ARTERIOVENOUS AND CAVERNOUS MALFORMATIONS

HANDBOOK OF CLINICAL NEUROLOGY

Series Editors

MICHAEL J. AMINOFF, FRANÇOIS BOLLER, AND DICK F. SWAAB

VOLUME 143

ELSEVIER

ARTERIOVENOUS AND CAVERNOUS MALFORMATIONS

Series Editors

MICHAEL J. AMINOFF, FRANÇOIS BOLLER, AND DICK F. SWAAB

Volume Editors

ROBERT F. SPETZLER, KARAM MOON, AND RAMI O. ALMEFTY

VOLUME 143

3rd Series

ELSEVIER

ELSEVIER
Radarweg 29, PO Box 211, 1000 AE Amsterdam, Netherlands
The Boulevard, Langford Lane, Kidlington, Oxford OX5 1GB, United Kingdom
50 Hampshire Street, 5th Floor, Cambridge, MA 02139, United States

British Library Cataloguing-in-Publication Data
A catalogue record for this book is available from the British Library

Library of Congress Cataloging-in-Publication Data
A catalog record for this book is available from the Library of Congress

ISBN: 978-0-444-63640-9

Publisher: Nikki Levy
Editorial Project Manager: Kristi Anderson
Production Project Manager: Sujatha Thirugnana Sambandam
Cover Designer: Alan Studholme

Typeset by SPi Global, India

Handbook of Clinical Neurology 3rd Series

Available titles

Vol. 79, The human hypothalamus: basic and clinical aspects, Part I, D.F. Swaab, ed. ISBN 9780444513571
Vol. 80, The human hypothalamus: basic and clinical aspects, Part II, D.F. Swaab, ed. ISBN 9780444514905
Vol. 81, Pain, F. Cervero and T.S. Jensen, eds. ISBN 9780444519016
Vol. 82, Motor neurone disorders and related diseases, A.A. Eisen and P.J. Shaw, eds. ISBN 9780444518941
Vol. 83, Parkinson's disease and related disorders, Part I, W.C. Koller and E. Melamed, eds. ISBN 9780444519009
Vol. 84, Parkinson's disease and related disorders, Part II, W.C. Koller and E. Melamed, eds. ISBN 9780444528933
Vol. 85, HIV/AIDS and the nervous system, P. Portegies and J. Berger, eds. ISBN 9780444520104
Vol. 86, Myopathies, F.L. Mastaglia and D. Hilton Jones, eds. ISBN 9780444518996
Vol. 87, Malformations of the nervous system, H.B. Sarnat and P. Curatolo, eds. ISBN 9780444518965
Vol. 88, Neuropsychology and behavioural neurology, G. Goldenberg and B.C. Miller, eds. ISBN 9780444518972
Vol. 89, Dementias, C. Duyckaerts and I. Litvan, eds. ISBN 9780444518989
Vol. 90, Disorders of consciousness, G.B. Young and E.F.M. Wijdicks, eds. ISBN 9780444518958
Vol. 91, Neuromuscular junction disorders, A.G. Engel, ed. ISBN 9780444520081
Vol. 92, Stroke – Part I: Basic and epidemiological aspects, M. Fisher, ed. ISBN 9780444520036
Vol. 93, Stroke – Part II: Clinical manifestations and pathogenesis, M. Fisher, ed. ISBN 9780444520043
Vol. 94, Stroke – Part III: Investigations and management, M. Fisher, ed. ISBN 9780444520050
Vol. 95, History of neurology, S. Finger, F. Boller and K.L. Tyler, eds. ISBN 9780444520081
Vol. 96, Bacterial infections of the central nervous system, K.L. Roos and A.R. Tunkel, eds. ISBN 9780444520159
Vol. 97, Headache, G. Nappi and M.A. Moskowitz, eds. ISBN 9780444521392
Vol. 98, Sleep disorders Part I, P. Montagna and S. Chokroverty, eds. ISBN 9780444520067
Vol. 99, Sleep disorders Part II, P. Montagna and S. Chokroverty, eds. ISBN 9780444520074
Vol. 100, Hyperkinetic movement disorders, W.J. Weiner and E. Tolosa, eds. ISBN 9780444520142
Vol. 101, Muscular dystrophies, A. Amato and R.C. Griggs, eds. ISBN 9780080450315
Vol. 102, Neuro-ophthalmology, C. Kennard and R.J. Leigh, eds. ISBN 9780444529039
Vol. 103, Ataxic disorders, S.H. Subramony and A. Durr, eds. ISBN 9780444518927
Vol. 104, Neuro-oncology Part I, W. Grisold and R. Sofietti, eds. ISBN 9780444521385
Vol. 105, Neuro-oncology Part II, W. Grisold and R. Sofietti, eds. ISBN 9780444535023
Vol. 106, Neurobiology of psychiatric disorders, T. Schlaepfer and C.B. Nemeroff, eds. ISBN 9780444520029
Vol. 107, Epilepsy Part I, H. Stefan and W.H. Theodore, eds. ISBN 9780444528988
Vol. 108, Epilepsy Part II, H. Stefan and W.H. Theodore, eds. ISBN 9780444528995
Vol. 109, Spinal cord injury, J. Verhaagen and J.W. McDonald III, eds. ISBN 9780444521378
Vol. 110, Neurological rehabilitation, M. Barnes and D.C. Good, eds. ISBN 9780444529015
Vol. 111, Pediatric neurology Part I, O. Dulac, M. Lassonde and H.B. Sarnat, eds. ISBN 9780444528919
Vol. 112, Pediatric neurology Part II, O. Dulac, M. Lassonde and H.B. Sarnat, eds. ISBN 9780444529107
Vol. 113, Pediatric neurology Part III, O. Dulac, M. Lassonde and H.B. Sarnat, eds. ISBN 9780444595652
Vol. 114, Neuroparasitology and tropical neurology, H.H. Garcia, H.B. Tanowitz and O.H. Del Brutto, eds. ISBN 9780444534903
Vol. 115, Peripheral nerve disorders, G. Said and C. Krarup, eds. ISBN 9780444529022
Vol. 116, Brain stimulation, A.M. Lozano and M. Hallett, eds. ISBN 9780444534972
Vol. 117, Autonomic nervous system, R.M. Buijs and D.F. Swaab, eds. ISBN 9780444534910
Vol. 118, Ethical and legal issues in neurology, J.L. Bernat and H.R. Beresford, eds. ISBN 9780444535016
Vol. 119, Neurologic aspects of systemic disease Part I, J. Biller and J.M. Ferro, eds. ISBN 9780702040863
Vol. 120, Neurologic aspects of systemic disease Part II, J. Biller and J.M. Ferro, eds. ISBN 9780702040870
Vol. 121, Neurologic aspects of systemic disease Part III, J. Biller and J.M. Ferro, eds. ISBN 9780702040887
Vol. 122, Multiple sclerosis and related disorders, D.S. Goodin, ed. ISBN 9780444520012
Vol. 123, Neurovirology, A.C. Tselis and J. Booss, eds. ISBN 9780444534880

Vol. 124, Clinical neuroendocrinology, E. Fliers, M. Korbonits and J.A. Romijn, eds. ISBN 9780444596024
Vol. 125, Alcohol and the nervous system, E.V. Sullivan and A. Pfefferbaum, eds. ISBN 9780444626196
Vol. 126, Diabetes and the nervous system, D.W. Zochodne and R.A. Malik, eds. ISBN 9780444534804
Vol. 127, Traumatic brain injury Part I, J.H. Grafman and A.M. Salazar, eds. ISBN 9780444528926
Vol. 128, Traumatic brain injury Part II, J.H. Grafman and A.M. Salazar, eds. ISBN 9780444635211
Vol. 129, The human auditory system: Fundamental organization and clinical disorders, G.G. Celesia and G. Hickok, eds. ISBN 9780444626301
Vol. 130, Neurology of sexual and bladder disorders, D.B. Vodušek and F. Boller, eds. ISBN 9780444632470
Vol. 131, Occupational neurology, M. Lotti and M.L. Bleecker, eds. ISBN 9780444626271
Vol. 132, Neurocutaneous syndromes, M.P. Islam and E.S. Roach, eds. ISBN 9780444627025
Vol. 133, Autoimmune neurology, S.J. Pittock and A. Vincent, eds. ISBN 9780444634320
Vol. 134, Gliomas, M.S. Berger and M. Weller, eds. ISBN 9780128029978
Vol. 135, Neuroimaging Part I, J.C. Masdeu and R.G. González, eds. ISBN 9780444534859
Vol. 136, Neuroimaging Part II, J.C. Masdeu and R.G. González, eds. ISBN 9780444534866
Vol. 137, Neuro-otology, J.M. Furman and T. Lempert, eds. ISBN 9780444634375
Vol. 138, Neuroepidemiology, C. Rosano, M.A. Ikram and M. Ganguli, eds. ISBN 9780128029732
Vol. 139, Functional neurologic disorders, M. Hallett, J. Stone and A. Carson, eds. ISBN 9780128017722
Vol. 140, Critical care neurology Part I, E.F.M. Wijdicks and A.H. Kramer, eds. ISBN 9780444636003
Vol. 141, Critical care neurology Part II, E.F.M. Wijdicks and A.H. Kramer, eds. ISBN 9780444635990
Vol. 142, Wilson disease, A. Członkowska and M.L. Schilsky, eds. ISBN 9780444636256

All volumes in the 3rd Series of the **Handbook of Clinical Neurology** are published electronically, on Science Direct: http://www.sciencedirect.com/science/handbooks/00729752.

Foreword

The varying clinical presentations of patients with vascular malformations of the central nervous system and its coverings can make their diagnosis challenging. Refinements in imaging procedures have facilitated the recognition of these lesions, however, and advances in surgical and endovascular techniques have led to remarkable changes in the management of affected patients. These developments have generated new questions concerning the need to treat these malformations in asymptomatic patients in whom they have been encountered incidentally by the sophisticated imaging procedures now in widespread use. A new understanding of the genetic underpinnings of various vascular malformations, their pathophysiology, and the manner in which these malformations affect neurologic function has also led to a more informed approach to patients harboring these lesions. Much has yet to be learned, of course, and a convenient up-to-date review might well be useful – we hoped – in pointing the way to future advances. For these various reasons, then, we felt the need to devote a volume of the *Handbook of Clinical Neurology* to the topic.

We were particularly glad that Professor Robert F. Spetzler, the J.N. Harber Chairman of Neurological Surgery and Director of the Barrow Neurological Institute in Phoenix, Arizona, agreed to edit this volume because he is an internationally recognized authority on these lesions, has contributed much to the relevant literature, and has unrivaled experience in their surgical management. He has been aided in this endeavor by two of his former trainees, Dr. Karam Moon and Dr. Rami O. Almefty, both also in Phoenix. Together, they have produced a most comprehensive and wide-ranging text.

The volume is divided into several parts. The first two-thirds, containing some 20 chapters, deals with arteriovenous malformations from many different viewpoints. Twelve chapters focus on intracranial lesions, covering both cerebral (eight chapters) and dural vascular anomalies (four chapters), and another eight chapters are devoted to intramedullary or dural spinal lesions. Different chapters discuss the epidemiology, clinical manifestations, and surgical and endovascular management of these arteriovenous anomalies. The last one-third of the volume contains 11 chapters devoted to all aspects of cranial and spinal cavernous malformations.

We are grateful to the volume editors, and to the various contributors whom they enlisted as coauthors, for crafting such a splendid and well-illustrated volume. As series editors, we reviewed all of the chapters for scope, substance, and style, making suggestions for improvement as needed. Based on our review, it is our belief that the volume will serve as a valuable reference work for neurologists, neurosurgeons, and interventional radiologists, as well as providing a practical guide to the management of patients with these lesions.

Elsevier has been the publisher of the *Handbook* series since its inception, and we are grateful for the continued support that we have received from the publisher. We acknowledge with particular pleasure, however, our personal indebtedness to Michael Parkinson in Scotland and to Mara Conner and Kristi Anderson in California for their assistance in seeing these volumes to fruition.

Michael J. Aminoff
François Boller
Dick F. Swaab

Preface

Vascular malformations, specifically arteriovenous and cavernous malformations, have long fascinated cerebrovascular surgeons with their complexity and management options. At Barrow Neurological Institute, we have studied the vast continuum of these malformations affecting our patients, and we have refined the techniques for managing these lesions. Although considerable knowledge and new treatment strategies have been added to our armamentarium in recent decades, many questions remain about the optimal management of individual vascular malformations.

Because these lesions can vary widely in size, location, and clinical presentation, the care of individual patients must be tailored to provide the least risky and most efficacious management available. For the best treatment options, the judicious use of microsurgery, radiosurgery, and embolization requires expertise from multiple specialists. However, nonintervention remains preferable for the nonsymptomatic patient who has a grade 5 arteriovenous malformation or a cavernous malformation under the floor of the fourth ventricle. Thus, the aims of this volume are to improve recognition of the pathology, awareness of the natural history, and understanding of the risks of treatment and nontreatment of these lesions.

In developing this volume for the *Handbook of Clinical Neurology,* we have drawn on the experience and expertise of established leaders in the field to update surgeons, neurologists, trainees, and others on the best current knowledge to diagnose and treat these often complex and challenging lesions. The chapters in this volume specifically address arteriovenous malformations and cavernous malformations. Their subject matter comprises the full spectrum of diagnostic modalities for the evaluation of patients, considerations crucial to clinical decision-making, the risks and benefits of treatment, and the latest neurosurgery literature on natural history and outcomes.

The volume is divided into two sections. The first section encompasses arteriovenous malformations, beginning with chapters that discuss genetics, natural history, and clinical presentation. The next several chapters discuss the intricacies of treatment of arteriovenous malformations, including the indications for treatment and the multiple treatment modalities that are available. This section also discusses arteriovenous fistulas, which are a close relative of arteriovenous malformations. Finally, discussions of spinal arteriovenous malformations and fistulas are included as well.

The second section of the book is devoted to cavernous malformations. The first few chapters are devoted to the natural history, epidemiology, clinical presentation, and pathophysiology of these lesions. A discussion of developmental venous anomalies, which are closely associated with cavernous malformations, is also included. Finally, because the presentation of cavernous malformations and the indications for their treatment depend on the location of these lesions, several chapters are dedicated to their surgical management.

We believe that readers will find this volume of the *Handbook of Clinical Neurology* to be a valuable contribution to the existing neurosurgery literature on these complex lesions. Although questions remain about spinal arteriovenous malformations and fistulas, ongoing research, from the molecular to the clinical level, is focused on providing more answers.

We sincerely thank all the contributors to this volume for the key roles they have played in bringing this endeavor to fruition. We hope that our readers glean much from the collective expertise within these pages.

Robert F. Spetzler, MD
Karam Moon, MD
Rami O. Almefty, MD

Contributors

I.J. Abecassis
Department of Neurological Surgery, University of Washington, Seattle, WA, USA

A.A. Abla
Department of Neurological Surgery, University of Arkansas for Medical Sciences, Little Rock, AR, USA

J.R. Adler
Department of Neurosurgery, Stanford University, Stanford, CA, USA

F.C. Albuquerque
Department of Neurosurgery, Barrow Neurological Institute, St. Joseph's Hospital and Medical Center, Phoenix, AZ, USA

R.O. Almefty
Department of Neurosurgery, Barrow Neurological Institute, St Joseph's Hospital and Medical Center, Phoenix, AZ, USA

S. Ambekar
Department of Neurosurgery, University of Miami, Miami, FL, USA

G.S. Atwal
Department of Neurosurgery, Barrow Neurological Institute, St. Joseph's Hospital and Medical Center, Phoenix, AZ, USA

N.C. Bambakidis
Department of Neurosurgery, University Hospitals, Case Western Reserve University, Cleveland, OH, USA

D.L. Barrow
Department of Neurosurgery, Emory University, Atlanta, GA, USA

S.C. Bir
Department of Neurosurgery, Louisiana State University Health Sciences Center, Shreveport, LA, USA

W. Brinjikji
Department of Radiology, Mayo Clinic, Rochester, MN, USA

P.A. Brown
Department of Radiology, Duke University, Durham, NC, USA

A. Can
Department of Neurosurgery, Brigham and Women's Hospital and Harvard Medical School, Boston, MA, USA

P.R. Chen
Department of Neurosurgery, University of Texas Health Science Center, Houston, Texas, USA

A.J. Clark
Department of Neurological Surgery, University of California, San Francisco, CA, USA

M.L. Cohen
Department of Pathology, University Hospitals, Case Western Reserve University, Cleveland, OH, USA

E.M. Cox
Department of Neurosurgery, University Hospitals, Case Western Reserve University, Cleveland, OH, USA

A.L. Day
Department of Neurosurgery, University of Texas Medical School at Houston, Houston, TX, USA

D. Ding
Department of Neurosurgery, University of Virginia, Charlottesville, VA, USA

R.L. Dodd
Department of Neurosurgery, Stanford University, Stanford, CA, USA

R. Du
Department of Neurosurgery, Brigham and Women's Hospital and Harvard Medical School, Boston, MA, USA

A.F. Ducruet
Department of Neurological Surgery, University of Pittsburgh Medical Center, Pittsburgh, PA, USA

M.S. Elhammady
Department of Neurosurgery, University of Miami, Miami, FL, USA

J.A. Ellis
Department of Neurosurgery, Emory University School of Medicine, Atlanta, GA, USA

C. Ene
Department of Neurological Surgery, Harborview Medical Center, University of Washington, Seattle, WA, USA

C. Gesteira Benjamin
Department of Neurosurgery, NYU Langone Medical Center, New York, NY, USA

H.E. Goldstein
Department of Neurosurgery, The Neurological Institute, Columbia University Medical Center, New York, NY, USA

L.F. Gonzalez
Department of Neurosurgery, Duke University, Durham, NC, USA

C. Grady
Department of Neurosurgery, NYU Langone Medical Center, New York, NY, USA

B.A. Gross
Department of Neurosurgery, Brigham and Women's Hospital and Harvard Medical School, Boston, MA and Department of Neurosurgery, Barrow Neurological Institute, St. Joseph's Hospital and Medical Center, Phoenix, AZ, USA

R.C. Heros
Department of Neurosurgery, University of Miami, Miami, FL, USA

O.R. Idowu
Division of Neuroradiology, Russell H. Morgan Department of Radiology and Radiological Science, Johns Hopkins University School of Medicine, Baltimore, MD, USA

B. Jian
Department of Neurological Surgery, University of California, San Francisco, CA, USA

M.Y.S. Kalani
Department of Neurosurgery, Barrow Neurological Institute, St. Joseph's Hospital and Medical Center, Phoenix, AZ, USA

A. Kaul
Department of Neurological Surgery, Harborview Medical Center, University of Washington, Seattle, WA, USA

L. Kim
Department of Neurological Surgery and Department of Radiology, University of Washington, Seattle, WA, USA

D. Kondziolka
Department of Neurosurgery and Center for Advanced Radiosurgery, NYU Langone Medical Center, New York, NY, USA

G. Lanzino
Department of Neurosurgery, Mayo Clinic, Rochester, MN, USA

M.T. Lawton
Department of Neurological Surgery, University of California, San Francisco, CA, USA

D.D.M. Lin
Division of Neuroradiology, Russell H. Morgan Department of Radiology and Radiological Science, Johns Hopkins University School of Medicine, Baltimore, MD, USA

T.K. Maiti
Department of Neurosurgery, Louisiana State University Health Sciences Center, Shreveport, LA, USA

P. McCormick
Department of Neurological Surgery, Columbia University Medical Center, Neurological Institute of New York, New York, NY, USA

C.G. McDougall
Department of Neurosurgery, Barrow Neurological Institute, St. Joseph's Hospital and Medical Center, Phoenix, AZ, USA

G. Mendez
Department of General Surgery, Rush University Medical Center, Chicago, IL, USA

K. Moon
Department of Neurosurgery, Barrow Neurological Institute, St. Joseph's Hospital and Medical Center, Phoenix, AZ, USA

M.A. Mooney
Department of Neurosurgery, Barrow Neurological Institute, St. Joseph's Hospital and Medical Center, Phoenix, AZ, USA

M.K. Morgan
Department of Clinical Medicine, Macquarie University, Sydney, New South Wales, Australia

C.B. Mulholland
Department of Neurosurgery, Barrow Neurological Institute, St. Joseph's Hospital and Medical Center, Phoenix, AZ, USA

P. Nakaji
Department of Neurosurgery, Barrow Neurological Institute, St. Joseph's Hospital and Medical Center, Phoenix, AZ, USA

A. Nanda
Department of Neurosurgery, Louisiana State University Health Sciences Center, Shreveport, LA, USA

M. Narayanan
Department of Neurosurgery, Barrow Neurological Institute, St. Joseph's Hospital and Medical Center, Phoenix, AZ, USA

J.W. Osbun
Department of Neurosurgery, Emory University, Atlanta, GA, USA

M. Otten
Department of Neurological Surgery, Columbia University Medical Center, Neurological Institute of New York, New York, NY, USA

A. Ozpinar
Department of Neurological Surgery, University of Pittsburgh Medical Center, Pittsburgh, PA, USA

M.R. Reynolds
Department of Neurosurgery, Emory University, Atlanta, GA, USA

R.F. Rudy
Department of Neurosurgery, Brigham and Women's Hospital and Harvard Medical School, Boston, MA, USA

M.J. Rutkowski
Department of Neurological Surgery, University of California, San Francisco, CA, USA

C.E. Sarris
Department of Neurosurgery, Barrow Neurological Institute, St. Joseph's Hospital and Medical Center, Phoenix, AZ, USA

J. Schramm
Department of Neurosurgery, University of Bonn, Bonn, Germany

J.P. Sheehan
Department of Neurosurgery, University of Virginia, Charlottesville, VA, USA

R.A. Solomon
Department of Neurosurgery, The Neurological Institute, Columbia University Medical Center, New York, NY, USA

R.F. Spetzler
Department of Neurosurgery, Barrow Neurological Institute, St. Joseph's Hospital and Medical Center, Phoenix, AZ, USA

R.M. Starke
Department of Neurosurgery, University of Virginia, Charlottesville, VA, USA

H. Sun
Department of Neurosurgery, Barrow Neurological Institute, St. Joseph's Hospital and Medical Center, Phoenix, AZ, USA

E.S. Sussman
Department of Neurosurgery, Stanford University, Stanford, CA, USA

A.H. Turkmani
Department of Neurosurgery, University of Texas Medical School at Houston, Houston, TX, USA

D.D. Wang
Department of Neurological Surgery, University of California, San Francisco, CA, USA

K.Y. Wang
Division of Neuroradiology, Russell H. Morgan Department of Radiology and Radiological Science, Johns Hopkins University School of Medicine, Baltimore, MD and Department of Radiology, Baylor College of Medicine, Houston, TX, USA

G.M. Weiner
Department of Neurological Surgery, University of Pittsburgh Medical Center, Pittsburgh, PA, USA

D.S. Xu
Department of Neurosurgery, Barrow Neurological Institute, St. Joseph's Hospital and Medical Center, Phoenix, AZ, USA

J.M. Zabramski
Department of Neurosurgery, Barrow Neurological Institute, St. Joseph's Hospital and Medical Center, Phoenix, AZ, USA

A.R. Zomorodi
Department of Neurosurgery, Duke University, Durham, NC, USA

Contents

PART I

ARTERIOVENOUS MALFORMATIONS

Section 1

Cerebral arteriovenous malformations

Handbook of Clinical Neurology, Vol. 143 (3rd series)
Arteriovenous and Cavernous Malformations
R.F. Spetzler, K. Moon, and R.O. Almefty, Editors
http://dx.doi.org/10.1016/B978-0-444-63640-9.00001-1

Chapter 1

Epidemiology, genetics, pathophysiology, and prognostic classifications of cerebral arteriovenous malformations

ALP OZPINAR[1]*, GUSTAVO MENDEZ[2], AND ADIB A. ABLA[3]

[1]*Department of Neurological Surgery, University of Pittsburgh Medical Center, Pittsburgh, PA, USA*

[2]*Department of General Surgery, Rush University Medical Center, Chicago, IL, USA*

[3]*Department of Neurological Surgery, University of Arkansas for Medical Sciences, Little Rock, AR, USA*

Abstract

Arteriovenous malformations (AVMs) are vascular deformities involving fistula formation of arterial to venous structures without an intervening capillary bed. Such anomalies can prove fatal as the high arterial flow can disrupt the integrity of venous walls, thus leading to dangerous sequelae such as hemorrhage. Diagnosis of these lesions in the central nervous system can often prove challenging as intracranial AVMs represent a heterogeneous vascular pathology with various presentations and symptomatology. The literature suggests that most brain AVMs (bAVMs) are identified following evaluation of the etiology of acute cerebral hemorrhage, or incidentally on imaging associated with seizure or headache workup. Given the low incidence of this disease, most of the data accrued on this pathology comes from single-center experiences. This chapter aims to distill the most important information from these studies as well as examine meta-analyses on bAVMs in order to provide a comprehensive introduction into the natural history, classification, genetic underpinnings of disease, and proposed pathophysiology. While there is yet much to be elucidated about AVMs of the central nervous system, we aim to provide an overview of bAVM etiology, classification, genetics, and pathophysiology inherent to the disease process.

INTRODUCTION

Vascular disease entities manifest in various forms, and are particularly dangerous when high-flow arterial systems erroneously communicate directly with venous systems. Arteriovenous malformations (AVMs) are vascular deformities involving fistula formation of arterial to venous structures without an intervening capillary bed. Such anomalies can prove fatal as the high arterial flow can disrupt the integrity of venous walls, thus leading to sequelae such as hemorrhage. The central nervous system (CNS) is particularly sensitive to such insults as intracranial hemorrhage secondary to AVM rupture can prove profoundly disabling or deadly. Diagnosis of these lesions can prove challenging, as intracranial AVMs represent a heterogeneous vascular pathology with various presentations and symptomatology, as well as a wide variation in severity of the effect of an AVM on a patient's overall well-being. The literature suggests that most brain AVMs (bAVMs) are identified following evaluation of the etiology of acute cerebral hemorrhage, or incidentally on imaging associated with seizure or headache workup (Hofmeister et al., 2000; Conger et al., 2015). As reinforced by Conger et al. (2015), subsequent to detection of the AVM, history and physical examination coupled with critical imaging, including computed tomography (CT), magnetic resonance imaging (MRI), and catheter angiogram, allow neurosurgeons to form a clinical picture of the AVM patient as well as visualize the structure and hemodynamics of the vascular malformation (Conger et al., 2015).

*Correspondence to: Alp Ozpinar, MD, Department of Neurological Surgery, University of Pittsburgh Medical Center, Suite B-400, 200 Lothrop Street, Pittsburgh PA, USA. Tel: +1-412-647-3685, E-mail: ozpinara@upmc.edu

To date, bAVM data have been derived from single-center experiences. From epidemiological studies, the incidence of AVMs ranges from 1.12 to 1.42 cases per 100 000 person-years, with 38–68% of new cases presenting as first-ever hemorrhage (Abecassis et al., 2014). Annual rates of hemorrhage in untreated bAVMs have been estimated at 2.10–4.12% (Abecassis et al., 2014). Various studies have tried to identify factors associated with increased risk of AVM rupture. Meta-analyses on the topic have found that increased risk of future rupture has been associated with factors such as previous rupture, location of AVM in deep brain structures, and exclusive deep venous drainage (Stapf et al., 2006; Gross and Du, 2013). Gross and Du (2013) calculated for observed AVMs an overall yearly risk of hemorrhage rate of 3.0%, with an annual rate of 2.2% for the unruptured subset and 4.5% for ruptured AVMs. Stapf et al. (2006) showed that the annual rate of rupture for an AVM is 35.5% when the AVM has the triad of deep venous drainage, deep location, and prior hemorrhage. The other large meta-analysis (Kim et al., 2014) examined four AVM cohorts and found an overall annual hemorrhage rate of 2.3%, with a rate of 1.3% for unruptured AVMs and 4.8% for the ruptured group. These estimates have held up as results of the multicenter prospective randomized trial ARUBA showed a 2.2% annual risk of rupture for unruptured bAVMs (Mohr et al., 2014).

Data for the natural history of ruptured AVMs are less prevalent as the decision to observe ruptured lesions is a riskier enterprise. Some studies (Kondziolka et al., 1995; Brown, 2000) have gone so far as to extrapolate lifetime risk of hemorrhage by using annual rates and applying the following equation:

$$\text{Annual rupture risk} = 1 - (\text{risk of no hemorrhage})^{(\text{life expectancy})}$$
$$\text{and rupture risk} = 105 - \text{patient age}$$

While the aforementioned studies have shown concordance in estimates of annual risk of rupture, variability among lesions and patient populations presents inherent challenges to generalizability and comprehensively predicting risk of rupture, as illustrated in various studies on the natural history of bAVMs (Pollock et al., 1996; Stapf et al., 2001; Stefani et al., 2002; Fullerton et al., 2005; Kim et al., 2010; Laakso et al., 2010; Gross and Du, 2013).

CLASSIFICATION SYSTEMS

Multiple scales have been formulated to predict the morbidity and mortality associated with AVMs and the associated risk of intervention. Spetzler and Martin (1986), in their seminal paper on bAVMs, proposed a grading system based on size, venous drainage, and eloquence of brain adjacent to the malformation. The five-point scale assigns a score as follows: for size, <3 cm = 1 point, 3–6 cm = 2 points, >6 cm = 3 points; for location, eloquent brain = 1 point, noneloquent = 0 points; for venous drainage, deep venous drainage = 1 point, not deep venous drainage = 0 points. Applying this scheme, higher scores translate to higher Spetzler–Martin grade, which corresponds to greater risk of morbidity and mortality with microsurgical removal of the lesion. (See Figure 1.1 for an example of Spetzler–Martin grade 3 AVM on preoperative imaging, intraoperative photos, and intraoperative fluorescence angiography processing.)

While the Spetzler–Martin grade is the most-quoted benchmark/standard in determining outcome and management, other grading systems have been introduced as adjuncts. Lawton et al. proposed a supplementary and complementary system to the Spetzler–Martin grade in 2010 that takes into account age, hemorrhagic presentation, and diffuseness of lesion, and yields improved accuracy in predictability of neurologic outcome when combined with the Spetzler–Martin scale (Table 1.1). This supplementary scale in conjunction with Spetzler–Martin is a 10-point scale that takes the five-point Spetzler–Martin scheme and, in addition, accounts for the following: for age, <20 years = 1 point, 20–40 = 2 points, >40 = 3 points; for hemorrhagic presentation, hemorrhage = 0 points, no hemorrhage = 1 point; for lesion nidus diffuseness, compact = 0 points, diffuse = 1 point. Lawton et al. (2010) have shown that application of their scale can help with preoperative risk prediction and extrapolation of outcome (Kim et al., 2014).

In 2011, Spetzler and Ponce suggested a consolidation of the Spetzler–Martin scale into a three-tiered system by combining grades I and II, and grades IV and V into classes A and C, respectively. Grade III lesions were classified as class B in the consolidated system. Grouping into these new classes was based on similar surgical results and was intended to provide simplified management recommendations as well as superior statistical power for comparative studies (Spetzler and Ponce, 2011).

Another predictive model of neurologic outcome following bAVM surgery is the University of Toronto Brain AVM Study Group's scale. Proposed by Spears and collaborators (2006), the scale is a nine-point stratified risk score where the predictive variable's influence is a function of its relative weight (eloquence = 4, diffuse nidus = 3, deep venous drainage = 2). Applying the score, the probability of suffering a disabling neurologic outcome with surgery is as follows: low risk (0–2 points) = 1.8%, moderate risk (3–5 points) = 17.4%, high risk (6–7 points) = 31.6%, very high risk (>7 points) = 52.9% (Spears et al., 2006).

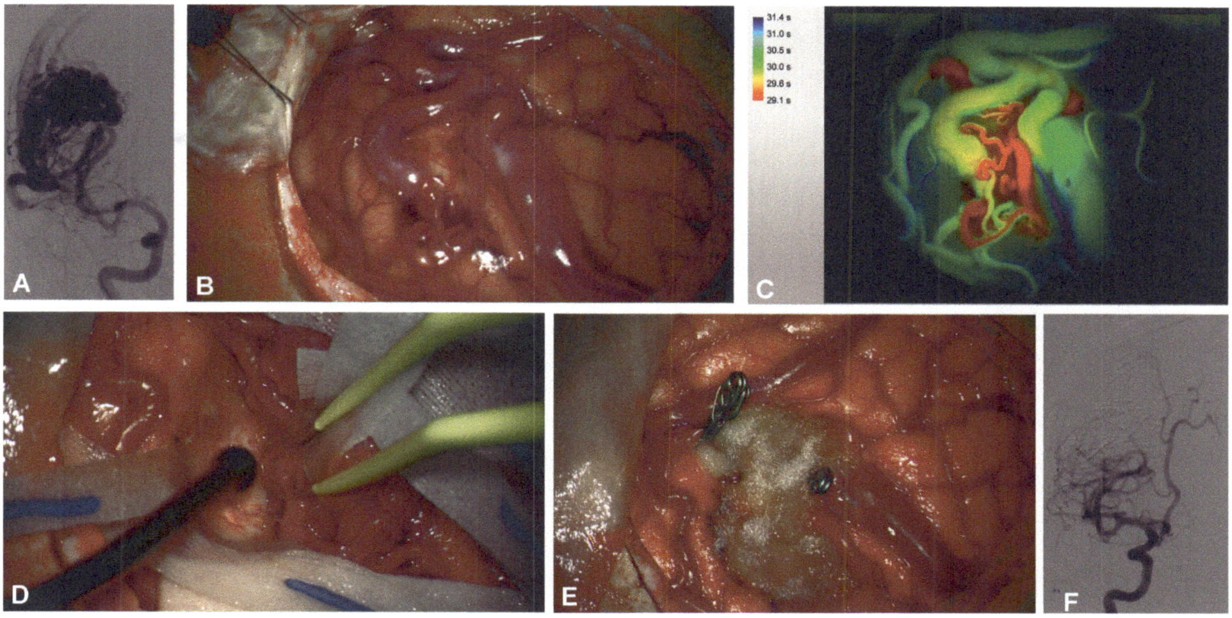

Fig. 1.1. (**A**) A representative Spetzer–Martin grade 3 arteriovenous malformation (AVM) (3 points for size) of the right frontal operculum and premotor region is demonstrated on anterior–posterior angiography. (**B**) Intraoperative photo demonstrates the dilated draining veins of Labbe and Trolard on the surface of the AVM with the AVM nidus located deep to the draining veins. (**C**) Intraoperative flow 800 processing of indocyanine green angiography demonstrates the arrival time of dye to the AVM with red demonstrating the vessels that fill first (feeders). Seconds in the figure key indicates latency in arrival time of dye. (**D**) AVM nidus seen deep to the draining veins along the anterior noneloquent border of the AVM. (**E**) Intraoperative photo demonstrating the surgical bed following resection. Note the change in color of the draining veins which are no longer arterialized. Change in color from red (before resection) to blue (following resection). (**F**) Postoperative anterior–posterior cerebral angiography demonstrating complete resection of AVM.

Table 1.1

Comparing the Spetzler–Martin scale and Lawton supplementary grade (Kim et al., 2014)

Spetzler–Martin grading	Points	Supplementary grading
Size (cm)		**Age (years)**
<3	1	<20
3–6	2	20–40
>6	3	>40
Venous drainage		**Bleeding**
Superficial	0	Yes
Deep	1	No
Eloquence		**Compactness**
No	0	Yes
Yes	1	No
Total	5	

Of note, the Toronto scale has been found to have a superior predictive ability to even the supplementary-Spetzler–Martin scale when looking at area under receiver operating characteristic curve, but is not as widely adopted as the multiple iterations of the Spetzler–Martin scheme.

Along with surgery, other methods have been adopted to treat bAVMs. Namely, radiosurgical modalities have become an established therapeutic option. Given that factors determining risk of procedure as well as variables associated with successful surgical resection of intracranial AVMs differ from those for radiosurgical management, development of a grading system for AVM radiosurgery was conceived. The radiosurgery-based AVM score (RBAS) was introduced following collaborative efforts between the University of Pittsburgh and the Mayo Clinic. Multiple validation studies have followed since then analyzing its application to lesions in various locations (including brainstem, deep structures, and all locations for AVMs), as well as with different radiosurgery techniques, including gamma knife, linac and CyberKnife (Pollock, 2013). The evolution of the grading scale has reflected a trend towards simplification, as the original Pittsburgh AVM radiosurgery scale (Pollock et al., 1997) takes into account AVM volume, patient age, AVM location, embolization status, and number of draining veins, while more recent iterations of the scale, namely the modified radiosurgery-based AVM score, focus on AVM location and volume, and patient age (Pollock and Flickinger, 2008) (See Fig. 1.2 for

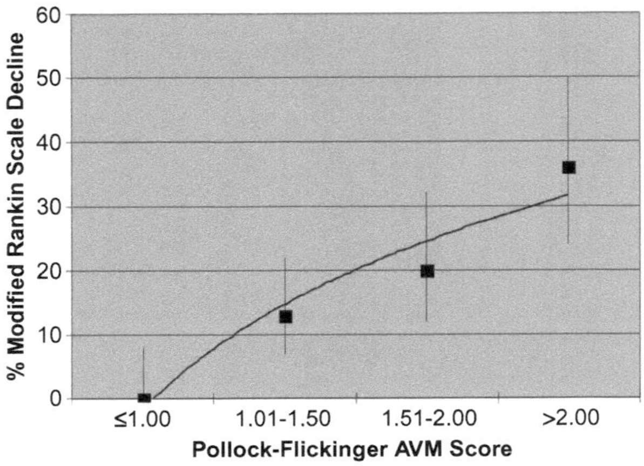

Fig. 1.2. Relationship of modified radiosurgery-based arteriovenous malformation (AVM) grading system and decline in modified Rankin Scale. Error bars illustrate 95% confidence interval for each point along the curve. (Adapted from Pollock and Flickinger, 2008.)

illustration of relationship between change in modified Rankin Scale and Pollock–Flickinger AVM score.)

The most recent grading scale in AVM management was created to address outcomes in endovascular treatment of intracranial AVMs. Dumont and colleagues (2015) devised the Buffalo score and retrospectively applied the metric to 50 bAVM patients treated with endovascular embolization, comparing accuracy of complication prediction to that seen with the Spetzler–Martin system. The proposed Buffalo score grade is determined by accounting for arterial pedicle number, arterial pedicle diameter, and eloquence of nearby cortex (Table 1.2; Dumont et al., 2015). Results from this initial retrospective study showed superior correlation of grade and complication incidence when compared to the Spetzler–Martin scale for this particular subset of AVM patients (Dumont et al., 2015). The authors note that part of the power of the new grading scale is in helping surgeons better stratify patients between surgical and endovascular treatment strategies. The Buffalo score represents a new tool with potential utility in guiding endovascular management of intracranial AVMs, which will likely undergo additional external validation moving forward.

Table 1.2

Buffalo grading system for endovascular treatment of brain arteriovenous malformations (Dumont et al., 2015)

Graded feature	Points assigned
Number of arterial pedicles	
1 or 2	1
3 or 4	2
5 or more	3
Diameter of arterial pedicles	
Most >1 mm	0
Most ≤1 mm	1
Nidus location	
Noneloquent	0
Eloquent	1

GENETICS

Animal models have allowed researchers to observe characteristic changes in nidal vessels of cerebral AVMs, including nonuniform changes in the thickness of vessel walls, lack of tight and adherent junctions, and splitting of elastic lamina (Tu et al., 2010). These observed changes illustrate the manifestation of molecular and cellular underpinnings that drive the development of cerebral AVMs. The genetic basis of intracranial AVMs is still being elucidated, but multiple candidate genes and pathways have been identified, both in syndromic and sporadic AVMs. Cerebral AVMs that result from associated syndromes provide insight into the etiology of AVM development. For example, the most common syndrome associated with bAVMs is Osler–Weber–Rendu (also known as hereditary hemorrhagic telangiectasia), which has been linked to haplo-insufficiency of transforming growth factor (TGF)-β pathway signaling genes like *ENG* and *SMAD4* (Rangel-Castilla et al., 2014). Another syndrome of note with increased incidence of AVM formation is Cobb's syndrome, in which patients are diagnosed with spinal AVMs and exhibit abnormal expression of platelet endothelial cell adhesion molecule (PECAM-1), vascular endothelial growth factor (VEGF), and matrix metalloproteinase (MMP)-9 (Rangel-Castilla et al., 2014).

Although there is no grand unifying signaling pathway for AVM proliferation in the context of associated syndromes, the dysregulation of angiogenesis, vasculogenesis, and inflammation have all been implicated (Sturiale et al., 2013). These recurring themes may also play a leading role in sporadic AVM formation and development. One proposed mechanism associated with AVM development is single-nucleotide polymorphisms (SNPs) of inflammatory factors like TGF-β and interleukin-6, and SNPs of vascular growth factors like angiopoietin-like 4 glycoprotein (Rangel-Castilla et al., 2014). In the same vein, studies have found overexpression of VEGF and angiopoietin-2 (a factor promoting vascular remodeling and destabilization) work in concert to assist in intracranial AVM development (Kim et al., 2009; Moftakhar et al., 2009). Notch4 signaling activation has also been characterized as sufficient to induce AVM phenotype in the developing mouse brain as it produces pathologically large vessels, and shunting physiology consistent with AVM (Murphy et al., 2008).

Along with AVM development, genetic anomalies have been found to predispose to AVM rupture. In particular, alterations in MMPs can lead to compromise of vascular stability and irregular angiogenesis. Studies have shown that even plasma levels of MMP-9 are elevated in AVM cases relative to controls prior to intervention (Starke et al., 2010). Authors from the University of California have also shown the potential for agents which block MMP activity; antibiotic medications, including doxycycline, may have the potential for inducing a static state in which AVM growth or activity can be mitigated (Frenzel et al., 2008).

Along with a better understanding of the etiology of bAVMs, genetic analysis allows for identification of targets for future molecular therapies, and therapeutic inhibition of aberrant pathways.

PATHOPHYSIOLOGY

The origin of AVMs remains unclear and is an area of ongoing investigation. Eliciting how these lesions arise can improve clinical management based on the presentation, especially in assessment of the risk of rupture leading to intracranial hemorrhage. Though intracranial hemorrhage is the most common presentation, others include seizure and neurologic deficit without underlying rupture (Zivin, 2012; Josephson et al., 2015). Norris et al. (1999) conducted a study on 31 patients showing that alterations in contrast dilution curves (decreased time to peak contrast versus increased time to peak contrast) are correlated with seizure or hemorrhage. The interplay of shear force and altered flow dynamics mixed with architectural vascular anomalies constitutes physiologic changes that shape the understanding of AVM development in addition to underlying molecular mechanisms (Moftakhar et al., 2009). Each respectively requires further validation.

Feeding artery pressure has been correlated with clinical presentation in conjunction with AVM size (Mckissock and Paterson, 1956; Henderson and Gomez, 1967; Houser et al., 1973; Waltimo, 1973; Guidetti and Delitala, 1980; Parkinson and Bachers, 1980; Itoyama et al., 1989; Spetzler et al., 1992). The relationship of size to risk of rupture has been postulated to be inversely proportional, or, in other words, smaller AVMs have greater risk of rupture resulting in intracranial hemorrhage (Norris et al., 1999). However, other studies have failed to support this theory (Gross and Du, 2013). One can infer feeding artery pressures are higher in AVMs that rupture compared to those that do not due to the wall stress exerted on the vessels. Spetzler et al. (1992) measured intraoperative perfusion pressure plus mean arterial pressure on 24 patients while documenting the AVM size. Key findings were a difference of 43.4% in mean arterial pressure between ruptured versus unruptured AVMs.

The nidus or functional unit of the AVM creates a tangled vascular network that connects feeding arteries and draining veins. Located proximal to the nidus are perinidal vessels that form a capillary network which some believe may contribute to postoperative hemorrhage or recurrence upon attempted surgical resection of the nidal unit (Ogilvy et al., 2001). It is important to note that endovascular embolization of an AVM, which is often done to prevent complications during surgical resection, can possibly increase the potential risk of rupture by altering local hemodynamics and re-routing blood flow within the AVM (Henkes et al., 2004). This may also be in part due to angiogenesis via upregulation of VEGF and hypoxia-inducible factor (HIF) following partial embolization (Meyer et al., 1999).

Compartmentalization of AVMs, first described by Yasargil (1987), is a concept pertaining to a single hemodynamic unit (i.e., compartment) served by one or more feeding pedicles with one or more draining veins. Intraoperatively, as feeding arteries are resected, the compartment will lose blood flow, causing it to collapse. Thus, theoretically, if feeders are removed and all compartments of the AVM collapse, hemorrhage would reasonably be avoided. However, Pellettieri et al. (1997) first proposed the idea of hidden compartments following surgery to explain the occurrence of postoperative hemorrhage and edema. If a compartment was unaccounted for, bleeding could be explained if that compartment became subsequently filled. There has been development of methods to detect the presence of underlying hidden compartments, including serial selective digital subtraction angiography and serial high-resolution MR angiography (Homan et al., 1986; Hashimoto and Nozaki, 1999). Yamada et al. (2004) established a protocol to potentially outline compartments to ensure preservation of surrounding brain tissue.

Abnormal venous architecture is a factor some have implicated in AVM pathogenesis (Mullan, 1994; Mullan et al., 1996a). Theories involve architectural changes that either give rise to *de novo* AVM production or perpetuate existing native arteriovenous connections. On histologic examination, endothelial cells seen within AVMs resemble those of early embryologic development despite arterial vasculature of feeding arteries lacking analogous primitive features (Deshpande and Vidyasagar, 1980). Though invalidated, embryologic anomalies leading to venous occlusion, stenosis, or agenesis seek to explain subsequent *de novo* AVM development due to venous hypertension (Bederson et al., 1991; Herman et al., 1995; Lawton et al., 1997). As the role of venous hypertension has yet to be definitively determined, an alternative hypothesis describes angiogenesis originating from venous structures due to venous hypertension causing transformation to AVM (Bederson et al., 1991; Wilson, 1992). Hypoxia resulting from vascular pressure changes may stimulate the formation of arteriovenous fistulas thought to arise from venous structures (Wilson, 1992). This would establish abnormal vascular architecture that increases venous pressure. Lastly, fistula formation of low-flow malformations has been implicated based on similar characteristics with select AVMs (abnormal surface and deep vein drainage plus the presence of deep collecting veins), accounting for the belief that low-flow malformations provide an architectural platform for AVMs to arise (Mullan, 1994; Mullan et al., 1996b; Pietilä et al., 2000). Abnormal venous drainage, specifically solely deep drainage, is a supported finding in the literature for possible formation of intracranial AVMs, and interestingly, *de novo* AVMs have often been found to drain into low-flow malformations (Nataf et al., 1997). It is thus hypothesized that low-flow malformations may contribute to the generation of AVMs. Previous authors have also demonstrated that AVMs, cavernous malformations, and capillary teleangiectasias may exist on a spectrum and may be related, as demonstrated by the sequential formation of all three entities in a single patient (Abla et al., 2008).

Native arteriovenous connections are thought to become pathogenic following a venous vaso-occlusive event progressing to AVM as blood flow is redirected through preformed connections from occluded vessels (Hasegawa et al., 1967). This postulate has been supported in animal studies, but is still in early investigational stages as to its validity in human models. In regard to rupture risk, those abnormalities that increase venous pressure (venous stenosis, predominant deep drainage, and venous reflux) have been associated with hemorrhage (Awad et al., 1993; al-Rodhan, 1995; Mullan et al., 1996a). Venous abnormalities are thus evidenced to have a role in AVM pathogenesis, yet other factors still remain pertinent. One such factor is the role of autoregulation.

The relationship between hemodynamics and AVM pathogenesis remains unclear. One postulate pertains to abnormal autoregulation, which may lead to changes in perfusion pressures. Spetzler et al. (1978) describe a theory whereby the presence of an AVM leads to local loss of autoregulation, described as normal perfusion pressure breakthrough theory. Those in opposition to this theory, however, believe paradoxically that autoregulation (or dysautoregulation) gives rise to the AVM rather than being caused by the presence of one (Quick et al., 2001). In this instance, loss of autoregulation causes venous hypertension that decreases perfusion pressure, ultimately resulting in reduction of blood flow and AVM development. In essence, as a mode of compensation following degeneration of normal vessels serving redundant areas, AVM production may occur as blood flow through shunted vessels increases. Related to the earlier discussion of perinidal vessels, hemodynamic overload state can result due to a hypervascular perinidal network potentially connected to the nidus termed "modja-modja" (Takemae et al., 1993). Solely following surgical resection of AVM, intravascular pressure can increase in this feeble vascular network, leading to rupture and hemorrhage.

Dilation of feeding arteries deprives alternate vessels of blood flow, shunting blood to the AVM under increased pressure (Taylor et al., 2002). The presence of vascular steal phenomenon, where blood is "stolen" from other regions and redirected toward the AVM, can be defined as such. Support for this theory comes from CT imaging studies, both single-photon emission and Xe-CT, that demonstrate decreased blood flow in areas adjacent to AVMs (Okabe et al., 1983; Homan et al., 1986). One observation describes cerebral calcifications arising, which is theorized to arise from steal phenomenon (Yu et al., 1987). Additionally, neuropsychologic studies have described a relationship between the potential occurrence of vascular steal phenomenon with lesion size, peripheral venous drainage, and the development of abnormal feeding arteries (Marks et al., 1991). In theory, steal may play a role in risk of cerebral edema postoperatively. Following lesion resection, increased blood flow through vessels previously deprived of fluid volume could account for this occurrence along with underlying molecular mechanisms of injury/ischemia (i.e., HIF and VEGF) (Moftakhar et al., 2009). This may be due to chronic hypoxia with loss of capillaries and hypoperfusion, leading to increased production of both HIF and VEGF (Meyer et al., 1999).

Some conflicting data surrounding the actual presence of AVM steal come from studies that provide no support for the presence of steal in cerebral vascular networks with AVMs (Mast et al., 1995; Meyer et al., 1998). Meyer and collaborators (1998) conducted a study that showed that blood flow is maintained at relatively normal levels, calling to question the existence of steal phenomenon in AVM physiology. Advanced imaging techniques have been used to further investigate these discrepancies, namely MR perfusion studies, which have described abnormal blood flow regulation (Bambakidis et al., 2001; Guo et al., 2004). Additional studies are still needed to provide further insight into the *in vivo* characteristics of "steal."

CONCLUSION

In conclusion, AVMs of the CNS represent a fascinating and complex disease entity with a presentation profile consisting of with hemorrhage, seizure, or headache.

Various classification systems are described, most notably the Spetzler–Martin grading system, the Lawton–Young supplementary grading scheme, the Pollock–Flickinger AVM score for radiosurgery, and the newly described Buffalo score for endovascular treatment. AVMs can occur in association with hereditary genetic syndromes; they can also involve sporadic development and formation, which may relate to abnormalities in cerebral vascular autoregulation as well as abnormalities in venous architecture. Finally, the entities of normal perfusion pressure breakthrough as part of a global process of loss of cerebral autoregulation and the process described as "steal" are fascinating pathophysiology described in AVM literature and are the subjects of ongoing study.

References

Abecassis IJ, Xu DS, Batjer HH et al. (2014). Natural history of brain arteriovenous malformations: a systematic review. Neurosurg Focus 37: E7.

Abla A, Wait SD, Uschold T et al. (2008). Developmental venous anomaly, cavernous malformation, and capillary telangiectasia: spectrum of a single disease. Acta Neurochir 150: 487–489; discussion 489.

Al-Rodhan NR (1995). Occlusive hyperemia remains the most logical explanation for the hemodynamic complications of resected intracerebral arteriovenous malformations. J Neurosurg Anesthesiol 7: 208–210.

Awad IA, Robinson JR, Mohanty S et al. (1993). Mixed vascular malformations of the brain: clinical and pathogenetic considerations. Neurosurgery 33: 179–188; discussion 188.

Bambakidis NC, Sunshine JL, Faulhaber PF et al. (2001). Functional evaluation of arteriovenous malformations. Neurosurg Focus. 11e2.

Bederson JB, Wiestler OD, Brüstle O et al. (1991). Intracranial venous hypertension and the effects of venous outflow obstruction in a rat model of arteriovenous fistula. Neurosurgery 29: 341–350.

Brown RD (2000). Simple risk predictions for arteriovenous malformation hemorrhage. Neurosurgery 46: 1024.

Conger A, Kulwin C, Lawton MT et al. (2015). Diagnosis and evaluation of intracranial arteriovenous malformations. Surg Neurol Int 6: 76.

Deshpande DH, Vidyasagar C (1980). Histology of the persistent embryonic veins in arteriovenous malformations of brain. Acta Neurochir 53: 227–236.

Dumont TM, Kan P, Snyder KV et al. (2015). A proposed grading system for endovascular treatment of cerebral arteriovenous malformations: Buffalo score. Surg Neurol Int 6: 3.

Frenzel T, Lee CZ, Kim H et al. (2008). Feasibility of minocycline and doxycycline use as potential vasculostatic therapy for brain vascular malformations: pilot study of adverse events and tolerance. Cerebrovascular Diseases (Basel, Switzerland) 25: 157–163.

Fullerton HJ, Achrol AS, Johnston SC et al. (2005). Long-term hemorrhage risk in children versus adults with brain arteriovenous malformations. Stroke; a Journal of Cerebral Circulation 36: 2099–2104.

Gross BA, Du R (2013). Natural history of cerebral arteriovenous malformations: a meta-analysis. J Neurosurg 118: 437–443.

Guidetti B, Delitala A (1980). Intracranial arteriovenous malformations. Conservative and surgical treatment. J Neurosurg 53: 149–152.

Guo W-Y, Wu Y-T, Wu H-M et al. (2004). Toward normal perfusion after radiosurgery: perfusion MR imaging with independent component analysis of brain arteriovenous malformations. AJNR Am J Neuroradiol 25: 1636–1644.

Hasegawa T, Ravens JR, Toole JF (1967). Precapillary arteriovenous anastomoses. "Thoroughfare channels" in the brain. Arch Neurol 16: 217–224.

Hashimoto N, Nozaki N (1999). Do cerebral arteriovenous malformations recur after angiographically confirmed total extirpation? Critical reviews in neurosurgery: CR 9: 141–146.

Henderson WR, Gomez RD (1967). Natural history of cerebral angiomas. Br Med J 4: 571–574.

Henkes H, Gotwald TF, Brew S et al. (2004). Pressure measurements in arterial feeders of brain arteriovenous malformations before and after endovascular embolization. Neuroradiology 46: 673–677.

Herman JM, Spetzler RF, Bederson JB et al. (1995). Genesis of a dural arteriovenous malformation in a rat model. J Neurosurg 83: 539–545.

Hofmeister C, Stapf C, Hartmann A et al. (2000). Demographic, morphological, and clinical characteristics of 1289 patients with brain arteriovenous malformation. Stroke; a Journal of Cerebral Circulation 31: 1307–1310.

Homan RW, Devous MD, Stokely EM et al. (1986). Quantification of intracerebral steal in patients with arteriovenous malformation. Arch Neurol 43: 779–785.

Houser OW, Baker HL, Svien HJ et al. (1973). Arteriovenous malformations of the parenchyma of the brain. Angiographic aspects Radiology 109: 83–90.

Itoyama Y, Uemura S, Ushio Y et al. (1989). Natural course of unoperated intracranial arteriovenous malformations: study of 50 cases. J Neurosurg 71: 805–809.

Josephson CB, Rosenow F, Al-Shahi Salman R (2015). Intracranial vascular malformations and epilepsy. Semin Neurol 35: 223–234.

Kim H, Pawlikowska L, Chen Y et al. (2009). Brain arteriovenous malformation biology relevant to hemorrhage and implication for therapeutic development. Stroke; a Journal of Cerebral Circulation 40: S95–S97.

Kim H, McCulloch CE, Johnston SC et al. (2010). Comparison of 2 approaches for determining the natural history risk of brain arteriovenous malformation rupture. Am J Epidemiol 171: 1317–1322.

Kim H, Al-Shahi Salman R, McCulloch CE et al. (2014). Untreated brain arteriovenous malformation: patient-level meta-analysis of hemorrhage predictors. Neurology 83: 590–597.

Kondziolka D, McLaughlin MR, Kestle JR (1995). Simple risk predictions for arteriovenous malformation hemorrhage. Neurosurgery 37: 851–855.

Laakso A, Dashti R, Juvela S et al. (2010). Natural history of arteriovenous malformations: presentation, risk of hemorrhage and mortality. Acta Neurochirurg Suppl 107: 65–69.

Lawton MT, Jacobowitz R, Spetzler RF (1997). Redefined role of angiogenesis in the pathogenesis of dural arteriovenous malformations. J Neurosurg 87: 267–274.

Lawton MT, Kim H, McCulloch CE et al. (2010). A supplementary grading scale for selecting patients with brain arteriovenous malformations for surgery. Neurosurgery 66: 702–713; discussion 713.

Marks MP, Lane B, Steinberg G et al. (1991). Vascular characteristics of intracerebral arteriovenous malformations in patients with clinical steal. AJNR Am J Neuroradiol 12: 489–496.

Mast H, Mohr JP, Osipov A et al. (1995). 'Steal' is an unestablished mechanism for the clinical presentation of cerebral arteriovenous malformations. Stroke; a Journal of Cerebral Circulation 26: 1215–1220.

Mckissock W, Paterson JH (1956). A clinical survey of intracranial angiomas with special reference to their mode of progression and surgical treatment: a report of 110 cases. Brain: A Journal of Neurology 79: 233–266.

Meyer B, Schaller C, Frenkel C et al. (1998). Physiological steal around AVMs of the brain is not equivalent to cortical ischemia. Neurol Res 20 (Suppl 1): S13–S17.

Meyer B, Schaller C, Frenkel C et al. (1999). Distributions of local oxygen saturation and its response to changes of mean arterial blood pressure in the cerebral cortex adjacent to arteriovenous malformations. Stroke; a Journal of Cerebral Circulation 30: 2623–2630.

Moftakhar P, Hauptman JS, Malkasian D et al. (2009). Cerebral arteriovenous malformations. Part 2: physiology. Neurosurg Focus 26: E11.

Mohr JP, Parides MK, Stapf C et al. (2014). Medical management with or without interventional therapy for unruptured brain arteriovenous malformations (ARUBA): a multicentre, non-blinded, randomised trial. Lancet 383: 614–621.

Mullan S (1994). Reflections upon the nature and management of intracranial and intraspinal vascular malformations and fistulae. J Neurosurg 80: 606–616.

Mullan S, Mojtahedi S, Johnson DL et al. (1996a). Cerebral venous malformation-arteriovenous malformation transition forms. J Neurosurg 85: 9–13.

Mullan S, Mojtahedi S, Johnson DL et al. (1996b). Embryological basis of some aspects of cerebral vascular fistulas and malformations. J Neurosurg 85: 1–8.

Murphy PA, Lam MTY, Wu X et al. (2008). Endothelial Notch4 signaling induces hallmarks of brain arteriovenous malformations in mice. Proc Natl Acad Sci U S A 105: 10901–10906.

Nataf F, Meder JF, Roux FX et al. (1997). Angioarchitecture associated with haemorrhage in cerebral arteriovenous malformations: a prognostic statistical model. Neuroradiology 39: 52–58.

Norris JS, Valiante TA, Wallace MC et al. (1999). A simple relationship between radiological arteriovenous malformation hemodynamics and clinical presentation: a prospective, blinded analysis of 31 cases. J Neurosurg 90: 673–679.

Ogilvy CS, Stieg PE, Awad I et al. (2001). AHA Scientific Statement: Recommendations for the management of intracranial arteriovenous malformations: a statement for healthcare professionals from a special writing group of the Stroke Council, American Stroke Association. Stroke; a Journal of Cerebral Circulation 32: 1458–1471.

Okabe T, Meyer JS, Okayasu H et al. (1983). Xenon-enhanced CT CBF measurements in cerebral AVM's before and after excision. Contribution to pathogenesis and treatment. J Neurosurg 59: 21–31.

Parkinson D, Bachers G (1980). Arteriovenous malformations. Summary of 100 consecutive supratentorial cases. J Neurosurg 53: 285–299.

Pellettieri L, Svendsen P, Wikholm G et al. (1997). Hidden compartments in AVMs – a new concept. Acta Radiologica (Stockholm, Sweden 1987) 38: 2–7.

Pietilä TA, Zabramski JM, Thèllier-Janko A et al. (2000). Animal model for cerebral arteriovenous malformation. Acta Neurochir 142: 1231–1240.

Pollock BE (2013). Development and testing of a radiosurgery-based arteriovenous malformation grading system. Prog Neurol Surg 27: 58–66.

Pollock BE, Flickinger JC (2008). Modification of the radiosurgery-based arteriovenous malformation grading system. Neurosurgery 63: 239–243; discussion 243.

Pollock BE, Flickinger JC, Lunsford LD et al. (1996). Factors that predict the bleeding risk of cerebral arteriovenous malformations. Stroke; a Journal of Cerebral Circulation 27: 1–6.

Pollock BE, Flickinger JC, Lunsford LD et al. (1997). The Pittsburgh arteriovenous malformation radiosurgery (PAR) grading scale. Radiosurgery 2: 137–146.

Quick CM, Hashimoto T, Young WL (2001). Lack of flow regulation may explain the development of arteriovenous malformations. Neurol Res 23: 641–644.

Rangel-Castilla L, Russin JJ, Martinez-Del-Campo E et al. (2014). Molecular and cellular biology of cerebral arteriovenous malformations: a review of current concepts and future trends in treatment. Neurosurg Focus 37: E1.

Spears J, Terbrugge KG, Moosavian M et al. (2006). A discriminative prediction model of neurological outcome for patients undergoing surgery of brain arteriovenous malformations. Stroke; a Journal of Cerebral Circulation 37: 1457–1464.

Spetzler RF, Martin NA (1986). A proposed grading system for arteriovenous malformations. J Neurosurg 65: 476–483.

Spetzler RF, Ponce FA (2011). A 3-tier classification of cerebral arteriovenous malformations. Clinical article. J Neurosurg 114: 842–849.

Spetzler RF, Wilson CB, Weinstein P et al. (1978). Normal perfusion pressure breakthrough theory. Clin Neurosurg 25: 651–672.

Spetzler RF, Hargraves RW, McCormick PW et al. (1992). Relationship of perfusion pressure and size to risk of hemorrhage from arteriovenous malformations. J Neurosurg 76: 918–923.

Stapf C, Mohr JP, Pile-Spellman J et al. (2001). Epidemiology and natural history of arteriovenous malformations. Neurosurg Focus. 11. e1.

Stapf C, Mast H, Sciacca RR et al. (2006). Predictors of hemorrhage in patients with untreated brain arteriovenous malformation. Neurology 66: 1350–1355.

Starke RM, Komotar RJ, Hwang BY et al. (2010). Systemic expression of matrix metalloproteinase-9 in patients with cerebral arteriovenous malformations. Neurosurgery 66: 343–348; discussion 348.

Stefani MA, Porter PJ, Terbrugge KG et al. (2002). Large and deep brain arteriovenous malformations are associated with risk of future hemorrhage. Stroke; a Journal of Cerebral Circulation 33: 1220–1224.

Sturiale CL, Puca A, Sebastiani P et al. (2013). Single nucleotide polymorphisms associated with sporadic brain arteriovenous malformations: where do we stand? Brain: A Journal of Neurology 136: 665–681.

Takemae T, Kobayashi S, Sugita K (1993). Perinidal hypervascular network on immediate postoperative angiogram after removal of large arteriovenous malformations located distant from the arterial circle of Willis. Neurosurgery 33: 400–405; discussion 405–406.

Taylor CL, Selman WR, Ratcheson RA (2002). Steal affecting the central nervous system. Neurosurgery 50: 679–688; discussion 688–689.

Tu J, Karunanayaka A, Windsor A et al. (2010). Comparison of an animal model of arteriovenous malformation with human arteriovenous malformation. Journal of Clinical Neuroscience: Official Journal of the Neurosurgical Society of Australasia 17: 96–102.

Waltimo O (1973). The relationship of size, density and localization of intracranial arteriovenous malformations to the type of initial symptom. J Neurol Sci 19: 13–19.

Wilson CB (1992). Cryptic vascular malformations. Clin Neurosurg 38: 49–84.

Yamada S, Brauer FS, Colohan ART et al. (2004). Concept of arteriovenous malformation compartments and surgical management. Neurol Res 26: 288–300.

Yasargil MG (1987). Pathological considerations. Microneurosurgery: AVM of the Brain, History, Embryology, Pathological Considerations, Hemodynamics, Diagnostic Studies, Microsurgical Anatomy, Thieme Verlag, Stuttgart.

Yu YL, Chiu EK, Woo E et al. (1987). Dystrophic intracranial calcification: CT evidence of 'cerebral steal' from arteriovenous malformation. Neuroradiology 29: 519–522.

Zivin J (2012). Hemorrhagic Cerebrovascular Disease. Goldman's Cecil Medicine, 24th ed. Saunders Elsevier, Philadelphia, PA.

Handbook of Clinical Neurology, Vol. 143 (3rd series)
Arteriovenous and Cavernous Malformations
R.F. Spetzler, K. Moon, and R.O. Almefty, Editors
http://dx.doi.org/10.1016/B978-0-444-63640-9.00002-3

Chapter 2

The natural history of cerebral arteriovenous malformations

ANIL CAN, BRADLEY A. GROSS, AND ROSE DU*
Department of Neurosurgery, Brigham and Women's Hospital and Harvard Medical School, Boston, MA, USA

Abstract

Cerebral arteriovenous malformations (AVMs) are composed of a complex tangle of abnormal arteries and veins and are a significant source of cerebral hemorrhage and consequent morbidity and mortality in young adults, representing a diagnostic and therapeutic challenge. Current natural-history studies of cerebral AVMs report overall annual rates of 1% and 3% for the risk of epilepsy and hemorrhage, respectively. Unruptured AVMs have an annual hemorrhage rate of 2.2% while ruptured lesions have an annual hemorrhage rate of 4.5%. These hemorrhage rates are can change over time, particularly for hemorrhagic lesions, with the rebleed rate ranging from 6% to 15.8% in the first year after rupture across several studies. Besides hemorrhage, other significant risk factors for AVM hemorrhage include deep location, deep venous drainage, associated aneurysms, and pregnancy. Other factors include patient age, sex, and small AVM size, which are not currently considered significant risk factors for AVM hemorrhage. In addition to hemorrhage risk and seizure risk, the natural history of an AVM also encompasses the daily psychologic burden that a patient must endure knowing that he or she possesses an untreated AVM. This chapter reviews the epidemiology, clinical features, and natural history of cerebral AVMs.

INTRODUCTION

Cerebral arteriovenous malformations (AVMs) are a complex of abnormal arteries and veins consisting of direct fistulous connections without normal intervening capillary beds or functional neural tissue (Steinheil, 1895; McCormick, 1966). Although their exact pathogenesis and pathophysiology remain poorly understood, AVMs are generally considered congenital lesions that arise from arrested vascular embryologic development resulting in atypical differentiation in the capillaries and subsequent abnormal communication between arteries and veins (McCormick, 1966; Gross et al., 2015). However, selected case reports of *de novo* AVM development and experimental models in animals showing *de novo* AVM formation have challenged congenital theories of AVM development (Gonzalez et al., 2005; Kilbourn et al., 2014). In addition, recurrent cerebral AVMs after complete surgical resection have been reported, as well as spontaneously disappearing AVMs (Waltimo, 1973; Pasqualin et al., 1985; Kader et al., 1996; Lee et al., 2002; Buis et al., 2004).

Cerebral AVMs are typically located in the cerebral hemispheres, but may be located in any region of the brain, including the cerebellum, brainstem, or spinal cord. Size of AVMs may vary from occult or cryptic, invisible on angiography or during surgery, to giant AVMs, involving one or several adjacent lobes, the entire cerebral hemisphere, or even the whole brain (Al-Rodhan et al., 1986). Due to this heterogeneity in size and location, cerebral AVMs may cause a wide range of clinical symptoms. Classically, the most frequent clinical symptoms include hemorrhage resulting from rupture, seizures, and focal neurologic deficits.

Treatment paradigms for AVMs continue to change as endovascular, microsurgical, and radiosurgical procedures evolve. However, essential for the management of these lesions is a thorough understanding of their natural history as it must be weighed against the risk of any anticipated treatment. Natural history not only includes

*Correspondence to: Rose Du, MD, PhD, Department of Neurosurgery, Brigham and Women's Hospital and Harvard Medical School, 75 Francis St, Boston MA 02115, USA. Tel: +1-617-732-6600, Fax: 617-734-8342, E-mail: rdu@bwh.harvard.edu

the anticipated hemorrhage risk and accompanying morbidity and mortality; it also includes the future risk of seizures as well as the underreported daily psychologic burden and consequent anxiety of harboring a life-threatening intracranial abnormality.

DEMOGRAPHICS

The true incidence and prevalence of cerebral AVMs remain unknown, due to the relative rarity of the disease and lack of large-scale epidemiologic studies. Across autopsy studies, the prevalence of AVMs has ranged from 5 to 613 cases per 100 000 (Courville, 1967; Stapf et al., 2001). Al-Shahi and colleagues (2002) found a crude AVM prevalence rate of 15–18 per 100 000 adults in a retrospective community-based study in Scotland. In the New York Islands study, a prospective population-based survey in which the incidence of AVM-related hemorrhage and the associated rates of morbidity and mortality were determined, the annual AVM detection rate was 1.34 per 100 000 person-years (95% confidence interval (CI) 1.18–49) (Stapf et al., 2001). Other studies reported detection rates of 1.12 (95% CI 0.90–1.37) and 1.11 (95% CI 0.6–1.8) per 100 000 person-years (Brown et al., 1996; Al-Shahi et al., 2003).

Despite their presumed congenital origin, cerebral AVMs are usually not hereditary lesions, and a recent study evaluating the prevalence of AVMs in first-degree relatives of patients harboring these lesions suggested against a familial risk factor (van Beijnum et al., 2014). However, aside from cases of hereditary hemorrhagic telangiectasia, rare cases of familial occurrence have been reported, supporting the hypothesis that genetic factors may play a role in the etiology (van Beijnum et al., 2007).

Six large natural-history studies with at least 200 patients with cerebral AVMs reported a mean age in a tight range of 32–39 years of age (Pollock et al., 1996; Stapf et al., 2006; Kim et al., 2007; Yamada et al., 2007; Hernesniemi et al., 2008; da Costa et al., 2009). Indeed, in a recent meta-analysis of cerebral AVMs, the mean age at presentation was reported to be 33.7 years (95% CI 31.1–36.2), underscoring that this is an abnormality discovered in younger patients (Gross and Du, 2013).

Most studies do not report a significant sex predisposition for patients with cerebral AVMs, including studies of pediatric patients (Stapf et al., 2006; Kim et al., 2007; da Costa et al., 2009; Darsaut et al., 2011).

CLINICAL PRESENTATION

Regardless of study design (natural history vs. surgical/interventional series), the most commonly reported presentation modality for patients with cerebral AVMs is hemorrhagic stroke. Population-based natural-history studies have reported an approximately 50% rate of hemorrhagic presentation, with intracerebral hemorrhage as the most common type of bleed, followed by intraventricular and subarachnoid hemorrhages (Stapf et al., 2006; Kim et al., 2007; da Costa et al., 2009; Gross and Du, 2013). In a recent meta-analysis, the overall rate of hemorrhagic presentation was 52% (95% CI 48–56%) (Gross and Du, 2013). AVM rupture typically presents in the third or fourth decade of life, and in a retrospective, hospital-based study of patients under 40 years of age with intracerebral hemorrhage, AVMs were the leading cause of intracerebral hemorrhage, affecting 33% (95% CI 40–85%) of patients (Ruiz-Sandoval et al., 1999). In addition to hemorrhage, seizures have been reported as the second most common presentation modality, seen in 27% of patients (95% CI 24–30%) (Gross and Du, 2013). Other less common presenting symptoms include headaches or focal neurologic deficits due to mass effect or, though rare, as sources of ischemic symptoms as a result of steal phenomena.

Headaches in the absence of hemorrhage are reported to occur in approximately 6–14% of patients with AVMs and Waltimo et al. (1975) reported that 58% of women with AVMs were found to have migraine with or without aura (Fults and Kelly, 1984; Ondra et al., 1990; Brown et al., 2005). The headaches are usually unilateral; however, more generalized headaches due to elevated venous pressures have also been reported (Al-Shahi and Warlow, 2001). However, the lack of prospective studies with validated diagnostic criteria hampers drawing definite conclusions regarding the association between headaches and AVMs, and more prospective longitudinal natural-history studies with standardized diagnostic headache criteria are needed.

Although rare, cerebral AVMs may cause focal neurologic deficits and signs of ischemia in the absence of prior or concomitant cerebral hemorrhage in approximately 3–10% of patients (Crawford et al., 1986; Itoyama et al., 1989; Kader et al., 1994; Mast et al., 1997; Al-Shahi and Warlow, 2001; Halim et al., 2004; Stapf et al., 2006; Yamada et al., 2007). Focal neurologic deficits can present with sudden onset and their course may vary from transient to progressive. The multifactorial pathophysiology includes vascular steal phenomenon, resulting from high-flow shunting through the AVM with consequent low blood pressure in the surrounding arteries and brain tissue (Mast et al., 1995). Other contributing factors are venous hypertension causing hypoperfusion of the surrounding brain parenchyma, and mass effect of the AVM or hydrocephalus. Other modes of AVM presentation that have been reported in the literature include cognitive dysfunction (Olivecrona and Riives, 1948), learning and behavioral disorders (Lazar et al., 1999), pulsatile tinnitus (Sabra, 1959), raised intracranial pressure (Chimowitz et al., 1990), and specific deficits associated with location of AVMs, such as movement disorders by AVMs located in the basal ganglia (Lobo-Antunes et al., 1974) or cranial nerve palsies (Hatori et al., 1991), trigeminal neuralgia (Johnson and Salmon, 1968), and hemifacial spasm caused by AVMs in the posterior fossa (Kim et al., 1991).

In addition, with advances in diagnostic radiology and the availability of high-resolution imaging techniques, asymptomatic AVMs are being detected more often, with an overall rate of 10% in modern series (Pollock et al., 1996; Stapf et al., 2006; Kim et al., 2007; Yamada et al., 2007; Hernesniemi et al., 2008; da Costa et al., 2009; Gross and Du, 2013).

SEIZURE RISK

The second most common presenting manifestation of cerebral AVMs is seizures (Thorpe et al., 2000), with approximately 17–30% of AVM patients presenting with this symptom (Galletti et al., 2014; Spetzler et al., 2015). However, the exact pathogenesis of seizures caused by cerebral AVMs remains unclear. Hemosiderin deposition, mass effect with cortical irritation, hemodynamic modifications, and/or vascular remodeling leading to steal, ischemia, and neuronal damage have all been suggested as pathophysiologic etiologies to epilepsy resulting from cerebral AVMs (Turjman et al., 1995a; Moftakhar et al., 2009).

Some studies have reported risk factors for seizure presentation in patients with AVMs. In one study, frontal, temporal, and/or superficial topography were significant risk factors for seizure presentation, with the strongest association for temporal-lobe location (odds ratio (OR) 3.48; 95% CI 1.77–6.85) (Galletti et al., 2014). Other studies also confirmed frontal and temporal location to be correlated with seizure presentation (Perret and Nishioka, 1966; Morello and Borghi, 1973) in addition to other angioarchitectural features such as cortical location of the AVM, feeding by the middle cerebral artery, cortical location of the feeder, absence of aneurysms, and presence of varix/varices (Turjman et al., 1995a). In Galletti et al.'s (2014) study, 31% of patients presented with seizures as the first clinical presentation of cerebral AVM, while 80% of these patients had epilepsy which led to the discovery of the lesion, and the remaining 20% had a single seizure at the time of diagnosis. Nearly half of the patients had simple partial seizures, 13% complex partial, 16% partial evolving to secondary generalized seizures, and only 3% generalized seizures (Galletti et al., 2014). In the remaining 23% of patients, the clinical pattern could not be established (Galletti et al., 2014). In another study, prior AVM hemorrhage, male sex, and frontotemporal lesion location were significant risk factors for seizures on presentation; deep artery perforators were associated with postoperative seizures (Englot et al., 2012).

Although several studies have demonstrated clear risk factors for seizures at presentation, this important feature is not as well represented in the prospective natural-history literature (Englot et al., 2012; Galletti et al., 2014). A recent observational analysis of 229 adults with a new diagnosis of an AVM reported a 72% 5-year risk of recurrent seizures in patients presenting with an initial seizure. Of patients presenting with seizures, by 2 years, 76% will develop epilepsy (Josephson et al., 2012). In patients presenting with hemorrhage, the 5-year risk of a first unprovoked seizure was 26%. In a natural-history study of 217 patients with AVMs, the risk of epilepsy for all-comers by 10-year follow-up was 11%, increasing to 18% by 20-year follow-up. This corresponds to an annual *de novo* seizure risk of 1% (Crawford et al., 1986). Twenty-two percent of patients presenting with hemorrhage developed epilepsy within 20 years of diagnosis, representing the greatest risk factor for *de novo* epilepsy. In addition, the younger the patient was at diagnosis, the higher the risk of developing epilepsy during follow-up, with rates ranging from 44% for patients aged 10–19 and 6% for patients over the age of 30. Size or depth of the AVM did not influence *de novo* epilepsy development (Crawford et al., 1986).

HEMORRHAGE RISK

Estimating the natural history of AVMs has been challenged by the heterogeneity of AVM lesions and diverse patient populations, with a wide range of clinical presentations and outcomes, in addition to surgeon or institutional bias toward or against treatment. Hemorrhage is often considered the most common source of morbidity and mortality from an AVM and is therefore the main focus of most natural-history studies that seek to identify risk factors predicting hemorrhage. In one long-term follow-up study of 168 patients with unruptured cerebral AVMs, morbidity and mortality from AVM hemorrhage were 35% and 29%, respectively (Brown et al., 1988). In another recent retrospective study of 51 patients sustaining a hemorrhage from a previously untreated AVM, 74% of patients who survived had neurologic deficits upon discharge, with 25% of patients having a severe deficit (Majumdar et al., 2015). On follow-up, 55% of patients were independent in their daily activities of living. The mortality rate was 8%. In a pediatric study of 115 patients, 68% of those sustaining a hemorrhage were independent (Gross et al., 2015). Overall, estimated hemorrhage-related mortality rates range from 10% to

30%, and morbidity rates from 20% to 30% (Crawford et al., 1986; Brown et al., 1988; Ondra et al., 1990).

Numerous studies have evaluated AVM hemorrhage risk, and there is general consensus that the overall annual hemorrhage rate for AVMs ranges from 2% to 4%. In a retrospective series of 238 patients with untreated AVMs and an average follow-up of 13.5 years, the average annual risk of hemorrhage from AVM was 2.4% (Hernesniemi et al., 2008). A recent meta-analysis confirmed these reports and defined the overall rate at 3.0% (95% CI 2.7–3.4%) (Gross and Du, 2013). However, this overall rate has little meaning for individualized counseling as it is well known that it varies widely depending on various risk factors.

Many early studies evaluated the influence of demographic and AVM angioarchitectural features on the risk of hemorrhagic presentation, a relatively simplistic approach as it does not require follow-up. Langer et al. (1998) found hypertension, small size, and deep venous drainage to be significantly associated with AVM hemorrhagic presentation. Other authors reported deep location (Stefani et al., 2002b; Stapf et al., 2006), nonborderzone/watershed location (Stapf et al., 2006), associated aneurysms (Stapf et al., 2006), infratentorial location (Khaw et al., 2004), small number of draining veins (Stefani et al., 2002b), high feeding artery pressure (Duong et al., 1998), and venous ectasias (Stefani et al., 2002b) as independent factors significantly associated with hemorrhagic AVM presentation. However, this type of analysis is largely influenced by presentation bias, as quiescent AVMs that may otherwise only present with hemorrhage, such as small lesions, were perceived as having a greater risk of hemorrhage (Itoyama et al., 1989; Turjman et al., 1995b; Stapf et al., 2002; Stefani et al., 2002a; Majumdar et al., 2015). Results from this approach have little validity for patient counseling as factors associated with hemorrhagic presentation are not equal to independent risk factors for future hemorrhage. Therefore, such decisions should be based on the perceived prospective risk of hemorrhage after presentation. Thus, recent studies have evaluated annual rates of hemorrhage and hemorrhage risk factors on this annual rate.

Most modern studies have not demonstrated a significant impact of demographic features such as patient age and sex on the risk of hemorrhage (Kim et al., 2007; Hernesniemi et al., 2008; da Costa et al., 2009; Gross and Du, 2013). Though studies based on hemorrhagic presentation suggested a greater risk of hemorrhagic presentation for small AVMs (Itoyama et al., 1989; Spetzler et al., 1992; Kader et al., 1994; Turjman et al., 1995b; Stapf et al., 2002; Stefani et al., 2002b), as discussed, this was likely an artifact of presentation bias as studies based on prospective annual hemorrhage rates have not demonstrated small size to be a significant risk factor for hemorrhage (Yamada et al., 2007; Hernesniemi et al., 2008; Gross and Du, 2013). This was confirmed in a recent meta-analysis where small AVM size was not associated with an increased risk of hemorrhage (Gross and Du, 2013). Significant risk factors for hemorrhage include prior hemorrhage, exclusively deep venous drainage, deep location, associated aneurysms, and pregnancy (da Costa et al., 2009; Gross and Du, 2012a, 2013).

Prior hemorrhage

Although the original Finnish and University of Toronto AVM natural-history studies did not report an increased risk of hemorrhage after AVM rupture (Ondra et al., 1990; Stefani et al., 2002b), subsequent reports from the same and other groups with longer follow-up clearly identified previous rupture to be the most significant risk factor for subsequent hemorrhage during follow-up (Forster et al., 1972; Crawford et al., 1986; Mast et al., 1997; Halim et al. 2004; Stapf et al., 2006; Yamada et al., 2007; Hernesniemi et al., 2008; da Costa et al., 2009). The prospective University of Toronto AVM study demonstrated an annual hemorrhage rate of 4.61% per year for the entire cohort of 678 patients and an annual hemorrhage rate of 7.48% for patients with initial hemorrhagic presentation, resulting in prior hemorrhage as a significant independent predictor of future hemorrhage (hazard ratio 2.15, $p = 0.07$) (da Costa et al., 2009). A recent meta-analysis of AVM natural-history studies confirmed these findings by demonstrating hemorrhage to be a statistically significant risk factor for subsequent bleeding, with a hazard ratio of 3.2 (95% CI 2.1–4.3) (Gross and Du, 2013). Unruptured AVMs had an annual hemorrhage rate of 2.2% (95% CI 1.7–2.7%) in this study while ruptured AVMs had an annual re-rupture rate of 4.5% (95% CI 3.7–5.5%) (Gross and Du, 2013). The former is consistent with the recent ARUBA trial results that demonstrated an annual hemorrhage rate of 2.2% for untreated, unruptured AVMs (Mohr et al., 2014).

Although rebleed rates are higher for AVMs, it is important to underscore that the rebleed rate is dynamic, with the highest hemorrhage rate during the first year of rupture, and declines thereafter. This may explain why studies with longer follow-up periods generally report lower average annual rupture rates than studies with shorter follow-up times. Re-rupture rates within the first year range from 6% to 15.8% across multiple studies (Mast et al., 1997; Yamada et al., 2007; Hernesniemi et al., 2008; da Costa et al., 2009; Gross and Du, 2012b), a fact that has important implications for both the timing of surgical treatment and the choice of radiosurgery for ruptured AVMs. In one study comparing the risk of hemorrhage in the clinical course of AVMs in 281 patients with and without initial hemorrhage, re-rupture occurred in 13% versus 2% in the nonhemorrhage group, leading to an annual risk of hemorrhage of 17.8% and 2.2%, respectively (Mast et al., 1997). However, in a subsequent report of the same cohort with 622 consecutive patients, with a mean follow-up of 829 days during which 39 patients developed AVM hemorrhage, the annual rupture rate for unruptured AVMs was 1.3% versus 5.9% for those after hemorrhagic AVM presentation (Stapf et al., 2006). In another study, two-thirds of AVMs that rebled within the first year were associated with aneurysms, suggesting that associated aneurysms are probably a contributing factor to this high re-rupture rate. The permanent morbidity from a rebleed was 45% (Gross and Du, 2012b). Interestingly, after reviewing the literature, this report illustrated that the rebleed rate within the first year was generally double the overall re-hemorrhage rate provided in each natural-history study (Gross and Du, 2012b).

Deep venous drainage

Deep drainage has been shown to be correlated with hemorrhagic presentation by multiple research groups (Marks et al., 1990; Kader et al., 1994; Turjman et al., 1995b). In one retrospective study, deep drainage was significantly associated with hemorrhagic presentation (odds ratio 5.77; $p=0.009$) and 25% of patients had additional intraventricular hemorrhage (Langer et al., 1998). In addition to these small retrospective studies with significant limitations, most natural-history studies consistently demonstrate that approximately half of AVMs possess deep venous drainage (Stapf et al., 2006; Yamada et al., 2007; Hernesniemi et al., 2008; da Costa et al., 2009). It has been postulated that the increased pressure of the deep venous system, and subsequent increase in the pressure gradient in the AVM nidus, is the etiologic cause of greater reported hemorrhage rates for AVMs with this feature (Hernesniemi et al., 2008; da Costa et al., 2009). Overall hemorrhage rates for AVMs with deep venous drainage range from 3.4% to 5.4% in the literature (Yamada et al., 2007; Hernesniemi et al., 2008; da Costa et al., 2009). Across two natural-history studies, annual hemorrhage rates for AVMs with deep venous drainage ranged from 2.4% to 2.6% for unruptured lesions and increased to 7.2–11.4% for ruptured AVMs (Stapf et al., 2006; Yamada et al., 2007). Deep venous drainage was reported as an independent predictor of subsequent hemorrhage in a prospective study of 622 patients with untreated AVMs from the Columbia AVM database (Stapf et al., 2006) and in an earlier reported analysis of the same cohort (Mast et al., 1997), with hazard ratios of 3.25 (95% CI 1.01–5.67) and 4.1 (95% CI 1.2–14.9), respectively. In a recent meta-analysis, the overall hazard ratio for hemorrhage from AVMs with deep venous drainage was 3.25 (95% CI 1.01–5.67) (Stapf et al., 2006).

Deep location

Approximately one-third of AVMs in natural-history studies are deep (Yamada et al., 2007; Hernesniemi et al., 2008; Gross and Du, 2013). It is important to underscore that deep location has been implicated as a risk factor for hemorrhage independent of the tendency of deep AVMs to possess deep venous drainage. Deep location may be an independent risk factor due to a greater tendency for deep AVMs to have supply from fragile perforators or potentially from a lack of "tamponade effect" from adjacent parenchyma for those presenting to a ventricular surface (Hernesniemi et al., 2008; Gross and Du, 2013).

Overall, annual hemorrhage rates for deep AVMs range from 4.1% to 8.6% in the literature (Yamada et al., 2007; Hernesniemi et al., 2008). Deep location was found to correlate with clinical presentation of hemorrhage in some retrospective reports (Turjman et al., 1995b; Mansmann et al., 2000), but not in others (Duong et al., 1998; Langer et al., 1998). Across two natural-history studies, the annual rupture rate for deep AVMs ranged from 3.1% to 4.5% for unruptured lesions and increased to 11.4–14.8% for ruptured AVMs (Stapf et al., 2006; Yamada et al., 2007). Deep location was an independent risk factor for subsequent AVM hemorrhage in several studies, with hazard ratios ranging from 3.07 (95% CI 1.28–7.40) (Yamada et al., 2007) to 3.25 (95% CI 1.30–8.16) (Stapf et al., 2006). In a recent meta-analysis, the overall meta-analytic hazard ratio of deep AVM location was 2.4 (95% CI 1.4–3.4) (Gross and Du, 2013). However, similar to small AVMs, deep lesions may be less likely to cause symptoms such as seizures and neurologic deficits due to their deep location, and are therefore prone to selection bias.

Associated aneurysms

The pathophysiology, incidence, and classification of intracranial aneurysms in relation to cerebral AVMs have been the topic of controversy and interest, and the subject of ongoing discussions in the medical literature. Aneurysms may be located in the nidus, on feeding arteries, or they may occur on peripheral arteries seemingly unrelated to the AVM lesion. They have been reported to occur in association with AVMs in approximately 10% of patients (Al-Shahi and Warlow, 2001), and across several natural-history studies, approximately 20% of cerebral AVMs were associated with aneurysms (Brown et al., 1988; da Costa et al., 2009; Gross and

Du, 2013). In these studies, approximately 50% of associated aneurysms were on feeding arteries while 25% were intranidal and 25% were in a remote location (Gross and Du, 2013).

Multiple studies found an association between clinical presentation of AVMs and the presence of related intracranial aneurysms (Marks et al., 1990; Turjman et al., 1995b). In a cross-sectional study of 463 prospectively enrolled patients, Stapf et al. (2002) demonstrated in a multivariate model an independent effect of feeding artery aneurysms on hemorrhagic AVM presentation (OR 2.11, 95% CI 1.18–3.78).

Although some of these studies demonstrated a clear correlation between hemorrhagic presentation and associated aneurysms, this important angiographic feature is underreported in the prospective natural-history literature. With intuitive preliminary evidence that AVMs associated with aneurysms have a greater risk of rupture (Turjman et al., 1995b; Stapf et al., 2002), and with additional evidence suggesting greater morbidity from rupture of AVMs associated with aneurysms (Gross et al., 2013), it is likely that these lesions are selected out early from natural-history studies and are therefore subject to selection bias, unless the aneurysms seem angiographically benign and/or small. Nevertheless, two studies reported relatively consistent annual hemorrhage rates ranging from 6.9% to 8.3% for AVMs associated with aneurysms. In the prospective Toronto AVM study, multivariate analysis did not reveal associated aneurysms as independent predictors for subsequent hemorrhage. However, in univariate analysis for specific aneurysm subtypes, both feeding artery aneurysms (HR 1.7, $p = 0.03$) and intranidal aneurysms (HR 2.1, $p = 0.03$) significantly increased the risk of hemorrhage, in contrast with remote aneurysms (da Costa et al., 2009). A subsequent recent meta-analysis reinforced associated aneurysms as a statistically significant risk factor for hemorrhage (Gross and Du, 2013).

Pregnancy

Although data regarding risk of AVM hemorrhage during pregnancy are sparse and inconclusive, pregnancy has long been implicated as a significant risk factor for AVM hemorrhage and the presence of an AVM significantly complicates the management of pregnant women (Crawford et al., 1986; Lanzino et al., 1994; Skidmore et al., 2001). In the early natural-history study by Crawford et al. (1986), 25% of women aged 20–29 years old presenting with AVM hemorrhage were pregnant. Pregnancy may be a risk factor for AVM hemorrhage due to several major physiologic changes that occur within the cardiovascular, hormonal, and hemostatic-thrombotic system, although the exact underlying mechanisms in relation to AVMs have yet to be clarified (Stieg et al., 2007).

Early studies report AVMs to be responsible for intracranial hemorrhage in 21–48% of cases of spontaneous intracranial hemorrhage during pregnancy (Fliegner et al., 1969; Amias, 1970; Robinson et al., 1972, 1974). In a study of 154 pregnant women with verified intracranial hemorrhage, 23% of the hemorrhages were secondary to ruptured AVMs (Dias and Sekhar, 1990). The overall maternal and fetal mortality from ruptured AVMs in patients that were observed were 32% and 23%, respectively. More than half of patients with ruptured AVMs were moribund or comatose at time of presentation.

In another report comprised of 24 women with AVMs, hemorrhage was associated with a 49% fetal complication rate and a 26% mortality rate, in contrast to a 33% fetal complication rate associated with aneurysmal subarachnoid hemorrhage (Robinson et al., 1974). A recent retrospective study of 54 women with an angiographic diagnosis of an AVM quantified a hemorrhage rate of 8.1% per pregnancy or an annual hemorrhage rate of 10.8% versus 1.1% in nonpregnant women. The statistically significant hazard ratio for hemorrhage during pregnancy was 7.91 in this study (Gross and Du, 2012a). These findings are consistent with another recent study that reported an annual hemorrhage rate of 11.1% for women who become pregnant during the 3-year latency interval between stereotactic radiosurgery and documented AVM obliteration (Tonetti et al., 2014). Among nonpregnant women, the annual hemorrhage rate was 2.5% during the latency interval (Tonetti et al., 2014). Timing of hemorrhage during pregnancy varies widely among studies, with reported cases during the first (Horton et al., 1990; Tonetti et al., 2014), second (Horton et al., 1990), and third trimester (Horton et al., 1990; Gross and Du, 2012a), and puerperium (Horton et al., 1990).

Due to preliminary studies that strongly suggest elevated hemorrhage risks during pregnancy and the obvious ethical dilemmas related to prospective natural-history studies in pregnant women harboring AVMs, no such studies have been performed to date and the annual hemorrhage rates for these patients are consequently lacking.

Age and sex

Although age was shown to be a significant predictor of AVM hemorrhage in some studies (Karlsson et al., 1997; Stapf et al., 2006), this finding was not confirmed by others (Yamada et al., 2007; da Costa et al., 2009; Gross and Du, 2013). In one prospective study, patients who experienced an AVM rupture were younger at admission (on average 5 years), but these results were not significant in univariate and multivariate Cox models (Hernesniemi et al., 2008). However, a recent meta-analysis of combined data from four large cohorts found increasing age to be significantly associated with an approximately 30% increase in risk for every 10-year increase in age (HR 1.34 per decade, 95% CI 1.17–1.53). However, the lack of consensus in the literature, combined with a higher risk of poor outcome after surgery in elderly patients (Ding et al., 2015), highlights the need for more well-designed larger prospective studies.

Similar to data regarding age, current studies do not provide clear associations between sex and risk of AVM-related hemorrhage. One clear exception was a natural-history study of 305 consecutive patients with AVMs that showed a significantly increased hemorrhage risk among female patients (HR 2.93 95% CI 1.20–7.16). However, the smaller portion of female patients and skewed distributions in this study compared with other studies may have influenced the results (Fults and Kelly, 1984; Mast et al., 1997; Stapf et al., 2003; Halim et al., 2004). Karlsson et al. (1997) reported that the risk of hemorrhage was higher in women during their fertile years as compared to males in the same age group. Although Mast et al. (1997) initially reported male sex to be significantly associated with subsequent hemorrhage (HR 9.2, 95% CI 2.1–41.3), these findings were not confirmed in a more recent report from the same group (Stapf et al., 2006). In line with the discussed differences in previously published reports, Kim et al. (2007) reported borderline associations between female sex and hemorrhage risk (HR 1.49, 95% CI 0.96–2.30) and a recent meta-analysis showed female sex not to be associated with hemorrhage (Gross and Du, 2013).

DISCUSSION

The wide variation in the clinical course of patients with AVMs and the heterogeneity of different patient populations hamper generalizations and make estimations of the natural history of AVMs the subject of controversy. The natural history of AVMs must incorporate a perceived, prospective risk of epilepsy, hemorrhage, and accompanying daily psychologic burden. The risk of epilepsy is considerably less explored than the risk of hemorrhage in the literature, and the importance of seizure control is often undervalued in the surgical treatment of AVMs; however, epilepsy may result in significant morbidity and decreased quality of life. Nevertheless, the decision to treat an AVM is often predicated on the presumed, prospective risk of hemorrhage from the lesion. This chapter gave an overview of risk factors for hemorrhage that should lower the threshold for treatment – specifically for hemorrhagic AVMs, deep AVMs, those with deep venous drainage, and those with associated aneurysms. In addition, women desiring to bear children with incidentally discovered AVMs should also be considered for curative treatment prior to planned conception.

Although the details of AVM treatment go beyond the scope of this chapter, the choice of management for patients harboring these lesions should consider treatment-related risks as related to patient age, location of the lesion, size of the AVM, and vascular topography, as well as the natural history of the individual patient. Therapeutic options for AVMs include operative obliteration or resection, endovascular embolization, and radiosurgery. Given the high risk of rebleeding within 1 year, ruptured AVMs are generally considered for surgical treatment as it provides the most expedient and definitive means for angiographic obliteration. Similarly, AVMs with deep venous drainage or associated aneurysms should also be considered for early surgery, if feasible at low morbidity. Deep AVMs (e.g., deep capsular, brainstem, or basal ganglia) may pose a challenge for surgery given their tendency to involve eloquent tissue; radiosurgery may thus serve as the optimal therapeutic option. Although one early prospective study of 73 consecutive patients with grades IV and V AVMs suggested that high-grade AVMs may in fact have a more benign natural history than their low-grade counterparts (Han et al., 2003), more recent studies have reported conflicting results. One retrospective study of 61 consecutive patients with high-grade AVMs reported an annual hemorrhage rate of 10.4% (95% CI 2.2–15.4) (Jayaraman et al., 2007), while another reported an overall annual hemorrhage rate of 3.3% that increased to 6.0% for ruptured, high-grade lesions (Laakso et al., 2008). Though they present a significant potential therapeutic challenge, these recent studies encourage creative, multimodality approaches when considering treatment of high-grade AVMs.

As an increasing number of unruptured cerebral AVMs are identified due to the increasing use of high-resolution imaging, this will lead to a shift in the AVM population, creating a diagnostic and treatment challenge for clinicians. For a better understanding of the natural history of cerebral AVMs, future studies with longer follow-up times are needed.

CONCLUSION

The natural history of cerebral AVMs encompasses overall annual rates of 2–4% for the risk of hemorrhage, and an annual rate of 1% for the development of *de novo* seizures. Significant risk factors for hemorrhage include prior rupture, deep location, deep venous drainage, associated aneurysms, and pregnancy. Patient age, sex, and small AVM size are not currently considered significant risk factors for AVM hemorrhage. In addition to hemorrhage risk and seizure risk, the natural history of an AVM also encompasses the daily psychologic burden that a patient must endure knowing he or she possesses an untreated AVM. All of the above factors need to be taken into consideration when determining the optimal treatment of an AVM.

References

Al-Rodhan NR, Al-Mefty O, Rifai A et al. (1986). Persistence of primitive cerebral vasculature in a newborn. A case report of whole brain AVM. Clin Neurol Neurosurg 88: 283–287.

Al-Shahi R, Warlow C (2001). A systematic review of the frequency and prognosis of arteriovenous malformations of the brain in adults. Brain 124: 1900–1926.

Al-Shahi R, Fang JS, Lewis SC et al. (2002). Prevalence of adults with brain arteriovenous malformations: a community based study in Scotland using capture-recapture analysis. J Neurol Neurosurg Psychiatry 73: 547–551.

Al-Shahi R, Bhattacharya JJ, Currie DG et al. (2003). Prospective, population-based detection of intracranial vascular malformations in adults: the Scottish Intracranial Vascular Malformation Study (SIVMS). Stroke 34: 1163–1169.

Amias AG (1970). Cerebral vascular disease in pregnancy. I. Haemorrhage. J Obstet Gynaecol Br Commonw 77: 100–120.

Brown Jr RD, Wiebers DO, Forbes G et al. (1988). The natural history of unruptured intracranial arteriovenous malformations. J Neurosurg 68: 352–357.

Brown Jr RD, Wiebers DO, Torner JC et al. (1996). Incidence and prevalence of intracranial vascular malformations in Olmsted County, Minnesota, 1965 to 1992. Neurology 46: 949–952.

Brown Jr RD, Flemming KD, Meyer FB et al. (2005). Natural history, evaluation, and management of intracranial vascular malformations. Mayo Clin Proc 80: 269–281.

Buis DR, van den Berg R, Lycklama G et al. (2004). Spontaneous regression of brain arteriovenous malformations – a clinical study and a systematic review of the literature. J Neurol 251: 1375–1382.

Chimowitz MI, Little JR, Awad IA et al. (1990). Intracranial hypertension associated with unruptured cerebral arteriovenous malformations. Ann Neurol 27: 474–479.

Courville CB (1967). Intracranial tumors. Notes upon a series of three thousand verified cases with some current observations pertaining to their mortality. Bull Los Angeles Neurol Soc 32 (Suppl 2): 1–80.

Crawford PM, West CR, Chadwick DW et al. (1986). Arteriovenous malformations of the brain: natural history in unoperated patients. J Neurol Neurosurg Psychiatry 49: 1–10.

da Costa L, Wallace MC, Ter Brugge KG et al. (2009). The natural history and predictive features of hemorrhage from brain arteriovenous malformations. Stroke 40: 100–105.

Darsaut TE, Guzman R, Marcellus ML et al. (2011). Management of pediatric intracranial arteriovenous malformations: experience with multimodality therapy. Neurosurgery 69: 540–556; discussion 556.

Dias MS, Sekhar LN (1990). Intracranial hemorrhage from aneurysms and arteriovenous malformations during pregnancy and the puerperium. Neurosurgery 27: 855–865; discussion 865–856.

Ding D, Xu Z, Yen CP et al. (2015). Radiosurgery for cerebral arteriovenous malformations in elderly patients: effect of advanced age on outcomes after intervention. World Neurosurg 84: 795–804.

Duong DH, Young WL, Vang MC et al. (1998). Feeding artery pressure and venous drainage pattern are primary determinants of hemorrhage from cerebral arteriovenous malformations. Stroke 29: 1167–1176.

Englot DJ, Young WL, Han SJ et al. (2012). Seizure predictors and control after microsurgical resection of supratentorial arteriovenous malformations in 440 patients. Neurosurgery 71: 572–580; discussion 580.

Fliegner JR, Hooper RS, Kloss M (1969). Subarachnoid haemorrhage and pregnancy. J Obstet Gynaecol Br Commonw 76: 912–917.

Forster DM, Steiner L, Hakanson S (1972). Arteriovenous malformations of the brain. A long-term clinical study. J Neurosurg 37: 562–570.

Fults D, Kelly Jr DL (1984). Natural history of arteriovenous malformations of the brain: a clinical study. Neurosurgery 15: 658–662.

Galletti F, Costa C, Cupini LM et al. (2014). Brain arteriovenous malformations and seizures: an Italian study. J Neurol Neurosurg Psychiatry 85: 284–288.

Gonzalez LF, Bristol RE, Porter RW et al. (2005). De novo presentation of an arteriovenous malformation. Case report and review of the literature. J Neurosurg 102: 726–729.

Gross BA, Du R (2012a). Hemorrhage from arteriovenous malformations during pregnancy. Neurosurgery 71: 349–355; discussion 355–356

Gross BA, Du R (2012b). Rate of re-bleeding of arteriovenous malformations in the first year after rupture. J Clin Neurosci 19: 1087–1088.

Gross BA, Du R (2013). Natural history of cerebral arteriovenous malformations: a meta-analysis. J Neurosurg 118: 437–443.

Gross BA, Lai PM, Du R (2013). Hydrocephalus after arteriovenous malformation rupture. Neurosurg Focus 34: E11.

Gross BA, Storey A, Orbach DB et al. (2015). Microsurgical treatment of arteriovenous malformations in pediatric patients: the Boston Children's Hospital experience. J Neurosurg Pediatr 15: 71–77.

Halim AX, Johnston SC, Singh V et al. (2004). Longitudinal risk of intracranial hemorrhage in patients with arteriovenous malformation of the brain within a defined population. Stroke 35: 1697–1702.

Han PP, Ponce FA, Spetzler RF (2003). Intention-to-treat analysis of Spetzler-Martin grades IV and V arteriovenous malformations: natural history and treatment paradigm. J Neurosurg 98: 3–7.

Hatori K, Urabe T, Kanazawa A et al. (1991). A case of brainstem vascular malformation with isolated trochlear nerve palsy as the initial symptom. No To Shinkei 43: 965–968.

Hernesniemi JA, Dashti R, Juvela S et al. (2008). Natural history of brain arteriovenous malformations: a long-term follow-up study of risk of hemorrhage in 238 patients. Neurosurgery 63: 823–829; discussion 829–831.

Horton JC, Chambers WA, Lyons SL et al. (1990). Pregnancy and the risk of hemorrhage from cerebral arteriovenous malformations. Neurosurgery 27: 867–871; discussion 871–872.

Itoyama Y, Uemura S, Ushio Y et al. (1989). Natural course of unoperated intracranial arteriovenous malformations: study of 50 cases. J Neurosurg 71: 805–809.

Jayaraman MV, Marcellus ML, Do HM et al. (2007). Hemorrhage rate in patients with Spetzler-Martin grades IV and V arteriovenous malformations: is treatment justified? Stroke 38: 325–329.

Johnson MC, Salmon JH (1968). Arteriovenous malformation presenting as trigeminal neuralgia. Case report. J Neurosurg 29: 287–289.

Josephson CB, Bhattacharya JJ, Counsell CE et al. (2012). Seizure risk with AVM treatment or conservative management: prospective, population-based study. Neurology 79: 500–507.

Kader A, Young WL, Pile-Spellman J et al. (1994). The influence of hemodynamic and anatomic factors on hemorrhage from cerebral arteriovenous malformations. Neurosurgery 34: 801–807; discussion 807–808.

Kader A, Goodrich JT, Sonstein WJ et al. (1996). Recurrent cerebral arteriovenous malformations after negative postoperative angiograms. J Neurosurg 85: 14–18.

Karlsson B, Lindquist C, Johansson A et al. (1997). Annual risk for the first hemorrhage from untreated cerebral arteriovenous malformations. Minim Invasive Neurosurg 40: 40–46.

Khaw AV, Mohr JP, Sciacca RR et al. (2004). Association of infratentorial brain arteriovenous malformations with hemorrhage at initial presentation. Stroke 35: 660–663.

Kilbourn KJ, Spiegel G, Killory BD et al. (2014). Case report of a de novo brainstem arteriovenous malformation in an 18-year-old male and review of the literature. Neurosurg Rev 37: 685–691.

Kim Y, Tanaka A, Kimura M et al. (1991). Arteriovenous malformation in the cerebellopontine angle presenting as hemifacial spasm – case report. Neurol Med Chir (Tokyo) 31: 109–112.

Kim H, Sidney S, McCulloch CE et al. (2007). Racial/Ethnic differences in longitudinal risk of intracranial hemorrhage in brain arteriovenous malformation patients. Stroke 38: 2430–2437.

Laakso A, Dashti R, Seppanen J et al. (2008). Long-term excess mortality in 623 patients with brain arteriovenous malformations. Neurosurgery 63: 244–253; discussion 253–255.

Langer DJ, Lasner TM, Hurst RW et al. (1998). Hypertension, small size, and deep venous drainage are associated with risk of hemorrhagic presentation of cerebral arteriovenous malformations. Neurosurgery 42: 481–486; discussion 487–489.

Lanzino G, Jensen ME, Cappelletto B et al. (1994). Arteriovenous malformations that rupture during pregnancy: a management dilemma. Acta Neurochir (Wien) 126: 102–106.

Lazar RM, Connaire K, Marshall RS et al. (1999). Developmental deficits in adult patients with arteriovenous malformations. Arch Neurol 56: 103–106.

Lee SK, Vilela P, Willinsky R et al. (2002). Spontaneous regression of cerebral arteriovenous malformations: clinical and angiographic analysis with review of the literature. Neuroradiology 44: 11–16.

Lobo-Antunes J, Yahr MD, Hilal SK (1974). Extrapyramidal dysfunction with cerebral arteriovenous malformations. J Neurol Neurosurg Psychiatry 37: 259–268.

Majumdar M, Tan LA, Chen M (2015). Critical assessment of the morbidity associated with ruptured cerebral arteriovenous malformations. J Neurointerv Surg 8 (2): 163–167.

Mansmann U, Meisel J, Brock M et al. (2000). Factors associated with intracranial hemorrhage in cases of cerebral arteriovenous malformation. Neurosurgery 46: 272–279; discussion 279–281.

Marks MP, Lane B, Steinberg GK et al. (1990). Hemorrhage in intracerebral arteriovenous malformations: angiographic determinants. Radiology 176: 807–813.

Mast H, Mohr JP, Osipov A et al. (1995). 'Steal' is an unestablished mechanism for the clinical presentation of cerebral arteriovenous malformations. Stroke 26: 1215–1220.

Mast H, Young WL, Koennecke HC et al. (1997). Risk of spontaneous haemorrhage after diagnosis of cerebral arteriovenous malformation. Lancet 350: 1065–1068.

McCormick WF (1966). The pathology of vascular ("arteriovenous") malformations. J Neurosurg 24: 807–816.

Moftakhar P, Hauptman JS, Malkasian D et al. (2009). Cerebral arteriovenous malformations. Part 2: physiology. Neurosurg Focus 26: E11.

Mohr JP, Parides MK, Stapf C et al. (2014). Medical management with or without interventional therapy for unruptured brain arteriovenous malformations (ARUBA): a multicentre, non-blinded, randomised trial. Lancet 383: 614–621.

Morello G, Borghi GP (1973). Cerebral angiomas. A report of 154 personal cases and a comparison between the results of surgical excision and conservative management. Acta Neurochir (Wien) 28: 135–155.

Olivecrona H, Riives J (1948). Arteriovenous aneurysms of the brain, their diagnosis and treatment. Arch Neurol Psychiatry 59: 567–602.

Ondra SL, Troupp H, George ED et al. (1990). The natural history of symptomatic arteriovenous malformations of the brain: a 24-year follow-up assessment. J Neurosurg 73: 387–391.

Pasqualin A, Vivenza C, Rosta L et al. (1985). Spontaneous disappearance of intracranial arterio-venous malformations. Acta Neurochir (Wien) 76: 50–57.

Perret G, Nishioka H (1966). Report on the cooperative study of intracranial aneurysms and subarachnoid hemorrhage. Section VI. Arteriovenous malformations. An analysis of 545 cases of cranio-cerebral arteriovenous malformations and fistulae reported to the cooperative study. J Neurosurg 25: 467–490.

Pollock BE, Flickinger JC, Lunsford LD et al. (1996). Factors that predict the bleeding risk of cerebral arteriovenous malformations. Stroke 27: 1–6.

Robinson JL, Hall CJ, Sedzimir CB (1972). Subarachnoid hemorrhage in pregnancy. J Neurosurg 36: 27–33.

Robinson JL, Hall CS, Sedzimir CB (1974). Arteriovenous malformations, aneurysms, and pregnancy. J Neurosurg 41: 63–70.

Ruiz-Sandoval JL, Cantu C, Barinagarrementeria F (1999). Intracerebral hemorrhage in young people: analysis of risk factors, location, causes, and prognosis. Stroke 30: 537–541.

Sabra F (1959). Observations on one hundred cases of cerebral angioma. J Am Med Assoc 170: 1522–1524.

Skidmore FM, Williams LS, Fradkin KD et al. (2001). Presentation, etiology, and outcome of stroke in pregnancy and puerperium. J Stroke Cerebrovasc Dis 10: 1–10.

Spetzler RF, Hargraves RW, McCormick PW et al. (1992). Relationship of perfusion pressure and size to risk of hemorrhage from arteriovenous malformations. J Neurosurg 76: 918–923.

Spetzler RF, Kalani Y, Nakaji P (2015). Neurovascular surgery, Thieme, New York.

Stapf C, Mohr JP, Pile-Spellman J et al. (2001). Epidemiology and natural history of arteriovenous malformations. Neurosurg Focus 11: e1.

Stapf C, Mohr JP, Pile-Spellman J et al. (2002). Concurrent arterial aneurysms in brain arteriovenous malformations with haemorrhagic presentation. J Neurol Neurosurg Psychiatry 73: 294–298.

Stapf C, Mast H, Sciacca RR et al. (2003). The New York Islands AVM Study: design, study progress, and initial results. Stroke 34: e29–e33.

Stapf C, Mast H, Sciacca RR et al. (2006). Predictors of hemorrhage in patients with untreated brain arteriovenous malformation. Neurology 66: 1350–1355.

Stefani MA, Porter PJ, terBrugge KG et al. (2002a). Angioarchitectural factors present in brain arteriovenous malformations associated with hemorrhagic presentation. Stroke 33: 920–924.

Stefani MA, Porter PJ, terBrugge KG et al. (2002b). Large and deep brain arteriovenous malformations are associated with risk of future hemorrhage. Stroke 33: 1220–1224.

Steinheil SO (1895). Ueber einen Fall von Varix aneurysmaticus im Bereich der Gehirngefaesse, F. Fromme, Wurzburg, pp. 1–56.

Stieg PE, Batjer HH, Samson DS (2007). Intracranial arteriovenous malformations, Informa Healthcare, New York.

Thorpe ML, Cordato DJ, Morgan MK et al. (2000). Postoperative seizure outcome in a series of 114 patients with supratentorial arteriovenous malformations. J Clin Neurosci 7: 107–111.

Tonetti D, Kano H, Bowden G et al. (2014). Hemorrhage during pregnancy in the latency interval after stereotactic radiosurgery for arteriovenous malformations. J Neurosurg 121 (Suppl): 226–231.

Turjman F, Massoud TF, Sayre JW et al. (1995a). Epilepsy associated with cerebral arteriovenous malformations: a multivariate analysis of angioarchitectural characteristics. AJNR Am J Neuroradiol 16: 345–350.

Turjman F, Massoud TF, Vinuela F et al. (1995b). Correlation of the angioarchitectural features of cerebral arteriovenous malformations with clinical presentation of hemorrhage. Neurosurgery 37: 856–860; discussion 860–862.

van Beijnum J, van der Worp HB, Schippers HM et al. (2007). Familial occurrence of brain arteriovenous malformations: a systematic review. J Neurol Neurosurg Psychiatry 78: 1213–1217.

van Beijnum J, van der Worp HB, Algra A et al. (2014). Prevalence of brain arteriovenous malformations in first-degree relatives of patients with a brain arteriovenous malformation. Stroke 45: 3231–3235.

Waltimo O (1973). The change in size of intracranial arteriovenous malformations. J Neurol Sci 19: 21–27.

Waltimo O, Hokkanen E, Pirskanen R (1975). Intracranial arteriovenous malformations and headache. Headache 15: 133–135.

Yamada S, Takagi Y, Nozaki K et al. (2007). Risk factors for subsequent hemorrhage in patients with cerebral arteriovenous malformations. J Neurosurg 107: 965–972.

Handbook of Clinical Neurology, Vol. 143 (3rd series)
Arteriovenous and Cavernous Malformations
R.F. Spetzler, K. Moon, and R.O. Almefty, Editors
http://dx.doi.org/10.1016/B978-0-444-63640-9.00003-5

Chapter 3

Arteriovenous malformations: epidemiology, clinical presentation, and diagnostic evaluation

JOSHUA W. OSBUN, MATTHEW R. REYNOLDS, AND DANIEL L. BARROW*
Department of Neurosurgery, Emory University, Atlanta, GA, USA

Abstract

Brain arteriovenous malformations (AVMs) represent an uncommon disease of the central nervous system characterized by an arteriovenous shunt in which one or multiple arterial pedicles feed into a vascular nidus, creating early drainage into a venous outflow channel. These lesions are considered to be congenital and can come to clinical attention in a variety of ways such as seizure, intracranial hemorrhage, chronic headache or progressive neurological deficit. We focus on the epidemiology, clinical presentation and diagnostic evaluation in this chapter.

INTRODUCTION

Arteriovenous malformations (AVMs) represent an uncommon disease of the central nervous system with important clinical implications. These lesions can occur anywhere in the brain and spinal cord and are a type of arteriovenous shunt in which one or multiple arterialized pedicles feed into a vascular nidus and drain in a variety of venous outflow channels. This gives the lesion its classic hallmark of "early venous drainage" on formal catheter-based cerebral angiography. These lesions are generally considered to be congenital and can come to clinical attention in a number of ways, such as seizure, chronic headache, progressive neurologic deficit, or intracranial hemorrhage. Some remain asymptomatic and are discovered only incidentally on cerebral imaging for unrelated reasons. Because AVMs are associated with a lifelong risk of hemorrhage (Ondra et al., 1990), these lesions require proper evaluation, diagnostic workup, and consideration for treatment based upon a variety of factors, including age, size, location, and drainage pattern.

EPIDEMIOLOGY

Brain AVMs are most likely congenital lesions that form for unclear reasons in the developing embryo. They can vary greatly from patient to patient in terms of angioarchitectural features related to size, location, and venous drainage pattern. Multiple AVMs in a single patient are uncommonly observed. No hereditary cause has been established and clustering in families is exceedingly infrequent in what is already a rare disease (Berman et al., 2000; van Beijnum et al., 2014). They are associated with a few syndromes, such as hereditary hemorrhagic telengiectasia (Osler–Weber–Rendu), Wyburn–Mason syndrome, von Hippel–Lindau disease, and Sturge–Weber syndrome. Although several mechanisms of AVM formation in the developing fetus have been proposed, the heterogeneity and lack of inheritance of these lesions cause their precise etiology to remain idiopathic.

A few population-based studies have been performed to aid in describing both the incidence and prevalence of brain AVMs. One of the earliest studies was data retrospectively recorded in Olmsted County, Minnesota, over a 27-year period, which captured an incidence of 0.82 AVMs per 100 000 persons based upon patients presenting with intracranial hemorrhage (Brown et al., 1996). The New York Islands AVM study is a registry and surveillance program of the population of the New York Islands (Manhattan, Staten Island, and Long Island) regarding patients with cerebrovascular anomalies (Stapf et al., 2003). The 19

*Correspondence to: Daniel L. Barrow, MD, 1365 Clifton Road Suite B6200, Atlanta GA 30322, USA. Tel: +1-404-778-5770, E-mail: Daniel_barrow@emoryhealthcare.org

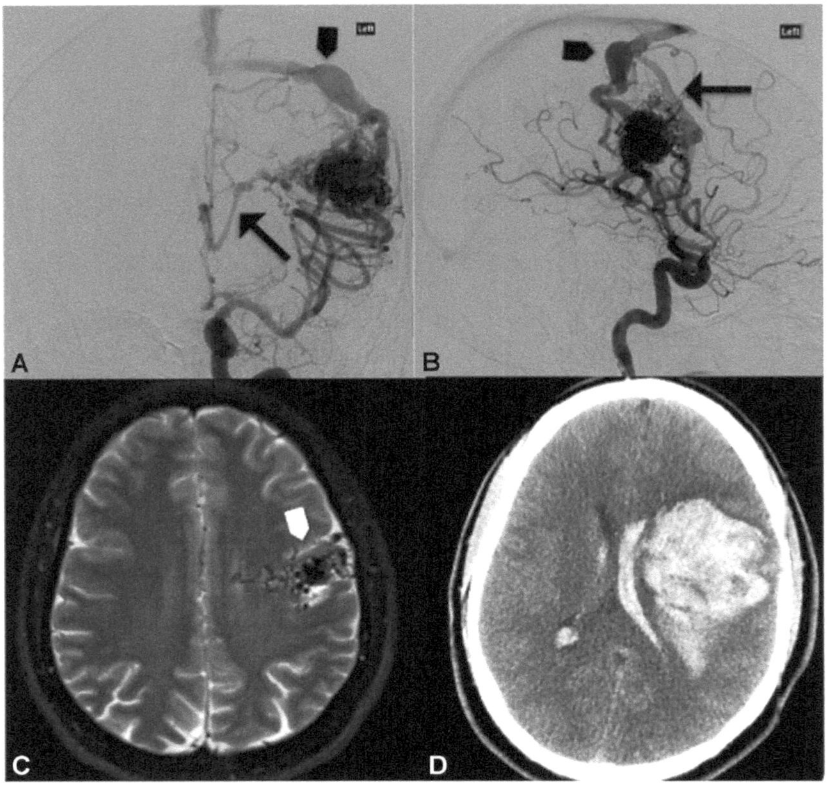

Fig. 3.1. A 34-year-old man who presented with a first-time seizure. (**A**) Anteroposterior and (**B**) lateral angiogram of the left internal carotid demonstrates a 4-cm arteriovenous malformation (AVM) nidus in the left frontal lobe fed by branches of the middle cerebral artery. Drainage is via both a superficial vein into the sagittal sinus (arrowheads) and a deep vein into the internal cerebral veins (arrows). (**C**) T2-weighted magnetic resonance demonstrated the appearance of "flow voids" (white arrowhead) in the frontal lobe signifying the AVM nidus. Due to these high-risk angiographic features, treatment was planned with stereotactic radiosurgery. Unfortunately, the patient suffered a devastating hemorrhage (**D**) while awaiting definitive treatment. Images courtesy of Emory University Hospital Department of Neurosurgery.

major area hospitals in the study captured 95% of the region's 9 429 541 inhabitants as of the 2000 Census. The study prospectively registered new AVM cases in the population with both hemorrhagic and nonhemorrhagic presentation. Initial results suggested an incidence of 1.34 per 100 000 and an incidence of first-time AVM hemorrhage of 0.51 per 100 000. A population study in Western Australia reported an incidence of 0.89 per 100 000 (ApSimon et al., 2002) and a similar Scottish study reported a rate of 1.12 per 100 000 (Al-Shahi et al., 2003). In summary, a collection of population-based studies from the literature approximate the incidence of brain AVMs at 0.69–1.32 per 100 000 (Brown et al., 1996; Berman et al., 2000; Stapf et al., 2001; ApSimon et al., 2002; Al-Shahi et al., 2003; Choi and Mohr, 2005; Laakso and Hernesniemi, 2012).

AVMs may present at any age and distribution among sexes is equal. Although AVMs are congenital, they most commonly come to clinical attention in the second through fourth decades of life. While no increase in the incidence or prevalence of AVMs is associated with any one race or ethnicity, a few studies have identified that Hispanic populations have a two- to threefold increase in the risk of intracranial hemorrhage from AVM (Kim et al., 2007; Yang et al., 2015).

CLINICAL PRESENTATION AND NATURAL HISTORY

AVMs can present with a variety of clinical symptoms. Intracranial hemorrhage, seizure, chronic headache, and progressive focal neurologic deficit are the most common presenting symptoms. In the modern era of magnetic resonance imaging (MRI), AVMs are being initially diagnosed incidentally on imaging performed for potentially unrelated reasons, such as headache, tinnitus, stroke, and/or transient ischemic attack symptoms, or other vague neurologic complaints.

Intracranial hemorrhage is the most common clinical presentation (See Fig. 3.1). AVMs are estimated to account for 2–4% of all hemorrhagic strokes and 1–2% of all total strokes (Choi and Mohr, 2005; Choi et al.,

2006; Laakso and Hernesniemi, 2012; Gross and Du, 2013). Population-based studies have reported that the incidence of hemorrhage from AVM is 0.51–1.11 per 100 000 per year. Natural-history studies report a risk of hemorrhage for known AVMs to be 2–5% per year (Ondra et al., 1990; Choi et al., 2006; van Beijnum et al., 2011; Gross and Du, 2013). In all, 5–25% of all AVM hemorrhages are fatal. Data from the UCSF Brain AVM Study Project have suggested that age, Hispanic race, hypertension, and previous intracranial hemorrhage are associated with increased risk of intracranial hemorrhage. A recent study from Johns Hopkins reported a threefold increased risk of hemorrhage in nonwhites. Particular angiographic and MR features have not definitively been demonstrated to increase the risk of hemorrhage, but in general there is a trend towards increased risk of hemorrhage in AVMs with multiple arterial feeders, deep-seated locations in the brain, exclusively deep drainage, and feeding-artery aneurysms (Stapf et al., 2006; Lv et al., 2011, 2013; Hetts et al., 2014). Smaller AVMS and those with limited venous drainage have been reported to have a higher risk of hemorrhage.

Patients who present with intracranial hemorrhage have symptoms that range from sudden severe headache, decreased level of consciousness, hemiparesis, neglect and gaze preference, aphasia, cranial neuropathy, visual field deficits, and coma. Approximately 25% of patients expire from their first hemorrhage.

Seizures are another common presentation of brain AVMs and occur in 20–29% of patients upon initial presentation (van Beijnum et al., 2011; Garcin et al., 2012; Laakso and Hernesniemi, 2012; Gross and Du, 2013; Lv et al., 2013; Galletti et al., 2014; Mohr et al., 2014; Ding et al., 2015a, b). A recent study from China found that 20.9% of patients presented with seizure out of a series of 3299 patients diagnosed with brain AVMs (Tong et al., 2016). Seizures are more common in male patients with AVMs in cortical locations, particularly the frontal and temporal lobe, and increase in frequency of occurrence with increasing AVM size (Garcin et al., 2012; Galletti et al., 2014; Ding et al., 2015b; Tong et al., 2016). For AVMs in deep location, up to 32% of insular AVMs may present with seizure, while other deep locations, such as the basal ganglia, thalamus, corpus callosum, brainstem, and cerebellum, present with seizure in less than 10% of cases (Ding et al., 2015b). With treatment, seizure control rates range from 49% to 85% depending on treatment strategy (surgery, radiosurgery, or embolization), with the highest seizure freedom being obtained in patients with confirmed angiographic obliteration after radiosurgery (Mohr et al., 2014; Ding et al., 2015b; Moon et al., 2015; Nerva et al., 2015, 2016).

The ARUBA trial was a randomized study that sought to compare the natural history of AVMs to intervention by surgery, embolization, or stereotactic radiosurgery (Mohr et al., 2014). The primary outcome measure was death or symptomatic stroke and the secondary outcome was clinical impairment (modified Rankin Scale of ≥ 2). During the study period, 1740 patients were screened, yet only 726 were deemed eligible to enroll in the study. A total of 323 patients refused to enroll and 177 had their management decision made by the treating team outside of the randomization process. The study was halted after randomizing 223 patients with 114 randomized to intervention and 109 to medical management. In the intervention group, 5 were treated with microsurgical resection, 30 with embolization, 31 with radiotherapy, and 28 with multimodality therapy. The primary endpoint was achieved in 35 (30.7%) of the 114 patients randomized to treatment and in 11 (10.1%) of the 109 patients in the medical management arm. The secondary outcome was achieved in 24 (46.2%) patients receiving intervention and 8 (15.1%) of the medically managed patients at 30 months. The authors concluded that medical management is superior to intervention in preventing death or stroke in patients with unruptured AVMs followed for 33 months. The design and interpretation of this study have raised a number of legitimate concerns. It is unclear why such a large number of eligible patients, who did not refuse enrollment, were not randomized. Secondly, the study design provides no standardization of the treatment arm. Anything less than complete obliteration of an AVM does nothing to protect patients from future hemorrhage. Many patients in ARUBA were treated with suboptimal therapy that did not accomplish the goal of complete obliteration. A recent large meta-analysis of treated AVMs reported obliteration rates of 96% for surgical resection, 38% for stereotactic radiosurgery, and 13% for embolization (van Beijnum et al., 2011). In the ARUBA study, only 5 patients received surgical resection alone when 76 patients in the treatment arm had grade Spetzler–Martin grade I or II AVMs. Embolization is generally not considered a curative procedure. It is concerning that 30 of the treatment patients were managed with embolization alone. Of the 114 patients randomized to treatment, 53 had not completed therapy and 20 had not even initiated therapy at the time of analysis. Given the lack of information and heterogeneity regarding therapies, it is impossible to understand what the treatment arm represents.

Many thoughtful critics of this study have pointed out that it is irresponsible to conclude superiority of medical management for AVMs when the treatment arm consisted of therapeutic options that are well below the current standard of care for elimination of AVMs. Finally, the 33-month follow-up for a disease with a lifelong risk of hemorrhage is clearly inadequate to make conclusions regarding the optimal treatment. The fact there was a 10% incidence of death or stroke over a time interval of less than 3 years in the medically managed group underscores the need to provide a low-risk cure for these patients and emphasizes the risk of untreated AVMs over the lifetime of the patient and warrants consideration of an appropriate treatment strategy.

While intracranial hemorrhage and seizures are the most common presenting features of a brain AVM, many present with less ominous symptoms such as chronic headache or tinnitus. As stated previously, these lesions are being discovered more and more frequently as incidental findings on noninvasive imaging. Regardless of the clinical presentation or route of discovery, any AVM harbors a risk of intracranial hemorrhage over the lifetime of the patient and warrants a full diagnostic workup so that an appropriate treatment strategy may be formed.

DIAGNOSTIC EVALUATION

AVMs have signature hallmarks on several noninvasive imaging studies. The appropriate workup will depend on the patient's presentation. AVMs are now presenting with increased frequency on noninvasive imaging studies. In such cases, larger AVMs can appear as a hyperdense mass that can even be confused for hemorrhage on noncontrast computed tomography (CT). CT angiography (CTA) can reveal a nidus or even arterial feeders and large tortuous veins. On plain magnetic resonance imaging (MRI), T2-weighted sequences often demonstrate a classic tangle of "flow voids" around the AVM nidus. Magnetic resonance angiography (MRA) may potentially delineate the nidus, arterial feeders, or draining veins. Even if an AVM is discovered incidentally on MRI, T2-weighted imaging is essential for understanding the exact anatomic location and size of the nidus for determining potential treatment decisions. Functional MRI testing can help determine the lesion's relationship to functionally important areas such as motor and language areas, and can be extremely valuable for determining the safety of surgical resection. A fine-cut volumetric T2-weighted study or T1-weighted study with contrast can be used with neuronavigation systems for intraoperative guidance.

Other patients will present in a more acute fashion. For patients presenting with seizures, often a noncontrast CT is performed to rule out intracranial hemorrhage, and occasionally a large AVM may be seen. For AVMs with all but a micronidus, MRI will usually demonstrate the AVM nidus and location. For the majority of patients who present with intracranial hemorrhage, CT scan will demonstrate an intraparenchymal hemorrhage adjacent to the AVM nidus. Another classic appearance is a "flame-shaped" hemorrhage emanating from the ventricle. In other cases an AVM may have an associated flow-related aneurysm and present with subarachnoid hemorrhage from aneurysm rupture. In the setting of acute hemorrhage, the diagnosis of AVM on noninvasive vascular imaging may not be possible because the hemorrhage will obscure the details of surrounding blood vessels. This is particularly true in smaller AVMs.

Regardless of how an AVM may be demonstrated on noninvasive imaging, the gold standard for diagnosis is a formal catheter-based cerebral angiogram. Since an angiogram is a time-resolved dynamic study, an AVM will demonstrate its hallmark of early venous drainage. Angiography will demonstrate an AVM's arterial feeders, nidus configuration, and pattern of venous drainage. It will demonstrate any flow-related aneurysms, intranidal aneurysms, or venous varices. Three-dimensional rotational angiography is crucial in understanding the precise angiographic architecture of the lesion, and greatly aids in treatment decisions such as the safety of embolization in regard to *en passage* vessels that simultaneously supply AVM nidus and normal cortex, and the technical feasibility of embolization in regard to microcatheterization of feeding arteries. Three-dimensional spatial resolution will also greatly assist in the formation of a surgical resection plan. Size, location, and drainage pattern are essential for understanding surgical risk as exhibited by the Spetzler–Martin grading scale, and angiography will help define most of these parameters. Angiography will therefore aid in surgical planning and help in determining whether preoperative embolization can be safe and effective.

CONCLUSIONS

AVMs are an uncommon vascular malformation of the brain and spinal cord with important clinical implications in terms of hemorrhagic stroke, epilepsy, and neurologic disability. An understanding of their epidemiology, clinical presentation, and natural history is imperative when forming a plan for full diagnostic workup. A full diagnostic workup, including both MRI and catheter-based cerebral angiography, is paramount in calculating a treatment modality that will maximize both AVM obliteration and patient safety.

REFERENCES

Al-Shahi R, Bhattacharya JJ, Currie DG et al. (May 2003). Prospective, population-based detection of intracranial vascular malformations in adults: the Scottish Intracranial Vascular Malformation Study (SIVMS). Stroke 34 (5): 1163–1169.

ApSimon HT, Reef H, Phadke RV et al. (Dec 2002). A population-based study of brain arteriovenous malformation: long-term treatment outcomes. Stroke 33 (12): 2794–2800.

Berman MF, Sciacca RR, Pile-Spellman J et al. (Aug 2000). The epidemiology of brain arteriovenous malformations. Neurosurgery 47 (2): 389–396; discussion 397.

Brown Jr RD, Wiebers DO, Torner JC et al. (Apr 1996). Incidence and prevalence of intracranial vascular malformations in Olmsted County, Minnesota, 1965 to 1992. Neurology 46 (4): 949–952.

Choi JH, Mohr JP (May 2005). Brain arteriovenous malformations in adults. Lancet Neurol 4 (5): 299–308.

Choi JH, Mast H, Sciacca RR et al. (May 2006). Clinical outcome after first and recurrent hemorrhage in patients with untreated brain arteriovenous malformation. Stroke 37 (5): 1243–1247.

Ding D, Quigg M, Starke RM et al. (Sep 2015a). Cerebral arteriovenous malformations and epilepsy, part 2: predictors of seizure outcomes following radiosurgery. World Neurosurg 84 (3): 653–662.

Ding D, Starke RM, Quigg M et al. (Sep 2015b). Cerebral arteriovenous malformations and epilepsy, part 1: predictors of seizure presentation. World Neurosurg 84 (3): 645–652.

Galletti F, Costa C, Cupini LM et al. (Mar 2014). Brain arteriovenous malformations and seizures: an Italian study. J Neurol Neurosurg Psychiatry 85 (3): 284–288.

Garcin B, Houdart E, Porcher R et al. (2012). Epileptic seizures at initial presentation in patients with brain arteriovenous malformation. Neurology 78 (9): 626–631.

Gross BA, Du R (Feb 2013). Natural history of cerebral arteriovenous malformations: a meta-analysis. J Neurosurg 118 (2): 437–443.

Hetts SW, Cooke DL, Nelson J et al. (Jul 2014). Influence of patient age on angioarchitecture of brain arteriovenous malformations. AJNR Am J Neuroradiol 35 (7): 1376–1380.

Kim H, Sidney S, McCulloch CE et al. (Sep 2007). Racial/ethnic differences in longitudinal risk of intracranial hemorrhage in brain arteriovenous malformation patients. Stroke 38 (9): 2430–2437.

Laakso A, Hernesniemi J (Jan 2012). Arteriovenous malformations: epidemiology and clinical presentation. Neurosurg Clin N Am 23 (1): 1–6.

Lv X, Wu Z, Jiang C et al. (Jul–Aug 2011). Angioarchitectural characteristics of brain arteriovenous malformations with and without hemorrhage. World Neurosurg 76 (1–2): 95–99.

Lv X, Li Y, Yang X et al. (Mar–Apr 2013). Characteristics of brain arteriovenous malformations in patients presenting with nonhemorrhagic neurologic deficits. World Neurosurg 79 (3-4): 484–488.

Mohr JP, Parides MK, Stapf C et al. (2014). Medical management with or without interventional therapy for unruptured brain arteriovenous malformations (ARUBA): a multicentre, non-blinded, randomised trial. Lancet 383 (9917): 614–621.

Moon K, Levitt MR, Almefty RO et al. (Dec 2015). Safety and efficacy of surgical resection of unruptured low-grade arteriovenous malformations from the modern decade. Neurosurgery 77 (6): 948–953.

Nerva JD, Mantovani A, Barber J et al. (May 2015). Treatment outcomes of unruptured arteriovenous malformations with a subgroup analysis of ARUBA (A Randomized Trial of Unruptured Brain Arteriovenous Malformations)-eligible patients. Neurosurgery 76 (5): 563–570; discussion 570; quiz 570.

Nerva JD, Kim LJ, Barber J et al. (2016). Outcomes of multimodality therapy in pediatric patients with ruptured and unruptured brain arteriovenous malformations. Neurosurgery 78: 695–707.

Ondra SL, Troupp H, George ED et al. (Sep 1990). The natural history of symptomatic arteriovenous malformations of the brain: a 24-year follow-up assessment. J Neurosurg 73 (3): 387–391.

Stapf C, Mohr JP, Pile-Spellman J et al. (2001). Epidemiology and natural history of arteriovenous malformations. Neurosurg Focus 11 (5): e1.

Stapf C, Mast H, Sciacca RR et al. (May 2003). The New York Islands AVM study: design, study progress, and initial results. Stroke 34 (5): e29–e33.

Stapf C, Mast H, Sciacca RR et al. (2006). Predictors of hemorrhage in patients with untreated brain arteriovenous malformation. Neurology 66 (9): 1350–1355.

Tong X, Wu J, Lin F et al. (2016). The effect of age, sex and lesion location on initial presentation in patients with brain arteriovenous malformations. World Neurosurg 87: 598–606.

van Beijnum J, van der Worp HB, Buis DR et al. (2011). Treatment of brain arteriovenous malformations: a systematic review and meta-analysis. JAMA 306 (18): 2011–2019.

van Beijnum J, van der Worp HB, Algra A et al. (Nov 2014). Prevalence of brain arteriovenous malformations in first-degree relatives of patients with a brain arteriovenous malformation. Stroke 45 (11): 3231–3235.

Yang W, Caplan JM, Ye X et al. (Aug 2015). Racial associations with hemorrhagic presentation in cerebral arteriovenous malformations. World Neurosurg 84 (2): 461–469.

Handbook of Clinical Neurology, Vol. 143 (3rd series)
Arteriovenous and Cavernous Malformations
R.F. Spetzler, K. Moon, and R.O. Almefty, Editors
http://dx.doi.org/10.1016/B978-0-444-63640-9.00004-7

Chapter 4

Seizures associated with cerebral arteriovenous malformations

JOHANNES SCHRAMM*
Department of Neurosurgery, University of Bonn, Bonn, Germany

Abstract

Various types of seizures and epilepsy are associated with 20–45% of cerebral arteriovenous malformations (AVMs). The necessity to differentiate between occasional seizures, epilepsy with repetitive seizures, and the much rarer drug-resistant epilepsy (DRE) is underlined. It is clear that where there is frequent seizures or DRE, vascular surgeons should take epilepsy surgery aspects into account. The epidemiology of AVM-associated seizures, assumed pathophysiologic mechanisms, most frequent seizures types, and medical treatment are described. Depending on the severity of the epilepsy, the diagnostic workup, including electroencephalogram (EEG), video-EEG, and, rarely, invasive evaluation, is explained. An invasive presurgical workup is only necessary in rare cases of DRE. The indication to extend the resection to more than just removal of the AVM is defined and the various specific resection techniques for this rare form are outlined. In the vast majority of AVM cases removal of the AVM with some adjoining gliotic or hemosiderotic rim of cortex will be sufficient, however. In the majority of cases with preoperative epilepsy, patients will be seizure-free after surgery. Patients who never had a seizure before AVM removal may develop *de novo* epilepsy postoperatively (5–15%). Rates of seizure freedom after different treatments (microsurgery, radiosurgery, endovascular) vary.

INTRODUCTION

Seizures are the second most frequent presentation of cerebral arteriovenous malformations (AVMs). Seizures are found in 20–45% of AVM cases seen for treatment, yet the major problem in cerebral AVM is the risk of hemorrhage which is around 2–3% annually. In large series hemorrhage is the most frequent presentation in up to 50% of cases and the true significance of this is the resulting morbidity and mortality. Thus, the major concern with cerebral AVM is prevention of bleeding or rebleeding with increased morbidity. However, having seizures or epilepsy brings about a set of problems for the patient which may be considerable – many professions off limits, more accidents, higher mortality, social stigma, no swimming, no car driving, no climbing, plus the need for medication with its side-effects. These factors make AVM-related epilepsy a significant factor to consider in case management to achieve a fully satisfactory outcome. This chapter deals with seizures and epilepsy types found in patients with cerebral AVM.

DEFINITIONS

Seizures in cerebral AVM may be induced by a first hemorrhage, but may be present in unruptured AVMs even for years. The World Health Organization (WHO) has defined epilepsy as a chronic neurologic disorder characterized by recurrent seizures, i.e., brief episodes of involuntary movement of a part or the entire body which may or may not be accompanied by loss of consciousness and loss of control of bowel and bladder function (World Health Organization, 2016). The International League Against Epilepsy (ILAE) prefers to call epilepsy a disease instead of disorder (International League Against Epilepsy, 2014). The WHO uses the term epilepsy if two or more unprovoked seizures are found, whereas a single seizure does not constitute epilepsy.

Seizures are caused by sudden synchronous electric discharges in large groups of neurons. They may be triggered by various stimuli (e.g., trauma, hemorrhage, blood–brain barrier disruption). The patient with AVM-associated seizures carries a double burden. Not

*Correspondence to: Johannes Schramm, M.D., Medical Faculty, Bonn University, Sigmund-Freud Str. 25, 53127 Bonn, Germany. E-mail: johannes.schramm@ukb.uni-bonn.de

only is there the fear of hemorrhage; epilepsy is also associated with many psychosocial problems and may need lifelong medication (with potential side-effects). Patients with chronic epilepsy are also endangered by seizure-associated injuries. Epilepsy patients have a higher mortality rate and patients with drug-resistant epilepsy (DRE) have in fact a shortened life expectancy. The older classification into symptomatic and kryptogenic epilepsies has been replaced by a new classification of epilepsies or syndromes: localization-related, generalized (idiopathic, special and other), and "unspecified" (World Health Organization, 2016). Single or occasional seizures are observed also in healthy brains associated with conditions such as seizure-provoking drugs or sleep deprivation: seizures in patients with epilepsy are frequently associated with structural lesions, cortical malformations, and biochemical alterations in cortical tissue.

It may be practical to look at AVM-associated seizures from the perspective of a neurosurgeon who sees the patient for the first time, and classify them into three groups:

1. Sporadic seizures group: patients who have had one or two seizures only.
2. Chronic epilepsy group: all patients who have had more than two seizures, but do not fulfill the criteria for DRE.
3. DRE: if seizures persist for over 2 years after failure of adequate trials of two tolerated and appropriately used antiepileptic drug (AED) schedules.

The mechanisms whereby epilepsy turns into DRE are unclear. DRE may include many or few seizures, e.g., four seizures per year, four seizures per month, or more. The definition of seizure freedom must be based somehow on the typical interval between seizures. The ILAE has therefore decided that sustained seizure freedom is achieved when the seizure-free period after a new drug regime or after surgery was longer than at least three times the longest interseizure interval (Kwan et al., 2010).

PATHOPHYSIOLOGY OF EPILEPTOGENESIS

AVMs consist of atypical and malformed blood vessels and do not contain neurons. Seizures are not generated by the AVM proper but they are induced by the effects the AVM exerts on the adjacent cortex. The basic rule is that only neurons can generate seizures and thus cortical lesions are most likely to produce seizures. In principle, seizures can develop if the balance between excitation and inhibition at the cellular level or synaptic level is disturbed. The synchronized neuronal discharges seen in seizures may be due to potentiation of excitatory transmissions or to the failure of inhibitory transmission.

The precise mechanisms are as yet unknown, but a number of mechanisms have been described in experimental setups, including gliosis, blood–brain barrier disruption, localized ischemia, hemosiderotic deposits, and scars following microbleeds (LeBlanc and Rasmussen, 1974; Stefan et al., 1987; Wolf et al., 1993; Englot et al., 2012). It is known that not one single cellular mechanism is involved in the generation of seizures. Numerous models of experimental epilepsy exist and seizures can be induced by topical application of chemical substances (kainate, penicillin, aluminum oxide, or compounds with metallic ions, e.g., hemosiderin). The primary inhibitory neurotransmitter gamma-aminobutyric acid (GABA) is presumed to play an important role in epileptogenesis. For temporal-lobe epilepsy GABA increases have been demonstrated by microdialysis technique in the extracellular space. GABA homeostasis is complicated and may be influenced by many factors, such as quantity of GABA transporter proteins, and by complicated astrocyte–pericyte interactions. Drugs effective in convulsive seizures can augment GABAergic neurotransmission (Macdonald and McLean, 1986) or can reduce the capability of cells to generate high-frequency discharges. Focal or generalized convulsions are associated with prolonged membrane deep polarizations.

A more recent discovery was the significant role of astroglia in epileptogenesis. Although astrocytes do not produce ictal discharges, they play a role in K^+ buffering and metabotropic glutamate receptor signaling (Kovács et al., 2012). Blood–brain barrier disruption has been found to play a role in epileptogenesis, especially the leaking of serum proteins, leading to inflammatory response, and impaired homeostasis of the extracellular space, leading to increased excitability of neurons, which results in an altered balance of excitatory and inhibitory elements (Friedman and Heinemann, 2012).

MECHANISMS IN AVM-RELATED EPILEPSY

Lesional epilepsies, such as AVM-related epilepsy, affect brain areas differently; they are more common in temporal and frontal lobe (Friedman and Heinemann, 2012). Several factors have been found to be associated with AVM-related seizures: hemorrhage from the AVM (Thorpe et al., 2000), size of the AVM, feeders from the middle cerebral artery territory, associated venous varices, superficial large AVMs (Morello and Borghi, 1973; Crawford et al., 1986b), and superficial venous drainage (Turjman et al., 1995; Garcin et al., 2012). Of 27 cases, histology of excised seizure foci contained gliosis in 26, hemosiderin deposits in 10, and focal hemorrhages in 4 (Yeh et al., 1990). In a recent series of 103 patients (von der Brelie et al., 2015) temporal and frontal localization were most frequent (37.9% and 33%). In that series unruptured AVMs were associated with seizures in 61%. A seizure as a first symptom was reported in that series in 43%; other authors report 15–35% of patients having a seizure as a first symptom (Perret and Nishioka, 1966; Morello and Borghi, 1973; Michelsen, 1979; Graf et al., 1983; Fults and Kelly, 1984; Wilkins, 1985; Brown et al., 1988; da Costa et al., 2009).

The storage of albumin in astroglial cells around vascular lesions has recently been implicated in AVM-associated epileptogenesis (Raabe et al., 2012). Since seizures are induced by the effects of AVM on the surrounding cortex via ischemia, microhemorrhages, gliosis, inflammatory processes, and albumin extravasation, it is no surprise that epilepsy may cease after removal of the AVM. On the other hand, there must be mechanisms that cause persistence of seizures despite removal of the AVM, most likely because these changes induced permanent epileptogenic areas in the neighboring cortex. One such factor may be the duration of the epilepsy until the removal of the primary lesion (e.g., AVM). The longer the epileptogenic process, brought on and maintained by the presence of an AVM, has existed, the more likely that the AVM-related epilepsy may persist after removal of the vascular malformation. Epilepsy occurs more frequently in larger AVMs (McKissock and Paterson, 1956; Hyun et al., 2012).

EXTENDED EPILEPTOGENIC ZONE AND SECONDARY FOCI

Epileptogenesis may take place in cortical areas distant from the immediate border of the AVM, where it normally has its primary location. This can occur in two different forms: (1) the primary ictal onset zone enlarges over time to form a larger ictogenic area in direct proximity to the AVM; and (2) a second ictogenic area develops away from the cortex surrounding the AVM – a process that usually takes time and happens only in some cases. These distant secondary foci are thought to be induced by exposure to repetitive seizure patterns from cortex close to the epileptogenic lesion. This kindling phenomenon is particularly prominent in the mesial temporal lobe but may also occur in the neocortex, especially in the central area. If kindling in a temporally located AVM occurs, it may lead to secondary hippocampal sclerosis, which may become an independent focus, then called dual pathology, with a higher risk for persisting seizures despite removal of the AVM. However, some secondary epileptogenic cortical areas may stop generating seizures after the primary lesion with its surrounding epileptogenic cortex has been removed. It may be warranted to opt for removal of the primary lesion and secondary focus in one procedure, or for a stepwise approach with invasive evaluation and secondary focus resection only done when seizures persist after lesionectomy.

The more frequently the seizures occur, the larger the lesion, the more cortex is affected, and the longer the duration of the epilepsy, the greater the chance that epileptogenic cortex is not only present in the immediate vicinity of the AVM. A more extensive cortical resection may be necessary, as demonstrated earlier (Yeh et al., 1990; Yeh and Privitera, 1991; von der Brelie et al., 2015). Seizure semiology, ictal electroencephalogram (EEG) findings, and the occasional invasive evaluation with grids or strips may pinpoint the area of a secondary focus (Yeh and Privitera, 1991). The big question is when to go for a more extensive epilepsy investigation, even with invasive recordings. DRE as opposed to chronic epilepsy or sporadic seizures is a definitive candidate for extensive presurgical evaluation (von der Brelie et al., 2015).

NATURAL HISTORY AND EPIDEMIOLOGY

Seizures are the presenting symptom in 20–40% of cases in larger series (Kida et al., 2000; Hoh et al., 2002; Josephson et al., 2011; von der Brelie et al., 2015). The most frequent seizure type in AVM patients is generalized tonic-clonic seizure (Kida et al., 2000; Hoh et al., 2002; Josephson et al., 2011; von der Brelie et al., 2015), although in the DRE subgroup initial focal seizures were more frequent (von der Brelie et al., 2015). Generalized tonic-clonic seizures were most frequent in a large series (Piepgras et al., 1993) and of the focally starting seizures half were associated with secondary generalization. Larger AVMs were found to be more likely to present with seizures, whereas AVMs <3 cm were more likely to present with hemorrhage (Piepgras et al., 1993). Temporally located AVMs have a tendency to present with complex partial seizures, many of which, however, later generalize.

The risk of developing seizures in an incidentally discovered AVM, where no seizures had occurred so far, is relatively low (8% within 5 years: Josephson et al., 2011, or 1.1% per person-year: Crawford et al., 1986a, b), so prophylactic AED therapy may not be indicated. If the patient had initially presented with hemorrhage the risk of a first seizure within the next 5 years increases to 23%. AVM patients who presented with a single seizure had a 5-year risk of developing epilepsy of 58% (Josephson et al., 2011). In a study of 153 nonoperated patients who were followed up for a 20-year period, 18% developed epilepsy (Crawford et al., 1986a, b). When diagnosed, younger patients aged 10–19 years were found to have a 44% risk compared to 6% risk in those who were over the age of 30, during a 20-year follow-up period (Crawford et al., 1986a). The same study also confirmed that temporal-lobe location carries the greatest risk of developing epilepsy (Crawford et al., 1986b).

Drug-resistant epilepsy

Whereas about 30% of all epilepsy patients suffer from DRE, the percentage of DRE in AVM-related seizures is much lower. A few studies have clearly described DRE cases (Yeh and Privitera, 1991; Yeh et al., 1993; Lim et al., 2006); however their percentage of DRE in the whole series was not given. It was 24 of 103 in the own AVM series (von der Brelie et al., 2015), and less than 1% AVM cases out of over 2000 resective epilepsy surgery cases in the Bonn epilepsy surgery program. In a very large series 18% of epilepsy cases had DRE (Englot et al., 2012).

In our own DRE group preoperative duration of epilepsy was significantly longer (156 months vs. 37 months for chronic epilepsy and 6 months for sporadic seizures). Initial intracerebral hemorrhage was less frequent and frontal localization of the AVM was most common; patients more frequently had multimodal treatment before surgery and they were twice as frequently associated with high Spetzler–Martin grades compared to the total cohort (von der Brelie et al., 2015). More extensive resections led to better results: 8 of 10 patients with extended lesionectomy were seizure-free, but only 6 of 14 with standard resections (no statistical significance because of low numbers). In another series factors predisposing to develop DRE were hemorrhage, frontotemporal lesion, male gender, and larger AVM size (3.1 cm vs. 2.4 cm in nonseizure cases) (Englot et al., 2012).

Few data on management of DRE in AVM cases are available (Yeh et al., 1993; Cao et al., 2011; von der Brelie et al., 2015). In another series with mostly DRE cases, 25 of 54 patients had additional cortical excisions, i.e., extended lesionectomies (Yeh et al., 1993); 12 of these were in a remote location, i.e., foci of secondary epileptogenesis. In our own series 13 of 24 DRE cases underwent epileptologic evaluation leading to an epilepsy surgery approach in 11 of those 13 cases (von der Brelie et al., 2015); in another series 10 of 17 temporal AVMs were associated with a remote seizure focus (Yeh et al., 1993).

DRUG TREATMENT

For patients presenting with a first ever seizure but no hemorrhage leading to the diagnosis of AVM, the first-line treatment is medical, unless AVM removal is indicated. Patients with primary generalized seizures need other AEDs than do patients with focal seizures. Since recommended drug regimes change over time it is wise to involve an epileptologist. The chance to achieve seizure freedom with AED is about 70% in any epilepsy, but was only 60% in a large observational population-based AVM study (Josephson et al., 2011). The same study described that immediate AED prescription increased the time to seizure recurrence and reduced the time to 2-year seizure freedom. After successful AVM removal in cases with previous chronic epilepsy or DRE it is recommended to continue the AED for at least 1 year, and preferably 2 years, as is standard practice in epilepsy surgery.

PRESURGICAL DIAGNOSTICS

The neuroradiologic routine examination is the same for AVM cases with and without seizures: four-vessel digital subtraction angiography and magnetic resonance imaging (MRI). The MRI examination may give hints with regard to gliotic areas, edematous zones, and microhemorrhages at the AVM border, but also of edema in the area of a more distended or stenotic draining vein as potential sources of distant epileptogenic cortex. The MRI in temporal AVM cases with chronic epilepsy or DRE should ideally be performed with sequences used for presurgical epilepsy workup. In order to obtain good visualization of the hippocampus and other mesiotemporal structures the axial slices should be inclined parallel to the length axis of the hippocampus.

No specific electrophysiologic techniques need be applied for AVMs with only sporadic seizures.

For chronic epilepsy, as defined above, preoperative video-EEG is advisable if seizure frequency is high and seizure types or frequency are very disturbing, for example, affecting professional activity.

Workup in DRE cases focuses on localization of the ictogenic area, since AVM removal is done in a standard fashion. DRE cases need to have video-EEG with the aim of recording at least two seizures. Video-EEG may not be available everywhere and should be done in collaboration with an epilepsy service. The seizure semiology and possible postictal neurologic deficits may give important hints regarding the presence and location of a so-called remote ictogenic focus.

The use of invasive evaluation with strip, depth, or grid electrodes will rarely be necessary. It will be indicated if there is a discrepancy between the MRI and the findings from the noninvasive evaluation, i.e., video-EEG recordings of two seizures, and the localizing signs obtained from careful history taking and the semiology of the seizures. A typical example would be a small AVM in the frontal lobe and a seizure semiology pointing to the temporal lobe. DRE from temporal AVM must be particularly carefully evaluated because of the well-known propensity of the temporal lobe to develop secondary epileptogenic areas (Yeh and Privitera, 1991; Cascino et al., 1992; Jooma et al., 1995; Moore et al., 1999). Intraoperative electrocorticography does not reliably indicate the epileptogenic zone or remote seizure foci because no seizures are recorded.

The implantation of strip and grid electrodes can be used for extraoperative mapping of both, epileptogenic and functional cortex. It requires a first craniotomy for the implantation and a second one for the resection, but offers the huge advantage that mapping of cortical function can be performed outside the operating room for lengthy periods if necessary. Video-EEG with implanted electrodes is the most useful EEG recording type for detection of seizure origin. It can help to plan the resection of AVM close to eloquent cortex and the localization of secondary epileptogenic areas and their topographic relationship to functional cortex. Again, collaboration with an epilepsy surgery group is very important. Awake craniotomy can only be used for functional mapping, not for precise localization of ictogenic areas, since intraoperative seizures are normally not present, and rather are to be avoided.

SURGICAL STRATEGIES

Considering surgery, prevention of hemorrhage is the classic indication, not treatment of seizures. Thus, with an unruptured AVM discovered because of one or a few seizures, the seizures do not constitute an indication for surgery. The primary treatment for rare seizures is pharmacotherapy. The main aim is complete removal of the AVM using the same microsurgical techniques as in cases without epilepsy. In patients with only a few seizures, the chance to obtain seizure freedom with simple AVM removal is good, so no specific resection strategy appears necessary, apart from including small areas of gliosis and hemosiderotic cortex immediately adjacent to the AVM border, provided eloquent cortex is not affected.

In patients with AVMs associated with chronic epilepsy where DRE is present, it is worth being aware of some concepts of epilepsy surgery. Ideally the vascular surgeon confronted with these two conditions should know about and include epilepsy surgery aspects in the planning of the AVM removal.

Rarely, patients who do not initially wish to have their AVM removed but suffer from uncontrolled seizures are candidates for surgery with the primary aim of achieving seizure freedom. For these cases planning should include considerations to optimize chances to achieve seizure freedom, i.e., potentially an invasive evaluation to detect extended areas of epileptogenicity. The indication for surgery may also be given for patients who cannot tolerate side-effects of anticonvulsants (Yeh et al., 1993).

Epilepsy surgery concepts applicable in AVM removals

The resection of an AVM which presented with one or a few seizures does not have to consider principles of epilepsy surgery, since the term epilepsy surgery is reserved for cases with DRE. AVM removal for chronic seizures and DRE may be considered lesional epilepsy surgery. Thus, the vascular surgeon should be aware of the definition of DRE (see section on definitions, above).

AVM removal alone (in epilepsy surgery terminology a "pure lesionectomy") may be supplemented by additional removal of a cortical rim, or adjacent scarred/gliotic cortical areas. For lesional epilepsy several options exist: simple lesionectomy (i.e., only the AVM is removed), lesionectomy with rim of cortex, and extended lesionectomy, i.e., lesionectomy with rim plus resection of secondary epileptogenic focus or dual pathology. The extended cortical removal may concern a defined cortical area in the vicinity of the lesion beyond the small cortical rim or in temporal-lobe epilepsy it may include amygdalohippocampectomy, or, rarely, even temporal lobectomy. Rasmussen already defined in 1983, for the purpose of resective surgery in epilepsy, primary and secondary localization. In cases of cerebral AVM the primary localization would be the irritated or disturbed cortex immediately bordering on the AVM, corresponding to the term lesionectomy with rim. The secondary localization would be the adjacent cortex that must be recruited to generate a clinical seizure. In a typical case these two regions would be lying side by side and the secondary location would correspond in some cases to a larger adjacent cortical area needing an extended lesionectomy. The rarer variant of extended lesionectomy is necessary if the recruited epileptogenic cortex lies distant from the immediate cortex around the AVM, such as, for example, the hippocampus in a case of lateral superficial temporal AVM.

During the very long debate on what procedure is best, some facts have evolved for lesion-related epilepsy in general, not necessarily restricted to AVM-related seizures. Although resection of the AVM alone may be associated with good postoperative control of seizure activity, simple lesionectomy for temporal-lobe lesions has poorer results than lesionectomy combined with resection of mesial structures (Cascino et al., 1992; Jooma et al., 1995; Moore et al., 1999). Findings that support a greater likelihood of being seizure-free after a simple lesionectomy were shorter duration of the seizure disorder, lower seizure frequency, or good response to antiepileptic treatment (Rassi-Neto et al., 1999). For several lesion types (benign tumors and vascular malformations) combined lesionectomy plus corticectomy were more successful (Weber et al., 1993).

For chronic epilepsy with disturbing seizure types or affecting professional activity it may be advisable in those cases where the AVM is distant from eloquent cortex to perform a lesionectomy with a wider rim, i.e., AVM removal, including areas of gliotic and hemosiderotic cortex in the immediate vicinity plus 0.5 cm of healthy-appearing cortex.

Although the superiority of an epilepsy surgery concept used in AVM removal with DRE or chronic epilepsy has been seen (Yeh et al., 1993; Jafar and Huaeng, 1999), but not proven so far, success in non-AVM cases supports its use.

POSTOPERATIVE SEIZURES AND OUTCOME ASSESSMENT

There are a variety of postoperative seizures. They may be a continuation of preoperative seizures, they may be *de novo* seizures in previously seizure-free patients, and they may constitute a deterioration of pre-existent seizures because of higher frequency or worse seizure type. To make assessment more complicated some patients have a single early postoperative seizure; for example, in the San Francisco series 16% had a single postoperative seizure (Englot et al., 2012).

It is difficult to interpret the available literature, since outcome scales common in epilepsy surgery are not used by many authors; only modern series have used the Engel outcome scale, sometimes considerably modified (Przybylowski et al., 2015). Many authors have used individualized outcome classifications. Unfortunately it is not always clear whether seizure freedom is meant according to the Engel classification or to the ILAE classification. In Engel I seizure-free patients can still have auras, whereas ILAE outcome class I is seizure free without aura. Outcome scores using different cutoffs vary widely. Sometimes a difference is made between one seizure versus more than two (Thorpe et al., 2000; Ghossoub et al., 2001; Schäuble et al., 2004; Cao et al., 2011) and others choose one to three seizures versus more than four (Piepgras et al., 1993; Yeh et al., 1993; Lim et al., 2006). In older references one may find terms like "improved" or combined outcome descriptions like "seizure-free or significantly improved." In one meta-analysis (Chen et al., 2014), the terms "seizure-free status" and "seizure control" were used, with the latter including "seizure improvement." Comparing seizure outcomes is difficult to impossible. It is also not easy to extract different outcome data for the various patient groups or for different durations of the epilepsy disorder, two factors known to influence seizure outcome.

In principle, one would expect best outcomes after the removal of the AVM with its surrounding damaged cortical rim. The AVM-derived factors known to induce epileptogenesis should ideally be removed in the course of AVM resection (e.g., hemosiderin deposits, gliosis) or disappear shortly after (e.g., edema, localized ischemia). Since some of these factors disappear only if they are mechanically removed, and others when the hemodynamic abnormalities around the AVM disappear, it becomes clear that complete obliteration of the AVM alone can lead to cessation of seizures. With microsurgery the hemodynamic effects will be gone but also the mechanical removal of structural changes has occurred, so a wider spectrum of ictogenic factors is dealt with when microsurgery is used. If edema, pulsatile compression, and relative ischemia are leading factors in seizure generation, total occlusion by embolization will significantly reduce the ictogenic potential of the AVM.

In microsurgical removal of the AVM by bipolar coagulation and use of sucker a small corridor of work around this lesion is created and a proportion of the structural changes very close to the AVM will automatically be included in the resected tissue volume. During microsurgery, hematomas and cortical zone overlying hematomas will additionally be removed It is thus no surprise that experience has demonstrated that simple AVM removal can lead to quite satisfactory seizure-free rates. One has to assume that, once the AVM is gone, the ictogenic factors originating from or produced by the AVM are so greatly reduced or have disappeared that they no longer irritate the surrounding cortex. Any treatment modality that does not obliterate or remove the AVM completely would therefore be expected to lead to less satisfactory seizure outcomes Some results support the concept that seizure-free rates likely improve when the AVM has been completely occluded (Ghossoub et al., 2001; Hoh et al., 2002; Lim et al., 2006; Hyun et al., 2012; Yang et al., 2012).

Early postoperative seizure versus *de novo* epilepsy

Unfortunately patients who never experienced seizures before AVM removal may have *de novo* seizures postoperatively. One or two early postoperative seizures do not constitute epilepsy. Occasional single seizures in the early postoperative period are not rare (Englot et al., 2012) and are not synonymous with persistent epilepsy. *De novo* seizures were seen in 6% of a large series (Piepgras et al., 1993); additionally in 2%, seizure patterns deteriorated, and there was a high rate of *de novo* seizures (17.6%) in another (Rohn et al., 2014). They appeared later than 1 year after surgery in 6.3% (Thorpe et al., 2000). Other authors reported that epilepsy worsened following a seizure (Forster et al., 1972; Parkinson and Bachers, 1980; Murphy, 1985).

The frequency of *de novo* seizures seems to be influenced by the size of the AVM: AVMs < 3 cm in 3%, 3–6 cm in 6%, and > 6 cm in 16% *de novo* seizures (Piepgras et al., 1993). Persisting and repetitive seizures, to be called epilepsy if they are more than two, should not be mixed up with one or two seizures in the first 7–10 days after surgery. They most likely stem from the surgical trauma to the cortex analogous to the occasional seizure after other types of cerebral surgery. In a meta-analysis 9.1% of 547 microsurgery patients, 5.4% of 568 stereotactic radiosurgery patients, and 39.4% in the endovascular group experienced *de novo* seizures (Baranoski et al., 2014).

SEIZURE OUTCOMES

Seizure outcomes after microsurgery

Approximately 45–80% of patients with AVM-associated seizures become seizure-free after microsurgery. Seizure-free outcomes were 43% (Yeh et al., 1990), 89.5% (Piepgras et al., 1993), 70.9% (Yeh et al., 1993), 70.4% (Thorpe et al., 2000), 77.4% (Heros et al., 1990), 78.1% (Cao et al., 2011), 83.3% (Hyun et al., 2012), 80.6% (Englot et al., 2012), 93% (Hoh et al., 2002). In our own cohort (von der Brelie et al., 2015), overall seizure freedom was 76.7% and 84.5% had a "favorable" outcome, corresponding to ILAE 1–3, which is Engel I and II, i.e., up to two seizures a year. Seizure outcomes varied considerably with duration of preoperative seizure disorder: 85.7% of patients with sporadic seizures were seizure-free, 80.5% with chronic epilepsy, and only 58.3% with DRE. Other series had better outcomes for DRE (Hoh et al., 2002), with 91% seizure-free. Interestingly, outcomes improved from 1 year (77%) to 79% at 5 years and 84% at 10 years (von der Brelie et al., 2015).

Seizure outcomes after embolization

Not many series with seizure outcomes after AVM embolization are available (Fournier et al., 1991; Hoh et al., 2002; Le Feuvre and Taylor, 2007; Lv et al., 2010; Los Reyes et al., 2011; Hyun et al., 2012). In some studies seizure outcome for different modalities is given (Hoh et al., 2002; Hyun et al., 2012; Baranoski et al., 2014). If embolization occluded the AVM totally the effect on epileptogenesis-inducing factors should be detectable early, in contrast to radiosurgery, where AVM occlusion may take 2–4 years. In a series of 30 patients who had endovascular treatment for their AVM, 70% became seizure-free during short-term follow-up (Lv et al., 2010). Other series found 50% seizure freedom after embolization (Hoh et al., 2002) or 70% seizure-free or occasional auras (Lv et al., 2010). Embolization, which is at present unable to totally occlude most AVMs – only smaller ones in up to 30% or 40% – leaves most of the cortex-disturbing factors behind. In one series, no patient became seizure-free (Fournier et al., 1991). In a meta-analysis endovascular treatment achieved 49.3% seizure freedom in a mean of five studies (Baranoski et al., 2014).

Seizure outcomes after radiosurgery

In two multidisciplinary series radiosurgery cases had lower seizure freedom rates than microsurgery: 43% vs. 81% (Hoh et al., 2002), and 78% vs. 66% (Hyun et al., 2012). A recent review article shows much lower rates of complete seizure freedom compared to microsurgically removed AVMs. Another review found a seizure freedom rate of 62.8% for all radiosurgically treated AVM versus 78.3% for microsurgery (Baranoski et al., 2014), and for completely obliterated AVM higher rates of seizure freedom (85.2%) with radiosurgery. Remarkably, only 41% of partially occluded cases were seizure-free versus 82% of totally occluded AVM (Chen et al., 2014). These numbers speak for complete obliteration being an important factor in achieving high rates of seizure freedom. In 73 of 253 patients with pretreatment seizures sufficient seizure outcome data were available in a series from Charlottesville, of which 36 were completely seizure-free (Engel class IA, 49%) or had auras or simple partial seizures (Engel class IB, $n = 29$, 40%) (Przybylowski et al., 2015). In a series of sylvian fissure AVM, total obliteration rates increased from 35% at 3 years to 60% at 4 and 5 years and 76% at 10 years. For patients with pretreatment seizures, rates of seizure improvement were 29% at 3 years, 36% at 5 years, 50% at 10 years, and 60% at 15 years – overall 53% Engel class I outcome. In a recent review of 19 radiosurgery case series (Chen et al., 2014), 997 patients with available seizure outcome were identified from 19 radiosurgery case series and had 43.8% seizure freedom and 68.7% "seizure freedom or improvement" and concluded that "improved seizure control is significantly more likely with complete AVM obliteration." However, it is difficult to prove beyond doubt that incomplete obliteration is the major factor in a poorer outcome since many other factors have been associated with influencing seizure outcome, e.g., short seizure disorder, infratentorial location, AVM size, hemorrhagic presentation.

On the one hand there is a presumed direct suppressive action of radiation on seizure generation in the cortex. On the other hand, when radiosurgery is applied for temporal-lobe epilepsy there is a typical considerable increase in the frequency of auras in the first 6–9 months, so there is no doubt that radiosurgery has the potential to induce seizures in the short term. Although radiosurgery has been used for the treatment of mesial temporal-lobe epilepsy, radiosurgery may cause delayed epilepsy via radiochemical damage (necrosis, edema) (Husain et al., 2001; Yang et al., 2012). Epilepsy is not expected to cease immediately after radiosurgery since obliteration of the AVM takes some time. In this period hemorrhages continue to occur at an annual rate of about 2% and these again may result in epilepsy (Yang et al., 2012).

SHORT SUMMARY OF MANAGEMENT STRATEGY

- AVM with no seizures: just treat with appropriate treatment modality, no prophylactic AEDs necessary. In case of microsurgery include microhemorrhages and areas of gliosis in adjacent cortex.
- AVM with sporadic seizures: choose appropriate treatment modality, no prophylactic AEDs necessary. In case of microsurgery include microhemorrhages and areas of gliosis in adjacent cortex.
- AVM with more than two seizures, with many seizures, or chronic epilepsy: if microsurgery is chosen, a small rim of adjacent normal cortex should be included when removing the AVM. In chronic epilepsy of long standing or in cases with significantly debilitating seizure types, video-EEG with recording of at least two seizures is recommended in order to detect a possible extended area of ictogenic cortex. In temporally located AVM, secondary epileptogenesis in mesial temporal-lobe structures should be excluded before surgery.
- AVM with DRE: efforts to get rid of the epilepsy are as important as total occlusion or removal of the AVM. A typical presurgical evaluation, as in any other case of DRE, is warranted in order to detect a possible extended area of ictogenic cortex. In temporally located AVM, secondary epileptogenesis in mesial temporal-lobe structures should be excluded before surgery. This should include at least video-EEG with recording of two seizures and possibly invasive presurgical evaluation with subdural or depth electrodes.

SUMMARY

It is obvious that all three treatment modalities can achieve seizure freedom. However, the literature does not easily allow us to conclude whether one of the available treatment modalities gives superior outcomes with regard to seizure freedom. Since complete AVM removal or obliteration yields better seizure outcomes, one would expect better seizure outcomes with microsurgery since this modality has the highest total occlusion rate. It is also frequently difficult to find out how outcomes vary for patients with sporadic seizures versus those with epilepsy of longer standing or even those with DRE.

The need for vascular surgeons to become acquainted with concepts of epilepsy surgery when they write up their AVM series and describe AVM-related seizures is obvious. It would be ideal, if vascular surgeons came to accept that chronic epilepsy or DRE associated with AVMs is a similarly important task to be dealt with like removal of the AVM. For this purpose vascular surgeons should become familiar with concepts from epilepsy surgeons on how to approach lesions (i.e., AVMs) associated with chronic or DRE. This includes the need to be aware that secondary epileptogenic foci may exist somewhat distant from the AVM. For those rare cases of DRE associated with AVM it may be most appropriate to seek collaboration with an epileptologist or epilepsy surgeon.

References

Baranoski JF, Grant RA, Hirsch LJ et al. (2014). Seizure control for intracranial arteriovenous malformations is directly related to treatment modality: a meta-analysis. Journal of neurointerventional surgery 6 (9): 684–690.

Brown RD, Wiebers DO, Forbes G et al. (1988). The natural history of unruptured intracranial arteriovenous malformations. J Neurosurg 68 (3): 352–357.

Cao Y, Wang R, Yang L et al. (2011). Bipolar electrocoagulation on cortex after AVMs lesionectomy for seizure control. The Canadian journal of neurological sciences. Le journal canadien des sciences neurologiques 38 (1): 48–53.

Cascino GD, Kelly PJ, Sharbrough FW et al. (1992). Long-term follow-up of stereotactic lesionectomy in partial epilepsy: predictive factors and electroencephalographic results. Epilepsia 33 (4): 639–644.

Chen C-J, Chivukula S, Ding D et al. (2014). Seizure outcomes following radiosurgery for cerebral arteriovenous malformations. Neurosurg Focus 37 (3): E17.

Crawford PM, West CR, Chadwick DW (1986a). Arteriovenous malformation of the brain: natural history in unoperated patients. J Neurol Neurosurg Psychiatry 49: 1–10.

Crawford PM, West CR, Shaw MD (1986b). Cerebral arteriovenous malformations and epilepsy: factors in the development of epilepsy. Epilepsia 27: 270–275.

Da Costa L, Wallace MC, Ter Brugge KG et al. (2009). The natural history and predictive features of hemorrhage from brain arteriovenous malformations. Stroke; a journal of cerebral circulation 40 (1): 100–105.

Englot DJ, Young WL, Han SJ et al (2012). Seizure predictors and control after microsurgical resection of supratentorial arteriovenous malformations in 440 patients. Neurosurgery 71 (3): 572–580; discussion 580

Forster DM, Steiner L, Håkanson S (1972). Arteriovenous malformations of the brain. A long-term clinical study. J Neurosurg 37 (5): 562–570.

Fournier D, Terbrugge KG, Willinsky R et al. (1991). Endovascular treatment of intracerebral arteriovenous malformations: experience in 49 cases. J Neurosurg 75 (2): 228–233.

Friedman A, Heinemann U (2012). Role of blood-brain barrier dysfunction in epileptogenesis. In: JL Noebels, M Avoli, MA Rogawski et al. (Eds.), Jaspers's Basic Mechanisms of the Epilepsies, 4th Edition. Oxford University Press, Oxford.

Fults D, Kelly DL (1984). Natural history of arteriovenous malformations of the brain: a clinical study. Neurosurgery 15 (5): 658–662.

Garcin B, Houdart E, Porcher R et al. (2012). Epileptic seizures at initial presentation in patients with brain arteriovenous malformation. Neurology 78 (9): 626–631.

Ghossoub M, Nataf F, Merienne L et al. (2001). Evolution des crises d'épilepsie associées aux malformations artérioveineuses cérébrales après traitement par radiochirurgie. Neurochirurgie 47 (2–3 Pt 2): 344–349.

Graf CJ, Perret GE, Torner JC (1983). Bleeding from cerebral arteriovenous malformations as part of their natural history. J Neurosurg 58 (3): 331–337.

Heros RC, Korosue K, Diebold PM (1990). Surgical excision of cerebral arteriovenous malformations: late results. Neurosurgery 26 (4): 570–577; discussion 577–578.

Hoh BL, Chapman PH, Loeffler JS et al. (2002). Results of multimodality treatment for 141 patients with brain arteriovenous malformations and seizures: factors associated with seizure incidence and seizure outcomes. Neurosurgery 51 (2): 303–309; discussion 309–311.

Husain AM, Mendez M, Friedman AH (2001). Intractable epilepsy following radiosurgery for arteriovenous malformation. J Neurosurg 95 (5): 888–892.

Hyun S-J, Kong D-S, Lee J-I et al. (2012). Cerebral arteriovenous malformations and seizures: differential impact on the time to seizure-free state according to the treatment modalities. Acta Neurochir 154 (6): 1003–1010.

International League Against Epilepsy (2014). Definition of epilepsy. Available online at http://www.ilae.org/Visitors/Centre/Definition-2014.cfm (accessed November 11, 2016).

Jafar JA, Huaeng P (1999). Intracranial vascular malformations: clinical decision and multimodality management strategies. In: JA Awad, R Rosenmeier (Eds.), Vascular Malformations of the Central Nervous System, Lippincott Williams and Wilkins, Philadelphia, pp. 219–232.

Jooma R, Yeh HS, Privitera MD et al. (1995). Lesionectomy versus electrophysiologically guided resection for temporal lobe tumors manifesting with complex partial seizures. J Neurosurg 83 (2): 231–236.

Josephson CB, Leach J-P, Duncan R et al. (2011). Seizure risk from cavernous or arteriovenous malformations: prospective population-based study. Neurology 76 (18): 1548–1554.

Kida Y, Kobayashi T, Tanaka T et al. (2000). Seizure control after radiosurgery on cerebral arteriovenous malformations. Journal of clinical neuroscience : official journal of the Neurosurgical Society of Australasia 7 (Suppl 1): 6–9.

Kovács R, Heinemann U, Steinhäuser C (2012). Mechanisms underlying blood-brain barrier dysfunction in brain pathology and epileptogenesis: role of astroglia. Epilepsia 53 (Suppl 6): 53–59.

Kwan P, Arzimanoglou A, Berg AT et al. (2010). Definition of drug resistant epilepsy: consensus proposal by the ad hoc Task Force of the ILAE Commission on Therapeutic Strategies. Epilepsia 51 (6): 1069–1077.

Leblanc FE, Rasmussen T (1974). Cerebral seizures and brain tumors. In: PJ Vinken, GW Bruyn (Eds.), Handbook of Clinical Neurology, vol. 15. North-Holland Publishing Co., Amsterdam, pp. 295–301.

Le Feuvre D, Taylor A (2007). Target embolization of AVMs: identification of sites and results of treatment. Interventional neuroradiology : journal of peritherapeutic neuroradiology, surgical procedures and related neurosciences 13 (4): 389–394.

Lim YJ, Lee CY, Koh JS et al. (2006). Seizure control of gamma knife radiosurgery for non-hemorrhagic arteriovenous malformations. Acta Neurochir Suppl 99: 97–101.

De Los Reyes K, Patel A, Doshi A et al. (2011). Seizures after onyx embolization for the treatment of cerebral arteriovenous malformation. Interventional neuroradiology : journal of peritherapeutic neuroradiology, surgical procedures and related neurosciences 17 (3): 331–338.

Lv X, Li Y, Jiiang C et al. (2010). Brain arteriovenous malformations and endovascular treatment: effect on seizures. Interventional neuroradiology : journal of peritherapeutic neuroradiology, surgical procedures and related neurosciences 16 (1): 39–45.

MacDonald RL, Mclean MJ (1986). Anticonvulsant drugs: mechanisms of action. Adv Neurol 44: 713–736.

Mckissock W, Paterson JH (1956). A clinical survey of intracranial angiomas with special reference to their mode of progression and surgical treatment: a report of 110 cases. Brain : a journal of neurology 79 (2): 233–266.

Michelsen WJ (1979). Natural history and pathophysiology of arteriovenous malformations. Clin Neurosurg 26: 307–313.

Moore J, Cascino GD, Trenerry MR (1999). A comparative study of lesionectomy versus corticectomy in patients with temporal lobe lesional epilepsy. J Epilepsy 6: 239–242.

Morello G, Borghi GP (1973). Cerebral angiomas. A report of 154 personal cases and a comparison between the results of surgical excision and conservative management. Acta Neurochir 28 (3): 135–155.

Murphy MJ (1985). Long-term follow-up of seizures associated with cerebral arteriovenous malformations. Results of therapy. Arch Neurol 42 (5): 477–479.

Parkinson D, Bachers G (1980). Arteriovenous malformations. Summary of 100 consecutive supratentorial cases. J Neurosurg 53 (3): 285–299.

Perret G, Nishioka H (1966). Report on the cooperative study of intracranial aneurysms and subarachnoid hemorrhage. Section VI. Arteriovenous malformations. An analysis of 545 cases of cranio-cerebral arteriovenous malformations and fistulae reported to the cooperative study. J Neurosurg 25 (4): 467–490.

Piepgras DG, Sundt TM, Ragoowansi AT et al. (1993). Seizure outcome in patients with surgically treated cerebral arteriovenous malformations. J Neurosurg 78 (1): 5–11.

Przybylowski CJ, Ding D, Starke RM et al. (2015). Seizure and anticonvulsant outcomes following stereotactic radiosurgery for intracranial arteriovenous malformations. J Neurosurg 122 (6): 1299–1305.

Raabe A, Schmitz AK, Pernhorst K et al. (2012). Cliniconeuropathologic correlations show astroglial albumin storage as a common factor in epileptogenic vascular lesions. Epilepsia 53 (3): 539–548.

Rasmussen TR (1983). Surgical treatment of complex partial seizures: results, lessons, and problems. Epilepsia 24 (Suppl 1): S65–S76.

Rassi-Neto A, Ferraz FP, Campos CR et al. (1999). Patients with epileptic seizures and cerebral lesions who underwent lesionectomy restricted to or associated with the adjacent irritative area. Epilepsia 40 (7): 856–864.

Rohn B, Haenggi D, Etminan N et al. (2014). Epilepsy, headache, and quality of life after resection of cerebral arteriovenous malformations. Journal of neurological surgery Part A, Central European neurosurgery 75 (4): 282–288.

Schäuble B, Cascino GD, Pollock BE et al. (2004). Seizure outcomes after stereotactic radiosurgery for cerebral arteriovenous malformations. Neurology 63 (4): 683–687.

Stefan H, Pawlik G, Böcher-Schwarz HG et al. (1987). Functional and morphological abnormalities in temporal lobe epilepsy: a comparison of interictal and ictal EEG, CT, MRI, SPECT and PET. J Neurol 234 (6): 377–384.

Thorpe ML, Cordato DJ, Morgan MK et al. (2000). Postoperative seizure outcome in a series of 114 patients with supratentorial arteriovenous malformations. Journal of clinical neuroscience: official journal of the Neurosurgical Society of Australasia 7 (2): 107–111.

Turjman F, Massoud TF, Sayre JW et al. (1995). Epilepsy associated with cerebral arteriovenous malformations: a multivariate analysis of angioarchitectural characteristics. AJNR Am J Neuroradiol 16 (2): 345–350.

von der Brelie C, Simon M, Esche J et al. (2015). Seizure outcomes in patients with surgically treated cerebral arteriovenous malformations. Neurosurgery 77 (5): 762–768.

Weber JP, Silbergeld DL, Winn HR (1993). Surgical resection of epileptogenic cortex associated with structural lesions. Neurosurg Clin N Am 4 (2): 327–336.

Wilkins RH (1985). Natural history of intracranial vascular malformations: a review. Neurosurgery 16 (3): 421–430.

Wolf HK, Campos MG, Zentner J et al. (1993). Surgical pathology of temporal lobe epilepsy. Experience with 216 cases. J Neuropathol Exp Neurol 52 (5): 499–506.

World Health Organization (2016). Epilepsy. Available online at http://www.who.int/mediacentre/factsheets/fs999/en/ (accessed November 11, 2016).

Yang S-Y, Kim DG, Chung H-T et al. (2012). Radiosurgery for unruptured cerebral arteriovenous malformations: long-term seizure outcome. Neurology 78 (17): 1292–1298.

Yeh HS, Privitera MD (1991). Secondary epileptogenesis in cerebral arteriovenous malformations. Arch Neurol 48 (11): 1122–1124.

Yeh HS, Kashiwagi S, Tew JM et al. (1990). Surgical management of epilepsy associated with cerebral arteriovenous malformations. J Neurosurg 72 (2): 216–223.

Yeh HS, Tew JM, Gartner M (1993). Seizure control after surgery on cerebral arteriovenous malformations. J Neurosurg 78 (1): 12–18.

Handbook of Clinical Neurology, Vol. 143 (3rd series)
Arteriovenous and Cavernous Malformations
R.F. Spetzler, K. Moon, and R.O. Almefty, Editors
http://dx.doi.org/10.1016/B978-0-444-63640-9.00005-9

Chapter 5

Surgical management

MICHAEL K. MORGAN*
Department of Clinical Medicine, Macquarie University, Sydney, New South Wales, Australia

Abstract

Surgical management includes selection of patients for surgery, performing the technical procedure of brain arteriovenous malformation (bAVM) resection and perioperative management that maximize the chance for the best outcome. In general the Spetzler–Ponce class (SPC) can divide patients into those with good evidence that surgery is appropriate in most cases (SPC A), those in whom surgery should only be considered occasionally with highly nuanced indications (SPC C), and surgery may be appropriate having made a detailed analysis of patient (including age), clinical (including mode of presentation), and AVM characteristics (including diffuseness), and a comparative analysis of outcomes with alternate management pathways for SPC B cases. The underlying competent performance of surgery must successfully achieve: consideration of the physiology; correct identification of vessel; protection of the arterial supply to normal brain; understanding of the expected anatomic relationship between feeding arteries and draining veins; and recognition and management of complex arterial feeding patterns from transdural and transosseous sources. Aggressive blood pressure management is required for bAVM with significant changes to brain vascular physiology as a consequence of surgery. For such cases, brain vascular remodeling will take approximately 1 week after surgery. During this period, protection against elevation of blood pressure must be strictly achieved.

BACKGROUND

There is almost nothing as satisfying in a neurosurgeon's career as the execution of a technically proficient excision of a brain arteriovenous malformation (bAVM). Nor is there likely to be anything as terrifying as a bAVM resection going wrong. Therefore, it is extremely important that the decision to recommend surgery for bAVM is not clouded by the desire to operate, nor fear, and is firmly based upon the needs of the patient and informed by the evidence. The complexity in establishing those patients for whom surgery is of benefit is confounded by the choices between surgery, embolization, radiosurgery, conservative management, and their combinations. Outcomes of combined management, where responsibilities towards the goal of obliteration of bAVM may vary between proceduralists and institutions, have been difficult to critically evaluate. Analyzing retrospectively surgical series almost always includes some series that perform preoperative embolization and difficulty may arise in identifying cases that failed to proceed to surgery because of complications sustained during the embolization procedure.

Establishing complications for bAVM management related to an intention to treat is presently limited to a single randomized control trial of unruptured bAVM (ARUBA) in which 18 cases had surgery as part of management, of whom only 5 cases were treated exclusively by surgery (Mohr et al., 2014). These small numbers prevent the ability to generalize outcomes from surgery and therefore ARUBA cannot be easily used as a guideline for treatment of unruptured bAVM (Korja et al., 2014b). Whether or not embolization is utilized prior to surgery may not influence outcomes in surgical series (Morgan et al., 2013; Korja et al., 2014a). The main conclusions that can be drawn from ARUBA are that the natural history of rupture for unruptured bAVM

*Correspondence to: Michael Kerin Morgan, Department of Clinical Medicine, Macquarie University, Sydney, New South Wales, Australia. Tel: +61-409-074-824, E-mail: michael.morgan@mq.edu.au

is likely to be 2.3% per year (confirming the meta-analysis results of Gross and Du, 2013), and that combined treatments and endovascular treatments may not be appropriate as a generalized standard treatment (this is not to say that there are not nuanced indications, such as targeting an aneurysm, for embolization other than as a presurgical treatment).

Having sufficient information to make a recommendation of treatment is different to explaining the estimated risk to a patient. Explaining the risks of a procedure simply requires the estimation of complication rate and to provide this information in a way that is understandable to the patient. Making a recommendation requires an understanding of likely outcomes from alternative management pathways and a synthesis of this knowledge. Knowing the results of surgical series is sufficient to inform the patient of what the risks of surgery are likely to be in that surgeon's hands but insufficient to make a judgment as to whether this is the best treatment for the patient. A surgeon's series reflects the surgeon's judgment as well as proficiency but sheds little light upon what is the recommended best treatment. Because of the complexity of bAVM management resulting from the variability in: time to complication depending on treatment; proportion with complications; time to bAVM obliteration; and proportion obliterated, many patients may request a recommendation from the treating team rather than traveling between specialists enquiring as to the estimated risks quoted from each team. That is, a synthesis needs to be made of the evidence from risks and efficacy from each management pathway.

For a recommendation that allows a comparison and synthesis of management options, the evidence must account for selection bias and efficacy of treatment over a period of time. Comparisons in outcomes between management pathways is possible if there is a common reasonable time point at which management pathways can be compared with adjustments made for bAVM ineffectively treated. Such an analysis can be considered in a complication–effectiveness analysis where a permanent new neurologic deficit over a defined period of time per effectively treated bAVM reduces the discrepancy between outcome variables comparing radiosurgery and surgery (Morgan et al., 2016b) (see below).

The best evidence with regard to recommending which cases may benefit from surgery currently rests with cohort studies. In order to apply conclusions from cohort studies to the general population of cases, the contextual institutional similarities and differences must be understood. Irrespective of the evidence that may support surgery, bAVM surgery (particularly for complex bAVM) requires a unique combination of: surgical skills; an operating team that is experienced and specialized; anesthesia that is experienced and specialized; radiologic support that is available, experienced, and specialized; intensive care that is experienced and specialized; a requirement that each team will work well with each other; and administrative support that this is of importance to the hospital. This makes it necessary for units performing bAVM surgery to measure their outcomes to ensure individual institutional expectations reflect the evidence.

Risk of hemorrhage and outcomes of hemorrhage in untreated bAVM

A meta-analysis, a combined database, and a randomized control trial report annual rates of first hemorrhage of 2.2% (95% confidence interval (CI): 1.7–2.7%), 1.3% (95% CI: 1.0–1.7%), and 2.2% (95% CI: 0.9–5.5%) respectively and a meta-analysis and a combined database reported annual rerupture rates of 4.5% (95% CI: 3.7–5.5%) and 4.8% (95% CI: 3.9–5.9%) (Gross and Du, 2013; Kim et al., 2014; Mohr et al., 2014).

The reported risk of permanent neurologic deficit as a consequence of hemorrhage varies widely in the literature because it is difficult to estimate. For most stroke studies, estimating outcomes from intracranial hemorrhage is reasonably reliable. However, there is a greater risk of rerupture with time from bAVM than many other sources of recurrent hemorrhage and there is a greater likelihood of treatment aimed at bAVM obliteration that confounds estimates. A simple 1-month outcome is less relevant for bAVM because of these factors. Permanent new neurologic deficits or deaths have been reported from approximately 35–85% (Ondra et al., 1990; van Beijnum et al., 2009).

Of relevance to management pathway selection is the incidence of mortality following hemorrhage after gamma-knife radiosurgery, reflecting the natural history of bAVM prior to obliteration. In a cohort series from Spetzler–Ponce class (SPC) A and B bAVM, of those known to have hemorrhage following radiosurgery there was a 55% mortality (95% CI: 41–69%) (Kano et al., 2012a, 2014). There may be a significant difference in reporting short-term outcomes from a single hemorrhage (e.g., 1-month morbidity and mortality) from long-term outcome from patients who may have multiple hemorrhages. Though interpreting the impact of permanent deficit arising as a consequence of hemorrhage is difficult, it is reasonable to assume an annual morbidity and mortality rate of 1% for first hemorrhage and 2% for subsequent hemorrhage. The impact of this with time is demonstrated in Figure 5.1.

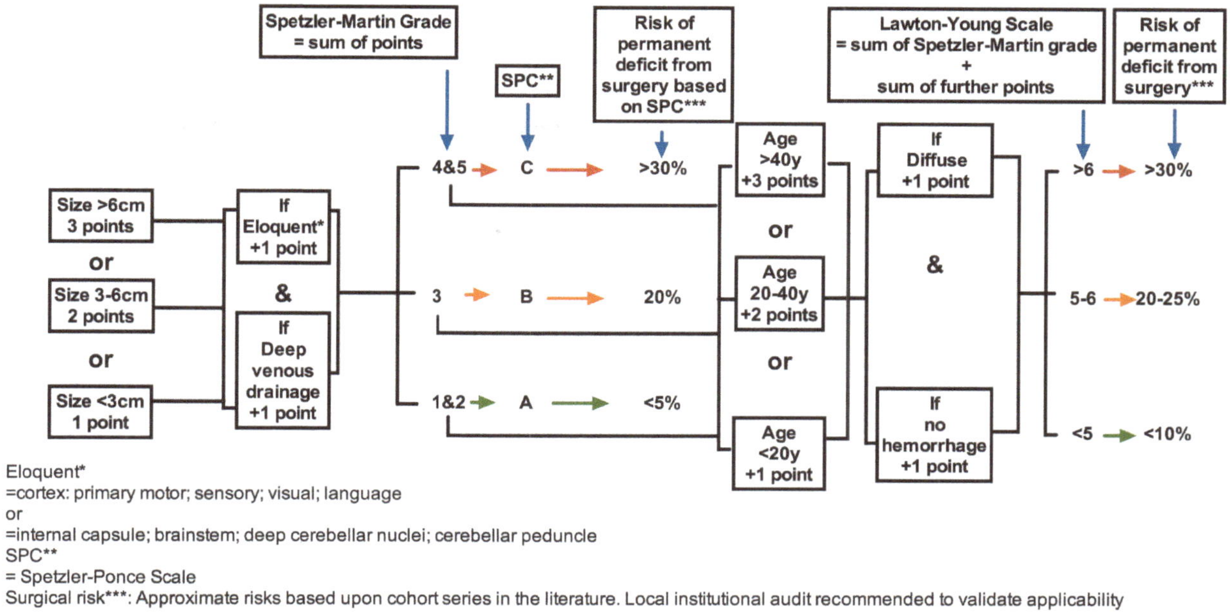

Fig. 5.1. Flow chart of decision-making guidelines commencing with brain arteriovenous malformations size and introducing other angiographic, patient, and clinical factors from left to right.

Surgery to reduce risk of seizure

Seizures account for 25% of presentations in untreated unruptured bAVM and there is reported to be a decline in functional outcomes due to the development of seizures or worsening frequency or severity of seizures in a small proportion of unruptured cases (Brown et al., 1988). Several reports have analyzed the outcome for seizures following surgery, confirming that seizure control is improved (Baronski et al., 2014). However, caution must be exercised against the expectation of cure for epilepsy following surgery for larger bAVM (O'Donnell and Morgan, 2015). Surgery for seizure control in larger bAVM should only be considered in highly nuanced cases that fail good medical management.

bAVM associated with proximal intracranial aneurysms

Associated proximal intracranial aneurysms (APIA) to bAVM may be identified in 6–27% of cases, due to long-term hemodynamic perturbations induced by arteriovenous shunting (Brown et al., 1990; Stapf et al., 2002, 2006; da Costa et al., 2009; Kano et al., 2012b; Platz et al., 2014; Morgan et al., 2016b). They are important to identify because unruptured bAVM with APIA are more likely to hemorrhage and the APIA may rupture perioperatively. APIA are more likely associated with posterior circulation supply to bAVM and have a greater prevalence in older patients (Morgan et al., 2016b). It is likely that APIA indicates a risk for bAVM first rupture rate approaching that of recently ruptured bAVM (Brown et al., 1990; da Costa et al., 2009; Kano et al., 2012b; Gross and Du, 2013; Platz et al., 2014). However, it is difficult to quantify these risks due to the confidence of the data around the natural history of unruptured bAVM and the variability in APIA identification.

The decision to leave an APIA untreated or repair the APIA is complex. APIA repair may be unnecessary because of anticipated remodeling of the artery after the bAVM has been removed. However, until the hemodynamic stress due to the AVM has been removed, and during the period that remodeling is occurring, there may be an increased risk of rupture due to the greater arterial wall tension arising from the increase in intravascular pressure. Furthermore, if repair is planned, whether this precedes, coincides with, or occurs after bAVM excision is also a complex decision. Small, round aneurysms at some distance from the bAVM may be watched with surveillance radiology for their anticipated involution after surgery. APIA within the surgical field of the bAVM resection can be dealt with at the time of the bAVM resection. Large irregular aneurysms that are likely to have significant increase in their parent artery pressure after arteriovenous shunt ablation should be eliminated prior to bAVM surgery. APIA that have been seen to increase in size after bAVM resection need to be repaired.

If APIA is deemed appropriate to treat, consideration of changes to the hemodynamics after bAVM resection needs to be anticipated. There will be a reduced shear stress within the parent artery that may increase the risk of thrombosis with stents or coils. The artery radius may be anticipated to reduce after arteriovenous shunt elimination and stenting the artery when it is larger may be inappropriate.

Risk of surgery

Concerns with regard to the lack of extant evidence for surgery from randomized control trials can be allayed from the overwhelming concordance of evidence from cohort studies with regard to strategies of surgical risk classification. The five-tier Spetzler–Martin grading (SMG) system (and the simplification into the three-tier SPC) is a reliable and accurate reflection of the risks of surgery. The SMG is established by allocating points for size (1 for less than 3 cm, 2 for size between 3 and 6 cm, and 3 for size greater than 6 cm), the presence of deep venous drainage (1 point), and location in "eloquent" brain (1 point if located in primary sensory cortex, motor cortex, language cortex, internal capsule, diencephalon, brainstem, deep cerebellar nuclei, or cerebellar peduncle) (Fig. 5.2) (Spetzler and Martin, 1986). The SPC categories A and C are derived from combining SMG 1 with 2 and SMG 4 with 5, respectively (Fig. 5.2). SMG 3 is SPC B (Spetzler and Ponce, 2011).

The Spetzler–Ponce scale has been validated by reviewing a number of cohort series. There is an additional nuancing of this grading system with the Lawton–Young scale that adds the SMG to points for age (1 point for less than 20 years of age, 2 points for ages between 20 and 40 years, and 3 points for age greater than 40 years), presentation without hemorrhage (1 point), and a bAVM considered diffuse (1 point) (Lawton et al., 2010). Combining cohort series have found this variation to be valid (Kim et al., 2015). For the Lawton–Young scale, a score of 7 or more is attended with a risk of surgery of greater than 25%. However, not all have identified age as a risk factor (independent of associated comorbidities that are associated with increasing age) and diffuse or compact bAVMs are readily identified when they are obvious. However, there is a gray area that may present difficulty allocating this point (Bervini et al., 2014; Korja et al., 2014a). To further refine the ability to determine whether a bAVM is diffuse, computer software can assist (Du et al., 2007). This may make application of the Lawton–Young scale more time consuming for some patients. Calculating the Lawton–Young score is useful for SPC B cases. However, it is likely to be unnecessary except for older, unruptured SPC A cases or young, ruptured SPC C bAVM. How these grading systems can inform surgical decisions as to risk can be seen in Figure 5.1.

The three components of the SMG have been independently validated as the most important criteria influencing surgical outcomes (including a sensitivity analysis adjusted for cases not undergoing surgery because of a perceived risk) (Korja et al., 2014a). Furthermore, the odds ratio for each of these factors is so large that a spread of surgical risks is extremely wide, and considerably

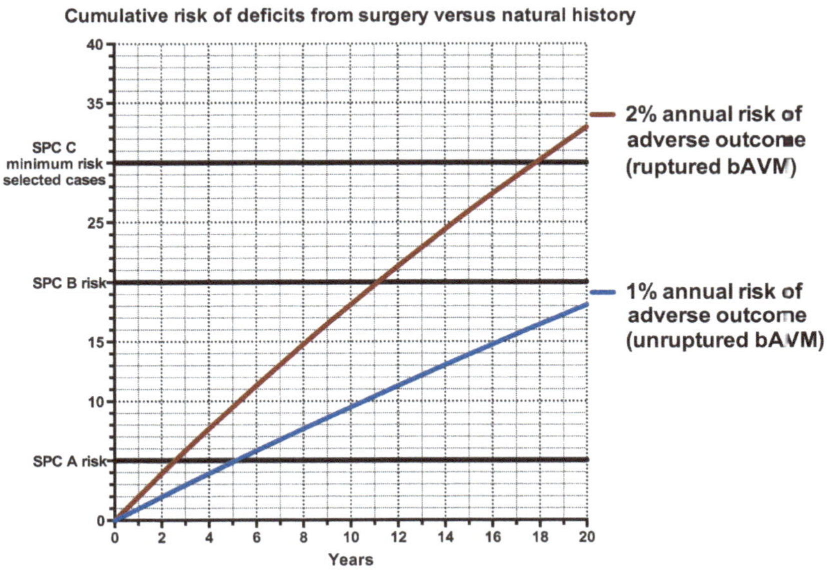

Fig. 5.2. The hypothetic projection of the average risks of deficit from natural history superimposed upon that of surgery (black lines). The calculated natural history has been calculated as being approximately half the natural history of next (2% curve in red) or first (1% curve in blue) hemorrhage. The natural-history risk has been calculated by the equation: cumulative risk % = (1 − *e* − annual rate × years) × 100. bAVM, brain arteriovenous malformation; SPC, Spetzler–Ponce class.

wider than the risks of future hemorrhage (influenced by presentation with hemorrhage or the presence of intranidal or feeding artery aneurysm) or the risks associated with radiosurgery. Therefore, commencing management decisions with SPC is reasonable and appropriate (Fig. 5.2).

A criticism of drawing conclusions from cohort studies is the unknown influence on the outcomes of the unoperated cases. This has been addressed by a complication effectiveness analysis of surgery for bAVM, allowing complications to be estimated having made adjustments for cases that were excluded from surgery because of perceived risk as well as those that underwent treatment but were not cured (Morgan et al., 2016a). Analysis by this methodology confirms the low risk for SPC A developing any permanent neurologic deficits (agreeing with surgical series), with a 95% likelihood that the risk is less than 4%, but increases the reported complication rates for both SPC B (the credible range of risks is likely to be 16–32%) and SPC C (the credible range of risks is likely to be 33–73%) (Morgan et al., 2016a). For SPC C in particular, the lowest end of this range is likely to be overly optimistic and only applies to cases that are highly selected, with the majority of cases deemed at much higher risk. For SPC C, the higher end of this range is likely to be more realistic for the group as a whole.

For the expediency of analyzing and comparing studies, it is necessary to report outcomes often in a dichotomized form that removes important information from the studies' intended audience. Although the risk of permanent neurologic deficit is a very important outcome measure, there are major categoric and qualitative differences that also should be considered when assessing the risks for some patients. There is clearly a marked difference in outcomes between a modified Rankin Scale (mRS) score of 2 and death (van Swieten et al., 1988). Therefore, individual patients and their families may be more focused upon the avoidance of any deficit, loss of independence, or death. The proportion of complications of surgery leading to permanent new neurologic deficits that result in disabling deficits, loss of independence, or death varies between the SPC. For SPC B, approximately one in five of complications result in outcomes with mRS greater than 2 (Morgan et al., 2015), whereas this proportion is likely to be two in three for SPC C (Bervini et al., 2014). The decision to trade a high likelihood of some immediate but mild deficit from surgery for a reduced risk of death or more profound deficit at an uncertain future time point is a consideration for some patients. This may suggest surgery is occasionally appropriate in some circumstances irrespective of the high risk of resulting in some deficit.

Who benefits from surgery?

SPC A bAVM have a low risk of surgery. This should be less than 10% of causing a permanent new neurologic deficit with an MRS score of more than 1 and maybe less than 5% in experienced centers (Gross and Du, 2013; Bervini et al., 2014; Korja et al., 2014a). This applies to all cases of SPC A bAVM that have no significant comorbidities (Morgan et al., 2016a). This is less than the natural history untreated or treatment by radiosurgery if a long-term view is considered (Bervini et al., 2014; Morgan et al., 2016a). These patients should benefit from surgery unless there are: comorbidities; too few years of expected life remaining to benefit; or bAVM or patient-specific factors that are highly nuanced. This benefit from surgery for SPC A should be apparent irrespective of the mode of presentation (ruptured, unruptured and no aneurysm, unruptured and APIA). However, the mode of presentation and associated angiographic features will influence how many years before a benefit is likely to occur from surgery, in comparison to the cumulative risks of untreated or radiosurgically treated bAVM (Figs 5.1 and 5.3).

Contrasting the risks of surgery with those of natural history has not yet been examined by a randomized control trial but, from cohort series and the natural history data discussed above, a figure can be constructed overlying these two data (Fig. 5.1). It can be seen that the risks of surgery for SPC A, in general, is likely to be less than 6 years of natural history.

Who does not benefit from surgery?

For SPC C, it is difficult to support a case for surgery (in general) for unruptured bAVM and only the very young (because of their normal life expectancy) are likely to benefit from surgery for selected ruptured bAVM. SPC C bAVM have a high risk of surgery. Although a subset of cases may be found for which a decision has been made to treat by surgery, even for these highly selected cases, the morbidity and mortality are significant. Such cohort series report a 30–40% risk of major morbidity by surgery (Han et al., 2003; Korja et al., 2014a). Considering that the minority of SPC C cases are recommended to undergo surgery (Han et al., 2003; Korja et al., 2014a), a sensitivity analysis suggests that the risk for such surgery may well be much greater than these figures would indicate (Morgan et al., 2016b). The decision to operate on SPC C cannot be generally recommended and when performed needs to be highly nuanced (e.g. less than 20 years of age, a compact bAVM, with recent hemorrhage and possibly a resection that is simple, such as lobectomy). Furthermore, trading a known deficit (e.g., homonymous hemianopia) for the

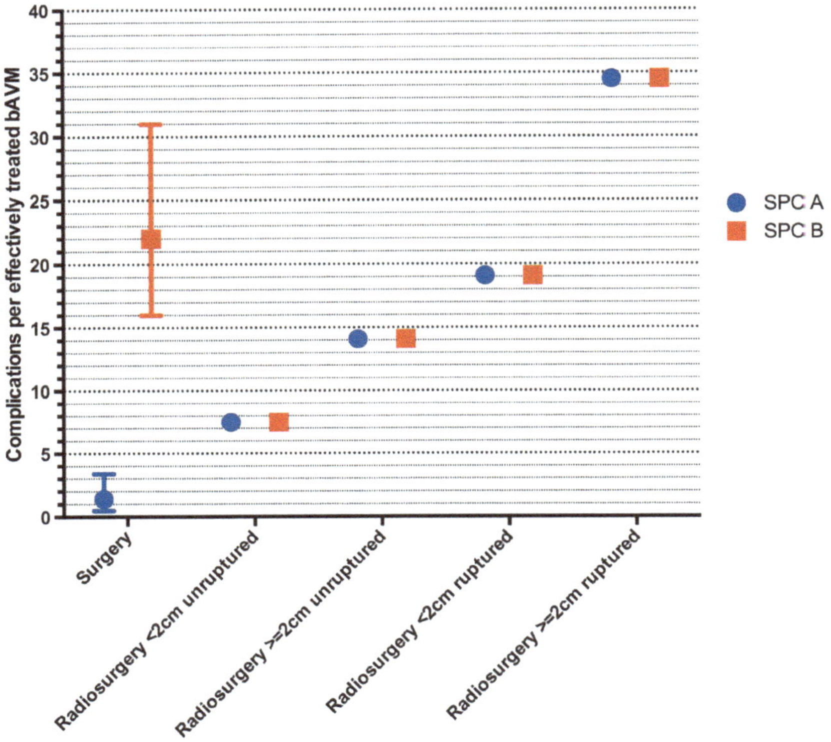

Fig. 5.3. Comparing the expected risk of new permanent neurologic deficits from surgery with that from radiosurgery. The surgical results are complications leading to a modified Rankin Scale > 1 at 12 months for all Spetzler–Ponce class (SPC) A (blue) and B (red) corrected for cases not undergoing surgery and corrected for cases not effectively cured at 9 years (the period of time that was considered reliable by the method of analysis) (Morgan et al., 2016b). The radiosurgery results for the 9-year period are derived from obliteration rates reported from SPC A and B brain arteriovenous malformations (bAVM) (Kano et al., 2012c, 2014) and the natural-history results (Gross and Du, 2013).

potential of a greater deficit may be seen as reasonable by patients and their family. However, before a decision is reached that surgical treatment is the pathway that should be followed, patient, family, and surgical team need to fully understand what is entailed in this pathway. Not only are there significant risks: there is a considerable gauntlet to pass through a successful hospital admission. It is these patients in particular who will likely require a prolonged stay in intensive care with aggressive blood pressure management and a prolonged hospital admission (Morgan et al., 2003).

Who may be considered for surgery, but in whom evidence for benefit is uncertain?

SPC B predominantly comprises this group. However, this may also apply to young (<20 years of age) compact ruptured SMG 4 cases and older (>40 years of age) diffuse unruptured SMG 2 cases. For these cases, surgery may be considered with a relatively high risk of a new permanent neurologic deficit (about one in five) (Gross and Du, 2013; Morgan et al., 2015), but with a relatively low risk of seriously disabling permanent deficit (mRS > 2) (about one in five who experience a deficit) and a low risk of mortality (Morgan et al., 2015). Such cases may be recommended for surgery after considerable discussion and an understanding of trading a deficit now for a greater risk of a worse deficit in the future. This discussion is complex and requires considerable experience to ensure that management pathways under consideration are fully understandable to the patients and that a decision to follow a surgical pathway is backed up by expertise. Consideration as to alternate management pathways needs to be understood, as small unruptured SPC B bAVM may be best managed by radiosurgery. However, a recently ruptured SPC B that is larger than 2 cm in maximum diameter may be better managed by surgery rather than radiosurgery by a complication–effectiveness analysis (Fig. 5.3).

PERFORMANCE OF SURGERY

Principles underlying competent surgery

The strategy for eliminating bAVM is not an elimination of a mass lesion but, rather, elimination of abnormal physiology. This distinction is important and needs to be explicitly stated because much of neurosurgical training emphasizes a neuro-oncologic-geographic approach to tumors, an approach that is found wanting in the management of bAVM. This is because the bAVM will continue to fill whilst ever there is even the smallest of arterial input, whereas in tumor surgery, leaving a small residual of brain tumor may be considered a successful operation and will not usually result in significant perioperative complications, as can happen with bAVM surgery.

The aim of surgery is the elimination of the abnormal physiology (the arteriovenous shunt) along with the removal of vasculature leading to this shunt assessed to be vulnerable to rupture (immediately after shunt removal) (see explanation for the vasculature to be at risk after shunt ablation in Fig. 5.4). Removing vulnerable vasculature leading to the arteriovenous shunt is of utmost importance. This is because elimination of the shunt increases pressure in the feeding arteries, increasing the tension within the progressively increasingly redundant (in terms of supplying either brain or AVM) small arterial walls, risking easy rupture and perioperative hemorrhage. By removing vessels that have been pathologically affected (to the extent that they may not withstand an increase in intravascular pressure at the time of shunt elimination), surgery can be successful.

The judgment as to when a vessel has transitioned from normal to pathologically abnormal, weak, and vulnerable to rupture with the increase in its pressure within (upon eliminating the arteriovenous shunt) may be difficult but is critical. The approach to this goal of excising bAVM can either be a geographic resection, within which is the region of shunting and pathologically altered arteries, or a marginal resection in which the arteries are ligated at a point as close to the arteriovenous shunting itself but far enough proximal to this point that the arteries left exposed to the systemic circulation can be confidently left to withstand the postoperative physiologic changes. This may be termed a marginal resection. To understand the important distinction between the principles of surgery for bAVM and that of a mass lesion the following interaction between physiology and pathology (pathophysiology) as well as anatomy should be understood:

1. There is a disproportionately thin wall for radius of artery feeding bAVM, in response to the chronically low mean and pulsatile pressures and the high shear stress.
2. With the increase in resistance (impedance) arising from progressive arteriovenous shunt occlusion, the pressure within the remaining arteriovenous shunt will increase within the residual bAVM and immediately proximal arteries. If the venous outflow has been occluded prematurely, any residual feeding to unresected bAVM will ultimately be subject to an increase in pressure within its interstices, the level of which is determined by the arterial pressure, the rate

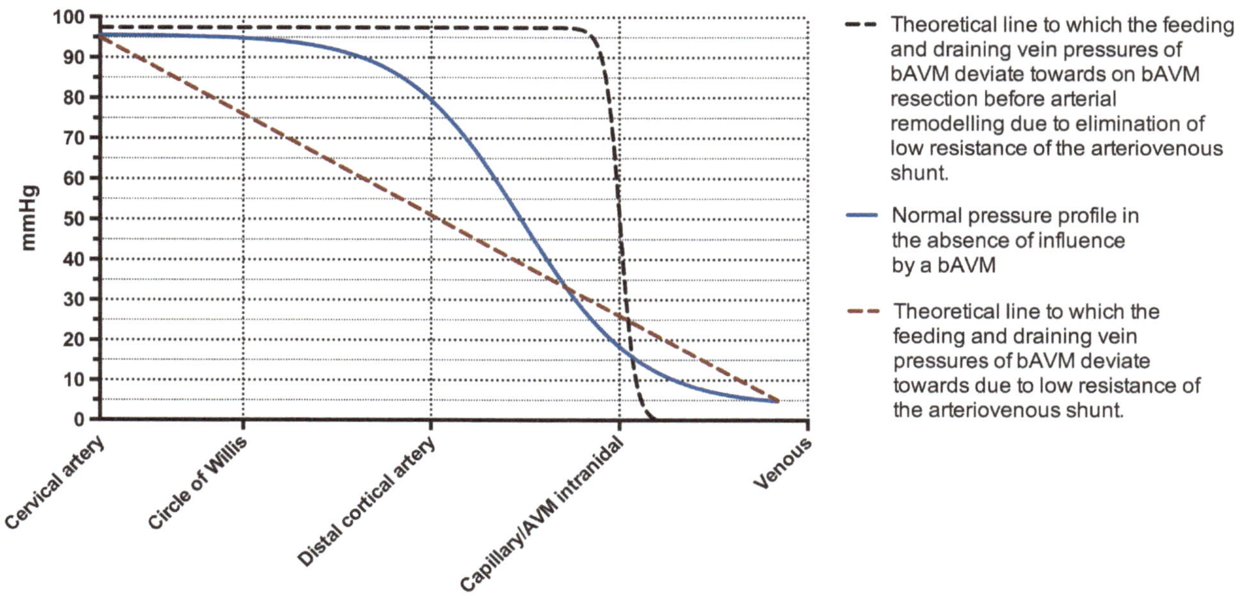

Fig. 5.4. The hypothetic pressure profile along the length of the vasculature involved in brain arteriovenous malformation (bAVM). The normal profile is illustrated in blue. With a reduced resistance the normal profile is lower. After bAVM elimination, due to the dilated arterial system, the pressure profile exceeds that expected in the normal arteries. This greater than normal pressure persists until remodeling of the arteries has occurred and the diameter is appropriate to ensure that wall shear stress has returned to normal.

of rise of which will be determined by size of the residual arterial supply. That is, even the smallest arterial input can lead to the pressure within the bAVM reaching that of the surrounding arterial pressure with venous outflow occlusion. Only when 100% of the arterial input has been secured and the immediately proximal vulnerable arteries are excluded from the inevitable blood pressure rise is a bAVM under control.

3. Because of the reflected pressure wave generated from the heart, and the substantial dilatation of the arterial system, before remodeling of the arteries has occurred after bAVM resection, pulsatility will increase in arteries (and their branches) immediately proximal to the bAVM with progressive shunt elimination.
4. A very low blood pressure, whilst reducing the potential of thin-walled vessels rupturing during surgery, may be too low and cause ischemic damage to the surrounding brain. Although the brain immediately surrounding the bAVM may have adjusted to the low perfusion pressure with arterial dilatation and altered microvasculature resistance (and therefore will tolerate reduced blood pressure as the shunt is eliminated), the more distant brain is likely to be unaffected by the shunt (Barnett et al., 1987). Therefore, the lower limits of autoregulation will vary from reduced (immediately adjacent to the bAVM) to normal (further from the bAVM).
5. There is no perivascular space surrounding the smallest veins, whereas there is surrounding the smallest artery. The dilated feeding arteries enlarge this perivascular space and the smallest amount of cerebrospinal fluid may give a clue that the smallest of the arterialized vessels is a feeding artery rather than an arterialized vein. The majority of these small arterialized vessels are arterialized veins (as opposed to arteries) that emerge from the surface of the bAVM and loop back to rejoin the bAVM at a slight distance. This penetration of the venous loops into the white matter results from the higher than normal venous pressure within the veins and the hemodynamic drive to reduce the resistance to flow by remodeling the veins into lengthened sweeping loops (even with the smallest vein).
6. Whether or not the bAVM has a supply from the deep arterial system (e.g., lenticulostriate and choroidal arteries), there is always a supply to the deepest point of the bAVM (the nose cone). When this is from superficially located arteries, they often reach this point by diving into the depths of the sulcus at least a small distance from what might be called the margin of the bAVM. This is not a surprise due to the principle that the circulation remodeling often achieves a minimization of work of the circulation, thus creating a widened angle between the arterial input and the axis of mass flow of the bAVM. This is restricted superficially because of the brain's covering and is therefore achieved by either sweeping at a distance superficially or by entering deeply from a superficial location.
7. Veins are generally located superficially to arteries in normal anatomy. This is no different with regard to bAVM, where there is often an arterial feeder located immediately deep to the main venous drainage of the bAVM. This makes it difficult to be sure that all arterial input has been controlled without exploring immediately beneath the draining veins.

Put into practice, these pathophysiologic and anatomic principles dictate that the following practice be followed:

1. Attention to the physiologic changes that may occur during anesthesia, particularly the anticipated increase in feeding arterial pressure (both as mean and pulsatility) during and after arteriovenous shunt elimination for larger bAVM. Anesthesia with significant blood pressure control to levels lower than normal (but avoiding ischemia) may require both the use of vasodilation and cardiac output suppression (e.g., beta-blockade and barbiturates). Cerebral protection needs to be ensured at low pressures (e.g., barbiturates). A benefit to the surgeons of reducing the cerebral metabolic activity of the brain will be brain relaxation. This reduces the need for retraction during surgery. This is particularly pertinent at low pressures, during which retraction may result in ischemia. In addition or instead of systemic hypotension, regional hypotension with temporary clips may be preferred. This author employs both practices.
2. Craniotomies need to be executed to provide optimum exposure of the bAVM, minimization of retraction (by maximizing the range of angles from which the bAVM can be approached), and optimum feeding artery exposure (both proximal and distal to the supply to the bAVM). Visualization needs to ensure minimal risk of misidentification of critical supply to the brain or confusion with draining veins. Distinguishing the arteries from arterialized veins may be difficult but is possible with experience. Intraoperative indocyanine green (ICG) and angiography may assist with identification. Accessing arteries from normal to abnormal provides confidence that arteries can be distinguished from arterialized veins. The craniotomy may need to consider exposure across the normal dural boundaries separating hemispheres or infratentorial from supratentorial compartments. As an example, a bAVM

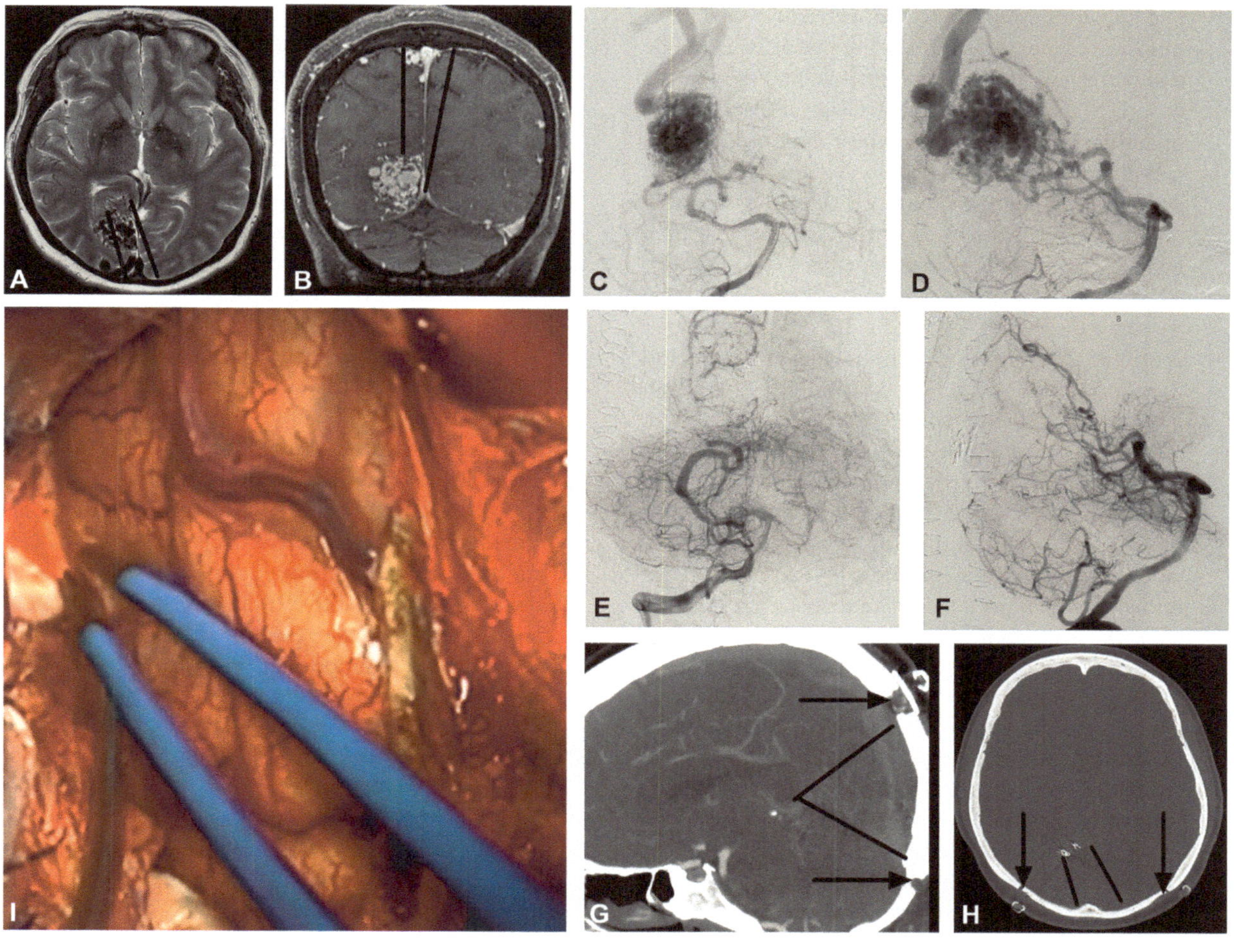

Fig. 5.5. A 41-year-old woman with an unruptured Spetzler–Ponce class B brain arteriovenous malformation (bAVM) of the right occipital lobe presenting with classic migraine and intact visual fields. To avoid excessive retraction both an ipsilateral and a contralateral interhemispheric approach is performed through a large occipital craniotomy with a large falcine window. To emphasize the unfavorable angle of approach, the corridor of access is bounded by the illustrated lines (**A**, **B**, **G**, and **H**). By accessing the bAVM bilaterally, access to the most problematic corner is achieved with less retraction than required from an ipsilateral approach. The large craniotomy was performed for this case in which the patient was positioned prone, a position in which it is more difficult to achieve relaxation due to higher venous pressures. The wider craniotomy minimizes the potential for brain to herniate at the margins during the approach and provides rapid direct access to the bAVM if uncontrolled hemorrhage becomes a problem. The preoperative axial T2 (**A**), coronal gadolinium-enhanced T1 (**B**), anteroposterior digital subtraction angiography (DSA) (**C**) and lateral DSA (**D**) demonstrate the bAVM. The postoperative DSA, in the same projection as preoperative images, was taken 6 days after surgery (**E** and **F**). The posterior cerebral artery is preserved after resection (**E** and **F**). Note the incomplete remodeling of the middle cerebral artery that will be expected to return to a normal diameter with time.

presenting on the medial surface may be accessed both from the ipsilateral interhemispheric approach as well as a contralateral approach through a wide window fashioned in the falx cerebri (Fig. 5.5). This may improve the angle at which the most lateral component of the bAVM may be more easily reached. Therefore, consideration as to positioning must, on occasions, include free access across dural boundaries (falx cerebri or tentorium cerebelli). Similarly, a bAVM on the superior surface of the cerebellum may be approached partially subtemporally with a large window fashioned in the tentorium cerebelli. For large or extensive superior cerebellar bAVM, this may be combined with a posterior fossa exposure with the possibility of dividing (if assessed safe in the individual case) the transverse sinus (performed in a way that allows reanastomosis at the end of the case) to allow the tentorium and posterior fossa to be opened as a single window. Therefore, consideration to positioning to ensure that the temporal lobe requires minimal retraction is required in such an approach.

3. Feeding arteries need to be dissected proximal to all branches to the bAVM (to ensure that arteries can be distinguished from arterialized veins) as well as ensuring the preservation of distal arterial supply to normal brain. This is often best achieved by commencing the dissection distally, beyond the bAVM, and progressing proximally. This allows correct identification of the brain critical arteries that may be much smaller than those branching to the bAVM. This requires extensive opening of the relevant sulci by sharp dissection.
4. Regional hypotension within the bAVM (and therefore reduced tension within the bAVM) can be facilitated by temporary arterial clips applied, to as many feeding arteries as possible, at a distance from the bAVM. Even when it is thought by the surgeon that all feeding arteries are controlled by this technique, in my experience, this can never be assumed until the bAVM is delivered on its venous umbilicus, from its bed. Irrespective of how well controlled the arterial feeders are believed to be prior to bAVM delivery, venous drainage should not be impeded. Embolization can be used to assist with regional hypotension, but may be unnecessary depending upon surgical technique (Morgan, 2013).
5. Retraction of the brain may exacerbate ischemia because perfusion pressure to the brain is low because of either low blood pressure of the anesthesia or due to pre-existing regional arterial hypotension coupled with venous hypertension (because of the arteriovenous shunt). Retraction of the bAVM itself may be safe, providing that venous outflow is not impeded. Impeding the venous outflow may result from coagulation of the surface vessels of the bAVM (that are most likely venous) coupled with the effects of bAVM retraction.
6. Division of the feeding arteries is particularly demanding. The use of bipolar is more difficult than usual because of the thinness of the walls of the arteries (or veins if distal outflow has been impeded) relative to the vessel radius. This can make coagulation difficult (even perilous) with apparent bursting of the vessels if not performed carefully. This can be assisted with very clean bipolar tips and coagulation settings lower than normally employed. Further assistance can be gained by the use of microclips (e.g., as invented by Dr. Sundt) to arrest flow (Sundt and Kees, 1986). Microclips can be employed permanently to ligate the smallest vessel and miniclips for larger vessels. Microclips can also be used to arrest flow allowing bipolar coagulation. Some surgeons prefer irrigating bipolars or disposable bipolar forceps. An additional technique suitable for some vessels includes incorporating adjacent brain with broader blades to spread the current and incorporate more proteinaceous material – "dirty coagulation" (Hernesniemi et al., 2010).
7. Because of the importance of securing all arterial input before compromising venous drainage, a useful strategy is to first secure all the accessible superficial feeding arteries followed by a partial and limited bAVM marginal arc dissection to the deepest component of the bAVM (including entry into the ventricles if the bAVM juxtaposes the ventricle) to secure arteries to the deep nose cone (Fig. 5.6). This differs from the circumferentially corkscrew spiraling dissection employed for brain tumor surgery, a technique that maintains an even depth of dissection on the full margin of the lesion. Such a corkscrew circumferential dissection is not recommended for bAVM until the deep feeders have been controlled. Having secured the deep feeding arterial supply, a circumferentially corkscrew deepening dissection of the resection margin can then be commenced. This approach is critical for cases that have not had the deep arterial supply embolized (Morgan, 2013).
8. It is necessary to follow small arterialized vessels, into the white matter, during the dissection. This is because small feeding arteries (that are usually few in number) need to be distinguished from the more frequent occurrence of arterialized venous loops that leave the imagined surface plane of the bAVM and loop in and out of the white matter. The cumulative ligation of venous loops, mistaken for arteries, will impede venous drainage from the depth of the bAVM. This may produce a problem for the resection of the final nose cone as it will become very tense if the nose cone has not been previously isolated from arterial input (as per point 7, above). The venous loops should be left unligated and only the infrequent arterial input ligated and divided (Hashimoto, 2001).
9. Before ligation and division of the main draining vein is attempted, the resected bAVM should be deliverable from the resection bed on its venous umbilical attachment.
10. A satisfactory resection bed is one with no arterialized bleeding. Any bleeding suggests the possibility of retained bAVM and needs to be further explored. Even when this bleeding is not due to retained bAVM, the small vessels responsible for persistent bleeding do not have the capacity to constrict due to the thinness of their muscular wall component. Bleeding needs to be specifically surgically arrested rather than allowing time to pass for the possibility of arterial vasoconstriction response for

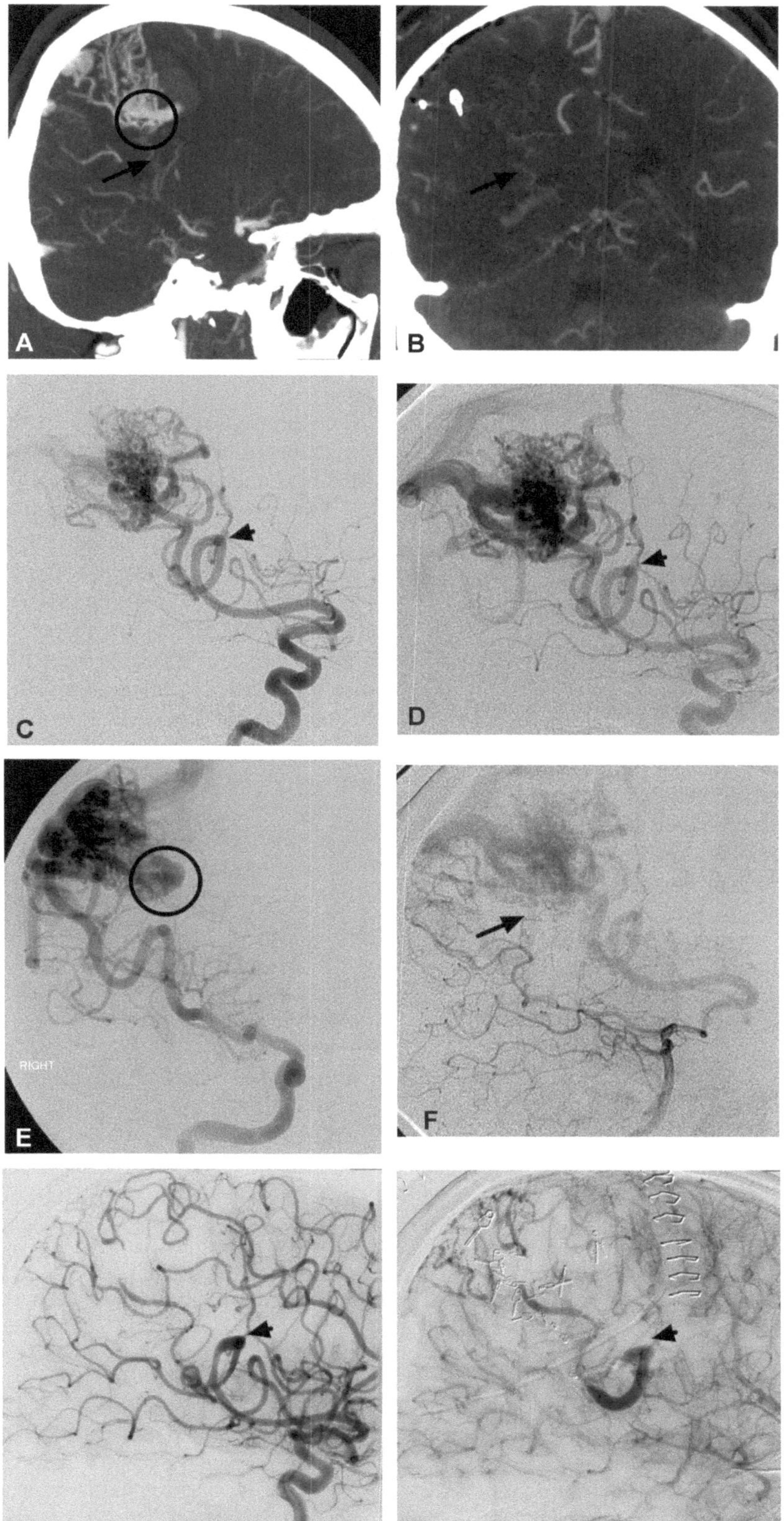

Fig.5.6. See legend on next page.

hemostasis control. It is important not to stress test the security of hemostasis by raising blood pressure during the surgery (for reasons stated above).

11. Occasionally, transdural arterial recruitment (TDAR) (from meningeal arteries) to the bAVM is significant. When TDAR occurs, the dura and brain must not be retracted away from each other (for fear of tearing the vascular connection and catastrophic hemorrhage). (This was the cause of the most catastrophic hemorrhage that the author has ever encountered or witnessed in bAVM surgery.) Specific strategies for dealing TDAR from tentorium or falx include creating an island of dura and leaving this on the brain over the bAVM. The tentorium or the falx can be incised at a considerable distance from the bAVM, securing the blood supply. Techniques, such as division of the tentorium from lateral to medial (to secure the meningeal arterial supply from the carotid artery) or opening of the falx from a contralateral approach at a distance from the nidus, can deal with feeders entering from the dura.
12. Occasionally, multiple transosseous arterial supply may be present and this can lead to potentially catastrophic bleeding on elevation of the craniotomy flap if performed in the normal way. Transosseous supply most often (but not always) occurs as a response to previous partial treatment of large-convexity bAVM (Fig. 5.7). When present, it is safer to do narrow-strip craniotomies, removing multiple narrow strips one at a time, securing bleeding at each stage. Although preoperative embolization may seem appealing in such cases, the result may interfere with wound healing, given the targeting of the extracranial supply, and is ill advised.

Technical execution of microsurgery

As stated above, the performance of surgery is dependent upon understanding the seven pathophysiologic and anatomic principles of surgery and following the 12 practices that follow from these principles without compromise. Advances in technology should not be seen as a substitute to relax these principles or practices. The most important instruments are the scalpel, microscissors, sucker, bipolars, and microscope. These must be appropriate and used appropriately for the situation. The following description is by necessity a personal approach. Unlike the principles and practices mentioned above (that must always be followed), there is undoubtedly many ways to effect surgery. This is, therefore, what I consider to be important.

Arachnoid dissection

The scalpel used for the fissure and larger sulci and releasing vessels from ensnaring arachnoid is an 11-blade that is changed often (to ensure that it remains sharp). For the efficiency of surgery it is important that an 11-blade is selected that can act as both a knife and a retractor. This is achieved with an 11-blade that has the spine (or blade flat) beveled as it approaches the tip, with the distal half of the tip incorporated in the grind of the cutting edge (or grind). This 11-blade can serve both as a knife, that divides arachnoid (the arachnoid is placed under slight tension with appropriate positioning of the sucker), and a retractor-dissector, by using the back (spine) of the blade (ensuring the tip or blade does not contact either vessels under tension or brain, facilitated by the beveled angle of the tip of the spine) with release of arachnoid tension previously induced by the sucker. For the smaller sulci as well as in the depths of sulci, an eye-knife (that is considerably smaller than the 11-blade) is used. However, these blades lack the distal bevel, making the dual purpose use of incision and retractor-dissector more difficult.

The selection of suckers is critical for the retraction role during the dissection. Because of the fragility of the vessels and the need to use the terminal part of the barrel of the sucker in retraction, a fine-tipped (although rounded to avoid sharp edges) sucker of the smallest sizes (#3 and #4) should be used. Suction pressure should be low. Pressure should be sufficient to clear the field of

Fig. 5.6. A 44-year-old man with sudden onset of distal left upper-limb weakness and subsequent diagnosis of ruptured Spetzler–Ponce class B brain arteriovenous malformation (bAVM) in right parietal lobe (**A**). Predominant arterial supply from the right middle cerebral artery (arrowhead at equivalent locations before surgery (**C** and **D**) and 6 days after surgery (**G**, **H**)). Small contribution from the right lateral posterior choroidal artery (arrow with long stem) (**F**). The lateral posterior choroidal arteries can be identified immediately after surgery (**B**). The nose cone (circled in **A** and **E**), that needs to be dissected and disconnected with deep arterial supply early in the surgery (before circumferential corkscrew dissection of bAVM) commences by reaching this point with a limited dissection of the bAVM margin at the convexity surface. Preoperative sagittal computed tomography angiogram (CTA) (**A**); postoperative coronal CTA (**B**) with evidence of microclips and absence of hematoma in cavity; preoperative right internal carotid artery (ICA) angiogram (early lateral arterial (**C**), later lateral arterial (**D**) and oblique (**E**)); lateral vertebral angiogram. Postoperative right lateral ICA angiogram (late arterial phase (**G**) and venous phase (**F**), with redundant dilated feeding arteries continuing to fill slowly). Note the incomplete remodeling of the right posterior cerebral artery that will be expected to return to a normal diameter with time.

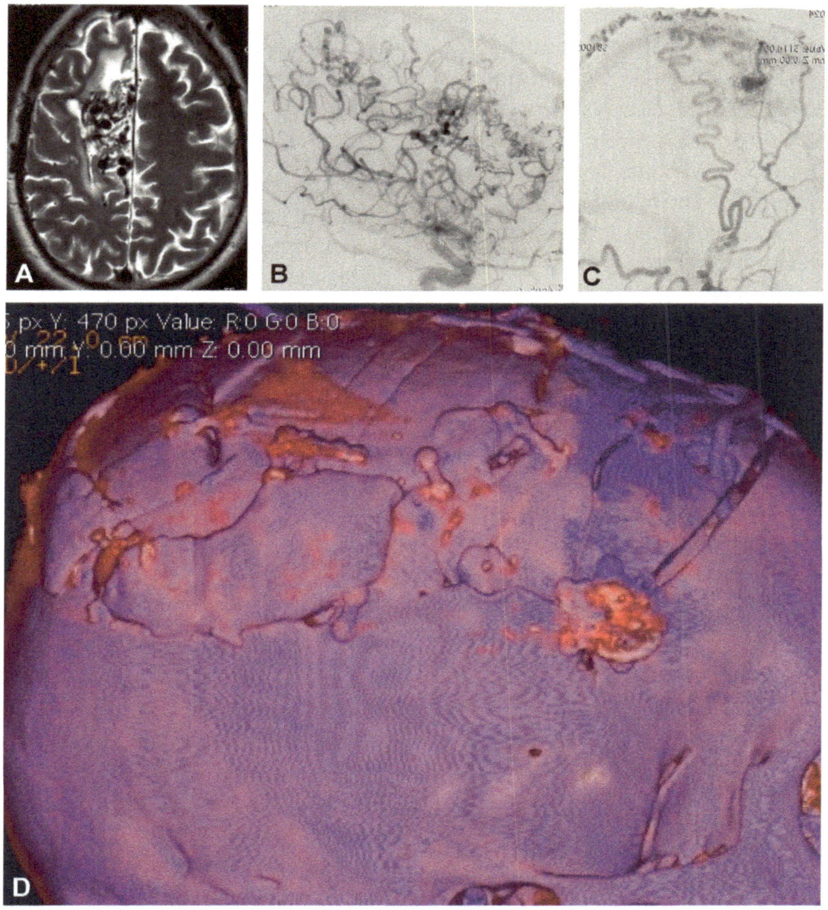

Fig. 5.7. A 54-year-old man presenting with recurrent hemorrhage from Spetzler–Ponce class B frontal-lobe brain arteriovenous malformation that had been previously treated by radiosurgery. (**A**) Magnetic resonance imaging, (**B**) right lateral internal carotid artery digital subtraction angiography (DSA), and (**C**) right lateral external carotid artery DSA. Total craniotomy exposure achieved by performing small adjacent strip craniotomies controlling bleeding from one small region at a time. (**D**) The postoperative computed tomography scan image of the repaired multiple bone plates.

cerebrospinal fluid and blood, but insufficient to disrupt the small fragile arteries.

Sucker tips damaged with drills should be avoided as they may catch arachnoid or vessels by sharp irregularities at their orifice. Because small suckers may easily block and need to be kept clean, it is important to have a number of spare suckers immediately available on the setup to prevent delays in surgery. In addition, as hemorrhage may be dramatic when it occurs, a variety of larger suckers on a separate suction line at higher pressure allows for immediate action if required. By the interplay between the sucker, as a retractor as well as keeping the field clear, and the 11-blade, as a knife dividing arachnoid and dissector, efficiency in minimizing instrument changes can be achieved. Tension to the arachnoid is provided by the sucker, allowing the caress of the 11-blade to divide the arachnoid, when tension is released; the 11-blade spine can be used as a dissector.

Controlling blood vessels in the sulci

Controlling all the arteries is imperative. This can start with temporary clips at a distance from the bAVM on arteries judged to have a normal wall thickness (although dilated). As the sulci and fissure are gradually opened the temporary clips can be advanced closer to the bAVM. Once an artery is judged to be terminating in the bAVM (without supply to critical brain) it can be ligated and divided. By dividing the artery, this point of bAVM anchoring is removed, reducing the tension when the bAVM needs to be manipulated. Ligation can be achieved by microclips or by bipolar coagulation. Where possible, bipolar should be used to reduce the cost of the procedure. The bipolar tips and current are both important considerations. For the small vessels in the sulci I favor a very fine-tipped bipolar applied with a low current.

Avoiding charring is critical as charring may cause uneven concentration of bipolar current, facilitating the bursting of the vessel wall. Having very clean mirror tips of the bipolar is essential. This is assisted, as with the scalpel, by having more than one bipolar. By applying the bipolar to a length of artery (much longer than the width of the tips of the bipolar) and taking time to ensure that coagulation has been achieved, the artery can then be divided with microscissors. This process may require a change in, or clean of, bipolar during the actual diathermy application of each individual artery. If the artery is considered too thin-walled to accept bipolar without rupturing prematurely, arresting the flow within the artery by microclips, before bipolar (or in the absence of bipolar), is reasonable.

Using a broader-tipped bipolar may assist with distribution of current along a greater length of artery, reducing the risk of premature vessel rupture. However, broader bipolars take up more room in this often very highly magnified field before the corticotomy is performed. I favor the use of the broader-tip bipolar when traversing the white matter when a far more agricultural (harvesting the adjacent brain) approach to dissection is practical and appropriate.

To reduce "sticking" and "charring" it is important to keep the tips of the bipolar clean. This may require exchanging bipolar mid-diathermy. The temptation to push on and try to complete the diathermy of the vessel with tips that are accumulating char during the application of diathermy must be resisted. Extending the time of diathermy application can be achieved by such techniques as using irrigation, disposable bipolars (thus keeping defects on the mirror surface of the bipolar tip to a minimum), or petroleum gel on the tips, but ultimately they need to be kept extremely clean.

Corticotomy

The decision as to the site of corticotomy is important. There are two corticotomy scenarios that need to be considered. The first is through a normal cortex. The second occurs where, during the process of the corticotomy, a number of small arterial feeders need to be divided. For the first scenario, the corticotomy should be in the deepest part of the sulcus, where possible, to make best use of the sulcal opening. This often affords a shorter course to the nose cone. The corticotomy in this scenario is the same as a corticotomy for other neurosurgical procedures. In the second technique, with a number of arterialized vessels crossing the cortical surface (below which access to normal white matter is anticipated), after incising pia, the broadest bipolar can be used to coagulate a broad width of vessels and cortex as one, followed by division of both with microscissors. In this second scenario, the fine bipolars are less helpful than the broader tips of the insulated bipolar and will slow down the surgery.

White-matter dissection

Dissection of the adjacent white matter can be performed quickly but needs to be performed in a way that allows arterialized vessels on the surface of the bAVM that sweep out (into the white matter) and into the peripheral margin of the bAVM to be swept back on to the bAVM. For this phase, the suction and the bipolar are used as both dissectors and retractors. The suction used at this point is often a little larger (#4 or 5) and a broader insulated bipolar with slightly more current. As the suction retracts and removes the peripheral white matter, the bipolar sweeps the immediate margin of the bAVM by lightly closing the bipolar blades, whilst coagulating, upon immediately adjacent white matter, with a movement that pulls a thin margin of white matter gently towards the bAVM. This allows arterialized loops to be brought back to the bAVM surface without placing them under undue tension. By using this sweeping motion, fewer of the arterialized vessels crossing the dissection void will be thought of as small feeding arteries to be diathermied and divided but rather as arterialized veins looping in and out of the plane of the bAVM's imagined surface. It is worth following such vessels, when they are identified, for a short distance to see whether they loop down and back. For the vessels that do cross the dissection void and need dividing, the use of broader bipolars or microclips is appropriate.

As discussed above, the initial white-matter dissection should be on a single side of the bAVM allowing access to the nose cone and dissection around the nose cone. The deepest target may be within the ventricle itself and entry into the ventricle should be anticipated and encouraged for any bAVM that approaches the ventricle. Identifying the choroidal feeders within the ventricle rather than through the white matter is preferred as the vessels are more readily controlled by bipolar diathermy (or microclip) within the ventricle and the circumferential dissection around the nose cone rendered more easily performed when incorporating ependyma.

Final surgical hemostasis

For those familiar with crystal-clear rock pools by the sea, this appearance is what needs to be achieved in the bed of the bAVM after resection with irrigation fluid in the bAVM cavity. No bleeding should be concealed by cottonoid or covering material. Nor should gentle compression be made to a bleeding point. Both these techniques, that might be appropriate in other neurosurgical contexts, can be disastrous with bAVM surgery. The small arteries do not have the muscular wall to constrict (the normal response to an artery opened) and there is a risk that bleeding that might start as a trickle into the bAVM bed can become torrential and, if concealed, spread widely within the brain parenchyma. Instead, if a point in the cavity wall is bleeding, by using the smallest sucker placed immediately over the point of bleeding, and perhaps a second small sucker to dissect around the point of bleeding, it can be ascertained if this is a solitary vessel or belongs to a nest of vessels under tension that need to be resected. In the case of a solitary vessel, applying a microclip should control the bleeding. These vessels, at this point in the operation, are very difficult to bipolar. A small square of Surgicel over the clip handles will nail or cage the microclip to the brain, preventing accidental dislodgement later by contacting the microclip with the sucker. If a nest of vessels is the source of bleeding, it should be assumed that there is "retained nidus" and that further dissection at a slight margin from this nest is appropriate to regain control and allow completion of the resection.

Preoperative functional imaging, intraoperative angiography, and indocyanine green angiography

These are all useful, to a varying extent, in the management of bAVM. However, they are of secondary importance to that of the surgical techniques and the recommended practices that follow the principles of bAVM surgery outlined above. The potential benefits from these techniques are obvious, so I will concentrate on the pitfalls.

1. Functional imaging may incorrectly suggest that brain adjacent to bAVM, that would be functionally critical in the absence of the bAVM, is nonfunctioning. Without an extremely detailed assessment by a neurocognitive scientist familiar with functional magnetic resonance imaging (fMRI) protocols, one should not trust that otherwise critical brain by anatomic location be regarded as noncritical by fMRI. One should remember that cerebrovascular neurosurgery differs from that of a neuro-oncologic approach to an intrinsic brain tumor. Whereas brain tumors may be indistinct on their margins blurring with brain, bAVMs are distinguished at a macroscopic level from the surrounding brain. To potentially benefit from brain tumor resection, in which the center of the tumor is easy to resect but the advancing edge is difficult to determine and is responsible for future progression, the extent of the resection is determined often by what is necessary to leave behind. For this reason, fMRI and awake craniotomy are of use in brain tumor surgery. However, this geographic approach to bAVM surgery is rarely appropriate and, therefore, these ancillary techniques seldom enhance the ability to deliver good surgery. As Dr. Sundt said, "where AVM is brain isn't" (TM Sundt Jr, personal communication, 1987).
2. Intraoperative angiography may be of assistance but it is important to have as good a quality study (as obtained in the angiography suite) and be aware of its limitation. Presence of residual bAVM by these studies should demand immediate surgery. However, a negative result may be misleading. This is because occlusion of outflow from "retained nidus" or fragile thin-walled feeding arteries may result in no flow (and therefore no angiographic filling) despite persistent continuity of the blood column, thus obscuring the presence of a risk for delayed hemorrhage. Given that usually only a single digital subtraction angiography (DSA) is warranted during hospital admission after bAVM surgery, the advantage of a delayed DSA (1 week after surgery for larger bAVM) is to allow the effects of propagated thrombosis, recanalization, and vascular remodeling to occur. The extent of vascular remodeling will also assist in deciding whether avoiding episodes of hypertension remains paramount (i.e., enabling relaxation of the blood pressure parameters) (Figs 5.5 and 5.6). Early postoperative computed tomography angiography (CTA) may be a satisfactory screening test to ensure that the surgical goal was likely to have been achieved. This may eventually prove to be as good as catheter DSA, but at this stage, I prefer to rely on a delayed (but during the hospital admission) DSA.
3. ICG. With experience, distinguishing the feeding arteries from arterialized veins at surgery is not problematic. However, for the less experienced, ICG may assist with this decision. The importance of securing all arterial input as early as possible whilst leaving intact venous outflow is critical to the success of surgery. Therefore, if doubt exists as to whether an arterialized vessel is artery or vein then ICG may assist. Of course, this does not eliminate the need to follow the artery proximally and distally to ensure that flow to normal brain remains secure. The absence of ICG enhancement should not be used to make a decision as to whether the bAVM is obliterated. This is because very little material needs to be over the vessel wall to obscure the ICG from the camera, giving a false-negative interpretation, as well as the explanation of a false-negative DSA, discussed above.

PERIOPERATIVE SURGICAL MANAGEMENT

For SPC A bAVM, the majority of patients can emerge from anesthesia at the end of surgery and be managed in an identical perioperative manner to that of most craniotomies, with the exception that high blood pressure should not be tolerated. For larger (>4 cm) SPC B and C bAVM, greater care needs to be taken to ensure that normal-low blood pressure is maintained until a degree of arterial remodeling has occurred and the patient is then capable of protecting the vasculature vulnerable to rupture in the early postoperative period (Morgan et al., 2003). Any episode of elevated blood pressure (examples of episodes include pain, seizures, and anxiety with upcoming procedures) needs to be absolutely avoided and anticipated. For some cases, where the institutional experience is appropriate, this may be best achieved by induced coma and rigorous continuous blood pressure control. The aim would be to keep systolic blood pressure below 70 mmHg and the cerebral perfusion pressure above 50 mmHg (necessitating intracranial pressure monitoring). The period of vulnerability is in that period before arterial remodeling provides protection and normally lasts no more than 7 days (Morgan et al., 2003).

Remodeling can be assessed both by CTA and DSA. A variety of agents are likely to be necessary to optimize blood pressure control, including calcium channel vasodilators, titratable agents (such as nitroprusside), beta-blockade, and barbiturates that impede the force of cardiac contraction, diuretics to prevent fluid overload, and angiotensin-converting enzyme inhibition to provide an element of blood pressure control, with the theoretic advantage of lowering the limit of autoregulation (Morgan and Winder, 2001). Because of the considerable time to emerge from barbiturate anesthesia (because of the reserves maintained within fat and myelin long after ceasing its administration), ceasing barbiturates after a short period is appropriate, well short of the expected time blood pressure protection is required. Because of the complexity of this management, there is considerable danger in this regimen without an intensive care team experienced in such management (Cordato et al., 2003). The more complex SPC B and C bAVM should only be operated on in institutions that have such intensive care experience, irrespective of the surgical or anesthetic skills.

References

Barnett GH, Little JR, Ebrahim ZY et al. (1987). Cerebral circulation during arteriovenous malformation operation. Neurosurgery 20: 836–842.

Baronski JF, Grant RA, Hirsch LJ et al. (2014). Seizure control for intracranial arteriovenous malformations is directly related to treatment modality: a meta-analysis. J NeuroIntervent Surg 6: 684–690.

Bervini D, Morgan MK, Ritson EA et al. (2014). Surgery for unruptured arteriovenous malformations of the brain is better than conservative management for selected cases: a prospective cohort study. J Neurosurg 121: 878–890.

Brown Jr RD, Wiebers DO, Forbes G et al. (1988). The natural history of unruptured intracranial arteriovenous malformations. J Neursurg 68: 352–357.

Brown Jr RD, Wiebers DO, Forbes GS (1990). Unruptured intracranial aneurysms and arteriovenous malformations: frequency of intracranial hemorrhage and relationship of lesions. J Neurosurg 73: 859–863.

Cordato DJ, Herkes GK, Mather LE et al. (2003). Barbiturates for acute neurological and neurosurgical emergencies – do they still have a role? J of Clinical Neuroscience 10: 283–288.

da Costa L, Wallace MC, terBrugge KG et al. (2009). The natural history and predictive features of hemorrhage from brain arteriovenous malformations. Stroke 40: 100–105.

Du R, Keyoung HM, Dowd CF et al. (2007). The effects of diffuseness and deep perforating artery supply on outcomes after microsurgical resection of brain arteriovenous malformations. Neurosurgery 60: 638–648.

Gross B, Du R (2013). Natural history of cerebral arteriovenous malformations: a meta-analysis. J Neurosurg 118: 437–443.

Han PP, Ponce FA, Spetzler RF (2003). Intention-to-treat analysis of Spetzler–Martin grades IV and V arteriovenous malformations: natural history and treatment paradigm. J Neurosurg 98: 3–7.

Hashimoto N (2001). Microsurgery for cerebral arteriovenous malformations: a dissection technique and its theoretical implications. Neurosurgery 48: 1278–1281.

Hernesniemi J, Romani R, Lehecka M et al. (2010). Present state of microneurosurgery of cerebral arteriovenous malformations. ACTA Neurochirurgica Suppl 107: 71–76.

Kano H, Kondzolka D, Flickinger JC et al. (2012a). Stereotactic radiosurgery for arteriovenous malformations, part 6: multistaged volumetric management of large arteriovenous malformations. J Neurosurg 116: 54–65.

Kano H, Kondziolka D, Flickinger JC et al. (2012b). Aneurysms increase the risk of rebleeding after stereotactic radiosurgery for hemorrhagic arteriovenous malformations. Stroke 43: 2586–2591.

Kano H, Lunsford LD, Flickinger JC et al. (2012c). Stereotactic radiosurgery for arteriovenous malformations, part 1: management of Spetzler-Martin grade I and II arteriovenous malformations. J Neurosurg 116: 11–20.

Kano H, Flickinger JC, Yang H-C et al. (2014). Stereotactic radiosurgery for Spetzler-Martin grade III arteriovenous malformations. J Neurosurg 120: 973–981.

Kim H, Al-Shahi Salman R, McCulloch CE et al. (2014). Untreated brain arteriovenous malformation. Patient-level meta-analysis of hemorrhage predictors. Neurology 83: 1–8.

Kim H, Abla AA, Nelson J et al. (2015). Validation of the supplemented Spetzler-Martin grading system for brain arteriovenous malformations in a multicenter cohort of 1009 surgical patients. J Neurosurg 76: 25–33.

Korja M, Bervini D, Assaad N et al. (2014a). The role of surgery in the management of brain arteriovenous malformations: a prospective cohort study. Stroke 45: 3549–3555.

Korja M, Hernesniemi J, Lawton MT et al. (2014b). Sacrifice of cerebrovascular neurosurgery at the altar of randomized controlled trial. Lancet 384: 27–28.

Lawton MT, Kim H, McCulloch CE et al. (2010). A supplementary grading scale for patients with brain arteriovenous malformations for surgery. Neurosurgery 66: 702–713.

Mohr JP, Parides MK, Stapf C et al. (2014). Medical management with or without interventional therapy for unruptured brain arteriovenous malformations (ARUBA): a multicenter, non-blinded, randomized trial. Lancet 383: 614–621.

Morgan MK (2013). AVM and embolization. J of Neurosurg 118: 967–968.

Morgan MK, Winder M (2001). Haemodynamics of arteriovenous malformations of the brain and consequences of resection. A review. J of Clinical Neuroscience 8: 216–224.

Morgan MK, Winder M, Little NS et al. (2003). Delayed hemorrhage following resection of an arteriovenous malformation in the brain. J Neurosurg 99: 967–971.

Morgan MK, Davidson AS, Koustais S et al. (2013). The failure of ethylene-vinyl alcohol copolymer embolization to improve outcomes in AVM management: case series. J of Neurosurg 118: 969–977.

Morgan MK, Assaad N, Korja M (2015). Surgery for unruptured Spetzler-Martin grade 3 brain arteriovenous malformations: a prospective surgical cohort. Neurosurgery 77: 362–370.

Morgan MK, Alsahli K, Wiedmann M et al. (2016a). Factors associated with proximal intracranial aneurysms to brain arteriovenous malformations: a prospective cohort study. Neurosurgery 78: 787–792.

Morgan MK, Wiedmann M, Assaad N et al. (2016b). Complication-effectiveness analysis for brain arteriovenous malformation surgery. A prospective cohort study. Neurosurgery 79: 47–57.

O'Donnell JM, Morgan MK (2015). Quality of life of patients affected by unruptured brain AVMs. Response. J Neurosurg 122: 985–986.

Ondra SL, Troupp H, George ED et al. (1990). The natural history of symptomatic arteriovenous malformations of the brain: a 24-year follow-up assessment. J Neurosurg 73: 387–391.

Platz J, Berkefeld J, Singer OC et al. (2014). Frequency risk of hemorrhage and treatment considerations for cerebral arteriovenous malformations with associated aneurysms. Acta Neurochir (Wien) 156: 2025–2034.

Spetzler RF, Martin NA (1986). A proposed grading system for arteriovenous malformations. J Neurosurg 65: 476–483.

Spetzler RF, Ponce FA (2011). A 3-tier classification of cerebral arteriovenous malformations. J Neurosurg 114: 842–849.

Stapf C, Mohr JP, Pile-Spellman J et al. (2002). Concurrent arterial aneurysms in brain arteriovenous malformations with haemorrhagic presentation. J Neurol Neurosurg Psychiatry 73: 294–298.

Stapf C, Mast H, Sciacc RR et al. (2006). Predictors of hemorrhage in patients with untreated arteriovenous malformations. Neurology 66: 1350–1355.

Sundt Jr TM, Kees G (1986). Miniclips and microclips for surgical hemostasis. J Neurosurg 64: 824–825.

van Beijnum J, Lovelock CE, Cordonnier C et al. (2009). Outcome after spontaneous and arteriovenous malformation-related intracerebral haemorrhage: population-based studies. Brain 132: 537–543.

van Swieten JC, Koudstaal PJ, Visser MC et al. (1988). Interobserver agreement for the assessment of handicap in stroke patients. Stroke 19: 604–607.

Handbook of Clinical Neurology, Vol. 143 (3rd series)
Arteriovenous and Cavernous Malformations
R.F. Spetzler, K. Moon, and R.O. Almefty, Editors
http://dx.doi.org/10.1016/B978-0-444-63640-9.00006-0

Chapter 6

Endovascular management of arteriovenous malformations

BRADLEY A. GROSS*, KARAM MOON, AND CAMERON G. McDOUGALL
Department of Neurosurgery, Barrow Neurological Institute, St. Joseph's Hospital and Medical Center, Phoenix, AZ, USA

Abstract

Arteriovenous malformation (AVM) embolization can serve as a crucial adjunct before surgical resection, a partial approach to target high-risk features, or, rarely, as a curative approach for high-risk, surgically inaccessible lesions. Specifically, embolization is a welcome surgical adjunct to reduce the size of medium to large AVMs, to target perforator supply, and/or to target the deep portion of a nidus. In addition, a crucial role for embolization is the targeting of associated aneurysms, particularly in the setting of a ruptured lesion, regardless of the subsequent therapeutic modality. Rarely, a deep, small ruptured AVM that cannot be accessed surgically may be embolized with intent to cure. This chapter will review patient selection, technical nuances, and published results for AVM embolization. With appropriate patient selection and well-defined goals of embolization, the risk of procedural morbidity can be outweighed by its benefit.

INTRODUCTION

Cerebral arteriovenous malformations (AVMs) are most commonly diagnosed in patients who present with intracranial hemorrhage (Crawford et al., 1986; Ondra et al., 1990; Hernesniemi et al., 2008; da Costa et al., 2009; Gross and Du, 2013). They are also a considerable source of neurologic morbidity as a result of seizures, focal neurologic deficits, and headaches (Crawford et al., 1986; Josephson et al., 2011; Gross and Du, 2013). They are commonly diagnosed in younger adults (in the third or fourth decade of life), and thus pose a lifelong threat of neurologic morbidity and psychologic burden (Crawford et al., 1986; van der Schaaf et al., 2002; Hernesniemi et al., 2008). In one study with a mean follow-up of 10.4 years, untreated patients had a 42% risk of hemorrhage, a 29% risk of death, an 18% risk of epilepsy, and a 27% risk of having a neurologic handicap by 20 years after diagnosis (Crawford et al., 1986). In another study with a mean follow-up period of 23.7 years, 23% of patients had died of AVM hemorrhage at the time of follow-up (Ondra et al., 1990). The study reported a combined annual rate of major morbidity and mortality of 2.7% per year and a mean interval of 7.7 years between initial presentation and subsequent hemorrhage (Ondra et al., 1990). Most modern natural-history studies are plagued by considerably shorter follow-up and thus attempt to compensate by tallying patient-years over large cohorts (Stapf et al., 2006; da Costa et al., 2009; Gross and Du, 2013). However, this approach does not account for gradual lesion morphologic changes (aneurysm development, venous drainage changes) that may cause the natural history of the AVM to worsen over an uncaptured follow-up period.

Except for patients with multiple medical comorbidities and short life expectancies, all patients with AVMs should be evaluated for potential treatment. Surgical excision provides an expedient curative option for AVMs and should be the treatment modality of choice for surgically accessible, low-risk lesions (Spetzler–Ponce grade A AVMs: Table 6.1) (Drake, 1979; Heros et al., 1990; Davidson and Morgan, 2010; Lawton et al., 2010; Spetzler and Ponce, 2011). Alternatively, small eloquent AVMs may be considered for radiosurgery at the discretion of the surgeon. Intermediate to larger-size AVMs (more than 3 cm) and/or Spetzler–Ponce grade B or C AVMs provide a greater therapeutic challenge and are

*Correspondence to: Bradley A. Gross, MD, c/o Neuroscience Publications, Barrow Neurological Institute, St. Joseph's Hospital and Medical Center, 350 W. Thomas Rd., Phoenix AZ 85013, USA. Tel: +1-602-406-3593, Fax: +1-602-406-4104, E-mail: Neuropub@dignityhealth.org

Table 6.1

Surgical grading schemes for arteriovenous malformations (AVMs)

AVM characteristics*	Spetzler–Martin grade	Spetzler–Ponce grade	Surgical risk
S1E0V0	1	A	Lowest
S1E1V0	2	A	Low
S1E0V1			
S2E0V0			
S1E1V1	3	B	Intermediate
S2E1V0			
S2E0V1			
S3E0V0			
S2E1V1	4	C	High
S3E1V0			
S3E0V1			
S3E1V1	5	C	Highest

*S1 denotes nidus size less than 3 cm, S2 3–6 cm size, and S3 size > 6 cm. E1 denotes eloquent nidus location, and E0 denotes non-eloquent location. V1 denotes deep venous drainage, and V0 denotes absence of deep venous drainage.

often best managed via multimodality approaches, including endovascular embolization (U et al., 1992; Lawton et al., 1995; Chang et al., 2003; Weber et al., 2007; Natarajan et al., 2008; Starke et al., 2009; Crowley et al., 2015). Other indications for embolization include the need to expediently obliterate a high-risk feature (associated aneurysm) and, rarely, to attempt cure in patients with surgically inaccessible, small lesions that require expedient treatment (ruptured, small, deep AVMs) (Haw et al., 2006; Krings et al., 2010; van Rooij et al., 2012; Crowley et al., 2015). In this chapter, we review these indications, along with endovascular techniques and outcomes.

PATIENT EVALUATION

Patient evaluation is based on known risk factors for AVM hemorrhage, as well as factors that influence the success and morbidity of select treatment approaches (Table 6.2). The primary goal of AVM treatment is the elimination of prospective hemorrhage risk; some studies have also shown a benefit of treatment for seizure risk (Yeh et al., 1993; Hoh et al., 2002).

Patient and clinical factors

Innumerable studies have demonstrated a greater risk of rehemorrhage after initial hemorrhage (Stapf et al., 2006; Hernesniemi et al., 2008; da Costa et al., 2009; Gross and Du, 2013). This feature should strongly prompt the consideration of AVM treatment. Importantly, some studies have shown that ruptured AVMs are associated with better surgical outcomes (Drake, 1979; Gross et al., 2008; Lawton et al., 2010); others have shown this feature to be associated with worse radiosurgical outcomes (Starke et al., 2013). The latter is in part intuitive because the risk of rebleeding is particularly greater within the first year after hemorrhage (Hernesniemi et al., 2008; Gross and Du, 2013), a period during which radiosurgical treatment does not provide protection. The impact of AVM rupture status on embolization risk is less clear.

Another potential risk factor for hemorrhage is pregnancy (Gross and Du, 2012a). Thus, young women with AVMs who wish to bear children should be counseled

Table 6.2

Pertinent factors from the patient evaluation and their impact on natural history and treatment approaches

Feature*	Natural history	Surgery	Radiosurgery	Embolization
Hemorrhagic AVM	Higher risk	Favored	Less favored	Treat/secure high-risk feature (aneurysm)
Younger age	No effect	Most favored	Favored	Favored
Female sex	Higher risk during pregnancy	No effect	No effect	No effect
Small AVM size	No effect	Favored	Favored	Favorable if needed standalone
Associated aneurysm	Higher risk	Favored	Less favored	Favorable to secure it
Perforator supply	No effect	Less favored	No effect	Desirable if feasible as adjunct to surgery
Diffuse nidus	No effect	Less favored	Least favored	Less favored
Deep venous drainage	Higher risk	Less favored	No effect	No effect
Deep location	Higher risk	Least favored	Less favored	Less favored

*Key factors to consider for embolization are in **bold**.
AVM, arteriovenous malformation.

and treatment should generally be advised. Pregnant patients with ruptured AVMs may undergo treatment before delivery (Gross and Du, 2012a).

Other patient-dependent factors that have not consistently been associated with hemorrhage risk are age and sex (Stapf et al., 2006; Hernesniemi et al., 2008; Gross and Du, 2013). Importantly, both surgical and radiosurgical studies have shown younger age to be associated with better treatment outcomes (Pollock and Flickinger, 2008; Raffa et al., 2009; Lawton et al., 2010; Starke et al., 2013). Similarly, endovascular treatment is generally more facile in younger adults than in older adults, as the former tend to have less atherosclerotic disease and proximal vessel tortuosity.

Angiographic factors

The crucial diagnostic study in the evaluation of patients before treatment is formal digital subtraction angiography. Several angiographic features require particular scrutiny when evaluating AVM hemorrhage risk and treatment risk: size, associated aneurysms, perforator supply, diffuse nidus angioarchitecture, deep location, and deep venous drainage.

Although early natural-history studies suggested a greater hemorrhage risk for smaller AVMs, they were largely based on rate of hemorrhagic presentation rather than on prospective hemorrhage risk and were thus subject to presentation bias (Kader et al., 1994; Langer et al., 1998). Specifically, because smaller AVMs would otherwise not garner clinical attention unless they ruptured, they came to be perceived as having a greater risk of hemorrhage. Modern natural-history studies have not demonstrated an association between AVM size and hemorrhage risk (Stapf et al., 2006; Gross and Du, 2013). Nevertheless, size is a crucial factor when an AVM is evaluated for treatment risk and feasibility. Smaller size has consistently been shown to be associated with higher obliteration and lower morbidity rates for both microsurgical and radiosurgical treatment (Spetzler and Martin, 1986; Heros et al., 1990; Pollock and Flickinger, 2008; Raffa et al., 2009; Lawton et al., 2010; Spetzler and Ponce, 2011; Gross and Du, 2012b; Starke et al., 2013). Although larger size is not a factor for endovascular treatment intended as an adjunct to other modalities, it is, of course, an adverse factor for obliteration and morbidity when considering potential curative embolization.

AVMs with associated aneurysms have been shown to have a greater prospective risk of hemorrhage (da Costa et al., 2009; Gross and Du, 2013). Thus, these types of AVMs are better managed with a more expedient means of treatment such as microsurgical excision (Natarajan et al., 2008; Crowley et al., 2015). Targeted embolization for high-risk features (associated aneurysms) is also an excellent treatment approach for these AVMs, particularly as a prelude to radiosurgery if microsurgical excision is not feasible (van Rooij et al., 2007; Weber et al., 2007; Crowley et al., 2015).

Although the presence of perforator supply and diffuse nidal angioarchitecture has not been clearly shown to impact natural history, both add difficulty in the surgical management of AVMs (Morgan et al., 1997; Du et al., 2007; Gross et al., 2008). Microsurgical coagulation of small, delicate perforators that retract into eloquent, deep tissue can be challenging. When feasible, adjunctive embolization of perforator supply to an AVM is a welcome approach for the neurosurgeon (U et al., 1992; Lawton et al., 1995; Gross et al., 2008); however, these vessels may pose a greater challenge for embolization compared to more capacious and less serpentine pial suppliers. Radiosurgical studies have not specifically shown an impact of perforator supply on results, although a diffuse nidus generally poses a greater challenge for effective AVM targeting.

Deep venous drainage and deep AVM location have both been shown to be independent risk factors for AVM hemorrhage (Stapf et al., 2006; Gross and Du, 2013). Both features also add risk to surgical treatment (Spetzler and Martin, 1986; Heros et al., 1990; Lawton et al., 2010; Spetzler and Ponce, 2011), whereas deep venous drainage has not been independently shown to adversely impact radiosurgical or endovascular results (Haw et al., 2006; Pollock and Flickinger, 2008; Raffa et al., 2009; Starke et al., 2013; Dumont et al., 2015). Yet deep location, correlative with eloquent location, adds to the morbidity of any therapeutic modality (Pollock et al., 2004; Andrade-Souza et al., 2005; Starke et al., 2013).

Summary of key features germane to embolization

Important angiographic features most germane to embolization include AVM size, location, associated aneurysms, and perforator supply. Small AVMs in surgically inaccessible locations that require an immediate therapeutic approach (ruptured lesions) may be considered for endovascular cure (Haw et al., 2006; Krings et al., 2010; van Rooij et al., 2012; Crowley et al., 2015). Larger AVMs that may seem unsuitable for surgical treatment may be “embolized down” to more compact, smaller, ideal targets (Fig. 6.1). If a lesion is being considered for surgery, perforator supply should be carefully perused and evaluated for the feasibility of embolization, which is a particularly welcome adjunct for the neurosurgeon. Similarly, embolization of deeper components of an AVM are

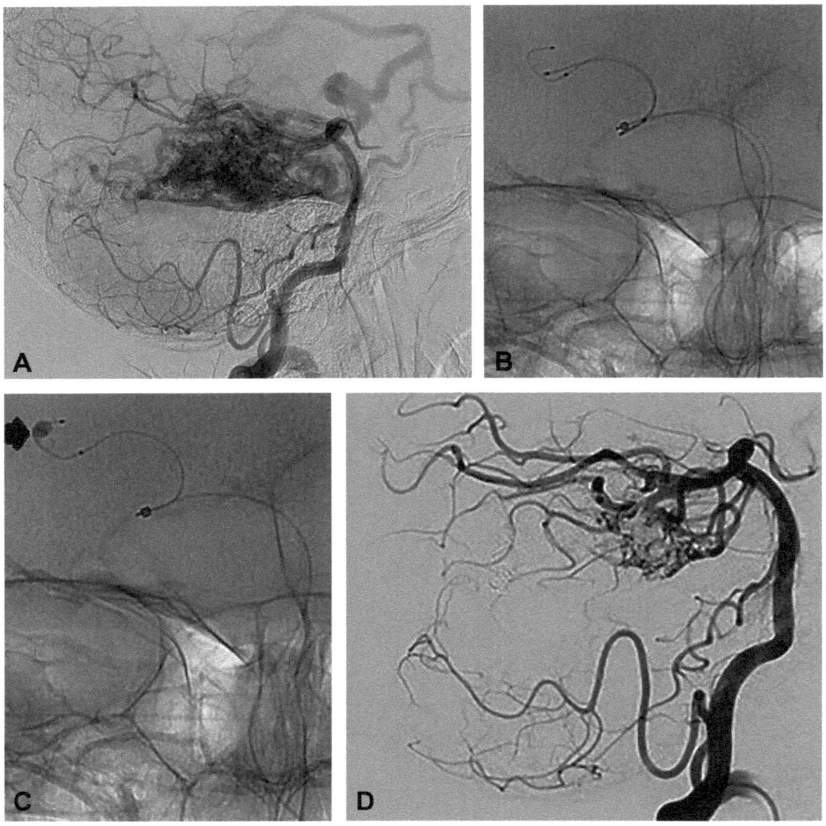

Fig. 6.1. A 21-year-old man presented with a medial temporal Spetzler–Martin grade 4 arteriovenous malformation (AVM) after a seizure (**A**, lateral view of vertebral artery injection). The treatment goal was to reduce it to a surgically amenable target, so two pedicles were embolized using the pressure-cooker technique. Two microcatheters were migrated into each respective pedicle (**B**, fluoroscopic image), using one to glue the other in place (**C**, lateral fluoroscopic image; arrow denotes glue plug). After embolization of the two pedicles, the AVM was markedly reduced in size (**D**, lateral view of control angiogram via the vertebral artery). The AVM was then resected without complication, and the patient was discharged neurologically intact. (Courtesy of Barrow Neurological Institute, Phoenix, Arizona.)

another welcome adjunct. Finally, associated aneurysms are high-risk features that should be considered for therapeutic embolization, a particularly important consideration for ruptured lesions (Fig. 6.2).

CLINICAL DECISION MAKING

An amalgamation of the aforementioned pertinent clinical and angiographic factors facilitates clinical decision making by the application of treatment grading schemes to help stratify treatment risk. The first issue to address is whether or not to treat the AVM; the second, for AVMs that can be treated, is how to treat. An AVM possessing at least one clinical feature denoted as "higher risk" for natural history should be considered for treatment (Table 6.2). In particular, patients with ruptured AVMs should be considered for treatment given the known greater risk of rehemorrhage (Stapf et al., 2006; Hernesniemi et al., 2008; da Costa et al., 2009; Gross and Du, 2013). Young women desiring to bear children should be given particular attention for treatment (Gross and Du, 2012a). Other features that should prompt treatment include associated aneurysms, deep venous drainage, and deep AVM location (Stapf et al., 2006; Hernesniemi et al., 2008; da Costa et al., 2009; Gross and Du, 2013). Even in the absence of these known risk factors for hemorrhage, AVMs that score well on surgical or radiosurgical grading schemes should be considered for treatment.

Surgical treatment

The most widely applied surgical grading scheme used to stratify surgical risk is the Spetzler–Martin grading scale (Spetzler and Martin, 1986; Spetzler and Ponce, 2011). This scale rates the three key factors of AVM size, deep venous drainage, and eloquent location (Table 6.1). The Lawton supplementary grade adds increasing age, unruptured presentation, and diffuse nidus as independent risk factors for worse outcomes after surgery (Lawton et al., 2010).

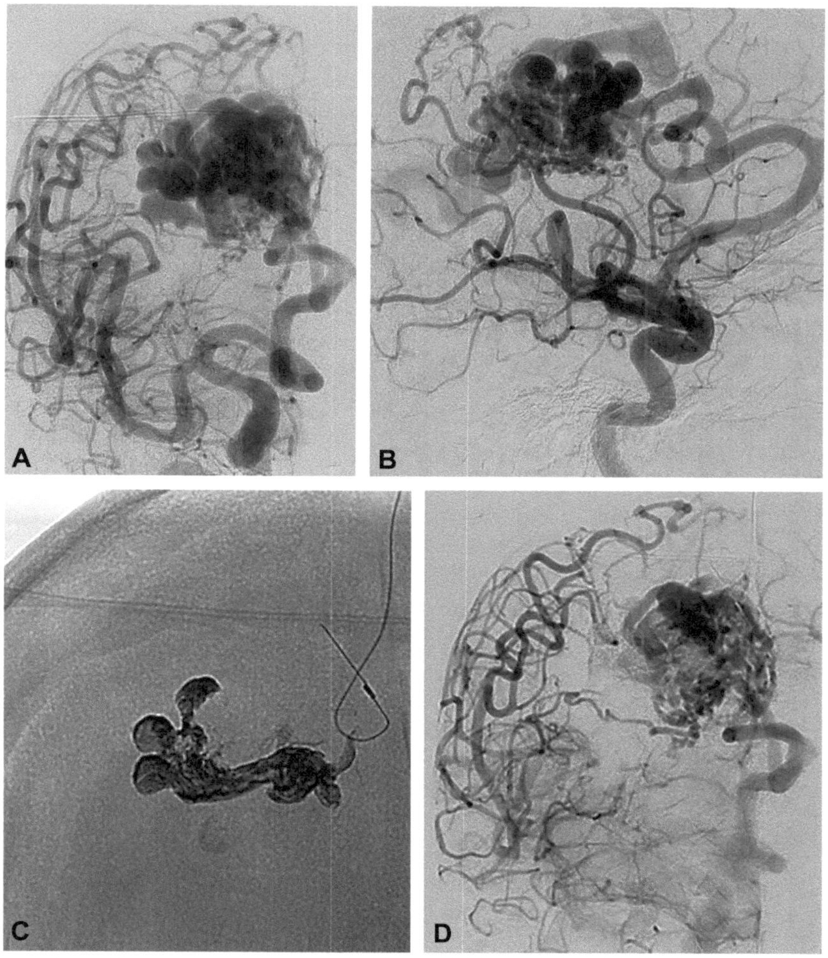

Fig. 6.2. A 54-year-old woman presented with subarachnoid hemorrhage from a medium-sized (3–6-cm) arteriovenous malformation in eloquent territory (right internal carotid artery injection, anteroposterior (**A**) and lateral (**B**) views). The patient's motor function was normal. After superselective catheterization and embolization of a single feeding pedicle from the anterior cerebral artery, Onyx can be seen casting three intranidal aneurysms (**C**, anteroposterior fluoroscopic image), without further opacification of these high-risk features on control angiography (**D**, anteroposterior view of right internal carotid artery control injection). (Courtesy of Barrow Neurological Institute, Phoenix, Arizona.)

Surgical treatment should be considered for any Spetzler–Martin grade I or II (Spetzler–Ponce grade A) AVMs (Spetzler and Martin, 1986; Heros et al., 1990; Davidson and Morgan, 2010; Lawton et al., 2010; Spetzler and Ponce, 2011; Gross and Du, 2012b). Unless precluded by the lesion being buried in eloquent parenchyma (applicable to grade II AVMs), surgical treatment should always be considered the first-line treatment given the expedient high rate of cure it affords. Generally, Spetzler–Martin grade I AVMs should be managed with surgery alone, as it is difficult to justify the added morbidity provided by multimodality approaches. This treatment approach also applies to select grade II lesions, particularly small AVMs (less than 3 cm) that are given a second point for only deep venous drainage. Spetzler–Martin grade III AVMs are often managed with multimodality therapy (Lawton et al., 1995; Natarajan et al., 2008; Spetzler and Ponce, 2011; Gross and Du, 2012b). Generally, unruptured grade IV or V AVMs should not be managed surgically (Spetzler and Ponce, 2011); in select cases, they can be treated initially with embolization or radiosurgery to reduce the size of the lesion to make it potentially amenable to subsequent surgical treatment (Fig. 6.1) (Chang et al., 2003; Sanchez-Mejia et al., 2009; Davidson and Morgan, 2010).

Embolization as an adjunct to surgical treatment

Embolization is often used as a preoperative adjunct for select Spetzler–Martin grade II and III AVMs (Lawton et al., 1995; Natarajan et al., 2008; Crowley et al., 2015).

It may also be used in select grade IV AVMs as a means to reduce them to surgically amenable targets (Fig. 6.1). In a recent report of our endovascular institutional experience, we noted that 78.9% of AVMs were embolized with intent for subsequent surgery (Crowley et al., 2015). With this embolization goal, Spetzler–Martin grade was not associated with endovascular procedural morbidity.

AVMs larger than 3 cm should generally be considered for preoperative embolization to mitigate the risk of postoperative normal perfusion pressure breakthrough. The immediate removal of an entire intermediate to large AVM places the adjacent normal parenchyma at risk for hyperperfusion as a result of the chronic loss of autoregulatory capabilities of preoperatively maximally dilated adjacent vasculature (Spetzler et al., 1978).

Given the added surgical risk of grade II AVMs in eloquent locales, embolization of the surgically more challenging or deep portion of the AVM is often a welcome adjunct for the surgeon. The same applies to eloquent grade III AVMs. In particular, the endovascular neurosurgeon should peruse perforator suppliers and attempt embolization of these pedicles as feasible in light of the added challenge they present to microsurgical excision (U et al., 1992; Lawton et al., 1995; Gross et al., 2008).

Radiosurgical treatment

Radiosurgery is an excellent option for small AVMs in surgically inaccessible or higher-risk locales (Pollock and Flickinger, 2008; Raffa et al., 2009; Starke et al., 2013). Larger lesions requiring treatment (ruptured AVMs) that cannot be embolized to surgically amenable targets may be managed with radiosurgery as a prelude to repeat radiosurgery or subsequent surgery (Sanchez-Mejia et al., 2009).

The success of radiosurgical treatment is largely predicated on AVM size and location (Pollock and Flickinger, 2008; Raffa et al., 2009; Starke et al., 2013). The most recent iteration of the radiosurgery-based AVM grading system calculates a score based on the following equation: $(0.1)\times(\text{volume in cc})+(0.02)\times(\text{age in years})+0.3$, if location is basal ganglia/thalamus/brainstem (Pollock and Flickinger, 2008). Patients with a score of ≤ 1 had an 89% rate of obliteration without new deficits; this rate fell to 46% for patients with a score of at least 2. The simpler Virginia Radiosurgery AVM Scale assigns 1 point for 2–4 cc AVMs, 2 points for those over 4 cc, 1 point for eloquent location, and 1 point for a history of hemorrhage (Starke et al., 2013). The percentage of patients with AVM obliteration without posttreatment hemorrhage or permanent complications was 83% and 79% for patients with scores of 0 or 1, respectively. This rate fell to 70% for patients with a score of 2, to 48% for patients with a score of 3, and to 39% for patients with a score of 4. In contrast to surgical results, this grading scheme emphasizes the adverse impact of a prior history of hemorrhage on overall radiosurgical success.

Embolization as an adjunct to radiosurgical treatment

Adjunctive embolization to anticipated radiosurgical treatment is a controversial approach. Multiple series, subgroup analyses of radiosurgical grading schemes, and even class II evidence have all shown worsened outcomes (obliteration and/or complications) for patients undergoing embolization and radiosurgery compared to radiosurgery alone (Andrade-Souza et al., 2007; Kano et al., 2012). Although these findings may be confounded by selection bias of patients with more complex or challenging lesions undergoing embolization and radiosurgery, it reinforces the maxim that the mere reduction in size of an AVM via embolization may not confer a benefit unless it is done in anticipation of subsequent surgical resection. Nevertheless, it is prudent to embolize AVMs with high-risk features (associated aneurysms), particularly ruptured AVMs, before radiosurgical treatment.

Embolization as sole therapy

Embolization as palliative or intended partial therapy should be considered for large lesions unamenable to either surgical or radiosurgical approaches with high-risk features (aneurysms). The goal of embolization is to target these features (Krings et al., 2010; Pollock et al., 2011; van Rooij et al., 2012). Embolization as sole curative therapy is a rare approach of consideration for high-risk (ruptured) AVMs in surgically inaccessible locations (van Rooij et al., 2012). In our recently published endovascular institutional experience, 2.3% of AVMs were embolized with curative intent (Crowley et al., 2015). Embolization should generally be considered in patients for whom a cure might be achieved, as merely reducing the size of the AVM has not been shown to be protective (Han et al., 2003; Lv et al., 2012). Thus, embolization is an option for small, deep, ruptured AVMs that would be considered at high risk for rerupture during the radiosurgical latency period (van Rooij et al., 2012). The recently proposed Buffalo score helps to quantify the endovascular risk for embolization with intent to cure (Dumont et al., 2015). One point is assigned for AVMs with one or two arterial pedicles; 2 points are assigned for three or four arterial pedicles, and 3 points are assigned for five or more pedicles. An additional point is given if the diameter of most pedicles is ≤ 1 mm and an additional point is given for eloquent location. No complications were observed for patients with grade 1 or 2 AVMs. The complication rate increased to 14% for patients with grade 3 AVMs, to 50% for those with grade 4, and to 75% for those with grade 5 AVMs.

TECHNICAL NUANCES

The endovascular treatment of AVMs has evolved considerably since the pioneering work in the 1960s by Luessenhop and Spence (1960), who used carotid puncture and particulate embolization. Innovative microcatheter designs by Serbinenko and Kerber and the popularization of the use of cyanoacrylates by Kerber in the late 1970s paved the way to distal intracranial catheterization and the introduction of more permanent embolisates (Kerber et al., 1978; Serbinenko, 1979; Kerber, 1980; Bank et al., 1981). Cyanoacrylate embolization as a crucial perioperative or intraoperative adjunct was championed by Drake in partnership with Debrun et al. (1982). Although cyanoacrylate glues are still used today, the introduction of Onyx was a crucial advance, allowing for potentially improved penetration and longer injection times from a single pedicle (Taki et al., 1990; van Rooij et al., 2007; Weber et al., 2007; Loh and Duckwiler, 2010). Unlike the liquid adhesives such as *N*-butyl cyanoacrylate (NBCA), Onyx is a copolymer of ethylene vinyl alcohol dissolved in dimethyl sulfoxide (DMSO) that precipitates to form a nonadhesive spongy solid.

Prior to embolization, the therapeutic goal should be defined. In any case, it is reasonable to consider partial embolization of high-risk features (associated aneurysms) (Fig. 6.2). This consideration applies to AVMs that may be inoperable or in patients who may go on for radiosurgical treatment. Additional treatment goals for patients with anticipated adjunctive embolization to surgical treatment include targeting surgically challenging features (perforator supply, deep component) and reducing the overall nidus size. In rare cases, embolization for cure of surgically inaccessible high-risk (ruptured) AVMs should be a goal defined before treatment. It is a rare approach that should be deemed safer than the natural history of the untreated AVM or surgical/radiosurgical alternatives.

With these goals in mind, we perform all procedures under general endotracheal anesthesia with neurophysiologic monitoring (somatosensory evoked potentials and electroencephalography). Meticulous hemodynamic monitoring, with strict avoidance of hypertension, is crucial throughout the case. All procedures are initiated in anticipation of a triaxial system of support with 6 F or 8 F common femoral artery access catheters, with the latter used in cases requiring a larger intermediate catheter. Patients are systemically heparinized. The guide catheter is typically navigated into the distal cervical internal carotid artery or the V2–3 junction of the vertebral artery. The intermediate catheter is navigated to the A1, M1, P1–2 junction, or distal V4, as needed, for the pedicles to be embolized. Triaxial systems allow for expedient access in anticipated multiple pedicle embolizations and in the rare circumstances of complications, such as vessel rupture requiring immediate recatheterization and management. They also allow for expedient runs after pedicle embolization. In our recent report of our institutional experience, we documented a mean of 3.5 embolized pedicles over a mean of 1.3 sessions (Crowley et al., 2015).

We typically use glue and DMSO-compatible microcatheters. For more distal embolization of smaller pedicles, we use detachable-tip, flow-directable microcatheters. Prior to embolization, a microcatheter run is performed to identify any opacification of pial or *en passage* vessels and to determine the rate of arteriovenous shunting. Ideally, angiographic views should have minimal overlap of the microcatheter tip and the AVM nidus, thus allowing the endovascular neurosurgeon to discern AVM nidus penetration and embolisate reflux. The location of draining veins is meticulously scrutinized to identify the maximal point of desired embolisate penetration. In circumstances of high-flow shunting and/or close proximity to the AVM nidus, we may use NBCA. In cases with very-high-flow shunting, balloon microcatheters may be used, with the balloon inflated at the time of embolization; alternatively, if the pedicle is adequately capacious, coils may be deposited first.

Most commonly, Onyx is used as the embolisate for added penetration and to allow for longer injection. The injection is performed under blank fluoroscopic road map control. In our recent account of our institutional experience, we reported that AVMs treated with Onyx had a greater mean number of arterial pedicles embolized over more treatment sessions (Crowley et al., 2015). However, the use of Onyx compared to NBCA was not associated with a difference in periprocedural morbidity.

For NBCA, the microcatheter is first flushed with 10 mL of D5 water before a mixture of Ethiodol and NBCA (typically 3:1) is injected in a single shot. After this injection, the microcatheter is briskly removed. For Onyx, the microcatheter is flushed with saline and then loaded with a volume of DMSO equal to the dead-space volume in the microcatheter. The DMSO is then displaced by injecting Onyx over 90 seconds into the microcatheter; then the embolization begins. Onyx is delivered until reflux at the microcatheter tip is observed. Injection then ceases for 1–2 minutes to allow the Onyx to precipitate around the microcatheter tip, forming a plug. Some neurosurgeons specifically prefer to perform this maneuver with Onyx-34, which is more dense. Injection then recommences under a new blank road map with increased chances of antegrade flow of the Onyx (plug-push technique). At the conclusion of the injection, the syringe is aspirated as it is withdrawn under traction. Although a greater amount of Onyx reflux may cause difficulty in withdrawing the microcatheter, reflux along a relatively straight segment is generally well tolerated while reflux around a vessel curve may increase the difficulty and the risk during withdrawal of the microcatheter.

As a means to augment penetration, the pressure-cooker technique may be used (Chapot et al., 2014). In this technique, two microcatheters, one a detachable-tip microcatheter, are used to catheterize the feeding pedicle (Fig. 6.1). The other microcatheter is then used to intentionally glue in the detachable tip of the microcatheter with NBCA. The glued-in microcatheter is then used to inject Onyx. This technique is analogous to the plug-push technique, but it allows for the swift, early penetration of Onyx with better control of reflux by the glue plug.

PATIENT OUTCOMES

In our recent report of our institutional experience with 342 AVMs embolized over 446 sessions, we documented neurologic morbidity and mortality rates of 9.6% and 0.3% (Crowley et al., 2015). A recent report from the University of Buffalo noted that 50 AVMs were managed with intent to cure, with a complete obliteration rate of 10% (Dumont et al., 2015). For a total of 194 arterial pedicles embolized in 177 procedures, there were 13 complications (nine minor, four major) (Dumont et al., 2015). On a per-patient basis, this emphasizes that embolization may have considerable associated morbidity. In another study of complications associated with cyanoacrylate glue embolization, the significant predictors of complications were eloquent nidus location, nidus with a fistulous component, and venous penetration of the glue cast (Haw et al., 2006).

A series depicting aggressive endovascular treatment of 350 AVMs with Onyx reported the highest reported angiographic cure rate of 51%; the rate of permanent morbidity and mortality was 7.1% and 1.4%, respectively (Saatci et al., 2011). From the aforementioned endovascular series and others reported in the medical literature, overall morbidity rates generally range from 7% to 11% and mortality rates range from 0% to 3% across these series (Mounayer et al., 2007; van Rooij et al., 2007; Weber et al., 2007; Katsaridis et al., 2008). Rates of catheter retention from Onyx usage have ranged from 1% to 8%, correlated to the general rates of attempted endovascular cure in respective series (Mounayer et al., 2007; van Rooij et al., 2007; Weber et al., 2007; Saatci et al., 2011).

An early comparative study of Onyx and NBCA concluded that the two embolisates were equivalent in regard to AVM volume reduction of at least 50%, serious adverse events, surgical resection time, and intraoperative blood loss (Loh and Duckwiler, 2010). A trend toward a greater rate of adverse events with Onyx was seen (9.2% vs. 4.8%); however, given the timing of this study, this finding likely reflects an early learning curve with Onyx usage. Our experience reflects a general equivalence in periprocedural complications with these embolisates (Crowley et al., 2015). However, our findings are consistent with those of other studies (Starke et al., 2013) in that a greater number of embolization sessions was associated with higher procedural morbidity.

Overall, NBCA and Onyx are complementary agents in the endovascular management of AVMs. Embolization remains a crucial adjunct in the surgical management of complex AVMs and can be a potentially life-saving approach in the management of patients with high-risk AVM features, particularly in the setting of hemorrhage. Proper patient selection and definition of the goals of therapy should be well delineated before treatment; with these goals, the risk of morbidity from embolization can be outweighed by its benefit.

References

Andrade-Souza YM, Zadeh G, Scora D et al. (2005). Radiosurgery for basal ganglia, internal capsule, and thalamus arteriovenous malformation: clinical outcome. Neurosurgery 56: 56–63; discussion 63–64.

Andrade-Souza YM, Ramani M, Scora D et al. (2007). Embolization before radiosurgery reduces the obliteration rate of arteriovenous malformations. Neurosurgery 60: 443–451; discussion 451–452.

Bank WO, Kerber CW, Cromwell LD (1981). Treatment of intracerebral arteriovenous malformations with isobutyl 2-cyanoacrylate: initial clinical experience. Radiology 139: 609–616.

Chang SD, Marcellus ML, Marks MP et al. (2003). Multimodality treatment of giant intracranial arteriovenous malformations. Neurosurgery 53: 1–11; discussion 11–13.

Chapot R, Stracke P, Velasco A et al. (2014). The pressure cooker technique for the treatment of brain AVMs. J Neuroradiol 41: 87–91.

Crawford PM, West CR, Chadwick DW et al. (1986). Arteriovenous malformations of the brain: natural history in unoperated patients. J Neurol Neurosurg Psychiatry 49: 1–10.

Crowley RW, Ducruet AF, Kalani MY et al. (2015). Neurological morbidity and mortality associated with the endovascular treatment of cerebral arteriovenous malformations before and during the Onyx era. J Neurosurg 122: 1492–1497.

da Costa L, Wallace MC, ter Brugge KG et al. (2009). The natural history and predictive features of hemorrhage from brain arteriovenous malformations. Stroke 40: 100–105.

Davidson AS, Morgan MK (2010). How safe is arteriovenous malformation surgery? A prospective, observational study of surgery as first-line treatment for brain arteriovenous malformations. Neurosurgery 66: 498–504; discussion 504–505.

Debrun G, Vinuela F, Fox A et al. (1982). Embolization of cerebral arteriovenous malformations with bucrylate. J Neurosurg 56: 615–627.

Drake CG (1979). Cerebral arteriovenous malformations: considerations for and experience with surgical treatment in 166 cases. Clin Neurosurg 26: 145–208.

Du R, Keyoung HM, Dowd CF et al. (2007). The effects of diffuseness and deep perforating artery supply on outcomes after microsurgical resection of brain arteriovenous malformations. Neurosurgery 60: 638–646; discussion 646–648.

Dumont TM, Kan P, Snyder KV et al. (2015). A proposed grading system for endovascular treatment of cerebral arteriovenous malformations: Buffalo score. Surg Neurol Int 6: 3.

Gross BA, Du R (2012a). Hemorrhage from arteriovenous malformations during pregnancy. Neurosurgery 71: 349–355; discussion 355–356.

Gross BA, Du R (2012b). Surgical and radiosurgical results of the treatment of cerebral arteriovenous malformations. J Clin Neurosci 19: 1001–1004.

Gross BA, Du R (2013). Natural history of cerebral arteriovenous malformations: a meta-analysis. J Neurosurg 118: 437–443.

Gross BA, Duckworth EA, Getch CC et al. (2008). Challenging traditional beliefs: microsurgery for arteriovenous malformations of the basal ganglia and thalamus. Neurosurgery 63: 393–410; discussion 410–411.

Han PP, Ponce FA, Spetzler RF (2003). Intention-to-treat analysis of Spetzler-Martin grades IV and V arteriovenous malformations: natural history and treatment paradigm. J Neurosurg 98: 3–7.

Haw CS, ter Brugge K, Willinsky R et al. (2006). Complications of embolization of arteriovenous malformations of the brain. J Neurosurg 104: 226–232.

Hernesniemi JA, Dashti R, Juvela S et al. (2008). Natural history of brain arteriovenous malformations: a long-term follow-up study of risk of hemorrhage in 238 patients. Neurosurgery 63: 823–829; discussion 829–831.

Heros RC, Korosue K, Diebold PM (1990). Surgical excision of cerebral arteriovenous malformations: late results. Neurosurgery 26: 570–577. discussion 577–578.

Hoh BL, Chapman PH, Loeffler JS et al. (2002). Results of multimodality treatment for 141 patients with brain arteriovenous malformations and seizures: factors associated with seizure incidence and seizure outcomes. Neurosurgery 51: 303–309; discussion 309–311.

Josephson CB, Leach JP, Duncan R et al. (2011). Seizure risk from cavernous or arteriovenous malformations: prospective population-based study. Neurology 76: 1548–1554.

Kader A, Young WL, Pile-Spellman J et al. (1994). The influence of hemodynamic and anatomic factors on hemorrhage from cerebral arteriovenous malformations. Neurosurgery 34: 801–807. discussion 807–808.

Kano H, Kondziolka D, Flickinger JC et al. (2012). Stereotactic radiosurgery for arteriovenous malformations after embolization: a case-control study. J Neurosurg 117: 265–275.

Katsaridis V, Papagiannaki C, Aimar E (2008). Curative embolization of cerebral arteriovenous malformations (AVMs) with Onyx in 101 patients. Neuroradiology 50: 589–597.

Kerber C (1980). Use of balloon catheters in the treatment of cranial arterial abnormalities. Stroke 11: 210–216.

Kerber CW, Cromwell LD, Sheptak PE (1978). Intraarterial cyanoacrylate: an adjunct in the treatment of spinal/paraspinal arteriovenous malformations. AJR Am J Roentgenol 130: 99–103.

Krings T, Hans FJ, Geibprasert S et al. (2010). Partial "targeted" embolisation of brain arteriovenous malformations. Eur Radiol 20: 2723–2731.

Langer DJ, Lasner TM, Hurst RW et al. (1998). Hypertension, small size, and deep venous drainage are associated with risk of hemorrhagic presentation of cerebral arteriovenous malformations. Neurosurgery 42: 481–486; discussion 487–489.

Lawton MT, Hamilton MG, Spetzler RF (1995). Multimodality treatment of deep arteriovenous malformations: thalamus, basal ganglia, and brain stem. Neurosurgery 37: 29–35. discussion 35–36.

Lawton MT, Kim H, McCulloch CE et al. (2010). A supplementary grading scale for selecting patients with brain arteriovenous malformations for surgery. Neurosurgery 66: 702–713; discussion 713.

Loh Y, Duckwiler GR (2010). A prospective, multicenter, randomized trial of the Onyx liquid embolic system and N-butyl cyanoacrylate embolization of cerebral arteriovenous malformations. J Neurosurg 113: 733–741.

Luessenhop AJ, Spence WT (1960). Artificial embolization of cerebral arteries. Report of use in a case of arteriovenous malformation. J Am Med Assoc 172: 1153–1155.

Lv X, Wu Z, Li Y et al. (2012). Hemorrhage risk after partial endovascular NBCA and ONYX embolization for brain arteriovenous malformation. Neurol Res 34: 552–556.

Morgan MK, Drummond KJ, Grinnell V et al. (1997). Surgery for cerebral arteriovenous malformation: risks related to lenticulostriate arterial supply. J Neurosurg 86: 801–805.

Mounayer C, Hammami N, Piotin M et al. (2007). Nidal embolization of brain arteriovenous malformations using Onyx in 94 patients. AJNR Am J Neuroradiol 28: 518–523.

Natarajan SK, Ghodke B, Britz GW et al. (2008). Multimodality treatment of brain arteriovenous malformations with microsurgery after embolization with Onyx: single-center experience and technical nuances. Neurosurgery 62: 1213–1225; discussion 1225–1226.

Ondra SL, Troupp H, George ED et al. (1990). The natural history of symptomatic arteriovenous malformations of the brain: a 24-year follow-up assessment. J Neurosurg 73: 387–391.

Pollock BE, Flickinger JC (2008). Modification of the radiosurgery-based arteriovenous malformation grading system. Neurosurgery 63: 239–243; discussion 243.

Pollock BE, Gorman DA, Brown PD (2004). Radiosurgery for arteriovenous malformations of the basal ganglia, thalamus, and brainstem. J Neurosurg 100: 210–214.

Pollock GA, Shaibani A, Awad I et al. (2011). Intraventricular hemorrhage secondary to intranidal aneurysm rupture—successful management by arteriovenous malformation embolization followed by intraventricular tissue plasminogen activator: case report. Neurosurgery 68: E581–E586; discussion E586.

Raffa SJ, Chi YY, Bova FJ et al. (2009). Validation of the radiosurgery-based arteriovenous malformation score in a large linear accelerator radiosurgery experience. J Neurosurg 111: 832–839.

Saatci I, Geyik S, Yavuz K et al. (2011). Endovascular treatment of brain arteriovenous malformations with prolonged intranidal Onyx injection technique: long-term results in 350 consecutive patients with completed endovascular treatment course. J Neurosurg 115: 78–88.

Sanchez-Mejia RO, McDermott MW, Tan J et al. (2009). Radiosurgery facilitates resection of brain arteriovenous malformations and reduces surgical morbidity. Neurosurgery 64: 231–238; discussion 238–240.

Serbinenko FA (1979). Six hundred endovascular neurosurgical procedures in vascular pathology. A ten-year experience. Acta Neurochir Suppl (Wien) 28: 310–311.

Spetzler RF, Martin NA (1986). A proposed grading system for arteriovenous malformations. J Neurosurg 65: 476–483.

Spetzler RF, Ponce FA (2011). A 3-tier classification of cerebral arteriovenous malformations. J Neurosurg 114: 842–849.

Spetzler RF, Wilson CB, Weinstein P et al. (1978). Normal perfusion pressure breakthrough theory. Clin Neurosurg 25: 651–672.

Stapf C, Mast H, Sciacca RR et al. (2006). Predictors of hemorrhage in patients with untreated brain arteriovenous malformation. Neurology 66: 1350–1355.

Starke RM, Komotar RJ, Otten ML et al. (2009). Adjuvant embolization with N-butyl cyanoacrylate in the treatment of cerebral arteriovenous malformations: outcomes, complications, and predictors of neurologic deficits. Stroke 40: 2783–2790.

Starke RM, Yen CP, Ding D et al. (2013). A practical grading scale for predicting outcome after radiosurgery for arteriovenous malformations: analysis of 1012 treated patients. J Neurosurg 119: 981–987.

Taki W, Yonekawa Y, Iwata H et al. (1990). A new liquid material for embolization of arteriovenous malformations. AJNR Am J Neuroradiol 11: 163–168.

U HS, Kerber CW, Todd MM (1992). Multimodality treatment of deep periventricular cerebral arteriovenous malformations. Surg Neurol 38: 192–203.

van der Schaaf IC, Brilstra EH, Rinkel GJ et al. (2002). Quality of life, anxiety, and depression in patients with an untreated intracranial aneurysm or arteriovenous malformation. Stroke 33: 440–443.

van Rooij WJ, Sluzewski M, Beute GN (2007). Brain AVM embolization with Onyx. AJNR Am J Neuroradiol 28: 172–177; discussion 178.

van Rooij WJ, Jacobs S, Sluzewski M et al. (2012). Endovascular treatment of ruptured brain AVMs in the acute phase of hemorrhage. AJNR Am J Neuroradiol 33: 1162–1166.

Weber W, Kis B, Siekmann R et al. (2007). Endovascular treatment of intracranial arteriovenous malformations with Onyx: technical aspects. AJNR Am J Neuroradiol 28: 371–377.

Yeh HS, Tew Jr JM, Gartner M (1993). Seizure control after surgery on cerebral arteriovenous malformations. J Neurosurg 78: 12–18.

Handbook of Clinical Neurology, Vol. 143 (3rd series)
Arteriovenous and Cavernous Malformations
R.F. Spetzler, K. Moon, and R.O. Almefty, Editors
http://dx.doi.org/10.1016/B978-0-444-63640-9.00007-2

Chapter 7

Radiosurgery for the management of cerebral arteriovenous malformations

DALE DING, ROBERT M. STARKE, AND JASON P. SHEEHAN*
Department of Neurosurgery, University of Virginia, Charlottesville, VA, USA

Abstract

Cerebral arteriovenous malformations (AVMs) are rare, unstable vascular lesions which spontaneously rupture at a rate of approximately 2–4% annually. Stereotactic radiosurgery is a minimally invasive treatment for AVMs, with a favorable risk-to-benefit profile in most patients, with respect to obliteration, hemorrhage, and seizure control. Radiosurgery is ideally suited for small to medium-sized AVMs (diameter <3 cm or volume <12 cm^3) located in deep or eloquent brain regions. Obliteration is ultimately achieved in 70–80% of cases and is directly associated with nidus volume and radiosurgical margin dose. Adverse radiation effects, which appear as T2-weighted hyperintensities on magnetic resonance imaging, develop in 30–40% of patients after AVM radiosurgery, are symptomatic in 10%, and fail to clinically resolve in 2–3%. The risk of AVM hemorrhage may be reduced by radiosurgery, but the hemorrhage risk persists during the latency period between treatment and obliteration. Delayed postradiosurgery cyst formation occurs in 2% of cases and may require surgical treatment. Radiosurgery abolishes or ameliorates seizure activity in the majority of patients with AVM-associated epilepsy and induces *de novo* seizures in 1–2% of those without preoperative seizures. Strategies for the treatment of large-volume AVMs include neoadjuvant embolization and either dose- or volume-staged radiosurgery.

INTRODUCTION

Cerebral arteriovenous malformations (AVMs) are rare, congenital vascular lesions with an incidence and prevalence of approximately 1 in 100 000 and 18 in 100 000, respectively (Al-Shahi and Warlow, 2001). AVMs are inherently unstable, and they rupture at a rate of approximately 2–4% annually (Graf et al., 1983; Crawford et al., 1986; Brown et al., 1988). However, the annual hemorrhage risk of an AVM has been reported to vary widely, from 1% to >30% depending on numerous factors such as patient age, prior AVM hemorrhage, deep location, nidus morphology, presence of an associated arterial aneurysm, and venous drainage pattern (Pollock et al., 1996a; Stapf et al., 2006; Gross and Du, 2013; Kim et al., 2014). Intracranial hemorrhage is the most frequent clinical manifestation of an AVM, and hemorrhage is the etiology of presentation in approximately 50% of AVM patients (Choi and Mohr, 2005).

The decision to treat an AVM is complex, and the primary goal of any management strategy is to completely occlude the AVM nidus, thereby abolishing the risk of hemorrhage. When intervention is deemed appropriate, AVMs can be treated with microsurgical resection, stereotactic radiosurgery, or endovascular embolization as standalone or combined therapies (Lawton et al., 2010; Saatci et al., 2011; van Beijnum et al., 2011; Ding and Liu, 2013; Ding et al., 2013b, 2016b, c; Chen et al., 2014; Yen et al., 2014; Cohen-Inbar et al., 2016a, b). Of the various AVM interventions, radiosurgery is the least invasive, with no significant immediate treatment risk. However, both the beneficial and adverse effects of radiosurgery are delayed, maturing over a course of months to years (Starke et al., 2014b; Chen et al., 2015). Therefore,

*Correspondence to: Jason P. Sheehan, M.D., Ph.D., University of Virginia, Department of Neurosurgery, P.O. Box 800212, Charlottesville VA 22908, USA. Tel: +1-434-924-8129, E-mail: jps2f@virginia.edu

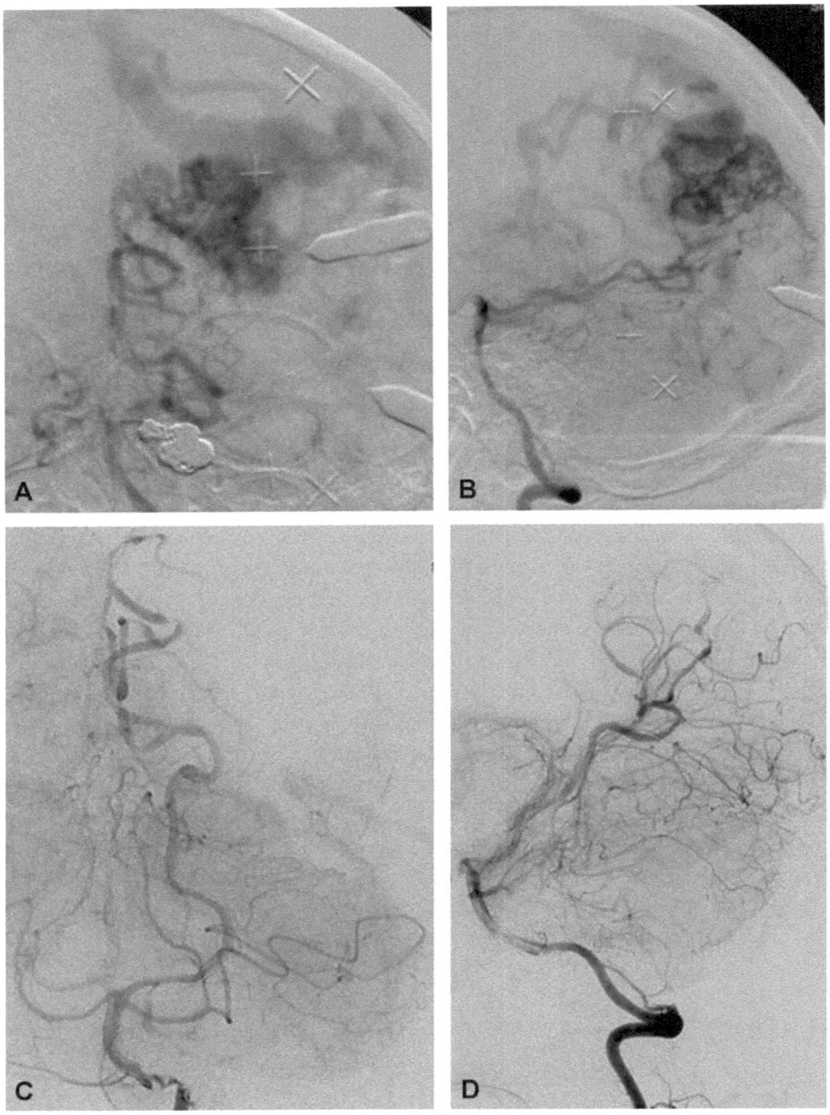

Fig. 7.1. A 43-year-old woman was incidentally diagnosed with a left parietal arteriovenous malformation (AVM) and four proximal prenidal aneurysms of the paraclinoid internal carotid artery. The paraclinoid aneurysms were treated with stent-assisted coil embolization 3 weeks prior to radiosurgery. Stereotactic cerebral angiography performed for gamma-knife radiosurgery planning: (**A**) anteroposterior (AP) and (**B**) lateral views of a left vertebral artery (VA) injection show a left parietal AVM, measuring 2.8 × 2.2 × 2.5 cm, supplied by distal branches of the middle cerebral artery (not shown) and posterior cerebral artery with exclusively superficial venous drainage into the superior sagittal sinus. This Spetzler–Martin grade II AVM was treated with a margin dose of 17 Gy. Follow-up angiography 2 years after radiosurgery: (**C**) AP and (**D**) lateral views of a left VA injection show complete AVM obliteration and occlusion of the flow-related aneurysms (not shown). At the most recent clinical follow-up 42 months after radiosurgery, the patient was asymptomatic.

a thorough understanding of the indications, efficacy, and complication profile of AVM radiosurgery is crucial to optimizing the care of patients harboring these complex lesions. The aims of this chapter are to: (1) summarize the radiologic and clinical outcomes after the treatment of brain AVMs with radiosurgery; (2) identify predictors of AVM radiosurgery outcomes; and (3) critically assess the role of radiosurgery in the contemporary management of AVM patients.

OBLITERATION AFTER AVM RADIOSURGERY

Prevention of intracranial hemorrhage is the most important long-term outcome of AVM intervention. Radiosurgery definitively achieves this goal when complete obliteration of the AVM nidus is achieved, as determined by a lack of flow voids on magnetic resonance imaging (MRI) and confirmed by the absence of anomalous arteriovenous

Table 7.1

Summary of arteriovenous malformation (AVM) obliteration rates after radiosurgery

Series (year)	Number of patients	AVM volume, margin dose,* delivery system	Obliteration rate† (%)	Adverse events	Predictors of obliteration‡
Touboul et al. (1998)	100	1.9 cm^3, 19 Gy, LINAC	51	Hemorrhage 10%, overall morbidity 8%, permanent ARE 1%, radionecrosis 1%	↑Margin dose, ↓SM grade
Chang et al. (2000)	128	12.1 cm^3, 16 Gy, GK and LINAC	79	Hemorrhage 7%, radiologic ARE 12%, symptomatic ARE 5%, permanent ARE 0.4%, cyst formation 0.4%	↓Diameter, single draining vein, compact nidus
Schlienger et al. (2000)	169	2.5 cm^3, 25 Gy, LINAC	64	Hemorrhage 2%, *de novo* seizures 2%, overall morbidity 2%, permanent morbidity 0.6%, radionecrosis 1%	Lack of prior embolization, mono isocentric irradiation
Flickinger et al. (2002)	351	5.7 cm^3, 20 Gy, GK	75	NR	↑margin dose (in-field and overall obliteration), male gender (in-field obliteration)
Friedman et al. (2003)	269	8.4 cm^3, NR, LINAC	53	Hemorrhage 10%, overall morbidity 5%, permanent morbidity 1%	↓SM grade, ↑margin dose
Bollet et al. (2004)	118	7.4 cm^3, 18 Gy, LINAC	54	Annual hemorrhage rate 1.7%, overall morbidity 7%, permanent morbidity 2%	Volume <7 cm^3
Shin et al. (2004)	400	1.9 cm^3, 20 Gy, GK	74 at 3 years, 88 at 5 years	Annual hemorrhage rate 1.9%, symptomatic ARE 7%, permanent ARE 2%	Prior AVM hemorrhage, ↓diameter, ↑margin dose
Liscak et al. (2007)	330	3.9 cm^3, 20 Gy, GK	92	Annual hemorrhage rate 2.1%, radiologic ARE 21%, symptomatic ARE 8%, permanent morbidity 3%, mortality 1%	Male gender, prior AVM hemorrhage, volume <10 cm^3, SM grade I or II, margin dose >19 Gy, maximum dose >35 Gy
Colombo et al. (2009)	102	5.2 cm^3, 19 Gy, CK	72	Hemorrhage 8%, transient morbidity 1%, mortality 1%	↓Volume, ↓SM grade, ↓RBAS, lack of prior radiosurgery, male gender§
Taeshineetanakul et al. (2012)	139	3.8 cm^3, 19 Gy, GK and LINAC	66	NR	↓Volume, noneloquent location, low transnidal flow, no or mild arterial enlargement, absence of perinidal angiogenesis
Fokas et al. (2013)	164	4.0 cm^3, 19 Gy, LINAC	61	Annual hemorrhage rate 1.3%, worsened neurologic deficit 3%, worsened seizure status 2%	Volume <4 cm^3, margin dose ≥19 Gy
Franzin et al. (2013)	127	2.7 cm^3, 22 Gy, GK	69	Annual hemorrhage rate 2.1%, permanent neurologic deficit 7%, radionecrosis 5%, cyst formation 2%, worsened seizure status 7%, mortality 4%	↓Volume
Starke et al. (2013)	1012	3.5 cm^3, 21 Gy, GK	69	Annual hemorrhage rate 1.1%, radiologic ARE 38%, symptomatic ARE 10%, permanent ARE 2%	↓Volume, age < 65 years, non-eloquent AVM location, lack of prior AVM hemorrhage, lack of prior AVM embolization
Hattangadi-Gluth et al. (2014)	248	3.5 cm^3, 15 Gy, PB	65	Hemorrhage 5%, *de novo* seizures 9%, major neurologic morbidity 1%, mortality 1%	↓Volume, noncritical/superficial location
Missios et al. (2014)	152	6.3 cm^3, 18 Gy, GK	46	Hemorrhage 4%, new neurologic symptoms 23%, neurologic worsening 2%, mortality 2%	↓Volume, ↓diameter, ↓SM grade, ↓RBAS, ↑margin dose, ↑maximum dose (all factors only significant in univariate analysis)
Paul et al. (2014)	662	3.6 cm^3, 19 Gy, GK	75	Annual hemorrhage rate 1.2%, overall morbidity 4%, mortality 2%	↓Volume, compact nidus, nondilated feeding arteries, ↑margin dose
Wang et al. (2014)	116	4.7 cm^3, NR, LINAC	82	Annual hemorrhage rate 1.9%, new neurologic deficit 1%, edema 5% radionecrosis 3%, cyst formation 5%, *de novo* seizures 6%	Prior AVM resection

*Mean values were preferentially reported; when a mean value was not available, the median was reported.
†Included obliteration determined by magnetic resonance imaging alone and confirmed by angiography.
‡Predictors based on multivariate analysis, when available.
§Unknown if significant factors were derived from univariate or multivariate analysis.
ARE, adverse radiation effects (perinidal T2-weighted hyperintensities on follow-up magnetic resonance imaging); CK, cyberknife GK, gamma knife; LINAC, linear accelerator; NR, not reported; PB, proton beam; RBAS, radiosurgery-based AVM score; SM, Spetzler–Martin.

shunting on cerebral angiography. Despite numerous published studies, the obliteration rates achieved after AVM radiosurgery vary from <50% to >90%. Based on our institutional experience and the available literature, obliteration can typically be achieved in approximately 70–80% of cases (Fig. 7.1). Table 7.1 summarizes the major AVM radiosurgery series in the literature (Touboul et al., 1998; Chang et al., 2000; Schlienger et al., 2000; Flickinger et al., 2002; Friedman et al., 2003; Bollet et al., 2004; Shin et al., 2004; Liscak et al., 2007; Colombo et al., 2009; Taeshineetanakul et al., 2012; Fokas et al., 2013; Franzin et al., 2013; Starke et al., 2013; Hattangadi-Gluth et al., 2014; Missios et al., 2014; Paul et al., 2014; Wang et al., 2014).

The postradiosurgery follow-up protocol for AVM patients treated at our center consists of an MRI every 6 months for the first 2 years after the procedure, and then annually thereafter. Angiography is performed to confirm obliteration after its diagnosis on MRI or to further characterize a residual AVM nidus for additional salvage therapy. Ideally, all cases of obliteration are confirmed by angiography. However, follow-up angiography is unavailable for approximately 10–20% of patients, and MRI is the only method to determine postradiosurgery obliteration for this subset of AVMs. In these cases, MRI has been found to be an acceptable accurate substitute for angiography. Pollock et al. (1996c) found MRI to have a sensitivity, specificity, and negative predictive value of 80%, 100%, and 91%, respectively, compared to angiography. O'Connor and Friedman (2013) reported MRI to have a predictive accuracy of 82% compared to angiography, which was greater for AVMs with a volume >2.8 cm^3 (90%) than for AVMs <2.8 cm^3 (70%). Most recently, we compared MRI to angiography with two independent observers and reported sensitivities of 85% and 77% and specificities of 89% and 95% (Lee et al., 2015).

Flickinger et al. (1996) analyzed the radiosurgery outcomes for 197 AVM patients with ≥3 years of angiographic follow-up to determine the relationship between dose and obliteration. The median target volume, radiosurgical margin dose (labeled minimum target dose in this study), maximum dose, isodose line, and number of isocenters were 4.1 cm^3, 20 Gy, 36 Gy, 50%, and 2, respectively. Of the 55 AVMs which remained patent (28%) after radiosurgery, 35 achieved in-field obliteration but had persistently filling portions of the nidus which were not targeted by the margin dose (64%). Upon detailed evaluation, the mechanisms for partial targeting of the nidus were inadequate angiographic definition of the AVM margins in 66%, imperfect angiographic technique in 14%, re-expansion of a compressed nidus compartment by hematoma in 11%, and recanalization of an embolized nidus compartment in 9%. A multivariate analysis for independent predictors of in-field obliteration identified only margin dose ($p = 0.04$), and a sigmoid dose–response curve correlated margin doses of 13 Gy, 16 Gy, 20 Gy, and 25 Gy with in-field obliteration rates of 50%, 70%, 90%, and 98%, respectively (Flickinger et al., 1996). However, neither margin nor maximum dose was significant in a multivariate model for overall obliteration; only AVM volume was independently associated with overall obliteration ($p < 0.001$).

AVMs which remain patent 3–4 years after radiosurgery can be treated with repeat radiosurgery. In a cohort of 140 AVMs patients treated with repeat radiosurgery, the rates of complete and subtotal obliteration on follow-up angiography were 55% and 6%, respectively. (Yen et al., 2010). An additional 11% had obliteration determined by MRI alone, and the rate of permanent neurologic deficits after radiosurgery was 4%. Surgical resection is an alternative to repeat radiosurgery for residual AVMs. Sanchez-Mejia et al. (2009) found that AVM patients who previously underwent radiosurgery had significantly better postoperative outcomes after resection compared to surgically treated patients with nonradiated AVMs.

ADVERSE RADIATION EFFECTS AFTER AVM RADIOSURGERY

Increasing radiosurgical margin dose, up to approximately 25 Gy, improves the likelihood of AVM obliteration. However, the risk of adverse radiation effects (ARE) also increases concomitantly with dose, and it is this balance between achieving obliteration and limiting the frequency and severity of ARE which determines the optimal radiosurgical plan for a specific AVM. ARE, otherwise known as postradiosurgery imaging changes and radiation-induced changes, are defined as perinidal T2-weighted hyperintensities that typically develop 6–18 months after radiosurgery. The pathophysiology of ARE is not well understood, with potential mechanisms including venous occlusion secondary to endothelial cell injury, demyelination secondary to glial cell injury, and vasogenic edema secondary to immune-induced breakdown of the blood–brain barrier (Yen et al., 2013; Ding, 2015). The onset of ARE precedes complete AVM obliteration and some studies have suggested that ARE development may be predictive of eventual obliteration (Ding et al., 2013a). The occurrence rate of ARE can be estimated by a "rule of thirds": one-third of all patients will develop ARE (30–40% of all AVM radiosurgery cases), one-third of cases of radiologic ARE result in neurologic symptoms (10% of all cases), and one-third of patients with symptomatic ARE suffer permanent neurologic deterioration (2–3% of all cases).

Flickinger et al. (1997) evaluated the ARE profile of 307 AVM patients who underwent radiosurgery and had ≥2 years of follow-up. The rates of radiologic and symptomatic ARE were 30% and 9%, respectively, with a median time to onset of 12 months for radiologic and 14 months for symptomatic ARE. The actuarial rate of resolution for radiologic ARE was 81% at 3 years (median time to resolution 12 months), and only 3% of patients developed radiologic ARE which persisted beyond 2 years. Patients with symptomatic ARE had significantly lower actuarial rates of ARE resolution at 3 years than those with asymptomatic ARE (95% vs. 53%, respectively; $p=0.03$). Only the 12-Gy volume (V12) was found to be an independent predictor of radiologic ARE ($p<0.0001$) in the multivariate analysis. Both 12-Gy volume ($p=0.001$) and brainstem location ($p=0.007$) were independently associated with symptomatic ARE.

We evaluated the ARE outcomes of 1426 gamma-knife radiosurgery procedures performed at our institution in 1309 AVM patients (Yen et al., 2013). Radiologic ARE (termed radiation-induced imaging changes in this study) were classified into grades I (mild; increased T2 signal less than 10 mm in thickness and not causing mass effect), II (moderate; increased T2 signal of 10 mm or greater in thickness causing local mass effect, evident as sulcal effacement or ventricular compression), and III (severe; increased T2 signal causing global mass effect, evident as midline shift). The median duration of MRI follow-up was 43 months (range 2–240 months). The rates of radiologic, symptomatic, and permanent ARE were 34%, 9%, and 2%, respectively. Radiologic AREs were stratified as grade I in 58%, II in 34%, and III in 8%. The median time from stereotactic radiosurgery to the appearance of radiologic ARE was 13 months, although 83% of these neuroimaging changes resolved at a median interval of 22 months after onset.

Cyst formation after AVM radiosurgery is substantially less common and develops in a more delayed fashion than ARE, occurring in approximately 2–5% of patients (Ding, 2014; Matsuo et al., 2014). Based on a histopathologic analysis of resected postradiosurgery cysts, Shuto et al. (2012) postulated that radiation-induced inflammatory changes in the perinidal brain tissue resulted in initial cyst development, with subsequent cyst enlargement mediated by neovascularization, hemorrhage, and osmotic expansion. An angiomatous lesion develops in some patients, and may provide an etiologic basis for a pathophysiologic spectrum which encompasses postradiosurgery cysts and chronic encapsulating hematomas. Specifically, rupture of the angiomatous lesion into the adjacent parenchyma results in cyst formation, whereas rupture of the lesion within itself results in the development of a chronic encapsulating hematoma. In many cases, postradiosurgery cysts can be managed conservatively, particularly if they are radiologically stable, do not exert significant local mass effect, or are asymptomatic. The treatment options for progressively enlarging or symptomatic cysts include stereotactic drainage, Ommaya reservoir placement, shunt placement, and fenestration/resection via craniotomy.

LATENCY-PERIOD HEMORRHAGE AFTER AVM RADIOSURGERY

Obliteration virtually eliminates the risk of AVM hemorrhage. However, until this radiologic endpoint is achieved, radiosurgically treated AVMs remain exposed to the risk of rupture. The relative risk of AVM hemorrhage during the latency period between radiosurgery and obliteration compared to that of an AVM's natural history is a topic of debate. Additionally, there have been infrequent reports of hemorrhage, and even complete AVM obliteration on angiography. Postobliteration hemorrhage has been observed in patients with and without evidence of AVM recurrence (Lindqvist et al., 2000).

Pollock et al. (1996b) evaluated the risks of pre- and postradiosurgery AVM hemorrhage in a cohort of 315 patients with a mean follow-up duration of 47 months. Prior to radiosurgery, the annual risk of AVM hemorrhage was 2.4%. The actuarial hemorrhage rate was 4.8% per year for the first 2 years after radiosurgery, and then 5.0% per for years 3–5 post treatment. The median interval from radiosurgery to hemorrhage was 8 months (range 1–60 months). The presence of a patent prenidal AVM-associated arterial aneurysm was an independent predictor of postradiosurgery hemorrhage in the multivariate analysis ($p<0.001$), with a relative risk of 4.6 (95% confidence interval (CI) 1.8–11.7), although no AVMs with a patent intranidal aneurysm ruptured during the latency period. The hemorrhage rates of AVMs treated with a margin dose ≥25 Gy were not significantly different from those treated with a margin dose <25 Gy. Notably, no patients with angiographic documentation of complete or subtotal (presence of an early draining vein without a residual nidus) AVM obliteration experienced a hemorrhage.

Maruyama et al. (2005) analyzed the AVM radiosurgery outcomes of a cohort of 500 patients and found that radiosurgery significantly reduced an AVM's hemorrhage risk, but their data did not support the complete abolition of this risk after obliteration. During an overall median observation period of 7.8 years, the proportion of patients who had an AVM hemorrhage was 8.4% in the interval from diagnosis to radiosurgery (median duration 0.4 years), 5.0% from radiosurgery to obliteration (median duration 2.0 years), and 2.4% after obliteration (median duration 5.4 years). All 6 patients with postobliteration hemorrhage underwent angiography, and none was found to have AVM recanalization. Of the 29 patients with postradiosurgery hemorrhage, 7 died, yielding a hemorrhage-related mortality rate of 24%. Compared to risk of hemorrhage from diagnosis to treatment, the hemorrhage risk decreased by 54% after radiosurgery (95% CI 20–74%, $p=0.006$) and by 88% after obliteration (95% CI 71–95%, $p<0.001$). Compared to the risk of latency-period hemorrhage from radiosurgery to obliteration, the hemorrhage risk decreased by 74% after obliteration (95% CI 34–90%, $p=0.006$). Radiosurgery reduced the hemorrhage risk of ruptured AVMs to a significantly greater extent compared to unruptured AVMs. The annual postradiosurgery hemorrhage rates of ruptured AVMs ($n=310$) at 1, 2, 3, and >3 years were 6.3%, 6.8%, 6.4%, and 6.3%, respectively.

We evaluated the rates of pre- and postradiosurgery hemorrhage in a cohort of 1204 AVMs treated with radiosurgery at our center (Yen et al., 2011). The annual AVM hemorrhage risk from birth to radiosurgery was 2.0% for the entire cohort and 3.7% for AVMs with prior hemorrhage. The annual hemorrhage risk from the time of diagnosis to radiosurgery was 6.6%, stratified as 10.4% and 3.9% for AVMs with and without prior hemorrhage. Until obliteration, the annual postradiosurgery hemorrhage risk was 2.5%, stratified as 2.8% and 2.2% for AVMs with and without prior hemorrhage. Only lower margin dose was identified as an independent predictor of postradiosurgery hemorrhage in the multivariate analysis ($p=0.046$). We found no cases of hemorrhage after obliteration. In another analysis of 159 AVMs with subtotal obliteration after radiosurgery, there were no cases of hemorrhage this cohort (Yen et al., 2007).

Kano et al. (2012a) analyzed the radiosurgery outcomes for a cohort of 407 ruptured AVMs with a median follow-up duration of 66 months (range 2–274 months). The annual hemorrhage risk between birth and radiosurgery was 3.4%, and the risk from diagnosis to radiosurgery was 16.5%. The annual postradiosurgery hemorrhage of the cohort was 1.3%. The mortality rate from a single hemorrhage was 41% (12/29 patients died at a median follow-up of 22 months) and from two hemorrhages was 100% (all 3 patients died). There were no cases of AVM hemorrhage after obliteration. The presence of a patent prenidal or intranidal arterial aneurysm ($p<0.0001$) and lower margin dose ($p<0.0005$) were independent predictors of postradiosurgery hemorrhage in the multivariate analysis. The annual postradiosurgery hemorrhage rate of AVMs with patent prenidal or intranidal aneurysms was significantly higher than that of AVMs with an associated aneurysm which was occluded (i.e., clipped or embolized) prior to radiosurgery (6.4% vs. 0.8%, $p=0.033$).

GRADING SYSTEMS FOR PREDICTING OUTCOMES AFTER AVM RADIOSURGERY

Table 7.2 summarizes major grading systems for predictors of AVM radiosurgery outcome: (1) Spetzler–Martin (SM) grading scale; (2) modified radiosurgery-based AVM score (RBAS); and (3) Virginia Radiosurgery AVM Scale (VRAS). In this section, we evaluate the predictive capability of each of these three major grading systems.

Spetzler–Martin grade

The original intent of the SM grading scale was to predict outcomes after AVM resection (Spetzler and Martin, 1986). However, the SM grade has also been shown to correlate reasonably well with AVM radiosurgery outcomes. Andrade-Souza et al. (2005) evaluated the predictive capability of the SM grading scale in 136 AVMs treated with radiosurgery. The rates of excellent outcome (complete AVM obliteration without new or worsening neurologic symptoms) for SM grade I, II, IIIB (defined in this study as a grade III AVM with diameter <3 cm), and IIIA (defined as a grade III AVM with diameter ≥3 cm) or IV AVMs were 89%, 70%, 62%, and 45%, respectively. Koltz et al. (2013) analyzed the long-term outcomes (follow-up duration ≥5 years) of 102 AVM patients treated with single-session or dose-staged radiosurgery and stratified their findings by SM grade. For SM grades I ($n=5$), II ($n=28$), III ($n=43$), IV ($n=22$), and V ($n=4$), the obliteration rates were 100%, 89%, 86%, 54%, and 0%, respectively; and the combined major morbidity and mortality rates were 20%, 11%, 9%, 18%, and 75%, respectively.

Due to the low operative morbidity associated with resection of SM grade I and II AVMs, surgery is the preferred intervention for these lesions (Morgan et al., 2004). However, patients who are unable, due to medical comorbidities, or unwilling to undergo a craniotomy can be afforded excellent outcomes with radiosurgery. We analyzed our radiosurgery outcomes for a cohort of 502 SM grade I and II AVMs, with median radiologic and clinical follow-up durations of 48 and 62 months, respectively (Ding et al., 2014f). Obliteration was achieved in 76%, with actuarial obliteration rates at 3, 5, and 10 years of 41%, 66%, and 80%, respectively. The rates of radiologic, symptomatic, and permanent ARE were 37%, 8%, and 1%, respectively, and the annual postradiosurgery hemorrhage rate was 1.4%. Kano et al. (2012b) reported the radiosurgery outcomes for a cohort of 217 SM grade I and II AVMs with a median follow-up duration of 64 months. The median volume and margin dose were 2.3 cm^3 and 22 Gy,

Table 7.2

Summary of grading systems for arteriovenous malformation (AVM) radiosurgery

Grading system	Series (year)	Number of patients	Components of grading system	Contribution of each component to grade
Spetzler–Martin (SM) grading scale	Spetzler and Martin (1986)	100	AVM maximum diameter, Eloquence* of AVM location, presence of deep venous drainage	SM grade (I–V) = AVM diameter (<3 cm = 1 point, 3–6 cm = 2 points, >6 cm = 3 points) + AVM location (noneloquent = 0 points, eloquent = 1 point) + venous drainage (superficial only = 0 points, deep component = 1 point)
Modified radiosurgery-based AVM Score (RBAS)	Wegner et al. (2011)	293	AVM volume, patient age, AVM location	RBAS = 0.1 * (AVM volume, cm^3) + 0.02 * (patient age, years) + 0.5 * (AVM location); AVM location = 0 for superficial, 1 for deep†
Virginia Radiosurgery AVM Scale (VRAS)	Starke et al. (2013)	1012	AVM volume, eloquence* of AVM location, prior AVM hemorrhage	VRAS (0–4) = AVM volume (<2 cm^3 = 0 points, 2–4 cm^3 = 1 point, >4 cm^3 = 2 points) + AVM location (noneloquent = 0 points, eloquent = 1 point) + prior AVM hemorrhage (no = 0 points, yes = 1 point)

*Eloquent locations = sensorimotor, language, and visual cortex, hypothalamus and thalamus, internal capsule, brainstem, cerebellar peduncles, and deep cerebellar nuclei.

†Deep locations = basal ganglia, thalamus, and brainstem.

respectively. The actuarial obliteration rates at 3, 5, and 10 years were 58%, 90%, and 93%, respectively. Symptomatic ARE occurred in only 2% of cases, none of which was permanent. The annual postradiosurgery hemorrhage rate was 2.3%, although AVMs with associated arterial aneurysms had a significantly greater hemorrhage risk.

SM grade III AVMs are the most angioarchitecturally heterogeneous lesions, comprising four distinct entities. The optimal management of SM grade III AVMs is controversial, since they straddle the boundary between grade I and II AVMs, which favor relatively aggressive management with intervention, and grade IV and V AVMs, which favor relatively conservative management. Radiosurgery has a reasonable risk-to-benefit profile for SM grade III AVMs, and should be considered a first-line therapy for these lesions, especially when they are unruptured and/or small (diameter <3 cm, eloquent location, and deep venous drainage). We analyzed our radiosurgery outcomes for a cohort of 398 SM grade III AVMs, with median radiologic and clinical follow-up durations of 54 and 68 months, respectively (Ding et al., 2014e). Obliteration was achieved in 69%, with actuarial obliteration rates at 3 and 5 years of 38% and 60%, respectively. The rates of radiologic, symptomatic, and permanent ARE were 35%, 12%, and 4%, respectively, and the annual postradiosurgery hemorrhage rate was 1.7%.

Kano et al. (2014) evaluated a cohort of 474 SM grade III AVMs treated with radiosurgery. The median volume and margin dose were 3.8 cm^3 and 20 Gy, respectively. After a mean follow-up duration of 89 months, the actuarial obliteration rates at 3, 5, and 10 years were 48%, 72%, and 77%, respectively. The rates of symptomatic and permanent ARE were 6% and 3%, respectively, and the annual postradiosurgery hemorrhage rate was 2.7%. The risk of latency-period hemorrhage was significantly higher in medium-sized (diameter 3–6 cm) grade III AVMs with deep venous drainage and noneloquent location than in medium-sized lesions with exclusively superficial venous drainage and eloquent location.

Due to their relatively larger volume and uniformly eloquent location, single-stage radiosurgery yields only modest obliteration rates for SM grade IV and V AVMs, with an elevated complication profile. In general, conservative management should be initially considered for most patients with unruptured SM grade IV and V AVMs. Ruptured SM grade IV and V AVMs and unruptured lesions resulting in progressive neurologic symptoms may be appropriate for intervention. Due to the high risk of surgical morbidity and mortality, radiosurgery is a reasonable treatment option for carefully selected lesions for which intervention is deemed necessary (Heros et al., 1990; Han et al., 2003). We analyzed our radiosurgery outcomes for a cohort of 110 SM grade IV and V AVMs, with median radiologic and clinical follow-up durations of 88 and 97 months, respectively (Ding et al., 2014d). Obliteration was achieved in 44%, with actuarial obliteration rates at 3 and 5 years of 10% and 23%, respectively. The rates of radiologic, symptomatic, and permanent ARE were 47%, 12%, and 3%, respectively, and the annual postradiosurgery hemorrhage rate was 3.0%. Transient and permanent clinical deterioration occurred in 9% and 10%, respectively.

Radiosurgery-based AVM score

The original RBAS was devised by Pollock and Flickinger (2002), and it was a weighted sum of AVM nidus volume, patient age, and AVM location. The intent of the RBAS was to predict excellent outcome, defined as complete nidal obliteration without new neurologic deficit, after AVM radiosurgery. The components of the RBAS were derived from a prior analysis by Pollock et al. (1998) of 220 AVM patients treated with radiosurgery, in which smaller AVM volume ($p=0.003$), fewer draining veins ($p=0.001$), younger patient age ($p=0.0003$), hemispheric AVM location ($p=0.002$), and lack of prior AVM embolization ($p=0.02$) were identified as independent predictors of excellent outcome in a multivariate model. The original RBAS was validated in a separate cohort of 136 patients treated at a different institution from that of the initial analysis (Pollock and Flickinger, 2002). The RBAS was subsequently modified such that the AVM location component was simplified from a three-tier variable in the original RBAS to a two-tier variable in the modified RBAS, comprising only superficial or deep (defined as basal ganglia, thalamus, and brainstem in this grading system) location (Pollock and Flickinger, 2008). Most recently, Wegner et al. (2011) modified the RBAS a second time by changing the coefficient multiplier of the AVM location component from 0.3 to 0.5, and verified its predictive capability in 293 AVM patients who underwent radiosurgery.

A number of centers have validated the RBAS in their own AVM cohorts (Andrade-Souza et al., 2005). Burrow et al. (2014) evaluated the radiosurgery outcomes of 80 AVM patients with a RBAS ≤ 1. The mean RBAS was 0.76, comprised of a mean AVM volume of 2.3 cm^3, mean patient age of 25 years, and superficial AVM location in 95%. There were no cases of AVM hemorrhage, ARE, or decline in modified Rankin Scale score after a mean follow-up duration of 68 months. Of the 76 patients with at least 3 years of follow-up, the obliteration rate was 92%. These impressive results led the authors to propose that, for younger patients with superficially located, small-volume AVMs without a mass-occupying hematoma requiring evacuation, intervention with radiosurgery may achieve similar outcomes to those of surgical extirpation.

Although the RBAS is a reasonably powerful tool for predicting AVM radiosurgery outcomes, we believe that much of its predictive capability is based on the AVM volume and location components, rather than patient age. In our experience, elderly patient age does not adversely affect AVM radiosurgery outcomes in the same way it has been found to portend worse AVM surgical outcomes (Lawton et al., 2010). In order to study the effect of advanced age on AVM radiosurgery outcomes, we performed a matched-cohort analysis of 66 elderly AVM patients (age >60 years), matched 1:1 with nonelderly AVM patients, who were treated with radiosurgery (Ding et al., 2015e). In comparison to the nonelderly AVM cohort, the elderly AVM cohort had a significantly higher mean age (67 vs. 36 years, $p<0.0001$), RBAS (1.70 vs. 1.11, $p<0.0001$), and number of radiosurgical isocenters (2.9 vs 2.7, $p=0.038$). Complete obliteration was achieved in 67% of cases in the elderly AVM cohort, with a significantly higher actuarial obliteration rate for those treated with a margin dose $\geq$22 Gy ($p=0.001$). In the elderly AVM cohort, radiologic and symptomatic ARE were evident in 36% and 11% (none experienced permanent symptoms), respectively, the annual postradiosurgery hemorrhage risk was 1.1%, and the AVM-related mortality rate was 1.5%. Elderly patient age was not significantly associated with AVM radiosurgery outcomes, including obliteration, ARE, and hemorrhage (Ding et al., 2015e).

Virginia Radiosurgery AVM Scale

The VRAS was developed at the University of Virginia, based on a 20-year institutional experience of more than 1400 patients who underwent AVM radiosurgery with the gamma knife (Starke et al., 2013). A total of 1012 patients. with $\geq$2 years of neuroimaging follow-up or <2 years of follow-up and who experienced treatment-related complications were selected for analysis. The mean age was 34 years, AVM volume was 3.5 cm^3 (stratified as <2 cm^3 in 20%, 2–4 cm^3 in 48%, and >4 cm^3 in 32%), and radiosurgical margin dose was 21 Gy. Prior AVM hemorrhage occurred in 56%, a component of deep venous drainage was identified in 52%, and nidus location was eloquent in 67%. The SM grade was $\geq$III in 48%, and the mean RBAS was 1.35. The mean follow-up duration was 8 years.

Favorable outcome, defined as AVM obliteration, no postradiosurgery hemorrhage, and no permanently symptomatic ARE, was achieved in 64% of patients after radiosurgery. A multivariate model comprising patient and AVM variables (i.e., excluding radiosurgical treatment factors) identified age <65 years ($p = 0.041$), smaller AVM volume ($p < 0.001$), noneloquent AVM location ($p < 0.001$), lack of prior AVM hemorrhage ($p < 0.001$), and lack of prior AVM embolization ($p < 0.001$) as independent predictors of favorable outcome. We devised a five-tiered grading system, the VRAS, based on this multivariate model. The VRAS ranges from 0 to 4 and comprises three AVM features: volume (<2 cm^3 = 0 points, 2–4 cm^3 = 1 points, >4 cm^3 = 2 points), location (noneloquent = 0 points, eloquent = 1 point), and prior hemorrhage (unruptured = 0 points, ruptured = 1 point). The rates of favorable outcome corresponding to a VRAS of 0–1, 2, and 3–4 were 80%, 70%, and 45%, respectively (Starke et al., 2013).

wCompared to the SM grading system and RBAS, the VRAS was developed relatively recently and has, therefore, yet to be tested and adopted to the same extent. Huo et al. (2016) evaluated the AVM radiosurgery outcomes of 162 partially embolized nidi and correlated the VRAS with obliteration and complications, including hemorrhage, seizure, and headache. VRAS was able to predict postradiosurgery outcomes in this cohort with reasonable accuracy. The obliteration rates for a VRAS of 0–1, 2, 3, and 4 were 89%, 68%, 51%, and 35%, respectively. The complication rates for a VRAS of 0–2, 3, and 4 were 8%, 24%, and 29%, respectively. Recently, we externally validated the VRAS in a multicenter cohort comprising over 2000 AVM patients who underwent radiosurgery (Starke et al., 2016). While the detailed findings of this analysis are forthcoming, among the three grading systems for AVM radiosurgery (i.e., SM grading scale, RBAS, and VRAS), the VRAS was determined to have the best predictive capability.

SEIZURE OUTCOMES AFTER AVM RADIOSURGERY

Although hemorrhage is the most feared and neurologically devastating component of an AVM's natural history, seizures represent the most common presentation of unruptured AVMs and can significantly impair a patient's quality of life. Radiosurgery reduces or abolishes AVM-associated epilepsy in the majority of patients, with a very low incidence of *de novo* seizures in patients without AVM-associated epilepsy, although the effect of AVM obliteration on seizure outcomes is inconsistent (Ding et al., 2015a). However, prospective studies do not support an improvement in seizure outcomes with intervention compared to medical therapy alone (Josephson et al., 2012).

A recent meta-analysis found that, for patients with AVM-associated epilepsy, microsurgery afforded the highest rate of seizure control (78%, 95% CI 70–86%), followed by radiosurgery (63%, 95% CI 55–70%) and embolization (49%, 95% CI 32–67%) (Baranoski et al., 2014). The rates of *de novo* seizures were highest after embolization (39%, 95% CI 8–68%), followed by microsurgery (9%, 95% CI 5–13%) and radiosurgery (5%, 95% CI 3–8%). Similarly, Wang et al. (2013) reported a significantly lower rate of persistent seizures (odds ratio 4.3, $p = 0.021$) after AVM resection compared to radiosurgery for patients with seizures at presentation. Conversely, the rate of *de novo* seizures was significantly lower (odds ratio 8.7, $p < 0.001$) after AVM radiosurgery compared to resection for patients without seizure presentation. The findings from this study indicate divergent seizure outcomes for radiosurgery versus resection, based on preoperative seizure status.

We analyzed a cohort of 1007 AVMs treated with radiosurgery at our center in order to identify predictors of seizure presentation and outcomes. The overall incidence of seizure presentation was 23% (Ding et al., 2015d). The lack of prior AVM hemorrhage ($p < 0.0001$), larger AVM diameter ($p < 0.0001$), and cortical AVM location ($p < 0.0001$) were found to be independent predictors of seizure presentation in the multivariate analysis. Cortical AVMs have a significantly higher incidence of seizure presentation than noncortical lesions (33% vs. 6%, respectively; $p < 0.0001$). In the cohort of 229 patients with AVM-associated seizures, the rates of seizure improvement and remission after radiosurgery were 57% and 20%, respectively (Ding et al., 2015b). Prior AVM hemorrhage ($p = 0.015$), longer follow-up duration ($p < 0.0001$), and a lack of postradiosurgery hemorrhage ($p = 0.048$) were found to be independent predictors of seizure improvement in the multivariate analysis. In the cohort of 778 patients without AVM-associated seizures, *de novo* seizures after radiosurgery occurred in only 2%. Prior AVM hemorrhage ($p = 0.001$) and higher SM grade ($p = 0.018$) were found to be independent predictors of *de novo* seizures in the multivariate analysis. Due to the low risk of *de novo* seizures after AVM radiosurgery, we do not recommend prophylactic antiepileptic drug (AED) therapy for patients without preoperative seizures. The seizure outcomes were not significantly different between obliterated and patent AVMs.

We also analyzed a cohort of 73 patients with AVM-associated epilepsy for whom detailed seizure and AED outcomes were available after radiosurgery (Przybylowski et al., 2015). Engel class I outcome was achieved in 89%. Presentation with a simple partial or secondarily generalized seizure was an independent predictor of Engel class I outcome in the multivariate analysis. AED therapy was administered to 86% of patients prior to radiosurgery, and of these patients, AEDs were tapered off completely in 33%. Although AVM obliteration was not predictive of Engel class I outcome, AED cessation was significantly more common in patients with obliterated AVMs ($p = 0.05$). We performed a systematic review of seizure outcomes after AVM radiosurgery (Chen et al., 2014). A total of 19 studies comprising 3971 patients were included in our analysis. The incidence of seizure presentation was 28%, and the rates of seizure remission and control were 44% and 69%, respectively. The rate of seizure remission was significantly higher in patients with obliterated AVMs (82%) than in those with patent AVMs (41%, $p = 0.0007$).

STAGED RADIOSURGERY FOR LARGE AVMS

The obliteration rates for large AVMs (volume >10–$15\ cm^3$) after single-session radiosurgery are relatively low. Alternative treatment strategies to conventional radiosurgery, which is delivered in a single dose during a single procedure, for large AVMs, are dose- and volume-staged radiosurgery. Dose-staged radiosurgery, also known as hypofractionated radiosurgery, treats the entire AVM nidus with multiple smaller doses, which are typically administered over a period of a few weeks. Volume-staged radiosurgery separates the nidus into two or more smaller compartments, so that each compartment can be treated with a higher margin dose than can be safely delivered to the entire AVM. In volume-staged radiosurgery, each treatment is typically separated by a period of 3–6 months. In a study of volume-staged AVM radiosurgery outcomes, Seymour et al. (2015) reported higher obliteration rates, with similar complication rates, when the treatment volume of each stage was limited to $\leq 8\ cm^3$, so that a higher margin dose (≥ 17 Gy) could be delivered to each portion of the AVM.

We performed a systematic review of dose- and volume-staged radiosurgery outcomes for large AVMs (volume $>10\ cm^3$), which included 15 studies comprising 401 patients (Moosa et al., 2014). The mean obliteration was higher for AVMs treated with volume-staged (48%, 95% CI 34–61%) compared to dose-staged (23%, 95% CI 4–41%) radiosurgery. The mean symptomatic ARE rates were 14% for each of the dose- and volume-staged cohorts. The mean frequency of postradiosurgery hemorrhage was slightly higher in the volume-staged cohort (18%, 95% CI 12–23%) compared to the dose-staged cohort (12%, 95% CI 1–25%). The mean mortality rates of the dose- and volume-staged cohorts were 3% and 5%, respectively. Therefore, volume-staged radiosurgery appears to be a superior approach to dose-staged radiosurgery, affording higher obliteration rates with a similar complication profile. Even when obliteration is not achieved, staged radiosurgery can frequently achieve sufficient volume reduction of a large AVM such that the shrunken residual nidus can be definitely treated with single-session radiosurgery or resection (Abla et al., 2015).

CURRENT CONTROVERSIES IN AVM RADIOSURGERY

Effect of prior embolization on AVM radiosurgery outcomes

The goals of preradiosurgery embolization are: (1) volume reduction of large AVMs (diameter >3 cm or volume $>12\ cm^3$); (2) occlusion of prenidal or intranidal AVM-associated arterial aneurysms; and (3) occlusion of high-flow intranidal arteriovenous fistulas. However, AVM embolization has been found to lower the obliteration rate after radiosurgery, although the etiology of this phenomenon is not well understood (Andrade-Souza et al., 2007; Ding et al., 2014a). The mechanisms by which AVM embolization has been postulated to reduce obliteration rates after radiosurgery include: (1) absorption or scattering of radiation beams by embolic agents; (2) increased difficulty in adequately targeting the remaining AVM after embolization due to obscuration of the borders of the nidus; (3) delayed recanalization of the embolized portion of the nidus during the latency period after radiosurgery; and (4) embolization-induced angiogenesis (Andrade-Souza et al., 2008; Valle et al., 2008; Bing et al., 2012; Buell et al., 2014).

We performed a matched cohort of analysis of 484 embolized and nonembolized AVMs (matched 1:1) treated with radiosurgery at our center (Oermann et al., 2015). The actuarial obliteration rate of the embolized AVM cohort was significantly lower than the nonembolized cohort ($p = 0.003$). However, prior embolization was not significant ($p = 0.744$) in the multivariate model for obliteration, due to a confounding effect of nidus angioarchitectural complexity (sum of the number of major feeding arteries and draining veins). Angioarchitectural complexity ($p < 0.001$) and radiologic evidence of ARE ($p = 0.016$) were independent predictors of obliteration in the multivariate analysis. Prior embolization ($p = 0.009$) and fewer draining veins ($p = 0.011$) were independent predictors of radiologic ARE in the multivariate analysis. Therefore, embolization appears to confer protection from the development of ARE, which may offset some of the procedural risks of embolization.

Radiosurgery for unruptured AVMs

The crux of the debate regarding the management of unruptured AVMs centers on two main issues: (1) the hemorrhage risk of untreated, unruptured AVMs is significantly lower than for ruptured AVMs (Pollock et al., 1996a; Stapf et al., 2006; Gross and Du, 2013; Kim et al., 2014); and (2) the treatment-related morbidity for unruptured AVMs may be higher than for ruptured AVMs (Lawton et al., 2005; Ding et al., 2013a, 2014b, c, 2015f). This controversy was brought to the forefront by two recent studies which found worse short-term outcomes after intervention compared to conservative management. A Randomized Trial of Unruptured Brain AVMs (ARUBA) was a multicenter, randomized controlled trial comparing intervention to conservative management for unruptured AVMs (Mohr et al., 2014). The initial interim analysis of ARUBA reported significantly higher rates of symptomatic stroke or death in patients assigned to intervention (31%) than to conservative management (10%; $p < 0.0001$). The Scottish Audit of Intracranial Vascular Malformations (SAIVM) prospective AVM cohort study also reported significantly worse outcomes, up to 4 years of follow-up, for unruptured AVMs which underwent intervention compared to conservative management (Al-Shahi Salman et al., 2014). A number of criticisms regarding both of the aforementioned studies have been appropriately raised by various cerebrovascular groups (Starke et al., 2014a).

At the time of analysis, 64% of patients in ARUBA who were assigned to intervention either had ongoing treatment plans or had not yet begun therapy, which suggests that at least half of patients considered to have treated AVMs had patent nidi at last follow-up (Mohr et al., 2014). When one considers the persistently elevated rate of symptomatic stroke or death over time and the excessive rate of hemorrhage in the interventional arm (22% as randomized, 25% as treated), the findings from ARUBA implicate a potentially increased rupture risk for partially treated, unruptured AVMs (Mohr et al., 2014). However, since treatment-specific radiologic and clinical outcomes from ARUBA have yet to be reported, the degree to which each treatment modality contributed to hemorrhage in the interventional arm is unknown (Starke et al., 2014a). Nevertheless, intervention likely yields a long-term benefit for the vast majority of patients harboring unruptured SM grade I and II AVMs, as well as many of those with small (diameter <3 cm) SM grade III AVMs.

A critical analysis of ARUBA, with suggests a potentially increased hemorrhage risk from incompletely obliterated AVMs, further emphasizes the importance of achieving complete nidal extirpation or occlusion (Mohr et al., 2014). Surgical resection should be the preferred intervention for most SM grade I and II AVMs, due to its high obliteration rate and relatively low risk of morbidity and mortality (Bervini et al., 2014; Potts et al., 2015). However, for SM grade II AVMs located in eloquent brain regions, patients who are medically unfit to tolerate the physiologic burden of resection, or patients who are unwilling to undergo an invasive neurosurgical procedure, radiosurgery should be considered a minimally invasive treatment option which affords an excellent risk-to-benefit profile for unruptured SM grade I and II AVMs (Kano et al., 2012b; Ding et al., 2013a, 2014f, 2016a). Additionally, based on the available literature and our institutional experience, we believe that radiosurgery should be the preferred treatment for unruptured SM grade III AVMs (Ding et al., 2014e; Kano et al., 2014).

When contemplating the management of an unruptured AVM patient, one should consider that the natural history of unruptured AVMs can vary considerably based on a lesion's angioarchitecture, and that most treatment-related morbidity is temporary (Stapf et al., 2006; Kim et al., 2014; Ding et al., 2015c; Mouchtouris et al., 2015). Thus, despite the controversial findings of ARUBA and the SAIVM prospective AVM study, intervention is likely warranted, and should remain the preferred management approach for most patients with unruptured SM grade I and II AVMs. However, further prospective, multicenter investigations are necessary to affirm the benefit of intervention over conservative management for appropriately selected unruptured AVMs.

CONCLUSIONS

Due to the variable natural history and heterogeneous outcomes after intervention, the proper management of an AVM patient is a multifactorial decision. In general, AVMs should be preferentially referred to high-volume cerebrovascular centers where experienced, multidisciplinary teams comprised of neurosurgeons, neurointerventional radiologists, radiation oncologists, and medical physicists are available to provide the latest microsurgical, radiosurgical, and endovascular techniques and technologies for the management of these challenging lesions. Radiosurgery has a unique and important role in the management of AVMs, and should be preferentially employed in patients with small to medium-sized nidi located in deep or functionally critical brain regions. For appropriately selected unruptured AVMs, radiosurgery affords a reasonable risk-to-benefit profile which likely results in outcomes which are superior to that of conservative management, although a prolonged follow-up duration may be necessary to fully realize this difference.

References

Abla AA, Rutledge WC, Seymour ZA et al. (2015). A treatment paradigm for high-grade brain arteriovenous malformations: volume-staged radiosurgical downgrading followed by microsurgical resection. J Neurosurg 122: 419–432.

Al-Shahi R, Warlow C (2001). A systematic review of the frequency and prognosis of arteriovenous malformations of the brain in adults. Brain 124: 1900–1926.

Al-Shahi Salman R, White PM, Counsell CE et al. (2014). Outcome after conservative management or intervention for unruptured brain arteriovenous malformations. JAMA 311: 1661–1669.

Andrade-Souza YM, Zadeh G, Ramani M et al. (2005). Testing the radiosurgery-based arteriovenous malformation score and the modified Spetzler–Martin grading system to predict radiosurgical outcome. J Neurosurg 103: 642–648.

Andrade-Souza YM, Ramani M, Scora D et al. (2007). Embolization before radiosurgery reduces the obliteration rate of arteriovenous malformations. Neurosurgery 60: 443–451; discussion 451–452.

Andrade-Souza YM, Ramani M, Beachey DJ et al. (2008). Liquid embolisation material reduces the delivered radiation dose: a physical experiment. Acta Neurochir (Wien) 150: 161–164; discussion 164.

Baranoski JF, Grant RA, Hirsch LJ et al. (2014). Seizure control for intracranial arteriovenous malformations is directly related to treatment modality: a meta-analysis. J Neurointerv Surg 6: 684–690.

Bervini D, Morgan MK, Ritson EA et al. (2014). Surgery for unruptured arteriovenous malformations of the brain is better than conservative management for selected cases: a prospective cohort study. J Neurosurg 121: 878–890.

Bing F, Doucet R, Lacroix F et al. (2012). Liquid embolization material reduces the delivered radiation dose: clinical myth or reality? AJNR Am J Neuroradiol 33: 320–322.

Bollet MA, Anxionnat R, Buchheit I et al. (2004). Efficacy and morbidity of arc-therapy radiosurgery for cerebral arteriovenous malformations: a comparison with the natural history. Int J Radiat Oncol Biol Phys 58: 1353–1363.

Brown Jr RD, Wiebers DO, Forbes G et al. (1988). The natural history of unruptured intracranial arteriovenous malformations. J Neurosurg 68: 352–357.

Buell TJ, Ding D, Starke RM et al. (2014). Embolization-induced angiogenesis in cerebral arteriovenous malformations. J Clin Neurosci 21: 1866–1871.

Burrow AM, Link MJ, Pollock BE (2014). Is stereotactic radiosurgery the best treatment option for patients with a radiosurgery-based arteriovenous malformation score </= 1? World Neurosurg 82: 1144–1147.

Chang JH, Chang JW, Park YG et al. (2000). Factors related to complete occlusion of arteriovenous malformations after gamma knife radiosurgery. J Neurosurg 93 (Suppl 3): 96–101.

Chen CJ, Chivukula S, Ding D et al. (2014). Seizure outcomes following radiosurgery for cerebral arteriovenous malformations. Neurosurg Focus 37.E17.

Chen CJ, Lee CC, Ding D et al. (2015). Stereotactic radiosurgery for intracranial dural arteriovenous fistulas: a systematic review. J Neurosurg 122: 353–362.

Choi JH, Mohr JP (2005). Brain arteriovenous malformations in adults. Lancet Neurol 4: 299–308.

Cohen-Inbar O, Ding D, Chen CJ et al. (2016a). Stereotactic radiosurgery for deep intracranial arteriovenous malformations, part 1: brainstem arteriovenous malformations. J Clin Neurosci 24: 30–36.

Cohen-Inbar O, Ding D, Sheehan JP (2016b). Stereotactic radiosurgery for deep intracranial arteriovenous malformations, part 2: basal ganglia and thalamus arteriovenous malformations. J Clin Neurosci 24: 37–42.

Colombo F, Cavedon C, Casentini L et al. (2009). Early results of CyberKnife radiosurgery for arteriovenous malformations. J Neurosurg 111: 807–819.

Crawford PM, West CR, Chadwick DW et al. (1986). Arteriovenous malformations of the brain: natural history in unoperated patients. J Neurol Neurosurg Psychiatry 49: 1–10.

Ding D (2014). Mechanisms of cyst formation after radiosurgery for intracranial arteriovenous malformations. Clin Neurol Neurosurg 124: 192–193.

Ding D (2015). Pathogenesis of radiosurgery-induced cyst formation in patients with cerebral arteriovenous malformations. Acta Neurochir (Wien) 157: 775–777.

Ding D, Liu KC (2013). Predictive capability of the Spetzler-Martin versus supplementary grading scale for microsurgical outcomes of cerebellar arteriovenous malformations. J Cerebrovasc Endovasc Neurosurg 15: 307–310.

Ding D, Yen CP, Xu Z et al. (2013a). Radiosurgery for patients with unruptured intracranial arteriovenous malformations. J Neurosurg 118: 958–966.

Ding D, Yen CP, Xu Z et al. (2013b). Radiosurgery for primary motor and sensory cortex arteriovenous malformations: outcomes and the effect of eloquent location. Neurosurgery 73: 816–824.

Ding D, Starke RM, Yen CP et al. (2014a). Radiosurgery for cerebellar arteriovenous malformations: does infratentorial location affect outcome? World Neurosurg 82: e209–e217.

Ding D, Yen C, Starke RM et al. (2014b). Effect of prior hemorrhage on intracranial arteriovenous malformation radiosurgery outcomes. Cerebrovasc Dis 39: 53–62.

Ding D, Yen CP, Starke RM et al. (2014c). Radiosurgery for ruptured intracranial arteriovenous malformations. J Neurosurg 121: 470–481.

Ding D, Yen CP, Starke RM et al. (2014d). Outcomes following single-session radiosurgery for high-grade intracranial arteriovenous malformations. Br J Neurosurg 28: 666–674.

Ding D, Yen CP, Starke RM et al. (2014e). Radiosurgery for Spetzler-Martin grade III arteriovenous malformations. J Neurosurg 120: 959–969.

Ding D, Yen CP, Xu Z et al. (2014f). Radiosurgery for low-grade intracranial arteriovenous malformations. J Neurosurg 121: 457–467.

Ding D, Quigg M, Starke RM et al. (2015a). Radiosurgery for temporal lobe arteriovenous malformations: effect of temporal location on seizure outcomes. J Neurosurg 123: 924–934.

Ding D, Quigg M, Starke RM et al. (2015b). Cerebral arteriovenous malformations and epilepsy, part 2: predictors of seizure outcomes following radiosurgery. World Neurosurg 84: 653–662.

Ding D, Starke RM, Liu KC et al. (2015c). Cortical plasticity in patients with cerebral arteriovenous malformations. J Clin Neurosci 22: 1857–1861.

Ding D, Starke RM, Quigg M et al. (2015d). Cerebral arteriovenous malformations and epilepsy, part 1: predictors of seizure presentation. World Neurosurg 84: 645–652.

Ding D, Xu Z, Yen CP et al. (2015e). Radiosurgery for cerebral arteriovenous malformations in elderly patients: effect of advanced age on outcomes after intervention. World Neurosurg 84: 795–804.

Ding D, Xu Z, Yen CP et al. (2015f). Radiosurgery for unruptured cerebral arteriovenous malformations in pediatric patients. Acta Neurochir (Wien) 157: 281–291.

Ding D, Starke RM, Kano H et al. (2016a). Radiosurgery for cerebral arteriovenous malformations in a randomized trial of unruptured brain arteriovenous malformations (ARUBA)-eligible patients: a multicenter study. Stroke 47: 342–349.

Ding D, Xu Z, Shih HH et al. (2016b). Stereotactic radiosurgery for partially resected cerebral arteriovenous malformations. World Neurosurg 85: 263–272.

Ding D, Xu Z, Starke RM et al. (2016c). Radiosurgery for cerebral arteriovenous malformations with associated arterial aneurysms. World Neurosurg 87: 77–90.

Flickinger JC, Pollock BE, Kondziolka D et al. (1996). A dose–response analysis of arteriovenous malformation obliteration after radiosurgery. Int J Radiat Oncol Biol Phys 36: 873–879.

Flickinger JC, Kondziolka D, Pollock BE et al. (1997). Complications from arteriovenous malformation radiosurgery: multivariate analysis and risk modeling. Int J Radiat Oncol Biol Phys 38: 485–490.

Flickinger JC, Kondziolka D, Maitz AH et al. (2002). An analysis of the dose–response for arteriovenous malformation radiosurgery and other factors affecting obliteration. Radiother Oncol 63: 347–354.

Fokas E, Henzel M, Wittig A et al. (2013). Stereotactic radiosurgery of cerebral arteriovenous malformations: long-term follow-up in 164 patients of a single institution. J Neurol 260: 2156–2162.

Franzin A, Snider S, Boari N et al. (2013). Evaluation of prognostic factors as predictor of AVMS obliteration after gamma knife radiosurgery. Acta Neurochir (Wien) 155: 619–626.

Friedman WA, Bova FJ, Bollampally S et al. (2003). Analysis of factors predictive of success or complications in arteriovenous malformation radiosurgery. Neurosurgery 52: 296–307; discussion 307–308.

Graf CJ, Perret GE, Torner JC (1983). Bleeding from cerebral arteriovenous malformations as part of their natural history. J Neurosurg 58: 331–337.

Gross BA, Du R (2013). Natural history of cerebral arteriovenous malformations: a meta-analysis. J Neurosurg 118: 437–443.

Han PP, Ponce FA, Spetzler RF (2003). Intention-to-treat analysis of Spetzler–Martin grades IV and V arteriovenous malformations: natural history and treatment paradigm. J Neurosurg 98: 3–7.

Hattangadi-Gluth JA, Chapman PH, Kim D et al. (2014). Single-fraction proton beam stereotactic radiosurgery for cerebral arteriovenous malformations. Int J Radiat Oncol Biol Phys 89: 338–346.

Heros RC, Korosue K, Diebold PM (1990). Surgical excision of cerebral arteriovenous malformations: late results. Neurosurgery 26: 570–577; discussion 577–578.

Huo X, Jiang Y, Lv X et al. (2016). Gamma knife surgical treatment for partially embolized cerebral arteriovenous malformations. J Neurosurg 124: 767–776.

Josephson CB, Bhattacharya JJ, Counsell CE et al. (2012). Seizure risk with AVM treatment or conservative management: prospective, population-based study. Neurology 79: 500–507.

Kano H, Kondziolka D, Flickinger JC et al. (2012a). Aneurysms increase the risk of rebleeding after stereotactic radiosurgery for hemorrhagic arteriovenous malformations. Stroke 43: 2586–2591.

Kano H, Lunsford LD, Flickinger JC et al. (2012b). Stereotactic radiosurgery for arteriovenous malformations, part 1: management of Spetzler–Martin grade I and II arteriovenous malformations. J Neurosurg 116: 11–20.

Kano H, Flickinger JC, Yang HC et al. (2014). Stereotactic radiosurgery for Spetzler–Martin grade III arteriovenous malformations. J Neurosurg 120: 973–981.

Kim H, Al-Shahi Salman R, McCulloch CE et al. (2014). Untreated brain arteriovenous malformation: patient-level meta-analysis of hemorrhage predictors. Neurology 83: 590–597.

Koltz MT, Polifka AJ, Saltos A et al. (2013). Long-term outcome of gamma knife stereotactic radiosurgery for arteriovenous malformations graded by the Spetzler–Martin classification. J Neurosurg 118: 74–83.

Lawton MT, Du R, Tran MN et al. (2005). Effect of presenting hemorrhage on outcome after microsurgical resection of brain arteriovenous malformations. Neurosurgery 56: 485–493; discussion 485–493.

Lawton MT, Kim H, McCulloch CE et al. (2010). A supplementary grading scale for selecting patients with brain arteriovenous malformations for surgery. Neurosurgery 66: 702–713; discussion 713.

Lee CC, Reardon MA, Ball BZ et al. (2015). The predictive value of magnetic resonance imaging in evaluating intracranial arteriovenous malformation obliteration after stereotactic radiosurgery. J Neurosurg 123: 136–144.

Lindqvist M, Karlsson B, Guo WY et al. (2000). Angiographic long-term follow-up data for arteriovenous malformations previously proven to be obliterated after gamma knife radiosurgery. Neurosurgery 46: 803–808; discussion 809–810.

Liscak R, Vladyka V, Simonova G et al. (2007). Arteriovenous malformations after Leksell gamma knife radiosurgery: rate of obliteration and complications. Neurosurgery 60: 1005–1014; discussion 1015–1016.

Maruyama K, Kawahara N, Shin M et al. (2005). The risk of hemorrhage after radiosurgery for cerebral arteriovenous malformations. N Engl J Med 352: 146–153.

Matsuo T, Kamada K, Izumo T et al. (2014). Cyst formation after linac-based radiosurgery for arteriovenous malformation: examination of predictive factors using magnetic resonance imaging. Clin Neurol Neurosurg 121: 10–16.

Missios S, Bekelis K, Al-Shyal G et al. (2014). Stereotactic radiosurgery of intracranial arteriovenous malformations and the use of the K index in determining treatment dose. Neurosurg Focus 37. E15.

Mohr JP, Parides MK, Stapf C et al. (2014). Medical management with or without interventional therapy for unruptured brain arteriovenous malformations (ARUBA): a multicentre, non-blinded, randomised trial. Lancet 383: 614–621.

Moosa S, Chen CJ, Ding D et al. (2014). Volume-staged versus dose-staged radiosurgery outcomes for large intracranial arteriovenous malformations. Neurosurg Focus 37. E18.

Morgan MK, Rochford AM, Tsahtsarlis A et al. (2004). Surgical risks associated with the management of grade I and II brain arteriovenous malformations. Neurosurgery 54: 832–837; discussion 837–839.

Mouchtouris N, Jabbour PM, Starke RM et al. (2015). Biology of cerebral arteriovenous malformations with a focus on inflammation. J Cereb Blood Flow Metab 35: 167–175.

O'Connor TE, Friedman WA (2013). Magnetic resonance imaging assessment of cerebral arteriovenous malformation obliteration after stereotactic radiosurgery. Neurosurgery 73: 761–766.

Oermann EK, Ding D, Yen CP et al. (2015). Effect of prior embolization on cerebral arteriovenous malformation radiosurgery outcomes: a case-control study. Neurosurgery 77: 406–417.

Paul L, Casasco A, Kusak ME et al. (2014). Results for a series of 697 arteriovenous malformations treated by gamma knife: influence of angiographic features on the obliteration rate. Neurosurgery 75: 568–583.

Pollock BE, Flickinger JC (2002). A proposed radiosurgery-based grading system for arteriovenous malformations. J Neurosurg 96: 79–85.

Pollock BE, Flickinger JC (2008). Modification of the radiosurgery-based arteriovenous malformation grading system. Neurosurgery 63: 239–243; discussion 243.

Pollock BE, Flickinger JC, Lunsford LD et al. (1996a). Factors that predict the bleeding risk of cerebral arteriovenous malformations. Stroke 27: 1–6.

Pollock BE, Flickinger JC, Lunsford LD et al. (1996b). Hemorrhage risk after stereotactic radiosurgery of cerebral arteriovenous malformations. Neurosurgery 38: 652–659; discussion 659–661.

Pollock BE, Kondziolka D, Flickinger JC et al. (1996c). Magnetic resonance imaging: an accurate method to evaluate arteriovenous malformations after stereotactic radiosurgery. J Neurosurg 85: 1044–1049.

Pollock BE, Flickinger JC, Lunsford LD et al. (1998). Factors associated with successful arteriovenous malformation radiosurgery. Neurosurgery 42: 1239–1244; discussion 1244–1247.

Potts MB, Lau D, Abla AA et al. (2015). Current surgical results with low-grade brain arteriovenous malformations. J Neurosurg 122: 912–920.

Przybylowski CJ, Ding D, Starke RM et al. (2015). Seizure and anticonvulsant outcomes following stereotactic radiosurgery for intracranial arteriovenous malformations. J Neurosurg 122: 1299–1305.

Saatci I, Geyik S, Yavuz K et al. (2011). Endovascular treatment of brain arteriovenous malformations with prolonged intranidal Onyx injection technique: long-term results in 350 consecutive patients with completed endovascular treatment course. J Neurosurg 115: 78–88.

Sanchez-Mejia RO, McDermott MW, Tan J et al. (2009). Radiosurgery facilitates resection of brain arteriovenous malformations and reduces surgical morbidity. Neurosurgery 64: 231–238; discussion 238–240.

Schlienger M, Atlan D, Lefkopoulos D et al. (2000). Linac radiosurgery for cerebral arteriovenous malformations: results in 169 patients. Int J Radiat Oncol Biol Phys 46: 1135–1142.

Seymour ZA, Sneed PK, Gupta N et al. (2015). Volume-staged radiosurgery for large arteriovenous malformations: an evolving paradigm. J Neurosurg: 1–12.

Shin M, Maruyama K, Kurita H et al. (2004). Analysis of nidus obliteration rates after gamma knife surgery for arteriovenous malformations based on long-term follow-up data: the University of Tokyo experience. J Neurosurg 101: 18–24.

Shuto T, Ohtake M, Matsunaga S (2012). Proposed mechanism for cyst formation and enlargement following gamma knife surgery for arteriovenous malformations. J Neurosurg 117 (Suppl): 135–143.

Spetzler RF, Martin NA (1986). A proposed grading system for arteriovenous malformations. J Neurosurg 65: 476–483.

Stapf C, Mast H, Sciacca RR et al. (2006). Predictors of hemorrhage in patients with untreated brain arteriovenous malformation. Neurology 66: 1350–1355.

Starke RM, Yen CP, Ding D et al. (2013). A practical grading scale for predicting outcome after radiosurgery for arteriovenous malformations: analysis of 1012 treated patients. J Neurosurg 119: 981–987.

Starke RM, Sheehan JP, Ding D et al. (2014a). Conservative management or intervention for unruptured brain arteriovenous malformations. World Neurosurg 82: e668–e669.

Starke RM, Yen CP, Chen CJ et al. (2014b). An updated assessment of the risk of radiation-induced neoplasia after radiosurgery of arteriovenous malformations. World Neurosurg 82: 395–401.

Starke RM, Kano H, Ding D et al. (2016). Stereotactic radiosurgery for cerebral arteriovenous malformations: evaluation of long-term outcomes in a multicenter cohort. J Neurosurg Mar 4: 1–9.

Taeshineetanakul P, Krings T, Geibprasert S et al. (2012). Angioarchitecture determines obliteration rate after radiosurgery in brain arteriovenous malformations. Neurosurgery 71: 1071–1078; discussion 1079.

Touboul E, Al Halabi A, Buffat L et al. (1998). Single-fraction stereotactic radiotherapy: a dose–response analysis of arteriovenous malformation obliteration. Int J Radiat Oncol Biol Phys 41: 855–861.

Valle RD, Zenteno M, Jaramillo J et al. (2008). Definition of the key target volume in radiosurgical management of arteriovenous malformations: a new dynamic concept based on angiographic circulation time. J Neurosurg 109 (Suppl): 41–50.

van Beijnum J, van der Worp HB, Buis DR et al. (2011). Treatment of brain arteriovenous malformations: a systematic review and meta-analysis. JAMA 306: 2011–2019.

Wang JY, Yang W, Ye X et al. (2013). Impact on seizure control of surgical resection or radiosurgery for cerebral arteriovenous malformations. Neurosurgery 73: 648–655; discussion 655–656.

Wang YC, Huang YC, Chen HC et al. (2014). Linear accelerator stereotactic radiosurgery in the management of intracranial arteriovenous malformations: long-term outcome. Cerebrovasc Dis 37: 342–349.

Wegner RE, Oysul K, Pollock BE et al. (2011). A modified radiosurgery-based arteriovenous malformation grading scale and its correlation with outcomes. Int J Radiat Oncol Biol Phys 79: 1147–1150.

Yen CP, Varady P, Sheehan J et al. (2007). Subtotal obliteration of cerebral arteriovenous malformations after gamma knife surgery. J Neurosurg 106: 361–369.

Yen CP, Jain S, Haq IU et al. (2010). Repeat gamma knife surgery for incompletely obliterated cerebral arteriovenous malformations. Neurosurgery 67: 55–64; discussion 64.

Yen CP, Sheehan JP, Schwyzer L et al. (2011). Hemorrhage risk of cerebral arteriovenous malformations before and during the latency period after gamma knife radiosurgery. Stroke 42: 1691–1696.

Yen CP, Matsumoto JA, Wintermark M et al. (2013). Radiation-induced imaging changes following gamma knife surgery for cerebral arteriovenous malformations. J Neurosurg 118: 63–73.

Yen CP, Ding D, Cheng CH et al. (2014). Gamma knife surgery for incidental cerebral arteriovenous malformations. J Neurosurg 121: 1015–1021.

Handbook of Clinical Neurology, Vol. 143 (3rd series)
Arteriovenous and Cavernous Malformations
R.F. Spetzler, K. Moon, and R.O. Almefty, Editors
http://dx.doi.org/10.1016/B978-0-444-63640-9.00008-4

Chapter 8

Multimodality management of cerebral arteriovenous malformations

MOHAN NARAYANAN, GURSANT S. ATWAL, AND PETER NAKAJI*
Department of Neurosurgery, Barrow Neurological Institute, St. Joseph's Hospital and Medical Center, Phoenix, AZ, USA

Abstract

Arteriovenous malformations (AVMs) of the brain are diverse lesions that vary widely in location, size, and complexity. Treatment options for AVMs are correspondingly complex. Complete elimination of an AVM is required to protect patients from future hemorrhage. Decisions about whether to treat and, if so, how to treat these lesions depend on the characteristics of the patient and the specific characteristics of the AVMs. The characteristics of AVMs are often summarized through grading systems. Some AVMs can be managed conservatively, whereas others can be managed with microsurgical resection, radiosurgical ablation, or endovascular embolization, either individually or in combination. Some AVMs may also be treated with partial therapy to reduce the risk of hemorrhage or to ameliorate symptoms. In this chapter, we review the key factors that influence whether and how to manage AVMs with multimodality treatment.

INTRODUCTION

Arteriovenous malformations (AVMs) of the brain are complex lesions characterized by a tortuous nidus of vessels interposed between one or more feeding arteries and one or more draining veins. The nidus functions as a direct connection between the arterial and venous sides without an intervening capillary bed, and as a result, it can be subject to atypical vascular pressures and forces, the formation of aneurysms, and, ultimately, hemorrhage or rupture. Cerebral AVMs are rare, and no standardized treatment guidelines have yet been created that encompass all AVMs. The available treatment modalities include microsurgical resection, radiosurgical ablation, and endovascular embolization (EE), which can be used individually or in combination, depending on the specific case. Outcomes of large cohorts of AVM patients have been published by leading treatment centers (Potts et al., 2014; Rutledge et al., 2014; Pollock, 2015). Better outcomes from recent studies reflect the benefit of modern technology and techniques not available at the time of older studies that documented poorer outcomes. Despite the availability of a variety of treatments, the treatment guidelines are not always agreed upon across grading systems. Ultimately, the correct treatment or combination of treatments must be chosen for each lesion to maximize patient benefit and to minimize morbidity and mortality, even if the best choice is conservative management.

NATURAL HISTORY AND PRESENTATION OF ARTERIOVENOUS MALFORMATIONS

Patients with brain AVMs most commonly present with intracranial hemorrhage (52%), seizures (27%), focal neurologic deficits, and headache (Gross and Du, 2013). Patients with AVMs should undergo expeditious and meticulous evaluation, including a focused investigation of past medical history and a thorough physical examination, although patients with clinically significant

*Correspondence to: Peter Nakaji, MD, c/o Neuroscience Publications, Barrow Neurological Institute, St. Joseph's Hospital and Medical Center, 350 W. Thomas Rd., Phoenix, AZ 85013, USA. Tel: +1-602-406-3593, Fax: +1-602-406-4104, E-mail: Neuropub@dignityhealth.org

neurologic impairment may not be able to provide a medical history or undergo a reliable physical examination. The age of the patients, their general medical condition, and the presence of clinically significant comorbidities will all influence the decision to intervene for a specific lesion. The timing of treatment relies on the acuity of presentation, and it differs for patients who have acute hemorrhage or rupture compared with patients who have symptoms without hemorrhage or incidentally found lesions.

AVMs carry an overall annual hemorrhage risk of approximately 3%, further stratified into risk ranges of 1–2% for unruptured lesions and 4.5–5% for ruptured lesions. Patients are at increased risk for hemorrhage if they have a history of prior hemorrhage, if the AVM has deep venous drainage (exclusive or partial), or if the aneurysm is a nidal or feeding artery aneurysm (Gross and Du, 2013; Kim et al., 2014). The most important risk factor in predicting future risk of hemorrhage is the existence of a previous rupture in the natural history of the AVM. Re-rupture of an AVM is more likely during the first year after the initial hemorrhage, which connotes a 6–15% increased risk. AVMs with deep venous drainage or in a deep nidal location may not be optimal for surgical resection. Patients with these lesions should therefore have longer follow-up and the natural history of the AVM should be monitored for progression of the pathology or symptoms.

The lifetime risk of hemorrhage can be calculated using the following equations (Kondziolka et al., 1995):

$$\text{Lifetime risk} = 1 - (0.97)^{\text{years of risk remaining}} \quad \text{assuming a 3\% rupture rate}$$

$$\text{Lifetime risk} = 1 - (0.96)^{\text{years of risk remaining}} \quad \text{assuming a 4\% rupture rate}$$

Noninvasive imaging is typically obtained when the patient presents with symptoms, and it can help identify hemorrhage or delineate the anatomy of the AVM. Catheter digital subtraction angiography is the preferred method for diagnosis of AVMs and definition of AVM characteristics that may direct or enhance the treatment plan. Noninvasive angiography has been found to be as helpful as catheter angiography for identifying AVMs after hemorrhage (Josephson et al., 2014). However, the preoperative angiogram is considered the gold-standard imaging modality, as it not only defines the anatomy with the best resolution but also provides information about flow within the lesion and the hemodynamics of the lesion. With invasive angiography, any necessary endovascular interventions can be performed at the time of lesion diagnosis and definition.

AVM grading scales

Many classification systems for AVMs can be found in the neurosurgery literature. The most widely used paradigm is the Spetzler–Martin grading system, which classifies lesions on a scale from grade I to V on the basis of their size, location in relation to eloquent areas of the brain, and venous drainage (Spetzler and Martin, 1986). The 1986 Spetzler–Martin grading system was revised in 2011, and lesions were reclassified into a new schema, the Spetzler–Ponce classification, which assigned letter groups to the previous Spetzler–Martin grades based on the recommended treatment options for each AVM class (Spetzler and Ponce, 2011). Spetzler–Martin grades I and II AVMs thus became class A, grade III became class B, and grades IV and V became class C. The authors of this updated paradigm recommended surgical treatment for patients with class A lesions, multimodality treatment (MMT) for those with class B lesions, and observation for those with class C lesions unless there was recurrent hemorrhage or progressive neurologic deficits. The reclassification of AVMs from five to three tiers maintained the ability of the neurosurgeon to use the classification system to guide the selection of therapy and to predict postoperative neurologic outcomes. Lawton et al. (2003) further recommended dividing Spetzler–Martin grade III or Spetzler–Ponce class B AVMs into four subgroups based on morbidity risk.

The modification proposed by Lawton et al. (2003) recommends treatment based on morbidity risk, with S1V1E1 (with S = size, V = venous drainage, and E = eloquent area) lesions similar to Spetzler–Martin grade I, S2V1E0 lesions similar to grade III lesions, and S2V0E1 lesions similar to grade IV and V lesions. Therefore, Spetzler–Martin grade III lesions require the highest amount of individualized planning to determine which aspects of MMT should be used for the patient. Further modification of the Spetzler–Martin scale, to more accurately predict neurologic outcome after open microsurgical resection, takes into account the age of the patient, the presence of hemorrhage, and the compactness of the lesion (Lawton et al., 2003; Kim et al., 2015). The Spetzler–Martin scale is still the most widely used classification system for AVMs, but, as discussed by Kim and colleagues (2015), no system or combination of systems for classifying AVMs can replace appropriate patient selection for microsurgical resection.

Radiosurgery-based AVM score

Grading scales for AVM treatment and outcomes have also been developed for radiosurgery-based applications, including the radiosurgery-based AVM score and the

Virginia Radiosurgery AVM Scale (Pollock and Flickinger, 2002; Pollock and Flickinger, 2008; Starke et al., 2013). Unlike the Spetzler–Martin grading system, which was initially developed to guide treatment and predict outcomes from microsurgical resection, these radiosurgery scores take into consideration the volume, age, and location of the AVM, and any history of hemorrhage.

MAJOR PARADIGMS IN ARTERIOVENOUS MALFORMATION MANAGEMENT

The risks of treatment-related morbidity and mortality should be carefully weighed when selecting one or more modalities to treat AVMs. Treatment is necessary if the future risk of hemorrhage is greater than the risk of open surgery, stereotactic radiosurgery (SRS), EE, or MMT. For microsurgical resection and EE, operator skill is an important variable in selecting the appropriate treatment modality, and some operators practice in specialty centers with resources better suited to specific techniques.

Whether to treat a ruptured AVM is rarely questioned. Patients who present with hemorrhage typically undergo noninvasive imaging and, in some centers, catheter angiography to determine the architecture of the lesion. These patients may then undergo either symptomatic treatment or both symptomatic and definitive treatment with microsurgical resection of the lesion, depending on the resources available for the patient.

When the primary goal is seizure control, the treatment of AVM-related epilepsy is controversial. Patients who have had a prior intracerebral hemorrhage or who have focal neurologic deficits as a result of an AVM are at higher risk for developing seizures (Crawford et al., 1986; Josephson et al., 2011). However, patients who have had a prior hemorrhage are more likely to experience improvement in seizure frequency over time (Ding et al., 2015a; Ding et al., 2015b).

Unruptured AVMs often present challenging clinical quandaries. The decision of whether to treat or not, and if so, then how, should be made jointly by the patient and the neurosurgeon or multidisciplinary team with the understanding that the data are often incomplete. The risk of intervention must be weighed against the risk of hemorrhage based on the natural history of each patient's individual lesion.

Microsurgical resection as monotherapy

Microsurgical resection of the AVM nidus is the classic mainstay of therapy. The aim of microsurgery is complete angiographic occlusion, which is immediate in open surgery, and which precludes future AVM rupture or hemorrhage. Microsurgery has the best outcomes in patients who have class A AVMs and long life expectancies that confer a higher lifetime risk of hemorrhage. Advances in intraoperative microscopy and navigation, and the use of tools such as intraoperative Doppler ultrasound aid in real-time definition and evaluation of lesions, help in identifying normal vasculature, and allow for a more complete AVM resection (Dempsey et al., 2004).

Microsurgical resection is not without its difficulties. AVMs are often located in deep or inaccessible areas of the brain not amenable to dissection. There may also be surrounding normal brain tissue that can be damaged during accessing, dissection, and/or removal of the AVM. Feeding arteries to surrounding brain tissue may be ligated as a result of complete AVM resection. Excessive intraoperative bleeding from the arterial supply or venous drainage can contribute to morbidity and mortality in patients who undergo microsurgical treatment. The reported rate of surgical morbidity and mortality of class A lesions is as low as 0%. In 2004, Morgan et al. reported a surgical morbidity and mortality rate of 1.4% for the same class of lesions and noted an even higher risk in large grade I and II AVMs in eloquent areas or on the threshold of grade III AVMs, requiring careful consideration of treatment strategies.

Radiosurgery as monotherapy

Radiosurgery is a technique of ablation of AVM using focused, high-dose radiation. The proposed mechanism of action is controlled injury to the vascular walls of the AVM, which will eventually lead to sclerosis of the nidus and the feeding and draining vessels (Ajiboye et al., 2014). It has traditionally been reserved for patients who have significant surgical morbidity, or for patients whose AVMs are not amenable to surgical resection (Starke et al., 2013). SRS is useful for lesions with a diameter of less than 3.5 cm, with reported obliteration rates of up to 80% after 2 years (Al-Shahi and Warlow, 2006), and 84% without risk of future hemorrhage (Yen et al., 2014). Various systems are available to deliver such radiation, including cobalt source systems (e.g., Gamma Knife (Leksell, Elekta)) and linear accelerator-based systems (e.g., CyberKnife (Accuray), XKnife (Integra LifeSciences), and Novalis (Brainlab)). The benefit of radiosurgery is not immediate; obliteration of the lesion in patients who respond to treatment occurs in 2–4 years after treatment. SRS can be used with excellent results in patients with a radiosurgery-based AVM score <1, and it may have similar outcomes for patients with small AVMs in eloquent areas that do not require craniotomy for evacuation of hematoma after AVM rupture (Burrow et al., 2014). Conger et al. (2015) reviewed the application of different grading scales to guide treatment and found that the Spetzler–Martin grading system is not useful for guiding treatment or predicting outcome for patients undergoing SRS, as the treatment plan requires volumetric information about the AVM.

Although SRS is technically a noninvasive therapeutic approach that does not require a craniotomy or invasive vascular access, it still has high latent risk. Open AVM resection has been associated with greater morbidity than SRS, and recurrent bleeding has been found to occur more often after SRS than after microsurgery (Pollock and Flickinger, 2008). SRS still carries a considerable risk of posttreatment hemorrhage (approximately 4–5% per year), and patients can sustain clinically significant neurologic deficits from damage to surrounding normal brain tissue, sensitivity to radiation, incomplete obliteration, and tumor formation that may occur secondary to radiation. Patient age influences the number of years the patient will be at risk and the decision making about the candidacy of the patient for surgery. One retrospective review demonstrated that, for elderly patients with AVMs, SRS does not cause significantly worse outcomes than microsurgical resection (Ding et al., 2015c).

SRS is used to treat large AVMs in dose-staged or volume-staged radiosurgery. It is primarily a tool to remove deep or unresectable parts of AVMs. However, AVMs can regrow after incomplete resection, or they can occur as *de novo* lesions after complete resection or treatment with radiosurgery (Moftakhar et al., 2009).

Endovascular embolization as monotherapy

Complete obliteration of AVMs has been difficult to achieve using EE alone, with angiographic cure ranging from 13% to 96% and being highly dependent on good selection of patients and lesions (Saatci et al., 2011; van Beijnum et al., 2011; van Rooij et al., 2012a; Consoli et al., 2014). Certain lesions, such as small and medium-sized superficial brain AVMs, are amenable to EE as monotherapy, especially in the acute phase of hemorrhagic stroke (van Rooij et al., 2012b). Appropriately administered EE may reduce the size of the nidus without abrupt changes in intraluminal pressure and minimize the risk of future arterial recruitment to the nidus (Krings et al., 2010). A transvenous approach has been suggested as a method to treat AVMs. This technique evolved from the treatment of dural arteriovenous fistulae and is technically challenging. Hypotension must be induced to reduce the nidal pressure and allow the embolisate to travel in a retrograde fashion across the nidus. This can be difficult to achieve because of the complicated anatomy of many AVMs. Additionally, the risk of venous perforation is not insignificant, and premature compromise of the venous drainage can lead to an abrupt increase in nidal pressure and subsequent rupture (Choudhri et al., 2015).

Risks associated with EE include showering embolic material in unintended parts of the vascular tree, hemorrhage, remote recanalization of embolized segments, contrast-induced kidney injury, retained foreign body (i.e., a piece of catheter gets stuck in embolisate), and inability to accomplish the goal of embolization. Newman et al. (2012) reviewed the literature related to retrieval of retained microcatheter pieces after EE, and described their own technique, but noted that retention of microcatheter in embolisate is often well tolerated by patients. Of note, EE has been described as an appropriate treatment for cranial dural arteriovenous fistulae (often called dural AVMs); however, these lesions represent a different type of pathology and a description of the treatment paradigm for them is outside the scope of this chapter.

Attempts have been made to create a standardized anatomic grading system to guide EE therapy. However, none of these systems has been validated.

Observation and conservative management

Patients who have Spetzler–Martin grade IV and grade V AVMs have a substantially high surgical risk and a short life expectancy (or low remaining lifetime risk of hemorrhage), and they may not be able to tolerate SRS or EE (or refuse those therapies). Conservative management is appropriate for these lesions. Recurrent hemorrhage or progressive neurologic deficits can be treated emergently, but the overall treatment plan is observation. Surveillance imaging can be performed, at the discretion of the managing physician, but no standard interval has been established for invasive or noninvasive imaging follow-up. Patients who pursue observation and conservative management should have close clinical follow-up and treatment for headaches, seizures, and hypertension (Ajiboye et al., 2014). Future improvements in current treatment modalities or the development of new techniques and technologies may allow for intervention in this group of complex lesions.

CLINICAL DECISION MAKING AND MULTIMODALITY TREATMENT

The primary goal in AVM intervention is resolution or obliteration of the AVM nidus, reduction in the risk of future hemorrhage, and minimization of any treatment-related morbidity or mortality. Symptom improvement can be considered a secondary endpoint, including the resolution of seizures and improvement in headaches or visual deficits. The decision to pursue a single modality of treatment or MMT should be determined by the best possible chance of achieving the primary endpoint and maximizing positive secondary endpoints.

Many single-institution studies have been conducted to analyze outcomes for single or multimodality AVM treatment. However, no clear treatment guidelines have been published and guidance in the literature remains obscure on this point.

EE followed by microsurgical resection

EE has traditionally been used as an adjunct to open microsurgery or SRS. As an adjunct to open surgery, EE can be used to reduce blood flow through AVM lesions by occluding feeding arteries to the nidus and helping to minimize intraoperative blood loss (van Beijnum et al., 2011; van Rooij et al., 2012a; Saatci et al., 2011; Consoli et al., 2014). Although in common use, adjunctive EE is still controversial and not universally recommended.

An early influential study by Morgan et al. (1997) did not recommend perioperative embolization in Spetzler–Martin grades III–V AVMs due to difficulty in catheterization of feeding arteries (specifically, lenticulostriate arteries), the unreliable prediction of neurologic deficits in patients after the Wada test, the augmentation of nonembolized arteries, and the significant incidence of postoperative neurologic deficits. Morgan et al. (2000, 2004) reported results from EE, prior to microsurgical resection, and noted that the use of embolization had declined in their institution as concern grew that there was no reduction in adverse events postoperatively when EE and microsurgery were used in combination compared to microsurgery alone. Weber and colleagues (2007) reported outcomes of preoperative EE and noted that it aids volume reduction and definition of AVM borders for microsurgical resection. However, postoperative deficits were seen in 51% of patients, the all-cause morbidity in their patients was high, and their findings highlight the considerable risk of EE.

The question of whether all AVMs treated with the liquid embolic agent Onyx should be resected also remains unresolved. Previous reports have documented partial resection of AVMs after EE with Onyx. Onyx can easily be visualized intraoperatively, which helps define the margins and boundaries of the AVM for extent of resection. Prior reports describe leaving behind small amounts of embolized vessels, which then undergo fibrosis and auto-obliterate. One retrospective study examined the use of preoperative EE in 28 patients over a total of 55 sessions (Natarajan et al., 2008). The authors noted that AVM feeders can be left in the resection bed with an Onyx plug effectively obliterating the vessel. However, the study documented observation of AVM filling intraoperatively in all lesions with previous angiographic cure by EE. A 2015 case report further cautioned neurosurgeons about recanalization and recurrence of AVMs (Bauer et al., 2015). The patient in that case underwent EE and partial resection of the nidus and Onyx cast but had recurrent seizures at 9 months after surgery. An angiogram revealed a recurrent AVM with flow in the nidus and Onyx volume loss in the cast, as well as recanalization in previously embolized vessels.

Incomplete obliteration of an AVM puts the patient at risk of bleeding and AVM recurrence. The goal of AVM treatment should be definitive obliteration. However, if embolic material in feeding vessels is left in the resection bed, risk remains for recanalization and subsequent hemorrhage. Bauer and colleagues (2015) recommended that every attempt should be made to resect the entirety of the Onyx cast and to obtain adequate surgical hemostasis. Additionally, a large patient cohort suggested that preoperative embolization was a risk factor for recurrence in adults (Ivanov et al., 2016). The authors warned that a false-negative postoperative angiogram (showing no residual AVM) might be caused by vessel spasm, a tamponade effect from adjacent brain edema, and temporary vessel thrombosis.

SRS followed by microsurgical resection

SRS can be used to shrink or prune dangerous sections of AVMs or AVMs in deep locations. SRS can be safely used in some patients with Spetzler–Martin grade IV AVMs to make the lesions more amenable to microsurgical resection (Fig. 8.1). In a 2016 series by Tong et al. (2016), good neurologic outcomes were obtained in patients who underwent SRS, had residual AVMs, and underwent microsurgery years later. SRS led to a reduction in volume and to lower Spetzler–Martin grades in this series of patients with mostly (80%) grade III and grade IV AVMs. SRS can aid in the resection of higher-grade AVMs in select patients, preventing perioperative neurologic deterioration. In another series of mostly (86%) grade II and grade IV AVMs by Sanchez-Mejia et al. (2009), SRS significantly reduced volume, required less EE, and decreased Spetzler–Martin grades. Additionally, in this series, surgical morbidity was higher in the group that did undergo SRS before surgery. However, there was a higher incidence of hemorrhages after SRS during the latency period. SRS can be a useful adjunct for treatment of complex AVMs, but it should be used judiciously in such patients.

Microsurgical resection followed by SRS

SRS can also be used as a salvage therapy for AVMs after microsurgical resection. In a recent retrospective study of Spetzler–Martin grade III AVMs treated with SRS, Ding

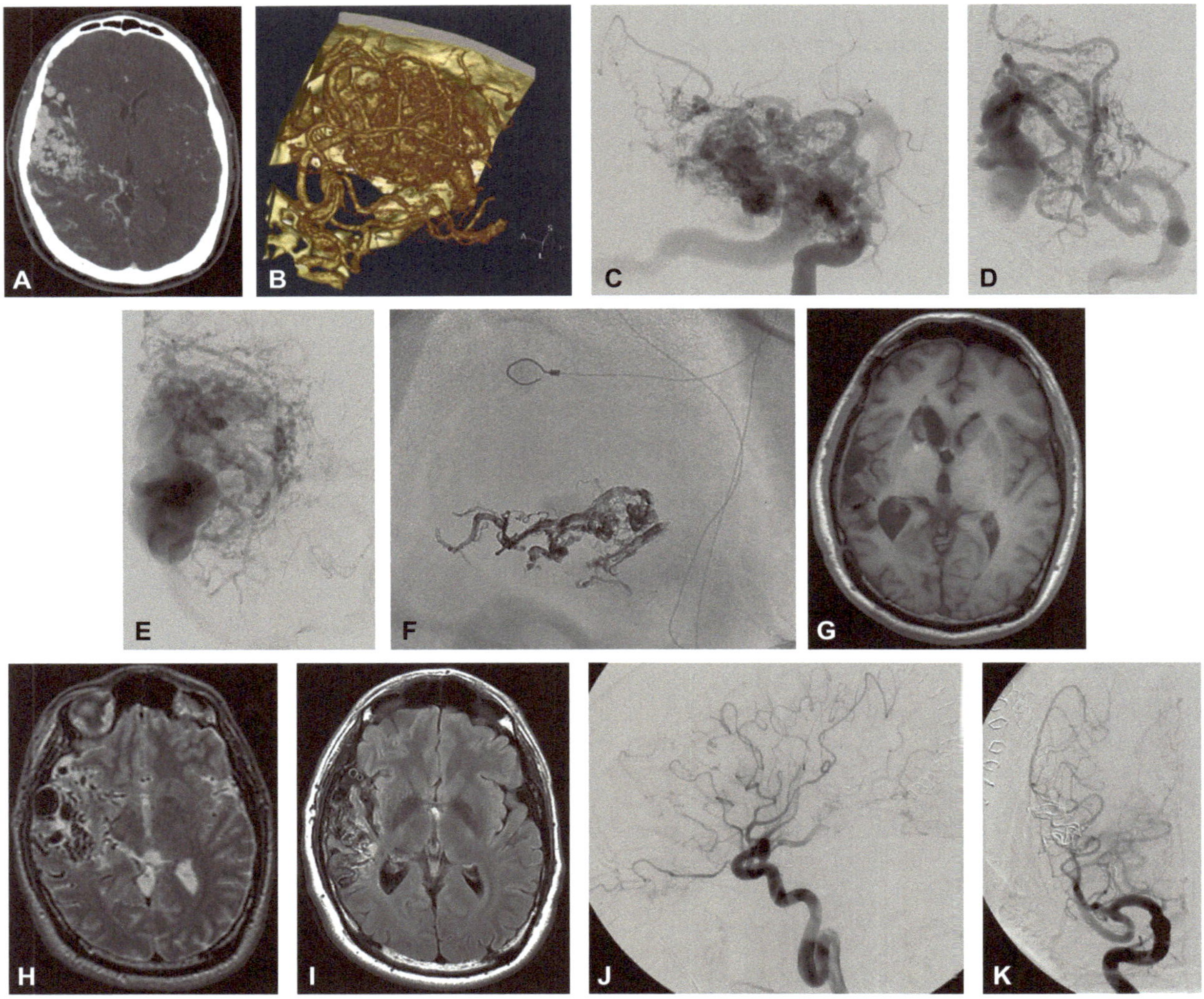

Fig. 8.1. A 28-year-old man who presented with seizures, a large right temporal arteriovenous malformation (AVM), and an intracranial hemorrhage was initially selected for partial endovascular embolization, followed by microsurgical resection. However, invasive angiography revealed a tortuous feeding architecture with a high risk for infarction, so a different multimodality treatment plan was developed. The patient underwent Gamma Knife stereotactic radiosurgery (SRS), and at the 4-year follow-up he had no further hemorrhage and his only reported neurologic deficit was seizures. Repeat imaging studies revealed a more compact AVM nidus, and the patient underwent microsurgical resection of the AVM with complete obliteration of the lesion. (**A**) Axial computed tomography angiogram (CTA) showing the large temporoparietal AVM. (**B**) Three-dimensional reconstruction of the CTA showing complicated angioarchitecture with both deep and superficial vasculature. (**C**) Digital subtraction angiography (DSA) of right internal carotid artery injection, lateral view, before embolization. (**D**) DSA of right internal carotid artery injection, anteroposterior view, before embolization. (**E**) DSA of right internal carotid artery injection, with emphasis on the venous phase of contrast, before embolization. (**F**) DSA showing partial nidal endovascular embolization. (**G**) T1-weighted magnetic resonance image (MRI) after microsurgical resection of the AVM. (**H**) T2-weighted MRI showing large-caliber vessels, before Gamma Knife SRS. (**I**) Fluid-attenuated inversion recovery MRI showing reduction of AVM burden, 4 years after Gamma Knife SRS. (**J**) DSA of right internal carotid artery injection, lateral view, after microsurgical resection of the AVM. (**K**) DSA of right internal carotid artery injection, anteroposterior view, after microsurgical resection of the AVM. (Courtesy of Barrow Neurological Institute, Phoenix, AZ.)

et al. (2017) identified several independent predictors of optimal outcome: smaller nidus volume, no previous hemorrhage, lack of previous embolization, absence of AVM-associated aneurysms, and higher radiation dose. In this series of 891 patients, 11% had symptomatic radiation-induced changes; 4% of these were permanent. The obliteration rates were 63% at 5-year follow-up and 78% at 10-year follow-up. Microsurgical resection can be used to reduce the volume of the nidus and thus make the lesion more amenable to complete obliteration with fewer radiation-induced changes after SRS. Additionally, if a small part of the nidus is left behind, EE can be a useful adjunct to reduce the volume further before SRS.

EE followed by SRS

As an adjunct to SRS, EE can be used to decrease the volume of the remaining nidus, and it can be used to occlude aneurysms in or around the AVM nidus (Flores et al., 2014; Ding et al., 2015d). Miyachi et al. (2000) examined a series of 37 patients who underwent EE before SRS; of these patients, 16 had further reduction in the size of the AVM after EE before SRS, 10 had no change, and 11 experienced regrowth. After SRS, only 5 patients achieved obliteration. The authors recommend a 2-month interval between EE and SRS to evaluate whether the embolization is adequate; if recanalization occurs, the risk of hemorrhage or regrowth of the AVM is high. The authors stressed the importance of superselective angiography, the use of a durable occlusive agent, and embolization of high-flow fistulae and meningeal feeding vessels.

In another series of 120 patients by Kano et al. (2012), EE before SRS reduced the rate of obliteration. EE had no effect on the hemorrhage rate during the latency period. However, if volume was reduced to 8 cm^3 with EE, the rate of obliteration was significantly improved. EE can be used with the goal of volume reduction to limit the radiation dose and help achieve a higher obliteration rate.

Dalyai et al. (2014) also reported their experience with a series of 95 patients undergoing EE followed by SRS to treat any residual AVM nidus. EE was performed with the goal of reducing AVM volume (not flow). With a combination of single-stage EE, multistage EE, single-stage SRS, and multistage SRS, they achieved complete obliteration in 40% ($n = 38$) of patients. The authors advise against SRS as a monotherapy for patients with larger AVMs.

Other MMT studies

In 2011, van Beijnum and colleagues conducted a meta-analysis of reported treatment modalities, and examined a large population in 142 cohorts examining microsurgery, EE, SRS, fractionated radiotherapy, and MMT. The primary outcome was defined as all-cause mortality. Younger patients and those with Spetzler–Martin grades I–III AVMs had lower associated mortality and lower rates of hemorrhage after microsurgery than after EE or SRS. Patients who underwent SRS had a low risk of acute complications, but a higher risk of late complications was found in patients who were older, who had larger AVMs, or who had ruptured AVMs. Higher obliteration rates in EE were associated with more recent cohorts, which may be due to improvement in EE operator skill and technology.

Nataraj et al. (2014) performed an extensive retrospective review of prospectively collected data on 290 patients who underwent various treatments for AVM from 1980 to 2008. Of the 290 patients, 265 had a single treatment or a combination of treatments. Their outcomes are summarized in Table 8.1, and their proposed treatment algorithm is shown in Figure 8.2. The patients who experienced the highest cure rates were in the EE

Table 8.1

Outcomes and complications based on type of treatment

Treatment	Minor deficits, no. (%)	Major deficits, no. (%)	Rehemorrhage, no. (%)	Death, no. (%)
Radiosurgery ($n = 37$)	2 (5)	0 (0)	5 (14)	4 (11)
Embolization ($n = 34$)	1 (3)	1 (3)	5 (15)	3 (9)
Surgery ($n = 57$)	1 (2)	0 (0)	0 (0)	1 (2)
Embolization and surgery ($n = 101$)	10 (10)	6 (6)	9 (9)	1 (1)
Embolization and radiosurgery ($n = 17$)	2 (12)	1 (6)	1 (6)	0 (0)
Surgery and radiosurgery ($n = 5$)	0 (0)	0 (0)	2 (40)	2 (40)
All modalities ($n = 14$)	0 (0)	1 (7)	2 (14)	0 (0)
No treatment ($n = 25$)	0 (0)	3 (12)	6 (24)	6 (24)
Total ($n = 290$)	16 (6)	12 (4)	30 (10)	17 (6)

Reproduced from Nataraj et al. (2014), with permission.

Disability and death include those due to rehemorrhage. All deaths were due to hemorrhage from residual arteriovenous malformations (except 2 in the surgery plus embolization group, which were postoperative hematomas, and 1 in the surgical group, which was a patient who presented in a coma and never regained consciousness).

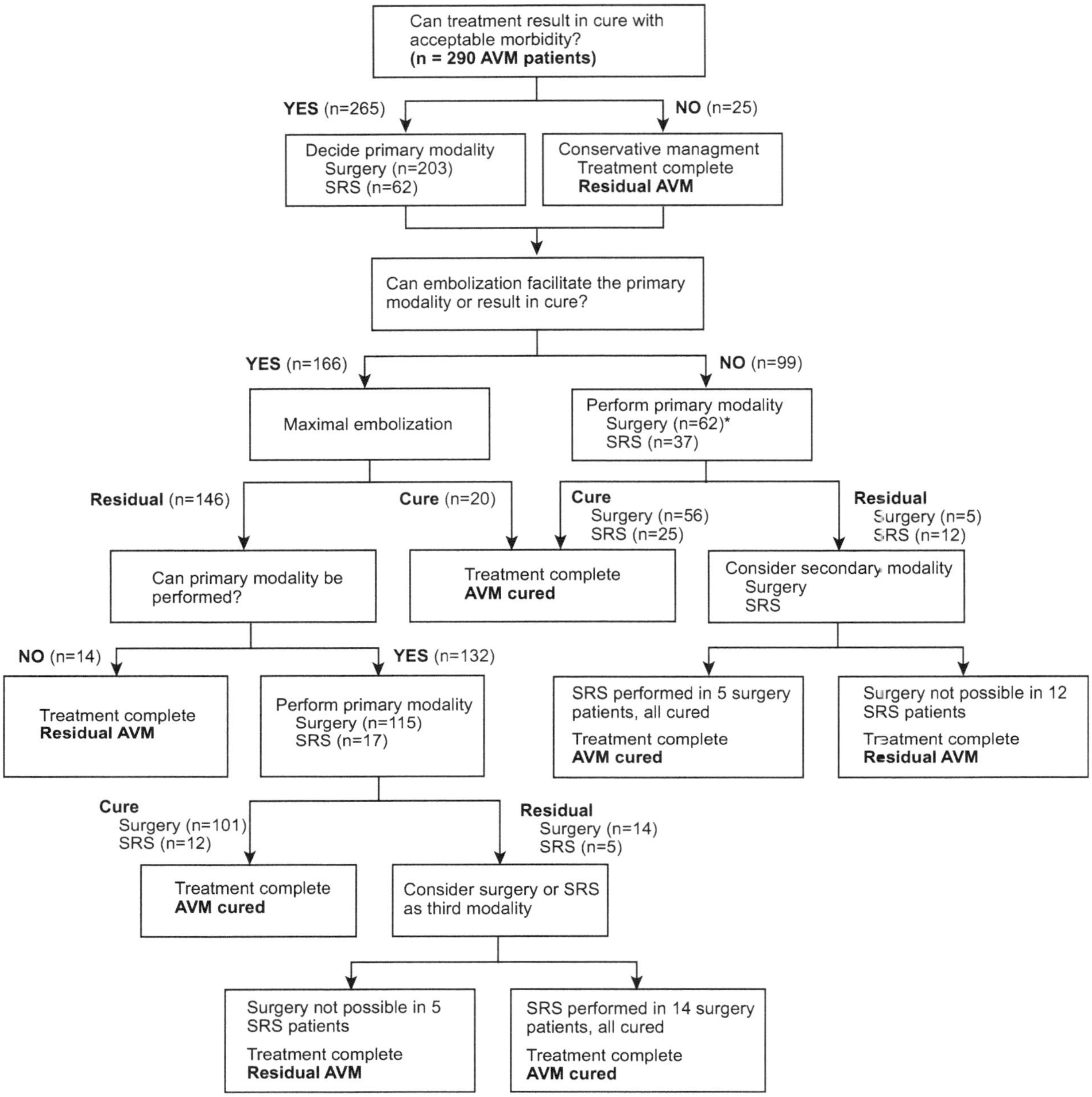

Fig. 8.2. Nataraj et al.'s treatment algorithm for arteriovenous malformations (AVMs) using a multimodal approach. SRS, stereotactic radiosurgery. *Of the 62 surgery patients, one patient died. (Adapted from Nataraj et al., 2014, with permission.)

group (99%) or the microsurgical resection group (98%). Fourteen of the 265 patients underwent a combination of EE, SRS, and microsurgical resection, and angiographic cure was achieved in all 14 of those patients. Five patients had microsurgical resection followed by postoperative SRS, and angiographic cure was achieved in all 5 of those patients. Nataraj et al. (2014) noted that the main benefit of EE is the elimination of deep feeding vessels, which does not really help to decrease the flow or volume of the AVM nidus.

Pandey et al. (2012) conducted a combined prospective and retrospective review of 100 patients with Spetzler–Martin grade III AVMs treated at their facility between 1984 and 2010. Twenty patients underwent single-modality treatment; 41 patients underwent EE followed by microsurgical resection, 25 patients underwent EE followed by SRS, and 11 patients underwent EE, resection, and SRS. Most of the patients had lesions classified as S2E1V0 subtype. When the authors compared the rates of complications in their patients over two discrete periods (1986–1999 and 2000–2009), they found decreases over time in the rates of overall complications and major complications and an increase over time in the rate of obliteration.

Potts et al. (2014) conducted a retrospective review of 108 patients with AVMs who were managed with MMT at their institution from 1997 to 2011. After 11 patents who had prior treatment elsewhere were excluded, 97 patients were studied. Most of these patients had Spetzler–Martin grade III and IV lesions (44% and 36%, respectively), and most presented with hemorrhage (64%). Forty-one patients underwent microsurgical resection, 44 underwent SRS, and 12 were observed. After initial treatment, radiographic cure was observed in 71% of the surgical patients and in 23% of the SRS patients; however, 21 patients who underwent resection had preoperative EE compared to 4 patients who underwent SRS. The authors concluded that the functional outcomes in this cohort were associated more with Spetzler–Martin grade of lesions than with actual obliteration rates.

Pediatric AVM MMT

Pediatric patients with AVMs are best treated at centers with extensive experience in management of young patients with these particular lesions. Darsaut et al. (2011) conducted a large retrospective review of prospectively collected data from 120 pediatric patients with AVMs who were treated in 1985–2009 at their institution. Seventy-six of these patients had Spetzler–Martin grade I–III AVMs. They recommend aggressive management for all pediatric AVMs, with resection of grade I–III lesions that are accessible via a safe surgical corridor. In these AVMs that are deemed inaccessible, SRS can be used with a high cure rate and a low risk of complications.

Darsaut et al. (2011) noted that many neurosurgeons are hesitant to use SRS in the developing central nervous system, although long-term follow-up of pediatric patients indicates good clinical outcomes. They report per-session complication rates for microsurgical resection of 9.6% per operative session for lesions with an average Spetzler–Martin grade of 2.22, for EE of 7.7% per embolization session with an average grade of 2.6, and for SRS of 7.5% per session.

Darsaut et al. (2011) emphasize that residual AVMs are commonly observed in delayed angiograms, so an angiogram should be obtained at 6-month follow-up to evaluate the efficacy of treatment. They also point out that their results show SRS to be ineffective as an initial treatment for high-grade lesions. For patients who have residual AVMs after initial therapy with EE, SRS, or resection, further microsurgical resection should be conducted when the residual AVM is accessible. If it is not accessible, the authors recommend SRS.

Eloquent location

AVMs in eloquent locations may prove to be a challenge to selection of the appropriate type of treatment. Institutions with a high volume of patients with AVMs and good outcomes for microsurgical resection, even in eloquent locations, should use this treatment if the risk of resection is lower than that of any other monotherapy or MMT. Neurosurgeons are less likely to use EE if they risk embolizing critical small vessels or *en passage* vessels supplying eloquent areas that cannot be sacrificed.

The deep gray-matter structures are sensitive to radiation exposure, and caution is advisable when SRS is used to treat AVMs adjacent to eloquent structures. Potts et al. (2013) reviewed a prospective registry of 48 patients treated surgically at their institution from 1997 to 2011 who had mostly Spetzler–Martin grade III AVMs located in deep areas. Of 34 patients who underwent complete resection, 14 also underwent SRS postoperatively. Forty-five patients had improved or unchanged Rankin scale scores at a mean follow-up of 1.6 years. The authors noted that deep lesions are more amenable to surgery than previously thought; however, this population was highly selected for hemorrhage (77% at presentation). The patients with hemorrhage were more likely to undergo microsurgical resection than other treatments initially. Since deep supratentorial AVMs are difficult to approach, most neurosurgeons would probably refer those patients for SRS and observation or to a specialized AVM center. For patients with AVMs located near the deep gray-matter structures, undergoing SRS is dangerous because of the sensitivity of this tissue to radiation and the risk of hemorrhage in critical structures during the latency period. In the Potts et al. (2013) cohort, patients had a 2.4% risk of mortality from surgical resection and a 5.3% risk of latent hemorrhage.

THE IMPACT OF ARUBA ON THE TREATMENT OF UNRUPTURED BRAIN AVMs

The ARUBA (A Randomized Trial of Unruptured Brain Arteriovenous Malformations) trial caused controversy about the appropriate management of unruptured AVMs. The study was one of the first attempts to compare the natural history of this complicated disease to the risk of treatment-associated morbidity (Wedderburn et al., 2008; Al-Shahi Salman et al., 2014; Mohr et al., 2014; Starke et al., 2014). At a mean follow-up of 33.3 months, ARUBA showed superior outcomes for patients undergoing conservative management ($n = 109$, of whom 11 (10.1%) had died or had a stroke) compared to patients undergoing intervention ($n = 114$, of whom 35 (30.7%) had died or had a stroke) (Mohr et al., 2014). This study has been criticized by many neurosurgeons (Gross et al., 2014; Lawton and Abla, 2014; Solomon and Connolly, 2014; Weiner et al., 2014), who argue that, while the short-term risks of treatment-related morbidity are real, it is also essential to consider the lifetime risk of hemorrhage for an unruptured lesion. The overall risk-to-benefit rates for microsurgery compared to observation would not become equivalent or beneficial until 10–15 years after surgery (Starke et al., 2014).

Steiger and colleagues (2015) subsequently published their own series of patients ($n = 97$) with AVM management, which demonstrated superior results for microsurgical resection in patients with Spetzler–Martin grade I and II AVMs but not in patients with grade III and IV AVMs compared to the natural history of the lesions and accounting for quality-adjusted life-years. Additionally, in that series, patients younger than age 39 years had better outcomes after treatment than older patients, as determined by modified Rankin scale scores.

Rutledge et al. (2014) reviewed the treatment and outcomes of 74 ARUBA-eligible patients at University of California, San Francisco, and noted that most of these patients ($n = 61$) underwent microsurgical resection with or without preoperative embolization compared to those included in ARUBA. These patients had significantly better outcomes than those reported by ARUBA in the intervention arm (14.8% (9 of 61) vs. 30.7% died or had a stroke). Only 1 patient had EE alone with the intent of pursuing surgery, but later declined surgery, and patients with higher-grade AVMs in inaccessible locations underwent SRS. These and other authors cite the importance of judiciously selecting the right patients, developing surgical expertise by operating at high AVM volumes, and meticulously reporting results and outcomes (Rutledge et al., 2014; White and Batjer, 2015).

In 2015, a large multicenter randomized controlled trial (Treatment of Brain AVMs (TOBAS)) was established to evaluate the effectiveness and outcomes of MMT for AVMs (Darsaut et al., 2015). This trial aims to address the design issues of ARUBA, and to use prospective data to compare conservative versus interventional management of patients with brain AVMs. The authors will also evaluate the effect of EE used before microsurgical resection or SRS for ruptured and unruptured AVMs. The results of the TOBAS trial will contribute significantly to the understanding and management of ruptured and unruptured AVMs.

CONCLUSIONS

MMT of AVMs may be a better treatment choice for some AVMs that are difficult to treat with single-modality therapy. For patients who do not fit into any particular discrete treatment category, neurosurgeons must rely on their experience and knowledge of the natural history of AVMs, their honest appraisal of their own technical skill with each modality, and their consideration of angiographic anatomy, patient risk factors, and the guidance of multiple grading scales. Doing so will allow the neurosurgeon to present information in an organized fashion to the patient so that they can work together to craft a reasonable treatment plan.

However, most authors agree that there is no good way to truly measure quality in healthcare delivery, especially when it is related to neurosurgical outcomes. Davies et al. (2012) report that the average risk of complications for patients after resection, EE, and SRS is 7.1%, 6.6%, and 5.1%, respectively. Patients with a ruptured AVM at presentation have the highest risk of mortality with EE alone (10.6%) or with EE followed by microsurgical resection (10.3%). Patients who undergo microsurgical resection have longer hospital stays, although the overall mortality was lower in patients undergoing microsurgical resection alone, when compared to endovascular monotherapy (1.0% vs. 1.6% unruptured, 6.6% vs. 10.6% ruptured).

The ultimate treatment decision should be reached in consultation with the patient, and ideally should include the goal of achieving resection or obliteration of the lesion, resolution of symptoms, and symptom-free survival with the least possible morbidity and mortality. All major treatment modalities carry associated risks of severe complications, but advances in technology, technique, and operator experience are improving outcomes after all types of treatment. Measurement of outcomes using the modified Rankin scale is crude, and more robust metrics should be implemented as a gauge of clinical outcomes. Longer-term follow-up is also necessary to make definitive statements about the benefits of intervention for unruptured AVMs. With commitment to rigorous self-evaluation and critical assessment of outcomes, we can ensure that we will continue to improve the care of patients with ruptured AVMs who are undergoing intervention with whichever modality or combination of modalities is ultimately chosen.

ACKNOWLEDGMENTS

We thank the staff of Neuroscience Publications at Barrow Neurological Institute for assistance with manuscript preparation.

References

Ajiboye N, Chalouhi N, Starke RM et al. (2014). Cerebral arteriovenous malformations: evaluation and management. Scientific World Journal 2014: 649036.

Al-Shahi R, Warlow CP (2006). Interventions for treating brain arteriovenous malformations in adults. Cochrane Database Syst Rev: CD003436.

Al-Shahi Salman R, White PM, Counsell CE et al. (2014). Outcome after conservative management or intervention for unruptured brain arteriovenous malformations. JAMA 311: 1661–1669.

Bauer AM, Bain MD, Rasmussen PA (2015). Onyx resorbtion with AVM recanalization after complete AVM obliteration. Interv Neuroradiol 21: 351–356.

Burrow AM, Link MJ, Pollock BE (2014). Is stereotactic radiosurgery the best treatment option for patients with a radiosurgery-based arteriovenous malformation score $\leq$ 1? World Neurosurg 82: 1144–1147.

Choudhri O, Ivan ME, Lawton MT (2015). Transvenous approach to intracranial arteriovenous malformations: challenging the axioms of arteriovenous malformation therapy? Neurosurgery 77: 644–651; discussion 652.

Conger A, Kulwin C, Lawton MT et al. (2015). Diagnosis and evaluation of intracranial arteriovenous malformations. Surg Neurol Int 6: 76.

Consoli A, Scarpini G, Rosi A et al. (2014). Endovascular treatment of unruptured and ruptured brain arteriovenous malformations with Onyx18: a monocentric series of 84 patients. J Neurointerv Surg 6: 600–606.

Crawford PM, West CR, Shaw MD et al. (1986). Cerebral arteriovenous malformations and epilepsy: factors in the development of epilepsy. Epilepsia 27: 270–275.

Dalyai R, Theofanis T, Starke RM et al. (2014). Stereotactic radiosurgery with neoadjuvant embolization of larger arteriovenous malformations: an institutional experience. Biomed Res Int 2014: 306518.

Darsaut TE, Guzman R, Marcellus ML et al. (2011). Management of pediatric intracranial arteriovenous malformations: experience with multimodality therapy. Neurosurgery 69: 540–556; discussion 556.

Darsaut TE, Magro E, Gentric JC et al. (2015). Treatment of Brain AVMs (TOBAS): study protocol for a pragmatic randomized controlled trial. Trials 16: 497.

Davies JM, Yanamadala V, Lawton MT (2012). Comparative effectiveness of treatments for cerebral arteriovenous malformations: trends in nationwide outcomes from 2000 to 2009. Neurosurg Focus 33: E11.

Dempsey RJ, Moftakhar R, Pozniak M (2004). Intraoperative Doppler to measure cerebrovascular resistance as a guide to complete resection of arteriovenous malformations. Neurosurgery 55: 155–160; discussion 160–161.

Ding D, Starke RM, Quigg M et al. (2015a). Cerebral arteriovenous malformations and epilepsy, part 1: predictors of seizure presentation. World Neurosurg 84: 645–652.

Ding D, Quigg M, Starke RM et al. (2015b). Cerebral arteriovenous malformations and epilepsy, part 2: predictors of seizure outcomes following radiosurgery. World Neurosurg 84: 653–662.

Ding D, Xu Z, Yen CP et al. (2015c). Radiosurgery for cerebral arteriovenous malformations in elderly patients: effect of advanced age on outcomes after intervention. World Neurosurg 84: 795–804.

Ding D, Sheehan JP, Starke RM et al. (2015d). Embolization of cerebral arteriovenous malformations with silk suture particles prior to stereotactic radiosurgery. J Clin Neurosci 22: 1643–1649.

Ding D, Starke RM, Kano H et al. (2017). Stereotactic radiosurgery for Spetzler–Martin grade III arteriovenous malformations: an international multicenter study. J Neurosurg 126: 859–871.

Flores BC, Klinger DR, Rickert KL et al. (2014). Management of intracranial aneurysms associated with arteriovenous malformations. Neurosurg Focus 37: E11.

Gross BA, Du R (2013). Natural history of cerebral arteriovenous malformations: a meta-analysis. J Neurosurg 118: 437–443.

Gross BA, Scott RM, Smith ER (2014). Management of brain arteriovenous malformations. Lancet 383: 1635.

Ivanov AA, Alaraj A, Charbel FT et al. (2016). Recurrence of cerebral arteriovenous malformations following resection in adults: does preoperative embolization increase the risk? Neurosurgery 78: 562–571.

Josephson CB, Leach JP, Duncan R et al. (2011). Seizure risk from cavernous or arteriovenous malformations: prospective population-based study. Neurology 76: 1548–1554.

Josephson CB, White PM, Krishan A et al. (2014). Computed tomography angiography or magnetic resonance angiography for detection of intracranial vascular malformations in patients with intracerebral haemorrhage. Cochrane Database Syst Rev: CD009372.

Kano H, Kondziolka D, Flickinger JC et al. (2012). Stereotactic radiosurgery for arteriovenous malformations after embolization: a case-control study. J Neurosurg 117: 265–275.

Kim H, Al-Shahi Salman R, McCulloch CE et al. (2014). Untreated brain arteriovenous malformation: patient-level meta-analysis of hemorrhage predictors. Neurology 83: 590–597.

Kim H, Abla AA, Nelson J et al. (2015). Validation of the supplemented Spetzler–Martin grading system for brain arteriovenous malformations in a multicenter cohort of 1009 surgical patients. Neurosurgery 76: 25–31; discussion 31–32.

Kondziolka D, McLaughlin MR, Kestle JR (1995). Simple risk predictions for arteriovenous malformation hemorrhage. Neurosurgery 37: 851–855.

Krings T, Hans FJ, Geibprasert S et al. (2010). Partial "targeted" embolisation of brain arteriovenous malformations. Eur Radiol 20: 2723–2731.

Lawton MT, Abla AA (2014). Management of brain arteriovenous malformations. Lancet 383: 1634–1635.

Lawton MT, Project UBAMS (2003). Spetzler–Martin grade III arteriovenous malformations: surgical results and a modification of the grading scale. Neurosurgery 52: 740–748; discussion 748–749.

Miyachi S, Negoro M, Okamoto T et al. (2000). Embolisation of cerebral arteriovenous malformations to assure successful subsequent radiosurgery. J Clin Neurosci 7 (Suppl 1): 82–85.

Moftakhar P, Hauptman JS, Malkasian D et al. (2009). Cerebral arteriovenous malformations. Part 1: cellular and molecular biology. Neurosurg Focus 26: E10.

Mohr JP, Parides MK, Stapf C et al. (2014). Medical management with or without interventional therapy for unruptured brain arteriovenous malformations (ARUBA): a multicentre, non-blinded, randomised trial. Lancet 383: 614–621.

Morgan MK, Drummond KJ, Grinnell V et al. (1997). Surgery for cerebral arteriovenous malformation: risks related to lenticulostriate arterial supply. J Neurosurg 86: 801–805.

Morgan MK, Zurin AA, Harrington T et al. (2000). Changing role for preoperative embolisation in the management of arteriovenous malformations of the brain. J Clin Neurosci 7: 527–530.

Morgan MK, Rochford AM, Tsahtsarlis A et al. (2004). Surgical risks associated with the management of grade I and II brain arteriovenous malformations. Neurosurgery 54: 832–837; discussion 837–839.

Nataraj A, Mohamed MB, Gholkar A et al. (2014). Multimodality treatment of cerebral arteriovenous malformations. World Neurosurg 82: 149–159.

Natarajan SK, Ghodke B, Britz GW et al. (2008). Multimodality treatment of brain arteriovenous malformations with microsurgery after embolization with Onyx: single-center experience and technical nuances. Neurosurgery 62: 1213–1225; discussion 1225–1226.

Newman CB, Park MS, Kerber CW et al. (2012). Over-the-catheter retrieval of a retained microcatheter following Onyx embolization: a technical report. J Neurointerv Surg 4: E13.

Pandey P, Marks MP, Harraher CD et al. (2012). Multimodality management of Spetzler–Martin grade III arteriovenous malformations. J Neurosurg 116: 1279–1288.

Pollock BE (2015). The alchemy of brain arteriovenous malformation management. World Neurosurg 83: 337–338.

Pollock BE, Flickinger JC (2002). A proposed radiosurgery-based grading system for arteriovenous malformations. J Neurosurg 96: 79–85.

Pollock BEE, Flickinger JC (2008). Modification of the radiosurgery-based arteriovenous malformation grading system. Neurosurgery 63: 239–243; discussion 243.

Potts MB, Young WL, Lawton MT et al. (2013). Deep arteriovenous malformations in the basal ganglia, thalamus, and insula: microsurgical management, techniques, and results. Neurosurgery 73: 417–429.

Potts MB, Jahangiri A, Jen M et al. (2014). Deep arteriovenous malformations in the basal ganglia, thalamus, and insula: multimodality management, patient selection, and results. World Neurosurg 82: 386–394.

Rutledge WC, Abla AA, Nelson J et al. (2014). Treatment and outcomes of ARUBA-eligible patients with unruptured brain arteriovenous malformations at a single institution. Neurosurg Focus 37: E8.

Saatci I, Geyik S, Yavuz K et al. (2011). Endovascular treatment of brain arteriovenous malformations with prolonged intranidal Onyx injection technique: long-term results in 350 consecutive patients with completed endovascular treatment course. J Neurosurg 115: 78–88.

Sanchez-Mejia RO, McDermott MW, Tan J et al. (2009). Radiosurgery facilitates resection of brain arteriovenous malformations and reduces surgical morbidity. Neurosurgery 64: 231–238; discussion 238–240.

Solomon RA, Connolly Jr ES (2014). Management of brain arteriovenous malformations. Lancet 383: 1634.

Spetzler RF, Martin NA (1986). A proposed grading system for arteriovenous malformations. J Neurosurg 65: 476–483.

Spetzler RF, Ponce FA (2011). A 3-tier classification of cerebral arteriovenous malformations. Clinical article. J Neurosurg 114: 842–849.

Starke RM, Yen CP, Ding D et al. (2013). A practical grading scale for predicting outcome after radiosurgery for arteriovenous malformations: analysis of 1012 treated patients. J Neurosurg 119: 981–987.

Starke RM, Sheehan JP, Ding D et al. (2014). Conservative management or intervention for unruptured brain arteriovenous malformations. World Neurosurg 82: e668–e669.

Steiger HJ, Fischer I, Rohn B et al. (2015). Microsurgical resection of Spetzler–Martin grades 1 and 2 unruptured brain arteriovenous malformations results in lower long-term morbidity and loss of quality-adjusted life-years (QALY) than conservative management—results of a single group series. Acta Neurochir (Wien) 157: 1279–1287.

Tong X, Wu J, Pan J et al. (2016). Microsurgical resection for persistent arteriovenous malformations following gamma knife radiosurgery: a case-control study. World Neurosurg 88: 277–288.

van Beijnum J, van der Worp HB, Buis DR et al. (2011). Treatment of brain arteriovenous malformations: a systematic review and meta-analysis. JAMA 306: 2011–2019.

van Rooij WJ, Jacobs S, Sluzewski M et al. (2012a). Curative embolization of brain arteriovenous malformations with Onyx: patient selection, embolization technique, and results. AJNR Am J Neuroradiol 33: 1299–1304.

van Rooij WJ, Jacobs S, Sluzewski M et al. (2012b). Endovascular treatment of ruptured brain AVMs in the acute phase of hemorrhage. AJNR Am J Neuroradiol 33: 1162–1166.

Weber W, Kis B, Siekmann R et al. (2007). Preoperative embolization of intracranial arteriovenous malformations with Onyx. Neurosurgery 61: 244–252; discussion 252–254.

Wedderburn CJ, van Beijnum J, Bhattacharya JJ et al. (2008). Outcome after interventional or conservative management of unruptured brain arteriovenous malformations: a prospective, population-based cohort study. Lancet Neurol 7: 223–230.

Weiner GM, Grandhi R, Friedlander RM (2014). Conservative management vs intervention for unruptured brain arteriovenous malformations. JAMA 312: 1057–1058.

White JA, Batjer HH (2015). Management of deep arteriovenous malformations. World Neurosurg 83: 339–340.

Yen CP, Ding D, Cheng CH et al. (2014). Gamma Knife surgery for incidental cerebral arteriovenous malformations. J Neurosurg 121: 1015–1021.

Section 2

Cerebral dural arteriovenous fistulas

Handbook of Clinical Neurology, Vol. 143 (3rd series)
Arteriovenous and Cavernous Malformations
R.F. Spetzler, K. Moon, and R.O. Almefty, Editors
http://dx.doi.org/10.1016/B978-0-444-63640-9.00009-6

Chapter 9

Epidemiology, clinical presentation, diagnostic evaluation, and prognosis of cerebral dural arteriovenous fistulas

MOHAMED SAMY ELHAMMADY*, SUDHEER AMBEKAR, AND ROBERTO C. HEROS
Department of Neurosurgery, University of Miami, Miami, FL, USA

Abstract

Dural arteriovenous fistulas are abnormal communications, within the dural leaflets, between meningeal arteries and dural venous sinuses and/or subarachnoid veins. Although many fistulas remain clinically silent and do not require treatment, presence of cortical venous reflux, intracranial bleed, and intolerable symptoms are the main indications for treatment. A thorough understanding of the natural history is of prime importance in the decision making and management of these lesions. In this chapter we discuss the epidemiology and natural history of intracranial dural arteriovenous fistulae.

INTRODUCTION

Dural arteriovenous fistulas (DAVFs) are abnormal communications, within the dural leaflets, between meningeal arteries and dural venous sinuses and/or subarachnoid veins. They account for approximately 10–15% of all intracranial vascular malformations (Newton and Cronqvist, 1969); however, a good percentage of DAVFs remain clinically silent or involute spontaneously and therefore the true incidence may be much higher (Luciani et al., 2001). Present evidence suggests that DAVFs are acquired lesions and present later in life as compared to parenchymal cerebral arteriovenous malformations (AVMs) (Borden et al., 1995). Although cerebral DAVFs frequently occur in proximity to dural venous sinuses, they can develop anywhere within the intracranial dura mater. Most common locations include the cavernous sinus, the cribriform plate, the transverse sigmoid sinus, and the tentorium (Zipfel et al., 2009). The fistulous connection is located within the dural leaflets. Their arterial supply arises from any of the dural branches of internal carotid artery (ICA), external carotid artery (ECA), and vertebral artery (VA), and in rare cases, may be fed by pial branches (Martins et al., 2005). Venous drainage occurs into dural veins/sinuses and/or cortical/pedimedullary veins.

In this chapter we will summarize the epidemiology, pathogenesis, diagnostic evaluation, clinical presentation, and natural history of DAVFs.

EPIDEMIOLOGY

DAVFs represent 10–15% of all intracranial AVMs (Newton and Cronqvist, 1969). In a Scottish population-based study, the detection rate of DAVFs in adults was 0.16 per 100 000 per year. In the same population, the detection rate of cerebral AVMs was 1.12 per 100 000 per year, and thus DAVFs accounted for approximately 12.5% of all intracranial AVMs. In another study, Brown et al. (1996) reported the detection rate of 0.15 per 100 000 per year among residents of Olmsted County, Minnesota, over a 27-year period. A Japanese survey showed a higher detection rate of 0.29 per 100 000 per year (Satomi and Satoh, 2008).

Although DAVFs can occur at any age, the mean age of presentation is between 50 and 60 years. DAVFs seen in adults are believed to be acquired lesions; however, those occurring in children may be congenital

*Correspondence to: Mohamed Samy A. Elhammady, MD, Department of Neurological Surgery, University of Miami, Lois Pope LIFE Center, 1095 NW 14th Terrace 2nd Floor, Miami FL 33136, USA. Tel: +1-305-710-2657, E-mail: samyelhammady@gmail.com

(Cognard et al., 1995; Davies et al., 1996). DAVFs can occur anywhere within the dura mater. The most common locations include the cavernous sinus, the cribriform plate, the transverse sigmoid sinus, and the tentorium (Zipfel et al., 2009). In DAVFs involving the transverse sigmoid sinus, there is a slight left-sided predominance (Zipfel et al., 2009). Most DAVFs are single lesions, but multiple lesions have been reported (Chaichana et al., 2012).

HISTORIC PERSPECTIVE

The earliest report of an abnormal communication between an artery and vein was made by Hunter in 1757 (Olivecrona and Ladenheim, 1957). In 1873 Rizzoli described the presence of a dural-based arteriovenous aneurysm causing pulsatile tinnitus and seizures in a 9-year-old girl (Perrini et al., 2007). Autopsy revealed fistulous connection between the occipital artery and transverse sinus. The earliest observations of fistulas occurring in the brain substance were reported by Steinheil in 1895, Hoffman in 1898, and Isenschmid in 1912, followed by a clinicopathologic report by Cushing and Bailey in 1928 (Olivecrona and Ladenheim, 1957). The angiographic appearance of intracranial DAVFs was first described by Sachs in 1931, followed by Tönnis in 1936 (Aminoff, 1973). In 1951, Verbiest and Fincher introduced the concept of spontaneous DAVFs. The first cases of congenital high-flow DAVFs in children were reported in 1964 (Sarma and ter Brugge, 2003).

PATHOGENESIS AND THE ROLE OF VENOUS SINUS OCCLUSION/ THROMBOSIS

The exact etiology of DAVFs is not clear. Many factors have been implicated in the pathogenesis of these lesions, such as anomalies of the venous system, venous sinus occlusion by tumors, venous sinus or cortical vein thrombosis, increased systemic thrombotic activity (activated protein C resistance, antithrombin III deficiency, mutations in factor V Leiden and factor II G20210A), head trauma or cranial surgery, otitis and sinusitis, and hormonal changes associated with oral contraceptive use, pregnancy, delivery, and menopause (Berenstein et al., 2004).

Although venous sinus occlusion is frequently seen in patients with DAVFs, its role in the development and formation of DAVFs remains unclear. Some have argued that the DAVF precedes the venous sinus occlusion, which occurs secondarily as a result of gradual and progressive hypertrophy of the lining of the sinus due to exposure to high blood flow and/or pressure of the fistula akin to the venous-occlusive phenomenon seen in the cortical veins draining a cerebral parenchymal AVM. On the other hand, some have proposed that the venous sinus occlusion precedes and is directly responsible for development of the fistula. Venous sinus thrombosis results in the release of angiogenic factors from the organizing thrombus, which subsequently leads to the invasion of small dural arteries and formation of small dural arteriovenous shunts (Houser et al., 1979). Another theory suggests that DAVFs arise from naturally occurring dormant channels between dural arteries and sinuses, which open when the sinus is occluded and venous pressure is increased (Piton et al., 1984; Mullan, 1994). Venous hypertension may lead to tissue hypoxia and increased production of angiogenic factors, which promote endothelial proliferation and neoangiogenesis (Lawton et al., 1997; Tirakotai et al., 2004; Kojima et al., 2007).

Histologic studies suggests that microscopic thrombosis is always present and plays an important role in the release of growth factors and the formation of DAVFs (Uranishi et al., 1999). In immunohistochemical studies, expression of basic fibroblast growth factor and vascular endothelial growth factor has been identified in the wall of the dural sinuses in patients with DAVFs (Terada et al., 1996; Uranishi et al., 1999).

DIAGNOSTIC EVALUATION AND CLASSIFICATION SYSTEMS

Noninvasive imaging, including computed tomography (CT), CT angiography (CTA), and CT venography as well as magnetic resonance imaging/angiography/venography (MRI, MRA, MRV), are first-line diagnostic imaging tools. CT and CTA can demonstrate presence of intracranial hemorrhage (ICH), hydrocephalus, cerebral edema related to venous congestion, venous sinus occlusion, or enlarged cortical veins. MRI, MRA, and MRV can detect flow voids, venous engorgement, T2 white-matter signal changes suggestive of venous congestion, venous sinus occlusion, or stenosis. Any such abnormalities suggesting the presence of a DAVF should warrant further workup with a formal catheter cerebral angiogram.

Digital subtraction angiography remains the gold standard for detecting and evaluating a suspected DAVF. A six-vessel angiogram, including both ICAs, both ECAs, and both VAs, is mandatory. The hallmark of DAVFs is the presence of early venous drainage due to an abnormal communication within the dura mater between meningeal arterial branches and a venous sinus and/or subarachnoid vein. A cerebral angiogram can provide information regarding the fistula location, the arterial supply, patency of the venous sinuses, and the pattern of venous drainage. In situations where a complex DAVF is supplied by multiple arterial feeders from different vascular territories, it may be difficult to ascertain the exact fistulous point. A simple trick is to superimpose the angiographic runs from different arterial injections; since the feeding arteries from different arterial injections are different but the venous drainage is identical regardless of the arteries injected, the point beyond which the venous drainage is identical marks the fistulous point.

Several classification and grading systems for cerebral DAVFs have been suggested. In early reports from 1960s, DAVFs were grouped by their location into cavernous sinus, transverse sigmoid sinus, and anterior fossa lesions (Hayes, 1963). In 1973, Aminoff proposed a classification based on topography and divided DAVFs into anteroinferior and superior posterior groups. The first classification distinguishing DAVFs based on cortical venous drainage (CVD) was by Castaigne et al., 1976. In 1977, Djindjian proposed the first classification of intracranial DAVFs based on angioarchitecture (Gupta and Periakaruppan, 2009).

Current classification systems are based on the pattern of venous drainage, which is now known to predict the clinical behavior of DAVFs, as we will discuss in the natural-history section. The most commonly used grading systems are the Cognard (Cognard et al., 1995) and the Borden (Borden et al., 1995) classifications.

The Borden classification system consists of three types (Table 9.1). Type I lesions drain directly into a dural venous sinus or meningeal vein. Type II lesions drain directly into a dural venous sinus/meningeal vein with retrograde flow into subarachnoid veins. Type III lesions drain primarily into subarachnoid veins and without drainage into a dural venous sinus or meningeal vein.

The Cognard classification is a modification of the Djindjian classification (Fig. 9.1 and Table 9.2). Type I DAVFs drain into a dural venous sinus with normal antegrade flow. Type IIa DAVFs drain into a dural venous sinus with retrograde venous drainage. Type IIb DAVFs drain into a dural venous sinus with antegrade flow as well as into cortical subarachnoid veins. Type IIa + IIb DAVF drain into a dural venous sinus with retrograde flow as well as into cortical subarachnoid veins. Type III DAVFs drain primarily into cortical veins without drainage into a dural venous sinus. Type IV DAVFs drain into cortical veins with venous ectasia. Type V DAVFs drain into spinal perimedullary veins.

Thus, Borden type I and Cognard type I and IIa DAVFs drain directly into venous sinuses, whereas in Borden type II and III, and Cognard types IIb, IIa+b, III, and IV fistulas there is direct or indirect reflux to cortical veins, causing more aggressive presentation, including hemorrhage and neurologic deficits. Cognard type V DAVFs drain into spinal perimedullary veins. It is important to note, as we will discuss in subsequent sections, that in both the Borden and Cognard classification systems, the higher the fistula grade the worse the natural history.

The Barrow classification is used for cavernous sinus DAVFs (Table 9.3) (Barrow et al., 1985). Barrow type A fistulas are direct high-flow shunts between ICA and cavernous sinus, usually caused by trauma or rupture of an intracavernous ICA aneurysm. They form a separate entity, referred to as direct carotid-cavernous fistulas (CCF), and should not be considered as true DAVFs. Barrow types B, C, and D are true DAVFs with indirect low flow from meningeal branches of ICA (type B) or ECA (type C), or from both the ICA and ECA (type D) to cavernous sinus.

Table 9.1

Borden classification of intracranial dural arteriovenous fistulas

Type	Description
Type I	Anterograde drainage into the dural sinus/ meningeal vein
Type II	Anterograde drainage into dural sinus and retrograde drainage into cortical veins
Type III	Isolated retrograde drainage: • Drains directly into cortical veins • Trapped segment of sinus with reflux into cortical veins • Venous varix/dural lake with reflux into cortical veins

CLINICAL PRESENTATION

The clinical presentation of DAVFs is determined by both the pattern of venous drainage as well as the location of the fistula (Lasjaunias et al., 1986; Awad et al., 1990; Borden et al., 1995; Tsai et al., 2004; Lucas et al., 2006) DAVFs that drain in antegrade fashion into a dural sinus (i.e., drainage into the transverse sinus with normal antegrade flow into the ipsilateral jugular vein without CVD) usually manifest with nonaggressive presentations such as headaches, pulsatile tinnitus, and bruits. If there is retrograde sinus drainage (i.e., the fistula drains into a transverse sinus with retrograde arterialization of the opposite transverse sinus across the torcula and drainage down the contralateral jugular vein), intracranial venous pressure and hence intracranial pressure are elevated and patients can suffer symptoms and signs of pseudotumor cerebri with severe headaches and papilledema and occasionally progressive cognitive decline leading to dementia.

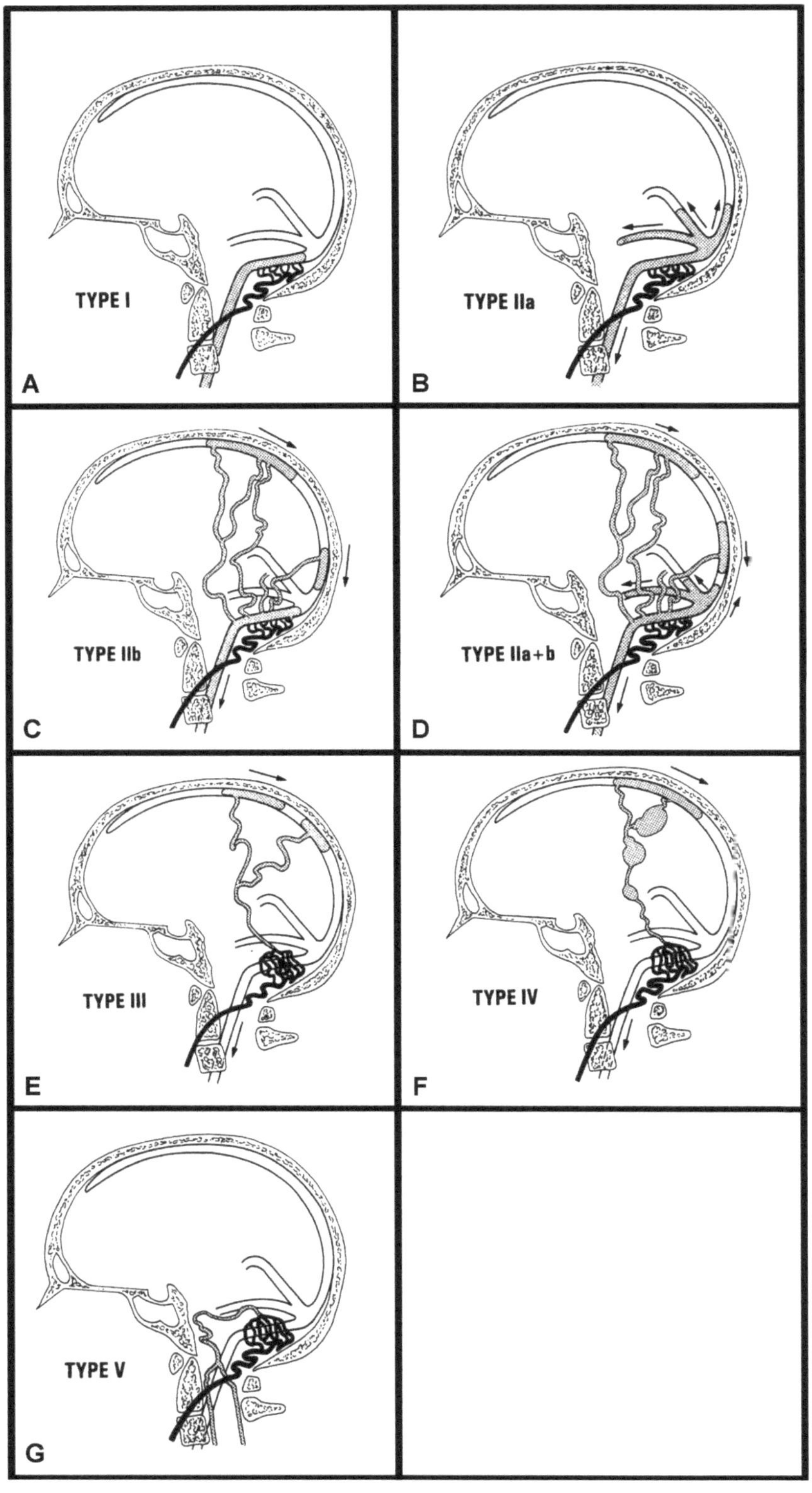

Fig. 9.1. **(A–G)** The Cognard classification. (Reproduced from Cognard et al. (1995), with permission from the Radiological Society of North America.)

Table 9.2

Cognard classification of intracranial dural arteriovenous fistulas

Type	Description
Type I	Anterograde drainage into venous sinus
Type II	
• IIA	• Drainage into main sinus with reflux into secondary sinus (retrograde sinus drainage)
• IIB	• Drainage into main sinus with reflux into cortical veins
• IIA + IIB	• Drainage into main sinus with reflux into secondary sinus and cortical veins
Type III	Direct cortical venous drainage without ectasia
Type IV	Direct cortical venous drainage with venous ectasia (>5 mm and 3 × larger than diameter of draining vein)
Type V	Drainage into the spinal perimedullary veins

DAVFs that acquire CVD can present with aggressive clinical manifestations, including either ICH or nonhemorrhagic neurologic deficits (NHNDs) such as progressive dementia, seizures, parkinsonism, or ataxia due to cerebral edema or ischemia (Barrow et al., 1985; Lasjaunias et al., 1986; Awad et al., 1990). In both the Borden and Cognard classification systems, the higher the fistula grade the worse the natural history (Borden et al., 1995; Cognard et al., 1995). DAVFs draining into perimedullary spinal veins may cause myelopathy and progressive tetraplegia (Hurst et al., 1999; Lv et al., 2011). The exception to the general rule that symptoms are benign unless there is cortical or perimedullary venous drainage occurs with carotid-cavernous dural fistulas, which can drain exclusively into orbital veins, yet can lead to severe ocular symptoms and even blindness if not treated due to increased orbital and intraocular venous pressure.

Mode of presentation and location of DAVFs are closely related. Patients with DAVFs of the transverse sigmoid junction typically suffer from pulsatile tinnitus due to the presence of a high-flow, low-pressure shunt in close proximity to the auditory apparatus. Similarly, middle fossa fistulas are more likely to present with pulsatile tinnitus due to increased drainage through the sigmoid and transverse sinuses. DAVFs in the region of cavernous sinus (indirect CCFs) arterialize the ophthalmic veins and typically present with exophthalmos, chemosis, and visual loss due to increased intraocular pressure (Suh et al., 2005). Anterior cranial fossa DAVFs often present with hemorrhage given the frequency of retrograde CVD with these fistulas, but they can also present with proptosis and chemosis if they acquire cavernous sinus drainage. Tentorial DAVFs have also been observed to

Table 9.3

Barrow classification of carotid-cavernous fistulas

Type	Description
Type A	Direct ICA-cavernous sinus high flow shunt (not considered a type of DAVF)
Type B	Dural ICA branches-cavernous sinus shunt
Type C	Dural ECA branches-cavernous sinus shunt
Type D	ECA/ICA dural branches shunt to cavernous sinus

ICA, internal carotid artery; DAVF, dural arteriovenous fistula; ECA, external carotid artery.

carry a high risk of intracerebral hemorrhage since they almost always acquire retrograde cortical vein drainage. DAVFs involving the superior sagittal sinus or those that drain into the deep venous system frequently present with NHNDs. Brainstem DAVFs are more likely to present with quadriparesis and lower cranial nerve palsies. Leptomeningeal drainage, variceal or aneurysmal venous dilatations, galenic drainage, and DAVFs of the petrosal and straight sinus have all been found to carry a high risk of intracerebral hemorrhage, although this has not been validated in larger studies (Zipfel et al., 2009).

ICH is believed to occur from rupture of fragile parenchymal veins as a result of exposure to increased pressure from retrograde venous reflux. On the other hand, parenchymal ischemia is thought to occur from venous congestion and hypertension, and prevents adequate arterial delivery of oxygen and removal of metabolic byproducts within the surrounding parenchyma. In exceedingly rare circumstances, venous hypertension can cause venous stasis within the DAVF and subsequent venous thrombosis, which can result in spontaneous obliteration and "cure" of the fistula.

NATURAL HISTORY AND PROGNOSIS

The natural history of DAVFs is influenced by both the pattern of venous drainage as well as the presenting symptoms. In a study on the natural history of DAVFs by Brown et al. (1994), an annual rupture and mortality rate of 1.6% and 2.3% respectively was detected during a mean follow-up period of 6.6 years. Angiographic features suggestive of increased hemorrhagic risk included petrosal or straight sinus location, presence of leptomeningeal venous drainage, and the presence of a venous varix on the draining vein.

DAVFs without CVD generally have a benign natural history, with the exception of CCFs, as discussed above. In a series of 117 benign (Borden type I) fistulas reported by Satomi et al. (2002) from the University of Toronto, no mortality was reported in 68 patients who were followed without treatment for a mean follow-up time of

27.9 months. A benign clinic course was observed in 98.5% of the patients, thus emphasizing the fact that such fistulas (those without CVD) should be managed conservatively and if treatment is contemplated it should not be too aggressive. An exception to this rule, as discussed above, is symptomatic DAVFs that drain into the cavernous sinus (indirect CCFs), where venous drainage of the eye is impaired and can result in visual loss. Only 1.5% of the patients in the series of Satomi et al. (2002) demonstrated angiographic progression to a more aggressive fistula with CVD and thus continued follow-up of patients with Borden type I fistulas is generally recommended.

DAVFs with CVD (Borden type II–III) typically behave more aggressively (Satomi et al., 2002; Soderman et al., 2008). The Toronto Brain Vascular Malformation Group analyzed a subgroup of 20 DAVFs with CVD that were followed over a 4-year period and found an annual hemorrhage rate of 8.1%, NHND rate of 6.9%, and a combined annual event rate of 15%. The annual mortality rate during the 87 person-year follow-up period was 10.4% (van Dijk et al., 2002). In a series of 20 patients with DAVFs presenting with hemorrhage, Duffau et al. (1999) found an exceeding high rebleed rate of 35% within 2 weeks and thus emphasized the importance of early treatment of DAVFs in such situations. It should be emphasized, however, that most of these patients presented with hemorrhage or a neurologic deficit.

The natural history of asymptomatic DAVFs with CVD is not well known but there is a suggestion that it is not as poor as in symptomatic patients. In a series of 85 patients with DAVFs with CVD, Soderman et al. (2008) reported an annual hemorrhage rate of 7.4% and 1.5% in patients presenting with and without hemorrhage respectively. Similar findings were reported by Strom and colleagues (2009) in a series of 28 patients with DAVFs with CVD. The annual ICH, NHND, and mortality rate was 7.6%, 11%, and 3.8% respectively in symptomatic patients. On the other hand, the annual ICH, NHND, and mortality rate was 1.4%, 0%, and 0% respectively in asymptomatic patients. In the study by Gross et al. (2012), asymptomatic patients or those with minimal symptoms had an annual hemorrhage rate of 2%, whereas patients who presented with NHND or ICH had an annual hemorrhage rate of 10% and 46% respectively.

The importance of CVD and validity of both the Borden and Cognard classification systems in determining the natural history and mode of presentation of DAVFs have been evaluated. In a series of 102 patients with intracranial DAVFs, Davies and colleagues (1996) from the University of Toronto found that an aggressive clinical presentation with either ICH or NHND was seen in 2%, 39%, and 79% of patients with Borden type I (Cognard type II and IIA), Borden type II (Cognard type IIB and type IIA + IIB), and Borden type III (Cognard type III, IV, and V) fistulas respectively. In a similar study, Gross et al. (2012) reported an annual hemorrhage rate of 0%, 6%, 10%, and 21% for Borden type I, type II, type III, and DAVFs with venous ectasia respectively.

CONCLUSION

Intracranial DAVFs are relatively uncommon lesions consisting of an abnormal connection, within the dural leaflets, between meningeal arteries and dural venous sinuses and/or subarachnoid veins. The association of DAVFs with venous sinus occlusion or thrombosis is common; however, its role in the development and formation of DAVFs remains unclear. Digital subtraction angiography remains the gold standard for detecting and evaluating a suspected DAVF. The natural history is influenced by the pattern of venous drainage as well as the mode of presentation. Intracranial DAVFs without CVD have a benign natural history and tend to remain "benign." Such lesions should be managed conservatively and if treatment is contemplated it should not be too aggressive. An exception to this rule is symptomatic DAVFs that drain into the cavernous sinus (indirect CCFs), where there is impairment of the venous drainage of the eye that can lead to visual loss. Intracranial DAVFs with CVD can have an aggressive natural history, particularly when symptomatic. Fistulas that drain into spinal veins can present with myelopathy. The early rebleeding rate following hemorrhage is extremely high and requires prompt treatment.

References

Aminoff MJ (1973). Vascular anomalies in the intracranial dura mater. Brain 96: 601–612.

Awad IA, Little JR, Akarawi WP et al. (1990). Intracranial dural arteriovenous malformations: factors predisposing to an aggressive neurological course. J Neurosurg 72: 839–850.

Barrow DL, Spector RH, Braun IF et al. (1985). Classification and treatment of spontaneous carotid-cavernous sinus fistulas. J Neurosurg 62: 248–256.

Berenstein A, Lasjaunias P, Brugge KG (2004). Dural arteriovenous shunts. Surgical neuroangiography, 2nd edn. Springer-Verlag, Berlin.

Borden JA, Wu JK, Shucart WA (1995). A proposed classification for spinal and cranial dural arteriovenous fistulous malformations and implications for treatment. J Neurosurg 82: 166–179.

Brown Jr RD, Wiebers DO, Nichols DA (1994). Intracranial dural arteriovenous fistulae: angiographic predictors of intracranial hemorrhage and clinical outcome in nonsurgical patients. J Neurosurg 81: 531–538.

Brown Jr RD, Wiebers DO, Torner JC et al. (1996). Incidence and prevalence of intracranial vascular malformations in

Olmsted County, Minnesota, 1965 to 1992. Neurology 46: 949–952.

Castaigne P, Bories J, Brunet P et al. (1976). Meningeal arterio-venous fistulas with cortical venous drainage. Rev Neurol (Paris) 132: 169–181.

Chaichana KL, Coon AL, Tamargo RJ et al. (2012). Dural arteriovenous fistulas: epidemiology and clinical presentation. Neurosurg Clin N Am 23: 7–13.

Cognard C, Gobin YP, Pierot L et al. (1995). Cerebral dural arteriovenous fistulas: clinical and angiographic correlation with a revised classification of venous drainage. Radiology 194: 671–680.

Davies MA, TerBrugge K, Willinsky R et al. (1996). The validity of classification for the clinical presentation of intracranial dural arteriovenous fistulas. J Neurosurg 85: 830–837.

Duffau H, Lopes M, Janosevic V et al. (1999). Early rebleeding from intracranial dural arteriovenous fistulas: report of 20 cases and review of the literature. J Neurosurg 90: 78–84.

Gross BA, Ropper AE, Du R (2012). Cerebral dural arteriovenous fistulas and aneurysms. Neurosurg Focus 32. E2.

Gupta A, Periakaruppan A (2009). Intracranial dural arteriovenous fistulas: a review. Indian J Radiol Imaging 19: 43–48.

Hayes GJ (1963). External carotid-cavernous sinus fistulas. J Neurosurg 20: 692–700.

Houser OW, Campbell JK, Campbell RJ et al. (1979). Arteriovenous malformation affecting the transverse dural venous sinus – an acquired lesion. Mayo Clin Proc 54: 651–661.

Hurst RW, Bagley LJ, Scanlon M et al. (1999). Dural arteriovenous fistulas of the craniocervical junction. Skull Base Surg 9: 1–7.

Kojima T, Miyachi S, Sahara Y et al. (2007). The relationship between venous hypertension and expression of vascular endothelial growth factor: hemodynamic and immunohistochemical examinations in a rat venous hypertension model. Surg Neurol 68: 277–284; discussion 284.

Lasjaunias P, Chiu M, ter Brugge K et al. (1986). Neurological manifestations of intracranial dural arteriovenous malformations. J Neurosurg 64: 724–730.

Lawton MT, Jacobowitz R, Spetzler RF (1997). Redefined role of angiogenesis in the pathogenesis of dural arteriovenous malformations. J Neurosurg 87: 267–274.

Lucas Cde P, Caldas JG, Prandini MN (2006). Do leptomeningeal venous drainage and dysplastic venous dilation predict hemorrhage in dural arteriovenous fistula? Surg Neurol 66 (Suppl 3): S2–S5; discussion S5-6.

Luciani A, Houdart E, Mounayer C et al. (2001). Spontaneous closure of dural arteriovenous fistulas: report of three cases and review of the literature. AJNR Am J Neuroradiol 22: 992–996.

Lv X, Yang X, Li Y et al. (2011). Dural arteriovenous fistula with spinal perimedullary venous drainage. Neurol India 59: 899–902.

Martins C, Yasuda A, Campero A et al. (2005). Microsurgical anatomy of the dural arteries. Neurosurgery 56: 211–251; discussion 211–251.

Mullan S (1994). Reflections upon the nature and management of intracranial and intraspinal vascular malformations and fistulae. J Neurosurg 80: 606–616.

Newton TH, Cronqvist S (1969). Involvement of dural arteries in intracranial arteriovenous malformations. Radiology 93: 1071–1078.

Olivecrona H, Ladenheim C (1957). Congenital arteriovenous aneurysms of the carotid and vertebral arterial systems, Springer, Berlin.

Perrini P, Nannini T, Di Lorenzo N (2007). Francesco Rizzoli (1809–1880) and the elusive case of Giulia: the description of an "arteriovenous aneurysm passing through the wall of the skull". Acta Neurochir (Wien) 149: 191–196; discussion 196.

Piton J, Guilleux MH, Guibert-Tranier F et al. (1984). Fistulae of the lateral sinus. J Neuroradiol 11: 143–159.

Sarma D, ter Brugge K (2003). Management of intracranial dural arteriovenous shunts in adults. Eur J Radiol 46: 206–220.

Satomi J, Satoh K (2008). Epidemiology and etiology of dural arteriovenous fistula. Brain Nerve 60: 883–886.

Satomi J, van Dijk JM, Terbrugge KG et al. (2002). Benign cranial dural arteriovenous fistulas: outcome of conservative management based on the natural history of the lesion. J Neurosurg 97: 767–770.

Soderman M, Pavic L, Edner G et al. (2008). Natural history of dural arteriovenous shunts. Stroke 39: 1735–1739.

Strom RG, Botros JA, Refai D et al. (2009). Cranial dural arteriovenous fistulae: asymptomatic cortical venous drainage portends less aggressive clinical course. Neurosurgery 64: 241–247; discussion 247–248.

Suh DC, Lee JH, Kim SJ et al. (2005). New concept in cavernous sinus dural arteriovenous fistula: correlation with presenting symptom and venous drainage patterns. Stroke 36: 1134–1139.

Terada T, Tsuura M, Komai N et al. (1996). The role of angiogenic factor bFGF in the development of dural AVFs. Acta Neurochir (Wien) 138: 877–883.

Tirakotai W, Bian LG, Bertalanffy H et al. (2004). Immunohistochemical study in dural arteriovenous fistula and possible role of ephrin-B2 for development of dural arteriovenous fistula. Chin Med J (Engl) 117: 1815–1820.

Tsai LK, Jeng JS, Liu HM et al. (2004). Intracranial dural arteriovenous fistulas with or without cerebral sinus thrombosis: analysis of 69 patients. J Neurol Neurosurg Psychiatry 75: 1639–1641.

Uranishi R, Nakase H, Sakaki T (1999). Expression of angiogenic growth factors in dural arteriovenous fistula. J Neurosurg 91: 781–786.

van Dijk JM, terBrugge KG, Willinsky RA et al. (2002). Clinical course of cranial dural arteriovenous fistulas with long-term persistent cortical venous reflux. Stroke 33: 1233–1236.

Zipfel GJ, Shah MN, Refai D et al. (2009). Cranial dural arteriovenous fistulas: modification of angiographic classification scales based on new natural history data. Neurosurg Focus 26. E14.

Handbook of Clinical Neurology, Vol. 143 (3rd series)
Arteriovenous and Cavernous Malformations
R.F. Spetzler, K. Moon, and R.O. Almefty, Editors
http://dx.doi.org/10.1016/B978-0-444-63640-9.00010-2

Chapter 10

Surgical management of cerebral dural arteriovenous fistulae

MARTIN J. RUTKOWSKI, BRIAN JIAN, AND MICHAEL T. LAWTON*
Department of Neurological Surgery, University of California, San Francisco, CA, USA

Abstract

Dural arteriovenous fistulae are high flow, low resistance intracranial vascular malformations defined by an aberrant connection between an artery and dural vein or sinus. Symptomatology and presentation are highly dependent on location, generally categorized as supratentorial, tentorial, or infratentorial, and consist primarily of sequelae secondary to local venous hypertension, insufficiency, and cortical venous reflux. Surgery is generally reserved for high risk or persistently symptomatic lesions that are unamenable or unresponsive to endovascular therapy.

For surgical lesions, familiarity with skull base approaches, specific fistula anatomy, and technical nuances based on fistula location offer patients the best chance of a favorable outcome.

INTRODUCTION

Intracranial dural arteriovenous fistulae (DAVF) are rare vascular lesions defined by an aberrant connection between an artery and a dural vein or sinus. Similar to arteriovenous malformations (AVMs), DAVFs are high-flow, low-resistance lesions, but are thought to be acquired rather than congenital lesions, and lack a parenchymal nidus. DAVFs have similar hemodynamics to brain AVMs, with low-resistance, high-flow arteriovenous shunting, with a higher annual rupture risk than AVMs. They are associated with venous hypertension and venous insufficiency, including cortical venous reflux, which is associated with a higher risk of spontaneous hemorrhage. Although the majority of DAVFs can be managed and cured through an endovascular approach, surgery is indicated for high-risk or persistently symptomatic DAVFs that have failed, or cannot be treated by endovascular embolization. The surgical management of DAVFs requires a team approach and is best managed in a collaborative setting by a dedicated neurovascular team.

CLINICAL PRESENTATION

Pathologic consequences of DAVF include arterialization of the venous system at the site of fistula, which can lead to venous hypertension and congestion and induce cerebral ischemia and hemorrhage. Some DAVFs can be benign, presenting as an incidental angiographic finding or with minor symptoms like headache or pulsatile tinnitus from turbulent blood flow through the shunt. Some DAVFs can be more insidious, presenting with progressive neurologic deterioration from venous ischemia or acute deterioration from intracranial hemorrhage. The variety of symptoms is wide and also includes double vision, visual obscuration, focal neurologic deficits, seizures, and progressive dementia (Lasjaunias et al., 1986; Datta et al., 1998; Matsuda et al., 1999; Tanaka et al., 1999; Yamakami et al., 2001; Bernstein et al., 2003).

Type of presentation and severity of symptoms are determined by the location of fistula, the anatomy of venous drainage, and resultant alterations in cerebral hemodynamics. Alterations in hemodynamics depend on the volume of shunt flow and the degree of

*Correspondence to: Michael T. Lawton, MD, Department of Neurological Surgery, University of California, San Francisco, 505 Parnassus Avenue, M780, Box 0112, San Francisco CA 94143-0112, USA. Tel: +1-415-353-3998, Fax: +1-415-353-3907, E-mail: lawtonm@neurosurg.ucsf.edu

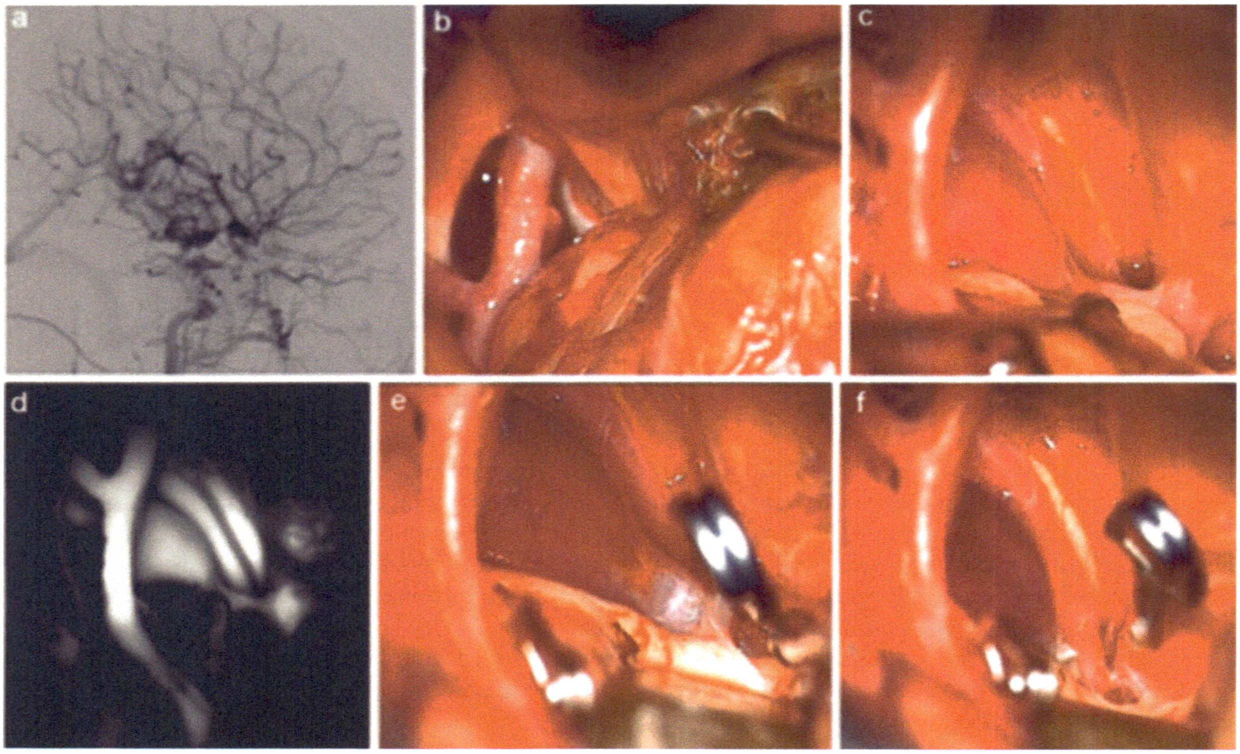

Fig. 10.1. Example of a carotid-cavernous fistula (CCF) that failed endovascular therapy. (**A**) This 8-year-old girl had multiple endovascular coiling treatments of her CCF and persistent deep venous drainage to the basal vein of Rosenthal remained (right internal carotid artery angiogram, lateral view). (**B**) A right orbitozygomatic craniotomy exposed the lateral wall of cavernous sinus, where coils were seen. (**C**) The arterialized vein was identified at the tentorial edge and (**D**) indocyanine green videoangiography confirmed shunt flow. (**E**) The dural arteriovenous fistula was interrupted with an aneurysm clip on the vein as it exited the tentorial dura, and (**F**) the vein was cut.

compromise of venous outflow. The exact mechanism has not been fully investigated and the relation to the patient's symptoms is not always clear (Lalwani et al., 1993; Cellerini et al., 1999; van Dijk, 2002). The annual risk of hemorrhage from a DAVF with cortical venous drainage is as high as 19% annually. Mortality and neurologic morbidity associated with hemorrhagic presentation have been estimated to be 20–30%. Cortical venous drainage and posterior fossa location of the DAVF are independent predictors of a hemorrhagic presentation (Singh et al., 2008).

DAVF patients present with mean ages around 50–60 years. No sex predilection has been established, although men appear more likely than women to present with hemorrhage. An underlying DAVF should be suspected in patients with nontraumatic intracerebral hemorrhage with a subarachnoid component (Singh et al., 2008). Women with DAVF more commonly present with pulsatile tinnitus than men, whereas men usually present with nontraumatic intraparenchymal and subarachnoid hemorrhage (Singh et al., 2008).

TYPES OF DAVFs

Our surgical experience with DAVFs indicates that they can be categorized as supratentorial, tentorial, and infratentorial. Supratentorial DAVFs include superior sagittal sinus (SSS) and sphenoparietal sinus DAVFs, ethmoidal DAVFs, and carotid-cavernous fistulae (CCFs). Tentorial DAVFs include Galenic, straight sinus, torcular, tentorial sinus, superior petrosal sinus, and incisural DAVFs. Infratentorial DAVFs include transverse–sigmoid sinus, marginal sinus, and inferior petrosal sinus DAVFs.

CLINICAL FEATURES AND TECHNICAL NUANCES BASED ON DAVF TYPE

Carotid-cavernous fistulae

A CCF is any abnormal communication between the internal carotid artery (ICA), external carotid artery (ECA), or any of their branches and the cavernous sinus (Fig. 10.1). CCFs can be classified hemodynamically into high-flow and low-flow fistulae, or anatomically into direct and indirect fistulae. The Barrow classification recognizes four types of CCF: type A is a direct fistula between the intracavernous ICA and cavernous sinus; type B is a dural AV shunt between intracavernous branches of ICA (like the branches from the meningohypophyseal trunk or inferolateral trunk) and the cavernous sinus; type C is a dural AV shunt between meningeal branches of ECA (like the accessory middle meningeal artery or ascending pharyngeal artery) and the cavernous sinus; and type D is a dural AV shunt between intracavernous branches of both ICA and ECA and the cavernous sinus. Type A CCFs usually occur in the setting of head trauma or as a result of a ruptured cavernous ICA aneurysm. These fistulae have high flow and produce symptoms related to venous engorgement of the orbit, including pulsatile exophthalmos, chemosis, cranial nerve III, IV, and VI paresis, and, rarely, glaucoma and visual loss. Types B, C, and D CCFs usually occur spontaneously or following minor trauma with symptoms that are less severe and more insidious in their onset. These indirect CCFs may occur at any age, but are more commonly present in women older than 40 years and may be associated with conditions such as vascular Ehlers–Danlos syndrome, hypertension, hypercoagulability, or pregnancy, among others (Desal et al., 2005; Miller, 2007). Their pathogenesis has been hypothesized in some cases to be related to prior, clinically silent, thrombosis of the cavernous sinus with subsequent formation of aberrant vascular connections during recanalization (Miller, 2007).

Type A CCFs require treatment to prevent progressive visual loss, reverse unsightly exophthalmos, or eliminate intolerable bruits or pain. The treatment is endovascular, occluding the fistula with a detachable balloon deployed from a microcatheter in the ICA and inflated in the fistula. These CCFs can also be occluded transvenously with coils in the cavernous sinus, deployed using microcatheters that reach the fistula via the internal jugular vein and inferior petrosal sinus.

Cavernous sinus DAVFs (types B, C, and D) have a benign natural history and often resolve spontaneously. Carotid self-compression (20 seconds, 3–4 times per hour when awake, using the contralateral hand) may promote thrombosis of the DAVF, but is contraindicated in patients with high-grade carotid stenosis or ulcerative plaques. Endovascular intervention is indicated in patients with visual deterioration, progressive intraocular hypertension, obtrusive diplopia, proptosis with corneal exposure, or intolerable bruit. ECA feeding arteries can be embolized safely in patients with type C and D CCFs, usually resulting in cures. Persistent lesions can be treated additionally with transvenous embolization. Selective catheterization of ICA feeding arteries is more difficult and embolization is associated with increased risk of cerebral emboli, but type B lesions are rare. Refractory CCFs that fail endovascular therapy and remain with retrograde cortical venous drainage may require surgical intervention to occlude the vein and interrupt the fistula.

Superior sagittal sinus DAVF

SSS DAVFs are uncommon, but the difficulty in curing these lesions endovascularly makes them more prevalent in surgical series. These lesions can present with pulsatile tinnitus, signs of raised intracranial pressure, or signs of raised cavernous sinus pressure from diversion of venous outflow. Local symptoms, such as focal seizures or focal neurologic deficits, can result in venous engorgement or varix formation and mass effect.

SSS DAVFs are supplied by middle meningeal arteries or branches thereof. Scalp arteries (superficial temporal and occipital arteries) can also supply these DAVFs, as can cortical arteries. Venous drainage is primarily to SSS, but these can be Borden type II or III lesions with cortical venous drainage through convexity veins. SSS DAVFs can be embolized transarterially with ease, but numerous meningeal feeding arteries make them difficult to cure. Fistulae with cortical venous drainage can also be difficult to cure endovascularly because there is typically poor access from the venous side.

SSS DAVFs requiring surgery are exposed with a midline craniotomy over the sinus and the fistula. Cutting the craniotomy flap over the sinus requires careful preservation of the dura, particularly in older patients with adherent dura. Two separate flaps may be needed to first expose ipsilateral dura and allow the sinus to be dissected from the inner table of skull before crossing the midline with the second flap. Surgical management of the fistula depends on its Borden type. Type I SSS DAVFs are treated by skeletonizing the SSS extensively, eliminating arterial supply while preserving normal flow in the sinus. Type III SSS DAVFs are treated by interrupting the fistula on its venous side as it exits the sinus, usually by clipping the vein, cauterizing it, and dividing it. Type II SSS DAVFs are treated with a combination of skeletonization of the sinus and interruption of arterialized veins.

Sphenoparietal sinus DAVFs

Like SSS DAVFs, sphenoparietal sinus DAVFs are supplied by middle meningeal arteries or branches thereof, and by scalp arteries (Fig. 10.2). Venous drainage is primarily into Sylvian veins that can course posteriorly towards the vein of Labbé, inferiorly under the temporal lobe, or superiorly towards the SSS.

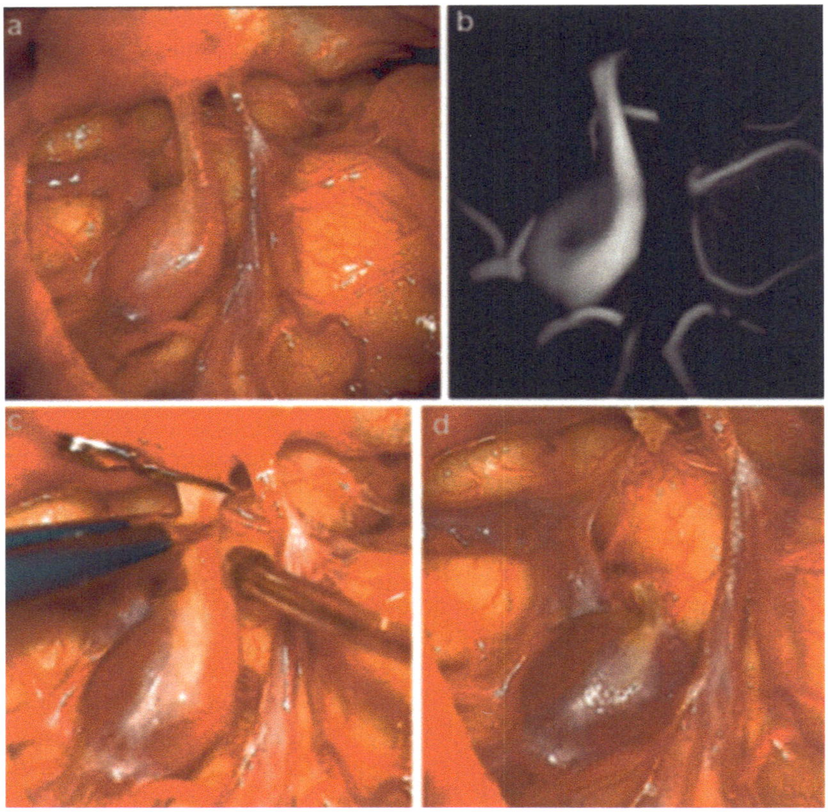

Fig. 10.2. Example of a sphenoparietal sinus dural arteriovenous fistula (DAVF). (**A**) A standard left pterional craniotomy exposed the Sylvian fissure where the arterialized vein draining the fistula was identified. (**B**) Indocyanine green videoangiography confirmed shunt flow. (**C**) The DAVF was interrupted with an aneurysm clip on the vein as it exited the pterional dura, and (**D**) the vein was cut. Note the immediate color change of blood in the venous varix, confirming complete obliteration.

These lesions are exposed with a standard pterional craniotomy. In our experience, these DAVFs are typically Borden type III lesions that can be treated with simple interruption of the arterialized draining vein as it exits the dura over the Sylvian fissure.

Ethmoidal DAVFs

Ethmoidal DAVFs are located on the floor of the anterior cranial fossa, adjacent to the cribriform plate. The fistula is supplied by anterior ethmoidal arteries and drains directly into cortical veins at the frontal pole. Unlike all the other DAVFs, ethmoidal DAVFs are not associated with a venous sinus. This unusual anatomy accounts for their propensity to hemorrhage.

We previously reported an experience with 16 patients with ethmoidal DAVFs, the largest in the literature (Lawton et al., 1999). Half of our patients presented with hemorrhage. Nonhemorrhagic presentations included headaches, decreasing visual acuity, and diminished sense of smell and taste. Male predominance is also striking, with 11 men in our series (69%). The mean age was 62 years.

The anterior ethmoidal artery that feeds these DAVFs is a branch from the ophthalmic artery, which is in turn a branch from the ICA. Feeding arteries can arise unilaterally (50% of our patients) or bilaterally (50% of our patients). Branches from ECA, namely the superficial temporal artery and internal maxillary artery, can also supply these lesions.

Endovascular treatment of ethmoidal DAVFs has not been widely performed for several reasons. First, selective catheterization of the ophthalmic and ethmoidal arteries is difficult due to the small caliber and tortuous course of these arteries. Second and most importantly, embolization of the central retinal artery, also a distal branch of the ophthalmic artery, can complicate the procedure and devastate a patient's vision. Third, unlike embolization of other DAVFs whose blood supply originates from the ECA, embolization of ethmoidal DAVFs through an ICA branch carries the risk of embolic agents refluxing into the cerebral circulation. These obstacles have deterred most endovascular surgeons from treating ethmoidal DAVFs. The endovascular results that have been published suggest that endovascular management of ethmoidal DAVFs has a small but clinically significant risk to vision, is rarely effective in curing the fistula, and does not eliminate the need for surgery (Agid et al., 2009; Gross et al., 2016).

These DAVFs are exposed through a bifrontal craniotomy and a subfrontal approach. The patient is positioned in the supine position with the head extended slightly. A bicoronal skin incision is needed to mobilize the scalp flap inferiorly enough for a low craniotomy. Pericranium is harvested during the approach to make it available for covering the frontal sinus, if it has been entered. Ethmoidal DAVFs are exposed optimally with a craniotomy flap whose medial border is either right on the SSS or crosses over to the contralateral side. Midline exposure enables the dura to be opened to the edge of the SSS and gives easy access to the interhemispheric fissure. The craniotomy flap is extended inferiorly as close to the anterior cranial fossa floor as possible. Crossing the midline with this cut sometimes requires drilling down the ridge of bone on the inner table in the midline.

Freeing the bone flap from the underlying dura can be difficult, given the age of many of these patients and the adherence of their dura. In these cases, the bone flap is divided adjacent to the SSS in order to visualize directly and dissect the sinus from the bone flap. The inner table of frontal bone in the inferior midline is drilled until flat, which increases visualization along the anterior cranial fossa floor and eliminates the need for frontal-lobe retraction.

The critical maneuver is the interruption of the fistulous connection between the arteries perforating the dura around the cribriform plate and the draining vein or veins. The cortical draining vein is identified early during the dissection as it enters the SSS and followed retrograde to the DAVF. When present, the venous varix is encountered during the dissection and is typically the site of hemorrhage in patients who bled. Excessive manipulation or evacuation of the hematoma is avoided to prevent bleeding from the varix before the fistula is controlled.

Numerous arteries in the dura of the anterior cranial fossa floor and falx typically merge into a dilated draining vein. The vein is coagulated and divided. Occasionally there are posterior ethmoidal feeding arteries that are obscured from view by the frontal lobe or the varix, but nonetheless contribute fistulous flow. If persistent arteriovenous shunting is observed in the draining vein, the vein just distal to the fistula should be dissected further to identify these posterior feeding arteries. Dissecting the varix is useful for this reason, but it does not need to be resected. Excision of dura involved with the fistula and adjacent bone is not necessary.

After the DAVF is occluded, any associated hematoma can be evacuated completely without risk of rupturing the varix. Manipulation of the olfactory bulbs and tracts is minimized to preserve olfaction. If the frontal sinus was entered during the exposure, sinus mucosa is exenterated and the sinus is packed with Gelfoam. A vascularized pericranial flap is then laid down over the defect and sutured to inferior frontal dura to block any communication with the intracranial compartment. The plated bone flap secures the pericranial flap's coverage of the defect.

In our experience, surgical management has low associated risk, no risk to vision, effects a cure in all cases, and can contribute to good clinical outcomes in most patients. For these reasons, surgical management of ethmoidal DAVFs remains the treatment of choice.

Galenic DAVFs

Galenic DAVFs are the most complex of the six types of tentorial fistulae. The Galenic region is the deepest location; the confluence of falx and tentorium creates awkward barriers and surgical blind spots; arterial inflow arrives from all directions; and venous outflow can be difficult to decipher, particularly when veins are tortuous or variceal. Consequently, a panoramic exposure is required that only the posterior interhemipheric approach can provide (Fig. 10.3) (Chi and Lawton, 2006). This approach is performed with the patient in the lateral position through a torcular craniotomy that exposes the SSS, both transverse sinuses, and torcula. After opening the dura, the dependent occipital lobe is retracted by gravity and the interhemispheric fissure opens widely without a retractor. Cuts in the falx above the straight sinus, and bilaterally in the tentorium parallel to the straight sinus, skeletonize the straight sinus and transform this unilateral, supratentorial exposure into a bilateral, supra-, and infratentorial exposure. Anatomy in front of the fistula in the quadrigeminal and ambient cisterns can be visualized clearly.

Skeletonizing the straight sinus also de-arterializes Galenic DAVFs. Transecting the tentorium obliterates supply from the tentorial arteries and ECA branches, while transecting the falx obliterates supply from MMA/falcine arteries. Supply from the occipital artery is already obliterated when the scalp flap is elevated, and the posterior meningeal artery can be interrupted along its course up the suboccipital dura to the torcular region. However, the goal of occluding the fistula is accomplished by interrupting the venous drainage, not by interrupting the arterial supply. Dural cuts are performed to widen the posterior interhemispheric corridor,

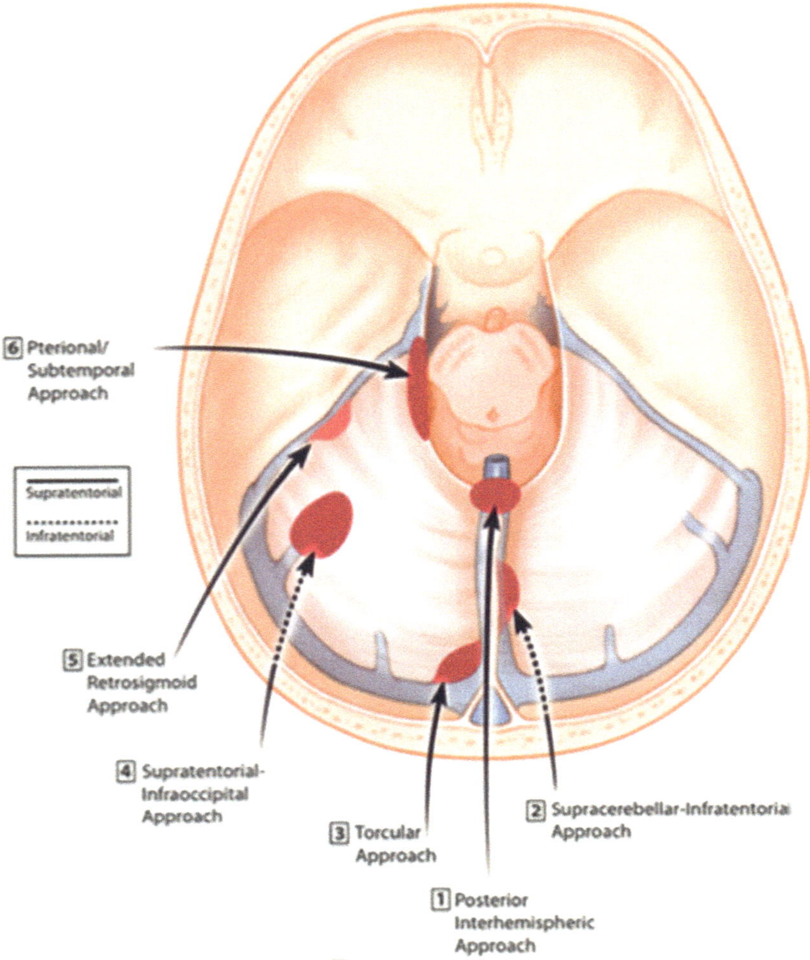

Fig. 10.3. Summary of surgical approaches to tentorial dural arteriovenous fistulae.

visualize the Galenic complex, and decipher the venous anatomy, not to de-arterialize the fistula.

Interruption of the fistula is completed by placing a clip on the vein draining the fistula, which requires meticulous microsurgical dissection of the venous anatomy. The vein of Galen and internal cerebral veins are most prominent from the perspective of the posterior interhemispheric approach, but the internal cerebral vein is rarely the target for the occluding clip. The basal vein of Rosenthal and the precentral cerebellar vein are more commonly the target for the clip, but are also more difficult to visualize. The basal vein of Rosenthal is identified by dissecting lateral and inferior to the internal cerebral veins into the ambient cistern. The precentral cerebellar vein lies in a blind spot in the surgical corridor beneath the falcotentorial dura, but this untethered dura can be mobilized aggressively to visualize the cerebellar veins. Veins that are red and dilated declare their participation in draining the fistula, whereas blue veins declare their participation in normal circulation. The connections between fistulous dura and its arterialized draining veins are not always obvious from gross inspection of extravascular anatomy.

Clip application is dictated by the straight sinus' patency, which must be determined preoperatively from the angiogram. All Galenic DAVFs in our experience were Borden type III fistulae with retrograde drainage in the vein of Galen or its tributaries. However, not all straight sinuses were occluded in patients with Galenic DAVFs; some straight sinuses had antegrade flow seen during the venous phase of the angiogram. A patent straight sinus requires vein of Galen preservation and clip application to occlude just the tributary vein draining the fistula, rather than the Galen trunk. Other tributary veins uninvolved with the fistulous outflow can continue to drain the deep cerebral circulation in an antegrade direction. In contrast, an occluded straight sinus allows clip application directly on the Galenic trunk to occlude fistulous outflow, which usually requires less dissection and is easier to decipher.

The supracerebellar-infratentorial approach is an alternative approach to Galenic DAVFs, which has the advantage of positioning the neurosurgeon on the same side of the tentorium as fistulae that drain inferiorly to cerebellar veins. However, the steep pitch of the tentorium at the vein of Galen creates a narrow attic with a very limited view, and a dilated, low-hanging vein of Galen can fill this small field.

Straight sinus DAVFs

Compared to Galenic DAVFs, straight sinus DAVFs are much simpler: they are not as deeply located; they usually drain out of a solitary vein; and they are exposed by opening the natural subarachnoid plane under the tentorium, without skeletonizing sinuses or dissecting venous complexes. The supracerebellar-infratentorial approach with the patient in the sitting position is the ideal operative approach. With the sitting position, gravity retracts the cerebellum and opens this plane to easily visualize the fistula, even in patients presenting with hemorrhage who have intraparenchymal clot, cerebellar swelling, and a tight surgical corridor. The supracerebellar-infratentorial approach can be performed in the prone position, but retractors are needed; this risks avulsing the draining vein and causing brisk bleeding from a fistula not yet in view. Neurosurgeons often prefer that prone position because it allows them to sit while operating. However, the surgeon can still sit while operating on patients in the sitting position. The patient's back is positioned nearly vertical and the head is slouched forward to align the plane of the tentorium nearly horizontal. The surgeon sits on a rolling stool and supports his or her elbows on a free-standing arm brace raised to shoulder level, which relaxes the arms and stabilizes the hands. The dissection and fistula occlusion are usually brief, so this somewhat awkward position for the neurosurgeon is easily tolerated.

The suboccipital craniotomy extends from above the transverse sinuses and torcula superiorly to just above, but not to, the foramen magnum. Exposure of the torcula removes the bony ledge that would otherwise obscure the infratentorial plane. Dural tears and venous sinus injury can be dangerous, particularly with patients in the sitting position. Older patients with adherent dura may have sinuses that are not safe to cross with the craniotome, and instead may require a suboccipital craniotomy first, then dissection of the dura from the inner table of skull under direct visualization. Alternatively, bone overlying the sinuses can be drilled away with a diamond drill bit until the inferior margins of the venous sinuses are seen. Once the torcula is exposed, dura is opened in a flap based on the transverse sinuses and tacking sutures that elevate the torcula. The arterialized vein draining the fistula is identified on the cerebellar surface by its red color and traced back to the fistula. Alternatively, the arterialized vein is seen in the subarachnoid space descending from the dura. The draining vein has a thickened wall and distinctive white color with red vasa vasorum. A clip is applied to the draining vein as it exits the tentorial dura, and it is coagulated and cut.

Torcular DAVFs

Treatment of torcular DAVFs depends on their Borden classification. Borden type II and type III torcular fistulae that drain exclusively to adjacent veins are treated simply by clipping the arterialized veins as they exit the sinus. These superficial fistulae are exposed using a torcular craniotomy with the patient positioned prone and with minimal subarachnoid dissection. In contrast, type II fistulae that drain to torcular sinuses and to adjacent veins are considerably more difficult to treat. Arterialized draining veins are occluded the same way as with type III fistulae, but shunt flow into the torcular sinuses cannot be interrupted without sacrificing a major sinus. Therefore, skeletonization of the torcula is also required with type II fistulae to interrupt arterial inflow (Sundt and Piepgras, 1983).

There are a total of eight dural leaflets around the torcula that can harbor arterial supply: falx cerebri, bilateral tentorium (two), bilateral occipital dura (two), bilateral suboccipital dura (two), and falx cerebelli. Complete skeletonization of the torcula requires a total of 12 cuts in these eight dural leaflets, with four of the leaflets requiring two cuts: the occipital dura must be cut along the tranverse sinus and the SSS, and the tentorium must be cut along the straight sinus and the transverse sinus. The torcular craniotomy exposes all of these leaflets, but some additional retraction is needed to make the tentorial incisions along the straight sinus.

Tentorial sinus DAVFs

We differentiate tentorial sinus DAVFs as a distinct but poorly described type of tentorial fistulae. This lesion has not been described in the neurosurgical literature in part because the tentorial sinus is an obscure entity that is often detected but ignored on angiography and magnetic resonance imaging (MRI). Furthermore, the anatomy of the tentorial sinuses is highly variable, which might make it difficult to recognize different tentorial sinus DAVFs as being the same type. Rhoton and colleagues classified the tentorial sinus into four groups depending on whether venous tributaries originated from the cerebrum (group I), cerebellum (group II), the tentorium itself (group III), or incisura (group IV) (Matsushima et al., 1989). In another anatomic study of 80 cadavers, Muthukumar and Palaniappan (1998) found tentorial sinuses in 86% of specimens and classified them into three types depending on their location (medial or lateral) and size (small or large). Miabi et al. (2004) used contrast-enhanced MRI to identify 104 tentorial sinuses in 55 patients and define yet another classification scheme: venous candelabra (type I), multiple independent veins (type II), and venous lakes within tentorium (type III). With so much confusion in the literature over a subtle and highly variable venous structure, it is no wonder that the association between the tentorial sinus and paramedian tentorial DAVFs has not been firmly established. Although we did not find any description of tentorial sinus DAVFs in the neurosurgical literature, it has been reported recently in the radiological literature (Horie et al., 2006).

Tentorial sinus DAVFs encountered in our experience had features implicating the tentorial sinus: none was associated with other dural sinuses (straight, superior petrosal, or transverse sinuses); some were laterally located, where the vein of Labbé might join the tentorial sinus under the temporal and occipital lobes; and others were medially located, where the medial variant of the tentorial sinus has been described.

The supratentorial-infraoccipital approach, as described by Smith and Spetzler (1995) for posteromedial temporal-lobe lesions, was used to expose tentorial sinus DAVFs. We prefer to position patients laterally rather than prone, with the head rotated downward to face the floor. A torcular craniotomy is important if the fistula is medially located, because a torcular craniotomy enables the dura to be opened widely and the occipital pole to be mobilized freely. More lateral DAVFs near the transverse–sigmoid junction may not need this midline exposure, and a unilateral temporal-occipital craniotomy taken down to or below the transverse sinus would suffice. Occlusion of the fistula is accomplished by interrupting the draining vein as it exits the tentorium.

Superior petrosal sinus DAVFs

Superior petrosal sinus DAVFs consistently drain infratentorially into the petrosal vein (Dandy's vein), making the extended retrosigmoid approach an ideal exposure (Quinones-Hinojosa et al., 2006). More radical transpetrosal approaches are not necessary because only the draining vein needs to be exposed to interrupt the fistula. The patient is positioned laterally with the head slightly flexed and angled downward toward the floor, optimizing the view along the angle between tentorium and petrous face. We use a C-shaped scalp incision behind the ear, a limited posterior mastoidectomy, skeletonization of the sigmoid sinus with a diamond-bit drill, craniotomy rather than craniectomy, and anterior mobilization of the sigmoid sinus with the dural flap. These maneuvers enhance the exposure of a conventional retrosigmoid approach that does not drill the sigmoid sinus (Quinones-Hinojosa et al., 2006). Feeding arteries from the ECA traverse the mastoid and petrous bones to make the drilling bloodier than with other lesions. These transosseus arteries are controlled easily with bone wax or drilling with a diamond bit.

If the fistula has ruptured and the cerebellum is swollen, cerebrospinal fluid should be released from the cisterna magna immediately after opening the dura to relax the cerebellum. Microsurgical dissection into the cerebellopontine angle leads to the arterialized petrosal vein, which is frequently variceal due to the high-flow nature of this fistula. The clip is applied as close to the petrous dura as possible, but not so close that closure of the blades avulses the vein. Draining veins distal to the clip should darken after the fistula is interrupted. The venous varices can sometimes hide an additional draining vein coursing medially towards the brainstem, so the varix should be mobilized and this medial territory inspected carefully. In our experience, half of these superior petrosal sinus DAVFs were Borden type II fistulae, with drainage medially into the superior petrosal sinus. Patency of this sinus can result in residual, but low-risk (Borden type I), shunting after occluding the draining petrosal vein, as in 1 patient in our experience.

Incisural DAVFs

Incisural DAVFs are the other type of tentorial fistula not well characterized in the literature. Picard et al. (1990) described a "marginal tentorial sinus" that courses along the free edge of the tentorial incisura, is not present in most people, and receives venous tributaries from the basal vein of Rosenthal and lateral mesencephalic veins (Tomak et al., 2003). We hypothesize that incisural DAVFs are associated with this marginal tentorial sinus, and the lack of a clear association between this fistula and venous sinus is due to the rarity of each entity. In our large experience, this subtype was the least common, with just 2 patients.

Incisural DAVFs and tentorial sinus DAVFs are similar because both are associated with intrinsic tentorial sinuses, both have variable anatomy, and both drain supratentorially. However, we differentiated these two types because they require different surgical approaches. Tentorial sinus DAVFs are located infraoccipitally and are exposed through a torcular or occipital craniotomy. In contrast, exposure of incisural DAVFs calls for either a more anterior pterional-transsylvian approach or a lateral subtemporal approach. In our experience, the venous drainage from incisural DAVFs coursed posteriorly to the Galenic system, but the fistulae were quite anterior, at the level of the uncus or supraclinoid ICA. Incisural DAVFs are also near the superior petrosal sinus and can be mistaken for the more common petrosal DAVF. However, the extended retrosigmoid approach used for petrosal DAVFs positions the neurosurgeon on the opposite side of the tentorium from an incisural DAVF, and even incising the tentorium was not sufficient to widen the exposure. Therefore, incisural DAVFs can be deceptive and require careful analysis of the venous drainage pattern on preoperative angiograms to be certain of their type and location along the incisura. We suspect that some of the difficulty in surgically obliterating superior petrosal sinus DAVFs relates to unrecognized differences between superior petrosal sinus and incisural DAVFs, and the erroneous selection of an infratentorial approach for incisural DAVFs.

INFRATENTORIAL DAVFs

Transverse–sigmoid sinus DAVFs

Symptoms associated with these DAVFs include pulsatile tinnitus, headache, visual obscuration, seizures, and hemorrhage. In low-risk DAVFs with low-flow AV shunts and no cortical venous drainage, symptoms are limited to pulsatile tinnitus. In high-risk DAVFs with increased shunt flow and retrograde cortical venous drainage, symptoms related to venous hypertension ensue with increased intracranial pressure, venous ischemia, focal neurologic deficits, and seizures.

Transverse–sigmoid sinus DAVFs are often amenable to endovascular therapy. Transvenous and transarterial approaches can be used alone or in combination to obliterate the fistulae. When treated surgically, transverse–sigmoid sinus DAVFs are exposed with an extended retrosigmoid craniotomy that widely exposes the junction of these two sinuses by removing overlying bone with a diamond drill bit, as described above for superior petrosal sinus DAVFs. Dura is opened to identify the arterialized draining vein, which is clipped, cauterized, and divided. Skeletonization of the transverse, sigmoid, and superior petrosal sinuses is also required for Borden type II lesions. Patients with Borden type I DAVFs are managed conservatively.

Marginal sinus DAVFs

The marginal sinus resides at the foramen magnum, encircling the cervicomedullary junction and draining into the jugular veins or sigmoid sinuses. Marginal sinus DAVFs are supplied by branches from the vertebral artery, including the posterior meningeal artery and direct branches from the vertebral artery trunk. Ascending muscular branches from the vertebral artery and ECA branches can also contribute to these fistulae. These DAVFs are rare. Patients present with hemorrhage, signs of venous hypertension, or symptoms of medullary or spinal cord compression from dilated veins.

Marginal sinus DAVFs are amenable to endovascular therapy, using transvenous or transarterial approaches to obliterate the fistulae, and in our experience, surgical intervention is rare. When necessary, these DAVFs are exposed through a far lateral craniotomy. The dural opening often occludes feeding arteries from the posterior meningeal artery. The approach provides access to the vertebral artery to identify additional feeders that arise from its subarachnoid segment. The arterialized vein is clipped, cauterized, and divided as it exits the dura overlying the occipital condyle.

Inferior petrosal sinus DAVFs

Inferior petrosal sinus DAVFs are rare lesions. They tend not to have leptomeningeal venous drainage, giving them a benign natural history. They have a low hemorrhage risk and can present with pain and cranial neuropathies. Management is often conservative. When treatment is necessary, endovascular therapy is highly successful. There have been no cases in our experience where surgical therapy has been needed for these lesions.

TREATMENT DECISION MAKING

The diverse clinical spectrum and variable natural history of these lesions demand an individualized approach to clinical decision making. Due to limited natural history data on these uncommon lesions, the decision to aggressively manage is usually solely based on the severity of neurologic symptoms and angiographic characteristics of DAVF. Emerging literature suggests that DAVFs with cortical venous drainage and those located in the posterior fossa should be aggressively treated to prevent hemorrhage (Singh et al., 2008). Besides hemorrhagic risk, the pathophysiology of chronic neurologic decline is thought to be due to venous hypertension based on computed tomography, MRI, and single-photon emission computed tomography findings. Concern about hemorrhage drives clinical management decisions after the diagnosis of DAVF, but the injurious effects of venous hypertension are underappreciated and poorly understood.

References

Agid T, TerBrugge K, Rodesch G et al. (2009). Management strategies for anterior cranial fossa (ethmoidal) dural arteriovenous fistulas with an emphasis on endovascular treatment. J Neurosurg 110 (1): 79–84.

Bernstein R, Dowd CF, Gress DR (2003). Rapidly reversible dementia. Lancet 361 (9355): 392.

Cellerini M, Mascalchi M, Mangiafico S et al. (1999). Phase-contrast MR angiography of intracranial dural arteriovenous fistulae. Neuroradiology 41 (7): 487–492.

Chi JH, Lawton MT (2006). Posterior interhemispheric approach: surgical technique, application to vascular lesions, and benefits of gravity retraction. Neurosurgery 59 (1 Suppl 1): ONS41–ONS49; discussion ONS41–ONS49.

Datta NN, Rehman SU, Kwok JC et al. (1998). Reversible dementia due to dural arteriovenous fistula: a simple surgical option. Neurosurg Rev 21 (2–3): 174–176.

Desal HA, Toulgoat F, Raoul S et al. (2005). Ehlers-Danlos syndrome type IV and recurrent carotid-cavernous fistula: review of the literature, endovascular approach, technique and difficulties. Neuroradiology 47 (4): 300–304.

Gross BA, Moon K, Kalani MYS et al. (2016). Clinical and anatomic insights from a series of ethmoidal dural arteriovenous fistulas at the Barrow Neurological Institute. World Neurosurg 93: 94–99.

Horie N, Morikawa M, Kitigawa N et al. (2006). 2D Thick-section MR digital subtraction angiography for the assessment of dural arteriovenous fistulas. AJNR Am J Neuroradiol 27 (2): 264–269.

Lalwani AK, Dowd CF, Halbach VV (1993). Grading venous restrictive disease in patients with dural arteriovenous fistulas of the transverse/sigmoid sinus. J Neurosurg 79 (1): 11–15.

Lasjaunias P, Chiu M, ter Brugge K et al. (1986). Neurological manifestations of intracranial dural arteriovenous malformations. J Neurosurg 64 (5): 724–730.

Lawton MT, Chun J, Wilson CB et al. (1999). Ethmoidal dural arteriovenous fistulae: an assessment of surgical and endovascular management. Neurosurgery 45 (4): 805–810; discussion 810–811.

Matsuda S, Waragai M, Shinotoh H et al. (1999). Intracranial dural arteriovenous fistula (DAVF) presenting progressive dementia and parkinsonism. J Neurol Sci 165 (1): 43–47.

Matsushima T, Suzuki SO, Fukui M et al. (1989). Microsurgical anatomy of the tentorial sinuses. J Neurosurg 71 (6): 923–928.

Miabi Z, Midia R, Rohrer SE et al. (2004). Delineation of lateral tentorial sinus with contrast-enhanced MR imaging and its surgical implications. AJNR Am J Neuroradiol 25 (7): 1181–1188.

Miller NR (2007). Diagnosis and management of dural carotid-cavernous sinus fistulas. Neurosurg Focus 23 (5): E13.

Muthukumar N, Palaniappan P (1998). Tentorial venous sinuses: an anatomic study. Neurosurgery 42 (2): 363–371.

Picard L, Bracard S, Islak C et al. (1990). Dural fistulae of the tentorium cerebelli. Radioanatomical, clinical and therapeutic considerations. J Neuroradiol 17 (3): 161–181.

Quinones-Hinojosa A, Chang EF, Lawton MT (2006). The extended retrosigmoid approach: an alternative to radical cranial base approaches for posterior fossa lesions. Neurosurgery 58 (4 Suppl 2): ONS-208–ONS-214; discussion ONS-214.

Singh V, Smith WS, Lawton MT et al. (2008). Risk factors for hemorrhagic presentation in patients with dural arteriovenous fistulae. Neurosurgery 62 (3): 628–635; discussion 628–635.

Smith KA, Spetzler RF (1995). Supratentorial-infraoccipital approach for posteromedial temporal lobe lesions. J Neurosurg 82 (6): 940–944.

Sundt Jr TM, Piepgras DG (1983). The surgical approach to arteriovenous malformations of the lateral and sigmoid dural sinuses. J Neurosurg 59 (1): 32–39.

Tanaka K, Morooka Y, Nakagawa Y et al. (1999). Dural arteriovenous malformation manifesting as dementia due to ischemia in bilateral thalami. A case report. Surg Neurol 51 (5): 489–493; discussion 493–494.

Tomak PR, Cloft HJ, Kaga A et al. (2003). Evolution of the management of tentorial dural arteriovenous malformations. Neurosurgery 52 (4): 750–760; discussion 760–762.

van Dijk JM (2002). terBrugge KG, Willinsky RA, et al. Clinical course of cranial dural arteriovenous fistulas with long-term persistent cortical venous reflux. Stroke 33 (5): 1233–1236.

Yamakami I, Kobayashi E, Yamaura A (2001). Diffuse white matter changes caused by dural arteriovenous fistula. J Clin Neurosci 8 (5): 471–475.

Handbook of Clinical Neurology, Vol. 143 (3rd series)
Arteriovenous and Cavernous Malformations
R.F. Spetzler, K. Moon, and R.O. Almefty, Editors
http://dx.doi.org/10.1016/B978-0-444-63640-9.00011-4

Chapter 11

Endovascular management of intracranial dural arteriovenous fistulas

CELENE B. MULHOLLAND, M. YASHAR S. KALANI, AND FELIPE C. ALBUQUERQUE*
Department of Neurosurgery, Barrow Neurological Institute, St. Joseph's Hospital and Medical Center, Phoenix, AZ, USA

Abstract

Dural arteriovenous fistulas are a heterogeneous group of lesions that comprise 10–15% of intracranial vascular malformations. The treatment strategy is devised after careful consideration of the arterial supply, venous drainage, clinical presentation, and risk of progression, hemorrhage, or neurologic decline. With recent advancements in endovascular technology, the majority of dural arteriovenous fistulas can be treated with either transarterial or transvenous embolization. Those that cannot be fully treated by endovascular means are approached with either adjuvant surgery or radiotherapy.

INTRODUCTION

Intracranial dural arteriovenous fistulas (DAVFs) account for 10–15% of all intracranial vascular malformations and about 6% of supratentorial arteriovenous malformations (AVMs) (Santillan et al., 2013). The arterial supply is generally attributed to the meningeal arteries. The hemorrhage risk for DAVFs is approximately 1.5% per year but varies by location and type of DAVF (Santillan et al., 2013).

Although there are different classification systems for intracranial DAVFs, the most commonly used are the Cognard and Borden classification systems. The Cognard classification system of four types of DAVFs is based on the venous drainage pattern: (1) patients who have type I DAVFs present with immediate drainage into a dural sinus or meningeal vein; (2) patients who have type II DAVFs present with initial drainage into the sinus and retrograde flow into other sinuses or cortical veins; (3) patients with type III DAVFs present with initial drainage into a cortical vein; and (4) patients with type IV DAVFs present with initial drainage into a cortical vein with venous ectasia (Table 11.1 and Figs 11.1–11.3) (Borden et al., 1995; Cognard et al., 1995; Santillan et al., 2013). In contrast, the Borden classification system simply stratifies DAVFs into type I, type II, and type III. In this classification system, type I lesions drain into dural venous sinuses or meningeal veins, and type II lesions have retrograde drainage into subarachnoid veins. The third type of DAVFs, type III lesions, drain into subarachnoid veins and do not have dural sinus or meningeal venous drainage.

AGGRESSIVE DAVFS

Some DAVFs can be quite aggressive, in particular those with cortical or leptomeningeal venous drainage and/or an associated varix. In these cases, the source of hemorrhage is not the arterial component but instead is the distended leptomeningeal venous varices (Liu et al., 2009). In a meta-analysis by Awad et al. (1990), leptomeningeal venous drainage, variceal or aneurysmal (venous) dilations, and galenic drainage were found to be significantly associated with an aggressive presentation (Tabrizi et al., 2003). In fact, patients with DAVFs involving retrograde venous drainage have been found to have an estimated risk of 35% (7 of 20) for rebleeding within 2 weeks after the initial hemorrhage (Duffau et al., 1999).

There are two types of high-grade lesions. One type of DAVF drains into a large dural sinus, with retrograde filling of the leptomeningeal veins. The other type

*Correspondence to: Felipe C. Albuquerque, MD, c/o Neuroscience Publications, Barrow Neurological Institute, St. Joseph's Hospital and Medical Center, 350 W. Thomas Rd, Phoenix AZ 85013, USA. Tel: +1-602-406-3593, Fax: +1-602-406-4104, E-mail: Neuropub@dignityhealth.org

Table 11.1

Classification system for intracranial dural arteriovenous fistulas

Cognard type	Drainage	Flow
I	Dural venous sinus	Anterograde sinus flow
IIa	Dural venous sinus	Retrograde sinus flow
IIb	Dural venous sinus	Anterograde sinus flow and retrograde leptomeningeal venous drainage
IIa+b	Dural venous sinus	Retrograde sinus flow and retrograde leptomeningeal venous drainage
III	Retrograde leptomeningeal venous drainage	No venous ectasia
IV	Retrograde leptomeningeal venous drainage	Venous ectasia
V	Retrograde leptomeningeal venous drainage	

Modified from Santillan et al. (2013).

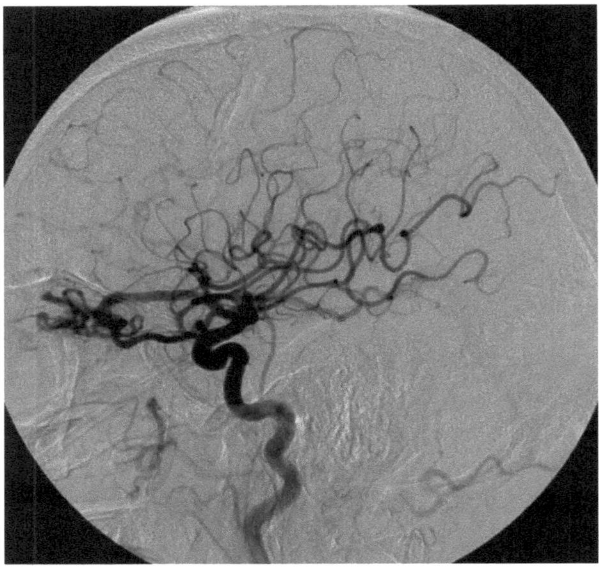

Fig. 11.1. Type I ethmoidal dural arteriovenous fistula. Type I fistulas demonstrate lack of retrograde flow and direct sinus drainage. (Courtesy of Barrow Neurological Institute, Phoenix, AZ.)

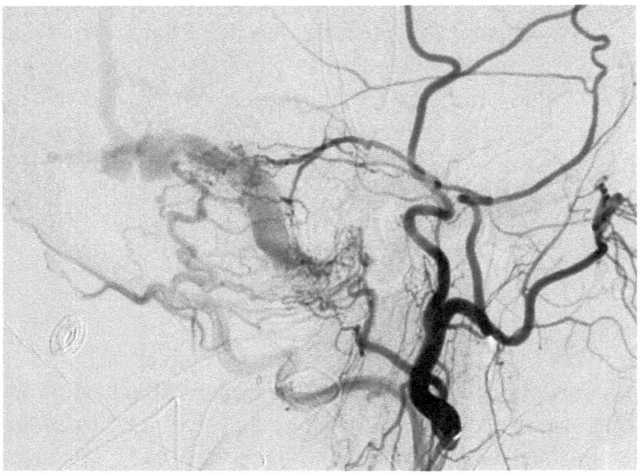

Fig. 11.2. Type II dural arteriovenous fistula. Type II fistulas demonstrate retrograde venous flow without leptomeningeal venous drainage. (Courtesy of Barrow Neurological Institute, Phoenix, AZ.)

of high-grade DAVF drains directly into the leptomeningeal veins without a large venous sinus (Collice et al., 2000).

The neurologic manifestations concomitant with DAVFs need not be caused solely by hemorrhage. Parenchymal venous congestion, neurovascular compression, and mass effect by dilated varices can also cause neurologic deficits.

ENDOVASCULAR MANAGEMENT

Patient evaluation

The decision to proceed with endovascular treatment should be based on the presentation of the patient, the patient's history of any previous hemorrhage, and the patient's age and medical comorbidities. Other key considerations that influence the decision include the venous drainage pattern, angiographic features, and the location of the DAVF. A history of a previous hemorrhage and low operative risk push the provider toward intervention. Low-risk features, such as a lack of cortical venous drainage and a lack of venous ectasia, can be managed conservatively if the patient is asymptomatic (Miller and Gandhi, 2015). Nonetheless, fistulas with high-risk angiographic features, such as cortical venous drainage, should be treated on a more emergent basis (van Dijk, 2002; Gandhi et al., 2012; Miller and Gandhi, 2015). The clinical indications are varied for emergent management, which some cases may require. These indications include leptomeningeal drainage, intracranial hemorrhage, vision loss, increased intracranial pressure, neurologic deficits, and neurovascular compression. Examples of

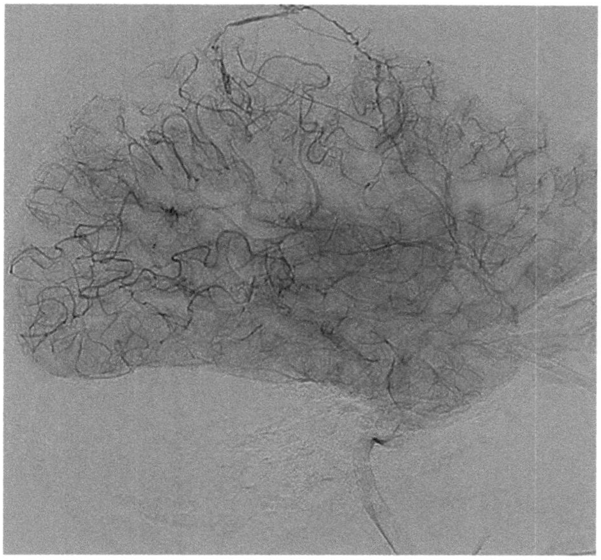

Fig. 11.3. Type III dural arteriovenous fistula. The characteristics of type III fistulas are significant for the lack of venous ectasia. (Courtesy of Barrow Neurological Institute, Phoenix, AZ.)

neurovascular compression include chemosis, proptosis, ophthalmoplegia, and decreased visual acuity resulting from DAVF of the cavernous sinus.

General tenets

The use of endovascular coils and liquid adhesive embolic agents has revolutionized the treatment of DAVFs. Endovascular management of DAVFs is considered to be the first-line therapy. However, this depends on the location of the DAVF. The goals of therapy are to close the shunt at the site of the fistula and to have embolic material extend into the venous side (Tabrizi et al., 2003; Santillan et al., 2013; Ambekar et al., 2015). One of the tenets of embolization of DAVFs is to avoid embolizing too proximally. If embolization is done too proximally, persistent arterial flow can recruit collateral blood supply and increase the risk of recanalization (Santillan et al., 2013; Ambekar et al., 2015). However, distal embolization can also result in undesirable outcomes, such as venous occlusion and venous hypertension (Santillan et al., 2013).

Although transarterial embolization can be performed, the transvenous approach is preferred, given the possible presence of external carotid artery-to-internal carotid artery or vertebral artery anastomoses. Transarterial embolization may be necessary when there is limited venous access because of thrombosis or stenosis (Santillan et al., 2013).

A variety of embolic materials can be used. The numerous types include glue (n-butyl cyanoacrylate (NBCA)), polyvinyl alcohol (PVA) particles, the Onyx liquid embolic system (ev3 Neurovascular/Covidien/Medronic), and detachable platinum coils. Onyx is generally preferred because it is more likely to reach the venous portion of the fistula, whereas PVA and NBCA are more likely to cause proximal vessel occlusion. Obliteration rates with Onyx range from 60% to 100% (Cognard et al., 2008; Stiefel et al., 2009; van Rooij and Sluzewski, 2010; Ambekar et al., 2015).

In preparation for surgery or transvenous embolization, transarterial embolization can be used to decrease blood flow and help delineate the angioarchitecture of the lesion and the draining arterialized veins (Woo et al., 2005; Liu et al., 2009). For treatment of aggressive DAVFs, transvenous embolization with coiling of the proximal venous outlet is an acceptable curative option (Santillan et al., 2013). Incompletely occluded DAVFs may recruit feeding vessels. In turn, this may result in more complex and aggressive lesions.

In high-grade aggressive DAVFs, the goal of treatment is to occlude the arterialized fistulous sinus and the proximal portion of the refluxed leptomeningeal vein. Doing so preserves the uninvolved sinus and distal veins that join the arterialized leptomeningeal veins functioning in the normal parenchymal venous drainage. For fistulas with direct leptomeningeal drainage, the goal of treatment is to occlude the arterialized veins at the most proximal segment as they exit the dura.

In cases where endovascular treatment is not feasible, a multidisciplinary approach is often required. Endovascular treatment can be used in preparation for open surgery or radiosurgery. Microsurgery is the first-line treatment for both ethmoidal and superior petrosal sinus DAVFs (Tabrizi et al., 2003). The success rate of open surgery for ethmoidal DAVFs has been reported by Agid et al. (2009) to be as high as 100% (15 of 15 patients).

Behavior and treatment by location

Transverse sigmoid DAVFs

The transverse sigmoid junction is the most common location for DAVFs. Patients with DAVFs in this location may present with pulsatile tinnitus, headaches, and, more rarely, intracranial hemorrhage. The main arterial supply is via the anastomosis branches of the occipital artery, the branches of the middle meningeal artery, the meningeal branches of the ascending pharyngeal artery, the branches of the vertebral artery, and the tentorial branches of the meningohypophyseal trunk (Santillan et al., 2013). It is especially important to note the direction of flow in the vein of Labbé (i.e., the inferior anastomotic vein) and where the vein inserts along the transverse sigmoid sinus junction. Anterograde flow through the vein of Labbé prohibits occlusion at its origin

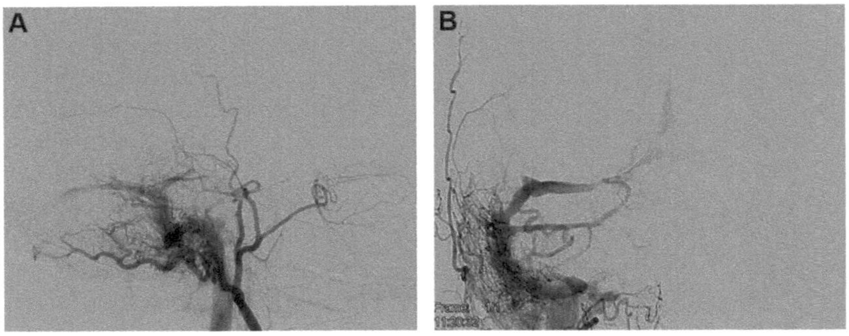

Fig. 11.4. Marginal sinus dural arteriovenous fistula. (A) Lateral view angiogram demonstrating contribution from the middle meningeal artery, and (B) anteroposterior view angiogram demonstrating shunting across the sigmoid sinus. (Courtesy of Barrow Neurological Institute, Phoenix, AZ.)

from a transvenous approach because of the risk of exacerbating venous hypertension (Santillan et al., 2013).

Fortunately, endovascular treatment of this type of DAVFs often leads to a high cure rate. Jiang et al. (2011) reported success in 5 of 5 patients (100%). Pelz et al. (1997) reported a lower cure rate of 31% (4 of 13) with use of a transarterial approach alone. Kirsch et al. (2009) reported a combined transarterial/transvenous treatment success rate of 54% (52 of 96). Similar high cure rates of 80% or more have been reported in other studies (Urtasun et al., 1996).

Case example

A 54-year-old woman presented with debilitating pulsatile tinnitus, and on magnetic resonance imaging was found to have a DAVF centered around the sigmoid sinus on the right. Angiography demonstrated that she had a Cognard type II A plus B marginal sinus DAVF, with extension of the shunt along the entire sigmoid sinus (Figs 11.4 and 11.5). The predominant supply was via the petrosquamosal branch of the right middle meningeal artery. There was shunting along the entire course of the sigmoid sinus, as well as a faint supply via clival and tentorial branches of the right internal carotid artery and a small tentorial branch of the right middle meningeal artery. This shunting was in addition to contribution from many muscular branches of the right vertebral artery and the posterior meningeal artery. There was retrograde flow in the right transverse and sigmoid sinuses, with reflux into the vein of Labbé and the cerebellar veins. Transarterial embolization was conducted of the right middle meningeal artery, the petrosquamosal branch, the right occipital artery branch, the right ascending pharyngeal artery, and the neuromeningeal trunk. Treatment included successful transvenous coil embolization of the right distal sigmoid sinus and jugular bulb. There was no obvious residual early shunting following transarterial and transvenous embolization. A postoperative angiogram performed at

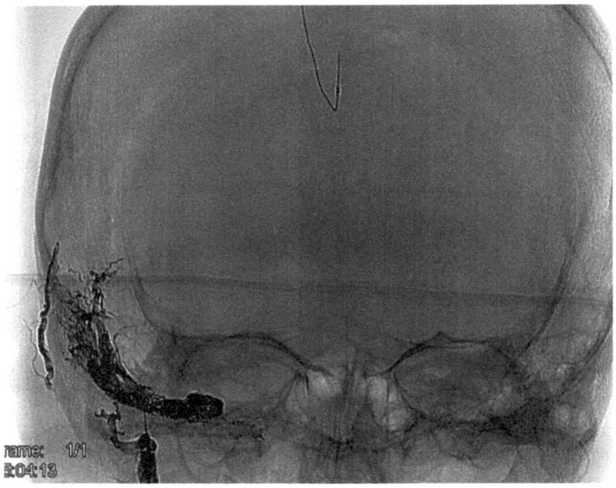

Fig. 11.5. Marginal sinus dural arteriovenous fistula. Postoperative anteroposterior view angiogram demonstrating cast after endovascular embolization. (Courtesy of Barrow Neurological Institute, Phoenix, AZ.)

5-month follow-up demonstrated no residual fistula and complete resolution of pulsatile tinnitus.

This case demonstrates complete cure following transarterial and transvenous embolization. As discussed above, the ability to perform the transvenous embolization was facilitated by the retrograde flow along the vein of Labbé.

Carotid-cavernous DAVFs

Carotid-cavernous fistulas are the second most common type of DAVF. These fistulas are categorized as direct (i.e., high-flow arterial blood passes directly through a defect in the artery, such as from head trauma) or indirect (i.e., spontaneous, low-flow) fistulas. They are classified using the Barrow et al. (1985) classification system for carotid-cavernous fistulas of type A to type D. Type A fistulas are direct high-flow shunts between the internal carotid artery and the cavernous sinus. Types B–D are indirect, low-flow dural shunts between the cavernous sinus and the meningeal branches of the internal carotid artery (type B), the external carotid artery (type C), or both internal and external carotid arteries (type D).

The most common signs and symptoms in a patient who presents with a carotid-cavernous fistula are chemosis, exophthalmos, oculomotor nerve (cranial nerve (CN III) and/or abducens nerve (CN XI)) palsy, decreased visual acuity, headache, and tinnitus (Klisch et al., 2003). Halbach et al. (1987) reported a total hemorrhage rate of 8.4% (13 of 155) in patients with direct carotid-cavernous fistulas but reported a 36% (4 of 11) rate of intracerebral hemorrhage in patients with cortical venous drainage.

The arterial supply to a carotid-cavernous fistula varies but can include branches of the inferolateral trunk, the meningohypophyseal trunk, the middle meningeal artery or the accessory meningeal artery, the artery of the foramen rotundum, and the ascending pharyngeal artery. The venous drainage is through the external jugular vein and the superior or inferior ophthalmic vein in the facial venous system.

The preferred therapy is transvenous treatment. It can be performed via the ipsilateral or contralateral inferior petrosal sinus, basilar plexus, circular sinus, facial vein, angular vein, superior ophthalmic vein, or pterygoid plexus (Klisch et al., 2003). If transvenous catheterization fails, either transarterial embolization or a surgical approach may be used.

Onyx, PVA, and NBCA are all useful agents for treating carotid-cavernous fistulas. Complete occlusion of these fistulas with transvenous treatment alone has been reported in up to 81% (17 of 21) of cases (Kirsch et al., 2009).

Superior sagittal sinus DAVFs

Although superior sagittal sinus DAVFs account for only 8–13% of all intracranial fistulas, the patients who have them are most likely to present with intracranial hemorrhage or neurologic deficits (Bertalanffy et al., 2001; Woo et al., 2005). This type of DAVF receives a bilateral arterial supply from branches of the middle meningeal artery, from the anterior falcine artery of the ophthalmic artery, or from the posterior meningeal artery. Given the presence of multiple arterial feeders and venous drainage, superior sagittal sinus DAVFs are more likely than other types of DAVFs to require aggressive treatment (Halbach et al., 1988). Transarterial embolization of superior sagittal sinus DAVFs, either alone or in combination with open surgery or radiosurgery, has also been described (Bertalanffy et al., 2001). Success rates for the transarterial approach alone have ranged from 30% to 100% (Halbach et al., 1988, 1990; Cognard et al., 2008).

Tentorial DAVFs

Tentorial DAVFs are relatively uncommon, encompassing only 4–8% of all intracranial fistulas (Santillan et al., 2013). Clinical manifestations include headache, ataxia, and trigeminal neuralgia; these symptoms are commonly present with hemorrhage secondary to leptomeningeal venous drainage (Awad et al., 1990; Kirsch et al., 2009). Aggressive clinical behavior is observed in 79–82% of cases. As might be expected, these lesions are commonly classified as Cognard type II or type III lesions. Their arterial supply comes from the first branch off the meningohypophyseal trunk (i.e., the posterior trunk, a branch of the proximal cavernous segment of the internal carotid artery) and from the petrosal and petrosquamosal branches of the middle meningeal artery. Treatment of tentorial DAVFs can be accomplished via transarterial or transvenous embolization with coils (Tomak et al., 2003).

Anterior cranial fossa (ethmoidal) DAVFs

Another rare lesion is the ethmoidal DAVF. This type of DAVF constitutes less than 10% of all DAVFs. Patients may present with headaches, seizures, and exophthalmos (Martin et al., 1990). Ethmoidal DAVFs often have a large venous pouch on the draining cortical vein and are associated with a high rate of intracranial hemorrhage because of the leptomeningeal drainage (Halbach et al., 1990; Gliemroth et al., 1999). Thus, these lesions are treated more aggressively than other DAVFs when patients are symptomatic.

The main arterial supply is from the posterior and anterior ethmoidal branches of the ophthalmic artery. However, ethmoidal DAVFs can also be supplied by branches of the internal maxillary artery, the superficial temporal artery, or the middle meningeal artery.

These lesions are better suited for surgical occlusion, given the risk of occluding the central retina artery (Tabrizi et al., 2003; Santillan et al., 2013). If transarterial embolization is performed, the microcatheter tip must be beyond the origin of the central retina artery (Pelz et al., 1997).

Superior petrosal sinus DAVFs

Superior petrosal sinus DAVFs are very rare lesions described largely in case reports (Ng et al., 2003). The clinical presentation of patients with this type of DAVF includes headaches, tinnitus, trigeminal neuralgia, and intracranial hemorrhage. Intracranial hemorrhage is especially notable in these patients because their venous drainage usually occurs to the superior petrosal sinus and to the pontine and perimesencephalic veins (Awad et al., 1990). The arterial supply can arise from the tentorial artery, from the first branch off the meningohypophyseal trunk, or from the petrosal and petrosquamosal branches of the middle meningeal artery (Woo et al., 2005).

Like ethmoidal DAVFs, superior petrosal sinus DAVFs are technically difficult to occlude via either transarterial or transvenous methods. One series demonstrated that 22% (4 of 18) of patients were able to be treated with endovascular therapy alone. The remaining DAVFs were treated with a combination of endovascular and open surgery (Ng et al., 2003).

SUMMARY

Endovascular management of intracranial DAVFs has made tremendous advancements in recent years. Of paramount importance is the assessment of the risk of hemorrhage or neurologic decline based on the clinical presentation of the patient and the type and location of the DAVF. Occlusion rates for the most common DAVFs are promising and, with continued technologic advancement, we may begin to see higher cure rates for DAVFs that are more technically difficult.

References

Agid R, Terbrugge K, Rodesch G et al. (2009). Management strategies for anterior cranial fossa (ethmoidal) dural arteriovenous fistulas with an emphasis on endovascular treatment. J Neurosurg 110: 79–84.

Ambekar S, Gaynor BG, Peterson EC et al. (2015). Long-term angiographic results of endovascularly "cured" intracranial dural arteriovenous fistulas. J Neurosurg: 1–5.

Awad IA, Little JR, Akarawi WP et al. (1990). Intracranial dural arteriovenous malformations: factors predisposing to an aggressive neurological course. J Neurosurg 72: 839–850.

Barrow DL, Spector RH, Braun IF et al. (1985). Classification and treatment of spontaneous carotid-cavernous sinus fistulas. J Neurosurg 62: 248–256.

Bertalanffy A, Dietrich W, Kitz K et al. (2001). Treatment of dural arteriovenous fistulae (dAVFs) at the superior sagittal sinus (SSS) using embolisation combined with micro- or radiosurgery. Minim Invasive Neurosurg 44: 205–210.

Borden JA, Wu JK, Shucart WA (1995). A proposed classification for spinal and cranial dural arteriovenous fistulous malformations and implications for treatment. J Neurosurg 82: 166–179.

Cognard C, Gobin YP, Pierot L et al. (1995). Cerebral dural arteriovenous fistulas: clinical and angiographic correlation with a revised classification of venous drainage. Radiology 194: 671–680.

Cognard C, Januel AC, Silva Jr NA et al. (2008). Endovascular treatment of intracranial dural arteriovenous fistulas with cortical venous drainage: new management using Onyx. AJNR Am J Neuroradiol 29: 235–241.

Collice M, D'Aliberti G, Arena O et al. (2000). Surgical treatment of intracranial dural arteriovenous fistulae: role of venous drainage. Neurosurgery 47: 56–66; discussion 66-67.

Duffau H, Lopes M, Janosevic V et al. (1999). Early rebleeding from intracranial dural arteriovenous fistulas: report of 20 cases and review of the literature. J Neurosurg 90: 78–84.

Gandhi D, Chen J, Pearl M et al. (2012). Intracranial dural arteriovenous fistulas: classification, imaging findings, and treatment. AJNR Am J Neuroradiol 33: 1007–1013.

Gliemroth J, Nowak G, Arnold H (1999). Dural arteriovenous malformation in the anterior cranial fossa. Clin Neurol Neurosurg 101: 37–43.

Halbach VV, Hieshima GB, Higashida RT et al. (1987). Carotid cavernous fistulae: indications for urgent treatment. AJR Am J Roentgenol 149 587–593.

Halbach VV, Higashida RT, Hieshima GB et al. (1988). Treatment of dural arteriovenous malformations involving the superior sagittal sinus. AJNR Am J Neuroradiol 9: 337–343.

Halbach VV, Higashida RT, Hieshima GB et al. (1990). Dural arteriovenous fistulas supplied by ethmoidal arteries. Neurosurgery 26: 816–823.

Jiang C, Lv X, Li Y et al. (2011). Transarterial Onyx packing of the transverse-sigmoid sinus for dural arteriovenous fistulas. Eur J Radiol 80: 767–770.

Kirsch M, Liebig T, Kuhne D et al. (2009). Endovascular management of dural arteriovenous fistulas of the transverse and sigmoid sinus in 150 patients. Neuroradiology 51: 477–483.

Klisch J, Huppertz HJ, Spetzger U et al. (2003). Transvenous treatment of carotid cavernous and dural arteriovenous fistulae: results for 31 patients and review of the literature. Neurosurgery 53: 836–856; discussion 856–857.

Liu JK, Dogan A, Ellegala DB et al. (2009). The role of surgery for high-grade intracranial dural arteriovenous fistulas: importance of obliteration of venous outflow. J Neurosurg 110: 913–920.

Martin NA, King WA, Wilson CB et al. (1990). Management of dural arteriovenous malformations of the anterior cranial fossa. J Neurosurg 72: 692–697.

Miller TR, Gandhi D (2015). Intracranial dural arteriovenous fistulae: clinical presentation and management strategies. Stroke 46: 2017–2025.

Ng PP, Halbach VV, Quinn R et al. (2003). Endovascular treatment for dural arteriovenous fistulae of the superior petrosal sinus. Neurosurgery 53: 25–32; discussion 32–33.

Pelz DM, Lownie SP, Fox AJ et al (1997). Intracranial dural arteriovenous fistulae with pial venous drainage: combined endovascular-neurosurgical therapy. Can J Neurol Sci 24: 210–218.

Santillan A, Nanaszko M, Burkhardt JK et al. (2013). Endovascular management of intracranial dural arteriovenous fistulas: a review. Clin Neurol Neurosurg 115: 241–251.

Stiefel MF, Albuquerque FC, Park MS et al. (2009). Endovascular treatment of intracranial dural arteriovenous fistulae using Onyx: a case series. Neurosurgery 65: 132–139; discussion 139–140.

Tabrizi P, Spetzler RF, McDougall C et al. (2003). Cerebral dural arteriovenous fistula. Interv Neuroradiol 9: 101–102.

Tomak PR, Cloft HJ, Kaga A et al. (2003). Evolution of the management of tentorial dural arteriovenous malformations. Neurosurgery 52: 750–760; discussion 760–762.

Urtasun F, Biondi A, Casaco A et al. (1996). Cerebral dural arteriovenous fistulas: percutaneous transvenous embolization. Radiology 199: 209–217.

van Dijk JM (2002). terBrugge KG, Willinsky RA, et al. Clinical course of cranial dural arteriovenous fistulas with long-term persistent cortical venous reflux. Stroke 33: 1233–1236.

van Rooij WJ, Sluzewski M (2010). Curative embolization with Onyx of dural arteriovenous fistulas with cortical venous drainage. AJNR Am J Neuroradiol 31: 1516–1520.

Woo HH, Masaryk TJ, Rasmussen PA (2005). Treatment of dural arteriovenous malformations and fistulae. Neurosurg Clin N Am 16: 381–393. x.

Handbook of Clinical Neurology, Vol. 143 (3rd series)
Arteriovenous and Cavernous Malformations
R.F. Spetzler, K. Moon, and R.O. Almefty, Editors
http://dx.doi.org/10.1016/B978-0-444-63640-9.00012-6

Chapter 12

Radiosurgery for dural arteriovenous malformations

CONOR GRADY[1], CAROLINA GESTEIRA BENJAMIN[1], AND DOUGLAS KONDZIOLKA[1,2]*
[1]*Department of Neurosurgery, NYU Langone Medical Center, New York, NY, USA*
[2]*Center for Advanced Radiosurgery, NYU Langone Medical Center, New York, NY, USA*

Abstract

Intracranial dural arteriovenous malformations (DAVFs) are relatively uncommon vascular lesions characterized by the direct connection of dural arteries into dural venous sinuses or leptomeningeal veins. Strategies for the treatment of these complex lesions have evolved significantly over the past three decades, and include open surgical disconnection, endovascular embolization, stereotactic radiosurgery (SRS), or a combination of these approaches. Radiosurgical intervention is unique in offering significant benefits to patients while exposing them to few of the risks associated with more invasive interventions. In this chapter we provide an overview of DAVFs and discuss the features of these lesions that affect management. We focus, in particular, on radiosurgical management of these lesions, describing present treatment paradigms, the procedure for the treatment of DAVFs with SRS, and expected clinical outcomes using SRS.

INTRODUCTION

Intracranial dural arteriovenous fistulas (DAVFs) are relatively uncommon vascular lesions characterized by the direct connection of dural arteries into dural venous sinuses or leptomeningeal veins. They are believed to constitute 10–15% of all intracranial vascular lesions (Houser et al., 1972), and are distinct from other intracranial arteriovenous malformations (AVMs) in their pathogenesis and presentation. Strategies for the treatment of these complex lesions have evolved significantly over the past three decades, and include open surgical disconnection, endovascular embolization, stereotactic radiosurgery (SRS), or a combination of these approaches. Radiosurgical intervention is unique in offering significant benefits to patients while exposing them to few of the risks associated with more invasive interventions. In this chapter we will provide an overview of DAVFs and discuss the features of these lesions that affect treatment. We will focus, in particular, on radiosurgical management of these lesions, and expected clinical outcomes using SRS.

PATHOGENESIS

In contrast to the majority of intracranial AVMs, DAVFs are primarily believed to be acquired lesions (Houser et al., 1979; Chaudhary et al., 1982; Nishijima et al., 1992). Over the past four decades, clinical observation, histologic examination of resected specimens, and animal models have led to multiple hypotheses of the cardinal event leading to their formation. Houser et al. (1979) were amongst the first to define the angiographic characteristics of these lesions, and noted the relationship of their formation to the presence of dural sinus thrombosis. Subsequent studies continued to support the association of DAVF with sinus thrombosis, though controversy existed regarding the etiologic relationship of sinus thrombosis to fistula formation (Chaudhary et al., 1982; Kutluk et al., 1991).

The present leading hypothesis of DAVF formation posits a repetitive cycle of anatomic and molecular events triggered by venous hypertension. Venous hypertension itself may result from any number of factors, including, but not limited to, trauma and venous

*Correspondence to: Douglas Kondziolka, MD, MSc, FRCSC, FACS, Center for Advanced Radiosurgery, NYU Langone Medical Center, 530 First Avenue, Suite 8R, New York NY 10016, USA. Tel: +1-646-501-2360, E-mail: douglas.kondziolka@nyumc.org

thrombosis (Nishijima et al., 1992). Anatomically, venous hypertension leads to dilation of dural veins and enlargement of microscopic arteriovenous shunts normally present in dura. The enlarging of these shunts further contributes to venous hypertension, as increased arterial flow reaches new venous territories. Blood flow through these microscopic fistulae become increasingly turbulent as they enlarge, leading to stasis and promoting thrombosis, further contributing to venous hypertension. The outcome of this cascade of events has been demonstrated by histologic examination of resected DAVFs, which reveals dilation and thickening of microscopic arterial and venous vessels present in the dura adjacent to fistula sites, as well as partially canalized sinus thromboses (Houser et al., 1979; Nishijima et al., 1992). This theory has been additionally supported by animal studies in which both carotid to external jugular anastomoses and experimentally induced sinus thromboses have first resulted in venous hypertension, followed by the formation of DAVFs (Terada et al., 1994; Herman et al., 1995).

On a molecular level, venous hypertension is believed to cause local ischemia and endothelial injury, leading to the increased expression of angiogenic growth factors which contribute to the histologic findings described above. Immunohistochemical analysis of resected DAVFs has demonstrated markedly increased expression of basic fibroblast growth factor and vascular endothelial growth factor (Uranishi et al., 1999). Lawton et al. (1997) also successfully demonstrated increased expression of angiogenic growth factors and DAVF formation in a rat model of venous hypertension.

PRESENTATION

DAVFs present with a variety of symptoms which are related to the anatomic location of the fistula itself. The majority of DAVFs present with local symptoms arising from altered hemodynamics. In one of the early large case series looking at the presenting symptoms of DAVFs, Lasjaunias et al. (1986) reported that 83% of patients with cavernous sinus DAVFs presented with proptosis or visual changes, while 70% of patients with torcular, transverse sinus, or sigmoid sinus DAVFs presented with an audible bruit. More recently, Gross et al. (2016), in a retrospective review of a prospectively maintained database of 260 patients with DAVFs, found that 92% of patient with cavernous sinus lesions presented with ocular findings, while 75% of patients with transverse sinus, sigmoid sinus, marginal sinus, or torcular DAVFs presented with pulsatile tinnitus.

DAVFs can also present with more malignant symptoms, such as seizure, neurologic deficit, or hemorrhage. Fistula location is again one of the factors correlated with the risk for the development of these symptoms. Awad et al. (1990) observed that only 10–20% of cavernous or transverse sigmoid sinus DAVFs presented with hemorrhage, while greater than 90% of tentorial DAVFs had hemorrhage at presentation. The discrepancy of risk for hemorrhage of fistulae in different locations has been further illustrated by more recent case series, which have reported a 6–11% risk of hemorrhage as the presenting symptom for DAVFs of the cavernous and transverse sigmoid sinuses compared to a 54–79% risk for presentation with hemorrhage in DAVFs of the tentorium or petrosal sinuses (Singh et al., 2008; Gross et al., 2016). Although this indicates that there is a correlation between the location of the fistula and the likelihood of presenting with hemorrhage or other malignant symptoms, it is important to note that fistulae at any location may exhibit aggressive behavior.

Aside from location, other factors have been identified as predictors of presentation with hemorrhage. In a review of 402 patients with DAVFs, Singh et al. (2008) identified cortical venous drainage (CVD), focal neurologic deficit, male sex, location in the posterior fossa, and age greater than 50 at presentation as independent risk factors for hemorrhage. Of note, amongst these risk factors, the presence of CVD was the strongest predictor of presentation with hemorrhage, with an odds ratio of 10.5, more than twice the odds ratio of the next highest predictor amongst the identified risk factors. The unique role of CVD in determining risk of hemorrhage at presentation was further elucidated by Gross and Du (2012), who identified male sex and tentorial or petrous location of fistulae as strongly associated with the presence of cortical drainage. These studies reinforce the findings of previous case series and literature reviews (Awad et al., 1990), and lend support to the long-accepted practice of sorting DAVFs into high- or low-risk categories based on the presence of CVD.

CLASSIFICATION

Given its significance in determining risk for malignant presentation, multiple classification schemes have been proposed for DAVFs, categorizing lesions primarily based on angiographic evidence of CVD. The two most widely applied of these criteria were formulated by Borden et al. (1995) and Cognard et al. (1995). The Borden classification system divides lesions into three categories based on the type of arteriovenous anastomosis (i.e., arterial drainage into a dural sinus or vein versus direct anastomosis with a cortical vein) as well as the presence of CVD. Borden type I lesions are a direct anastomosis between a meningeal artery and a dural sinus or vein with purely anterograde drainage. Type II lesions are also a result of anastomosis between a meningeal artery and a dural vein or sinus, but with reversal of flow and reflux into cortical veins. Type III lesions occur when meningeal arteries directly drain into cortical veins. Type I lesions are considered benign, and their spontaneous remission has been well documented (Newton and Hoyt, 1970; Vinuela et al., 1984; Kutluk et al., 1991). Both type II and type III lesions are considered aggressive, and warrant evaluation for intervention.

The classification system proposed by Cognard and colleagues (1995) takes several other features of a DAVF into account. Based on a retrospective analysis of a single center's experience with 205 patients presenting with DAVF, the Cognard system divides lesions into five categories, corresponding with their risk for malignant presentation. Type I lesions drain directly into a dural sinus, with only anterograde flow. Type II fistulae drain into dural sinuses with resulting retrograde flow, and are subdivided based on whether the retrograde flow involves only dural sinuses (subtype a), only cortical veins (subtype b), or involves both sinuses and cortical veins (subtype a+b). Type III lesions are direct anastomosis of meningeal arteries with cortical veins without ectasia, while type IV lesions are direct cortical vein anastomoses with ectasia. Type V lesions drain directly into the spinal venous system. As in the Borden classification system, those types featuring CVD (i.e., types IIb–V) are considered to be aggressive and should be considered for intervention.

Though amenable to classification by the Borden and Cognard systems, DAVFs involving the cavernous sinus can additionally be classified by the Barrow system (Barrow et al., 1985). In this system, type A lesions are direct, high-flow fistulae between the internal carotid artery and the caverous sinus. Type B, C, and D lesions all represent low-flow fistulous connections between dural arteries and the cavernous sinus, and are distinguished by whether the dural vessels arise from the internal (type B), external (type C), or both (type D) carotid arteries. Barrow type A fistulae are traumatic lesions which are distinct from other forms of DAVFs, and their natural history and management are beyond the scope of this chapter. Type B–D lesions are analogous to DAVF occurring at other locations.

NATURAL HISTORY

Attempts to precisely characterize the natural history of DAVFs have been hindered by the relatively low incidence of these lesions. A prospective, population-based cohort estimated a detection rate of only 0.16 per 100 000 adults per year (Al-Shahi et al., 2003). Consequently, a recent literature review and meta-analysis discovered that all literature describing the natural history of DAVFs is comprised of small, hospital-based cohorts, the vast majority of which are retrospective in design, and which seldom account for the treatment of lesions that are symptomatic at presentation when reporting subsequent outcomes (Jolink et al., 2015). In the absence of high-quality data, quoted risks of morbidity and mortality vary widely.

Brown et al. (1994), in reviewing the records of 54 patients with untreated DAVF with average follow-up of 6.6 years, found that 5 patients had a DAVF-related hemorrhage, resulting in an overall 1.8% per year risk of hemorrhage. This is similar to the annual risk of rupture of 1.5% reported by Soderman et al. (2008), in their evaluation of 53 patients presenting with unruptured DAVFs. In the same study, patients presenting with hemorrhage were calculated to have a 7.4% per year risk of re-hemorrhage. This is significantly lower, however, than the 20–35% re-hemorrhage rate demonstrated by Duffau et al. (1999) in their review of 20 patients with previous hemorrhage from DAVF. Likewise, following 20 patients with long-term CVD, van Dijk et al. (2002) calculated an annual mortality rate of 10.4%, as well as a 15% annual rate of hemorrhage or focal neurologic deficit, rates significantly higher than reported in previous studies.

Looking at an endpoint other than symptomatic presentation, Hetts et al. (2015), found that only 3.2% of 574 patients with angiogram-confirmed DAVF demonstrated progressive or de novo fistulae. Of note, these progressive lesions demonstrated 26%, 32%, and 47% risk for hemorrhage, focal neurologic deficit, or cranial neuropathy, respectively, during the period of the study.

Taken in sum, the available literature on the natural history of DAVFs indicates that these lesions pose a variable risk to affected patients. In general, aggressive features, such as CVD or formation of de novo lesions, indicate that patients are at an increased risk for hemorrhage or significant neurologic symptoms compared to patients with stable lesions not featuring CVD.

RADIOSURGERY FOR THE MANAGEMENT OF DAVFS

Symptomatic lesions or high-grade lesions at increased risk for malignant presentation warrant intervention. Although endovascular embolization, surgical disconnection, or radiosurgery can each be used alone, single-modality treatment may not be effective, and thus many DAVFs require multimodality approaches (Kobayashi and Al-Shahi Shalman, 2014). Endovascular embolization, especially via the transvenous route, has emerged as the most commonly employed solitary treatment of DAVFs. However, studies have shown that embolization alone is not sufficient for complete obliteration of the fistula in 50% of patients (Yang et al., 2013). This is largely due to limitation in visualizing all the feeding vessel with angiography and the technical challenges of superselective catheterization of small feeding vessels. Therefore it is frequently combined with radiosurgery for higher obliterations rates (Kobayashi and Al-Shahi Salman, 2014).

SRS can also be employed as a solitary approach, but is more frequently used as a valuable adjunct to endovascular or open approaches to complex lesions. Scenarios in which radiosurgery would be used as a single modality of therapy include patients whose lesions may not be amenable to other intervention, patients with significant medical comorbidities, or patients with low-grade, minimally symptomatic lesions.

Similar to the experience with parenchymal AVMs, radiosurgery is expected to lead to progressive obliteration of DAVFs over 1–3 years following irradiation. Initial SRS followed by transarterial embolization either the same day or soon after has been shown to be an effective treatment paradigm, with embolization providing more immediate symptomatic relief and decreased flow through the lesion, and SRS providing long-term, durable closure with lower risk for recanalization (Yang et al., 2013). If treatment with both SRS and endovascular embolization is planned, SRS should be performed first to prevent embolization material from obscuring any portion of the nidus during planning and irradiation.

Procedure

In the majority of cases cited in the literature, SRS for the treatment of DAVFs is performed using gamma knife technique (Brown et al., 2005; Chen et al., 2015). The procedure begins with the application of a Leksell stereotactic head frame under conscious sedation with local scalp anesthesia. Given the peripheral location of many noncavernous sinus DAVFs, eccentric head frame placement used to be critical to prevent patient collision with the collimator helmet. This, however, has become less of a concern with new models of gamma knife, which feature fully automated patient positioning via movement of the table on which the patient is introduced into the collimator. Once the head frame is applied, target identification is performed using stereotactic angiography, often requiring injections of bilateral internal and external carotid arteries as well as the bilateral vertebral arteries. High-resolution stereotactic magnetic resonance imaging is then performed to further aid target acquisition and to identify adjacent structures which must be spared from the dose plan. All imaging is exported to a computer workstation, where dose planning is performed.

In the published experience of Yang et al. (2013), the median prescription dose delivered to the margin of DAVFs undergoing SRS was 20 Gy (range, 15–25 Gy). The median isodose volume was 2.0 cm^3 (range 0.2–8.6 cm^3), with prescription isodose of 50% in all but two cases, in which the prescription isodose was 60%. Median maximum dose was 40 Gy, with a range of 30–50 Gy. This is similar to the features of all series contained in a recent meta-analysis of outcomes in DAVFs treated by SRS; across all studies, reported median marginal doses ranged from 18 to 25 Gy, treatment volume ranged from 1 to 9.3 cm^3, and prescription isodose from 40 to 60% (Chen et al., 2015). Likewise, a large, single-center cohort of 105 patients not included in the review employed a median radiation dose delivered to the margin of 19 Gy (range, 15–25 Gy) with a maximum dose of 38 Gy (range, 22–50 Gy) (Yang et al., 2013). Figures 12.1 and 12.2 provide example dose plans for representative cases.

OUTCOMES FOLLOWING RADIOSURGERY

SRS provides reasonable rates of complete occlusion with low risk for serious adverse events. In their systematic review and meta-analysis of 19 studies examining the outcomes of a pooled total of 729 patients having undergone SRS for treatment of 743 DAVFs, Chen et al. (2015) reported a complete obliteration rate of 63%, while calculating post-SRS rates of hemorrhage, neurologic deficit, or death at last follow-up as 1.2%, 1.3%, and 0.3%, respectively.

This review further subdivided outcomes based on the location, angioarchitecture, and Borden classification of the lesions treated, given the relationship of these features to risk for hemorrhage and neurologic deficit. By location, 73% of cavernous sinus DAVFs treated with SRS demonstrated complete occlusion at follow-up, compared to 58% of DAVFs occurring at other locations. No patients in any study with a cavernous sinus DAVF experienced post-SRS hemorrhage while 1.3% of patients with noncavernous sinus locations did. Of patients with noncavernous sinus DAVFs, variability in outcome was seen based on Borden classification, with 67% of type I lesions experiencing complete occlusion, compared to 43% of type II lesions and 50% of type III lesions. The rate of post-SRS hemorrhage for patients with Borden type I was 0%, compared to 2.6% and 3.7% in patients with Borden type II and Borden type III lesions, respectively. The presence or absence of CVD, regardless of location or specific classification type, also demonstrated a relationship with outcome. Of lesions without CVD, 74.5% demonstrated complete occlusion on follow-up imaging, and none experienced post-SRS hemorrhage, compared to 56% occlusion and 4.6% post-SRS hemorrhage rates in lesions with CVD.

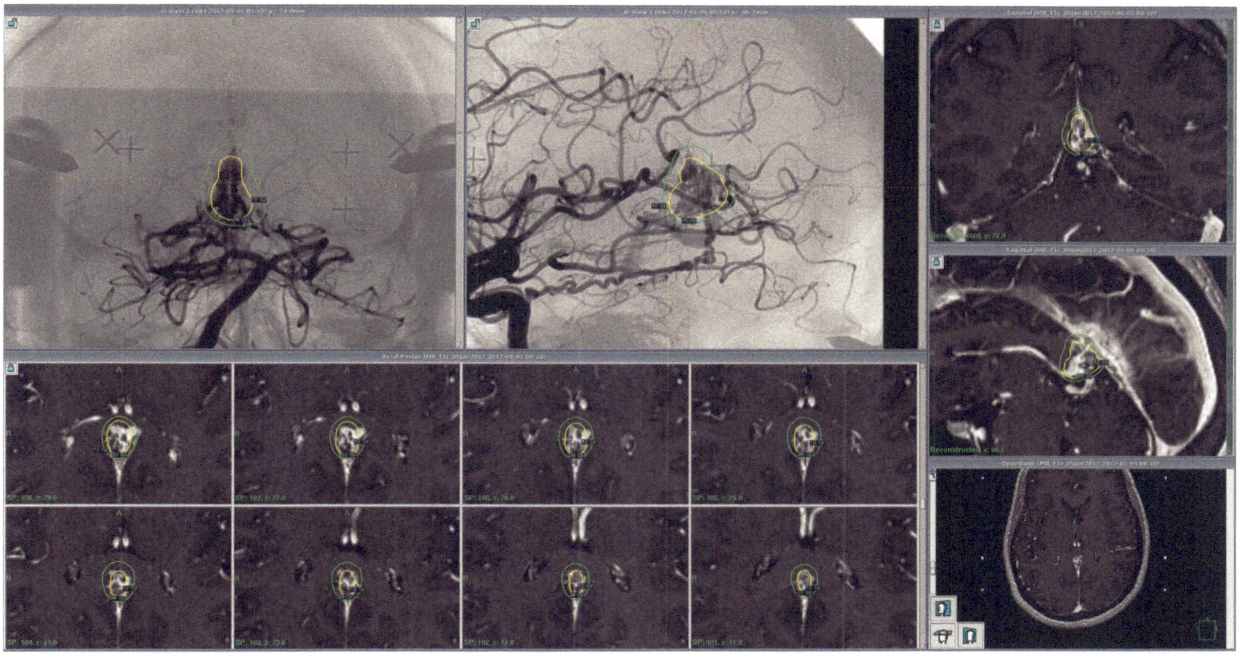

Fig. 12.1. A 46-year-old woman with memory dysfunction found to have a falcotentorial dural arteriovenous fistula. Radiosurgery was performed with 20 Gy to be followed by embolization to reduce venous hypertension.

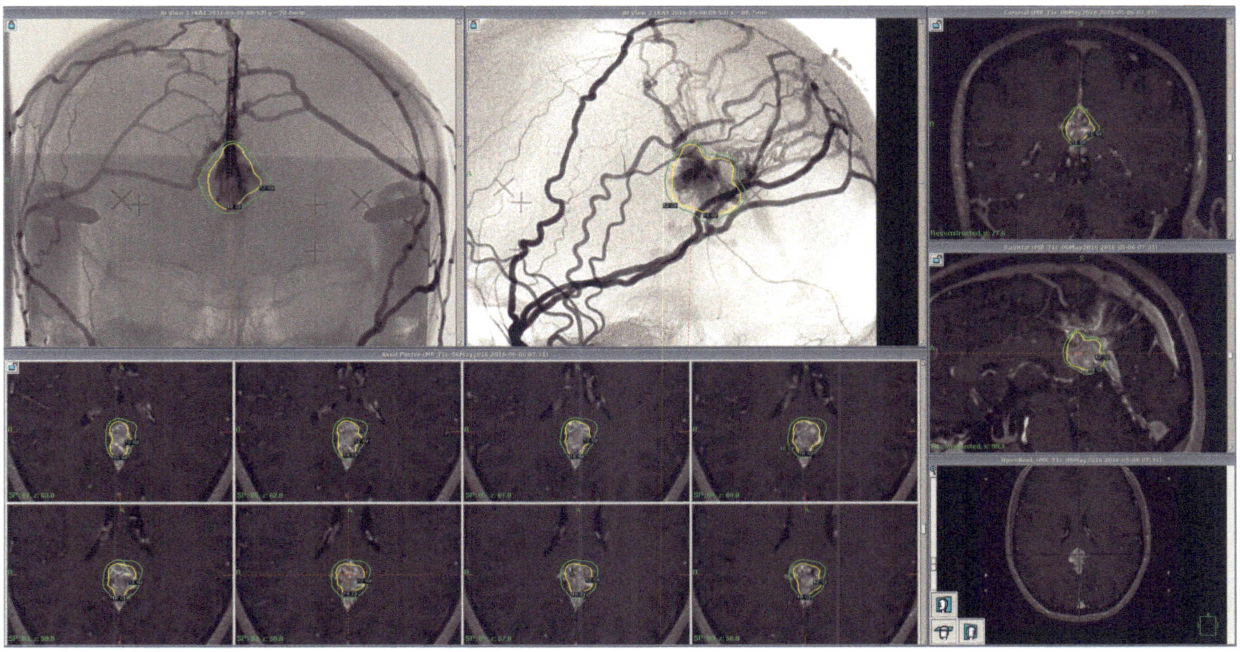

Fig. 12.2. A 36-year-old woman with a falcotentorial arteriovenous fistula associated with 2 years of headache. The radiosurgery plan was created to delivery 18 Gy to the 50% isodose margin.

A significant limitation in reporting outcomes for DAVFs treated by SRS is the previous or concurrent treatment of many patients with endovascular embolization. In the meta-analysis by Chen et al. (2015), at least 17% of all patients included in the analysis were reported to undergo endovascular intervention, while many of the studies included provided insufficient information to allow the authors to extract whether patients had previous treatment or were undergoing SRS alone. A recent retrospective series by Park et al. (2016) helps to address this, following 20 patients with DAVFs treated with SRS alone. The study included eight symptomatic lesions without CVD, 11 lesions with CVD but no symptoms, and one "high-risk" lesion which had both CVD and symptoms, but which was not amenable to endovascular or open intervention. Follow-up with thin-cut time-of-flight magnetic resonance angiography revealed complete occlusion in 18 of 20 lesions at a mean of 29 months' follow-up. All patients with symptomatic lesions experienced complete resolution of their symptoms, with a mean time of 3.5 months from SRS to symptom improvement.

CONCLUSION

DAVFs are complex lesions that pose variable risks to the patients who harbor them. SRS is a safe and effective treatment modality for these lesions, both as a solitary treatment and in conjunction with endovascular embolization or open surgical disconnection. SRS is especially effective in lesions of the cavernous sinus, and in lesions without CVD.

REFERENCES

Al-Shahi R, Bhattacharya J, Currie DG et al. (2003). Prospective, population-based detection of intracranial vascular malformations in adults: the Scottish Intracranial Vascular Malformation Study (SIVMS). Stroke 34: 1163–1169.

Awad I, Little J, Akarawi W et al. (1990). Intracranial dural arteriovenous malformations: factors predisposing to an aggressive neurological course. J Neurosurg 72: 839–850.

Barrow D, Spector R, Braun I et al. (1985). Classification and treatment of spontaneous carotid-cavernous sinus fistulas. J Neurosurg 62: 248–256.

Borden J, Wu J, Shucart W (1995). A proposed classification for spinal and cranial dural arteriovenous fistulous malformations and implications for treatment. J Neurosurg 82: 166–179.

Brown R, Wiebers D, Nichols D (1994). Intracranial dural arteriovenous fistulae: angiographic predictors of intracranial hemorrhage and clinical outcome in nonsurgical patients. J Neurosurg 81: 531–538.

Brown R, Flemming K, Meyer F et al. (2005). Natural history, evaluation, and management of intracranial vascular malformations. Mayo Clin Proc 80: 269–281.

Chaudhary M, Sachdev V, Cho S et al. (1982). Dural arteriovenous malformation of the major venous sinuses: an acquired lesion. AJNR Am J Neuroradiol 3: 13–19.

Chen C, Lee C, Ding D et al. (2015). Stereotactic radiosurgery for intracranial dural arteriovenous fistulas: a systematic review. J Neurosurg 122: 353–362.

Cognard C, Gobin YP, Pierot L et al. (1995). Cerebral dural arteriovenous fistulas: clinical and angiographic correlation with a revised classification of venous drainage. Radiology 194: 671–680.

Duffau H, Lopes M, Janosevic V et al. (1999). Early rebleeding from intracranial dural arteriovenous fistulas: report of 20 cases and review of the literature. J Neurosurg 90: 78–84.

Gross B, Du R (2012). The natural history of cerebral dural arteriovenous fistulae. Neurosurgery 71: 594–602.

Gross B, Albuquerque F, Moon K et al. (2016). Evolution of treatment and a detailed analysis of occlusion, recurrence, and clinical outcomes in an endovascular library of 260 dural arteriovenous fistulas. J Neurosurg Sep2: 1–10. http://dx.doi.org/10.3171/2016.5.JNS16331; E pub ahead of print.

Herman J, Spetzler R, Bederson J et al. (1995). Genesis of a dural arteriovenous malformation in a rat model. J Neurosurg 83: 539–545.

Hetts S, Tsai T, Cooke D et al. (2015). Progressive versus nonprogressive intracranial dural arteriovenous fistulas: characteristics and outcomes. AJNR Am J Neuroradiol 36: 1912–1919.

Houser O, Baker H, Rhoton A et al. (1972). Intracranial dural arteriovenous malformations. Radiology 105: 55–64.

Houser O, Campbell J, Campbell R et al. (1979). Arteriovenous malformation affecting the transverse dural venous sinus – an acquired lesion. Mayo Clin Proc 54: 651–661.

Jolink W, van Dijk J, van Asch C et al. (2015). Outcome after intracranial haemorrhage from dural arteriovenous fistulae; a systematic review and case-series. J Neurol 262: 2678–2683.

Kobayashi A, Al-Shahi Salman R (2014). Prognosis and treatment of intracranial dural arteriovenous fistulae: a systematic review and meta-analysis. Int J Stroke 9: 670–677.

Kutluk K, Schumacher M, Mironov A (1991). The role of sinus thrombosis in occipital dural arteriovenous malformations – development and spontaneous closure. Neurochirurgica 34: 144–147.

Lasjaunias P, Chiu M, ter Brugge K et al. (1986). Neurological manifestations of intracranial dural arteriovenous malformations. J Neurosurg 64: 724–730.

Lawton M, Jacobowitz R, Spetzler R (1997). Redefined role of angiogenesis in the pathogenesis of dural arteriovenous malformations. J Neurosurg 87: 267–274.

Newton T, Hoyt W (1970). Dural arteriovenous shunts in the region of the cavernous sinus. Neuroradiology 1: 71–81.

Nishijima M, Takaku A, Endo S et al. (1992). Etiological evaluation of dural arteriovenous malformations of the lateral and sigmoid sinuses based on histopathological examinations. J Neurosurg 76: 600–606.

Park K, Kang D, Park S et al. (2016). The efficacy of gamma knife radiosurgery alone as a primary treatment for intracranial dural arteriovenous fistulas. Acta Neurochir 158: 821–828.

Singh V, Smith W, Lawton M et al. (2008). Risk factors for hemorrhagic presentation in patients with dural arteriovenous fistulae. Neurosurgery 62: 628–635.

Soderman M, Pavic L, Edner G et al. (2008). Natural history of dural arteriovenous shunts. Stroke 39: 1735–1739.

Terada T, Higashida R, Halbach V et al. (1994). Development of acquired arteriovenous fistulas in rats due to venous hypertension. J Neurosurg 80: 884–889.

Uranishi R, Nakase H, Sakaki T (1999). Expression of angiogenic growth factors in dural arteriovenous fistula. J Neurosurg 91: 781–786.

van Dijk J, ter Brugge K, Willinsky R et al. (2002). Clinical course of cranial dural arteriovenous fistulas with long-term persistent cortical venous reflux. Stroke 33: 1233–1236.

Vinuela F, Fox A, Debrun G et al. (1984). Spontaneous carotid-cavernous fistulas: clinical, radiological, and therapeutic considerations. Experience with 20 cases. J Neurosurg 60: 976–984.

Yang H, Kano H, Kondziolka D et al. (2013). Stereotactic radiosurgery with or without embolization for intracranial dural arteriovenous fistulas. Prog Neurol Surg 27: 195–204.

Section 3

Spinal vascular malformations

Handbook of Clinical Neurology, Vol. 143 (3rd series)
Arteriovenous and Cavernous Malformations
R.F. Spetzler, K. Moon, and R.O. Almefty, Editors
http://dx.doi.org/10.1016/B978-0-444-63640-9.00013-8

Chapter 13

Classification and pathophysiology of spinal vascular malformations

ISAAC JOSH ABECASSIS[1], JOSHUA W. OSBUN[2], AND LOUIS KIM[1,3]*

[1]*Department of Neurological Surgery, University of Washington, Seattle, WA, USA*

[2]*Department of Neurosurgery, Emory University, Atlanta, GA, USA*

[3]*Department of Radiology, University of Washington, Seattle, WA, USA*

Abstract

Spinal arteriovenous malformations and fistulas comprise spinal vascular malformations (sVMs), a rare but challenging neurosurgic entity. A number of mechanisms have been proposed as explanations of neurologic decline in the setting of arteriovenous shunting, including venous hypertension, venous congestion, hemorrhage, vascular steal, or mass effect, which can be worsened with venous varices or aneurysmal dilations. Anatomic location and features dictate the categorization of sVMs. Two major systems are in place for classification of sVMs – (1) the American/French/English connection and (2) the Spetzler system – as well as a number of systems that have been described over the past century.

INTRODUCTION

Spinal arteriovenous malformations (sAVMs) and fistulas (sAVFs) comprise spinal vascular malformations (sVMs), a rare but challenging neurosurgical entity, comprising approximately 4% of all intradural spinal cord lesions (Bao and Ling, 1997; Kim et al., 2013; Osbun et al., 2015). Other spinal vascular lesions not included in the sVM category are aneurysms and vascular neoplasms (i.e., hemangioblastomas). In 1890, Berenbruch first observed a spinal lesion in surgery that would later be recognized as an sVM at autopsy, and Fedor Krause (1911) reported the first surgical encounter during a laminectomy of a spinal dural arteriovenous fistula (AVF) in 1910, though resection was not attempted. Two years later, Charles Elsberg (1916) made the first surgical attempt at resection, with no clinical improvement at 6-month follow-up. Foix and Alajounanine, two French physicians, reported in 1926 a case of rapid neurologic decline due to thrombosis of an sAVM, a syndrome they termed subacute necrotic myelopathy (Heros, 2009). Two years later, in 1928, Michon coined a separate term to describe a syndrome of sudden and abrupt back pain associated with subarachnoid hemorrhage related to an sAVM, a syndrome now referred to as *coup de poignard* of Michon. Historically, sVMs were either deemed amenable to surgical resection or not, with high morbidity and mortality naturally coupled to their course. The advent of both microsurgical technique and endovascular neurosurgery reformatted the landscape for both diagnosis and treatment of these lesions, and elegant multimodality treatment regimens are now standardly employed, including conservative observation, endovascular embolization, microsurgical resection, and radiosurgery.

ANATOMY

The anterior two-thirds of the spinal cord is supplied by the anterior spinal artery (ASA), which originates from the two branches of the vertebral arteries, travels along the anterior medial fissure, with additional contributions from approximately six to eight segmental radiculomedullary arteries (there are initially 31 pairs of segmental arteries at each level during the embryonic

*Correspondence to: Louis Kim, M.D., Department of Neurological Surgery, University of Washington, Seattle WA 98122, USA. E-mail: ljkim1@u.washington.edu

period with eventual regression of most of them). These typically include an artery at C3 (also referred to as the artery of cervical enlargement, from the vertebral artery), C6 (from the deep cervical artery), C8 (from the costocervical trunk), T4/T5, and the artery of Adamkiewicz, which usually originates on the left side between the T9 and L2 level (Taveras and Wood, 1976). This anterior blood supply is also referred to as the "sulcocommisural system" or centrifugal system, and gives rise to 200–400 sulcocommissural arteries within the ventral sulcus to supply the central gray matter and some peripheral white matter (Causin et al., 2013). There are extensive craniocaudal anastomoses both on the pial surface and within the substance of the cord.

The posterior one-third relies on a pair of posterior spinal arteries (PSA), which originate from the intradural portion of the vertebral artery or the posterior inferior cerebellar artery, with contributions from 10–23 radiculopial branches (Krings and Geibprasert, 2009). This posterior system is also referred to as the dorsolateral pial supply or centripetal system. This network gives off radial arteries that circumferentially run along the spinal cord, reaching anastomosis with the ventral system, and giving off axially oriented perforating branches into the white matter. Inferiorly on the spinal cord, the ASA and PSA systems are anatomized around the conus. The venous system begins with a set of radially oriented, intrinsic spinal cord veins and small superficial pial veins that lead into the superficial longitudinal median spinal cord veins. Eventually, these drain into a set of veins that mirror the arterial system (i.e., the anterior and posterior median spinal veins), with a vast interconnected network, including transmedullary anastomoses (Krings and Geibprasert, 2009). Ultimately, the venous blood reaches the epidural plexus and the extraspinal veins and plexus, with a reflux-impeding mechanism within the dura mater (Causin et al., 2013).

PATHOPHYSIOLOGY

A number of mechanisms have been proposed as explanations of neurologic decline in the setting of arteriovenous shunting in both sAVFs and sAVMs. With an abnormal fistulous connection there is resultant venous hypertension, and investigators in the past have documented intraoperatively elevations in venous pressure as high as 70% of the systemic arterial pressure (Hassler et al., 1989; Hassler and Thron, 1994). Venous congestion is now recognized as a primary source of neurologic disability with sVMs, and fortunately one that oftentimes is potentially reversible (Kataoka et al., 2001). More recently, animal model experiments have validated venous hypertension as a causative agent in myelopathy, and shown pathologic correlates, including hyalinized small blood vessels, perivascular/intraparenchymal lymphocytic infiltration, glial cell proliferation, and neuronal degeneration (Zhang et al., 2013). Hemorrhage can also precipitate neurologic decline: 25% of sVM patients will present with a ruptured lesion, as evidenced by subarachnoid hemorrhage, intraparenchymal hemorrhage, or a combination. Rupture at presentation is more frequently encountered in sAVMs (38%) as compared to sAVFs (9%) (Rangel-Castilla et al., 2014). Hemorrhage can injure neurologic structures either directly via disruption of fiber tracts, or indirectly with mass effect. Venous engorgement, venous varices, and aneurysmal dilatations can further contribute to mass effect and compress the spinal cord or nerve roots. Finally, sAVMs and sAVFs induce alterations in the normal spinal blood flow and perfusion pressures via aberrant fistulous connections, creating a phenomenon referred to as vascular steal. In fact, this was previously considered the major mechanism of neurologic decline prior to experimental investigation and understanding of venous congestion (Aminoff et al., 1974; Djindjian et al., 1978; Bandyopadhyay and Sheth, 1999).

CLASSIFICATION SYSTEMS

Anatomic location and features dictate the categorization of sVMs. There are two major systems in place for classification of sVMs – (1) the American/French/English connection and (2) the Spetzler system – though a number have been described and are discussed below.

The first classification system originated from Virchow in 1858 (Virchow, 1864): Virchow observed lesions at autopsy that he referred to as neoplasms or "vascular tumors." Lesions were either considered angioma cavernosum, if there was no parenchyma between the blood vessels, or angioma racemosum. This latter category was further subdivided into telangiectasias and angioma racemosum arteriale sive venosum (i.e., arteriovenous shunts). Elsberg proposed an updated scheme in 1916, to include: (1) aneurysms of spinal vessels; (2) angiomas; and (3) dilated posterior spinal veins, or "hemorrhoids" of the spinal cord. Cushing and Bailey (1928) published a system that separated hemangioblastomas from sVMs, including a plexus of dilated veins, aneurysmal varix, venous angioma, telangiectasis, and a separate category for those venous angiomas or telangiectasias that went on to develop an aberrant arteriovenous fistulous connection. Bergstrand et al. described a very similar system in 1936, as did Wyburn Mason in

1943, but, as with most of these early classification systems, there were limited gains with regard to understanding the pathogenesis or intricacies of the anatomy, as spinal angiography had not yet been described.

The American/English/French connection and the Borden system

During the 1960s, a constellation of advancements yielded tremendous progress in understanding and treating sVMs, including interventional neuroradiology with selective spinal angiography and embolization, the operating microscope, microsurgical instruments such as the bipolar forceps, and neuroanesthesiology (Black, 2006). Three major centers – including the National Institutes of Health in Bethesda (Di Chiro and Wener, 1973; Oldfield et al., 1983), the National Hospital for Neurology and Neurosurgery in London (Aminoff et al., 1974; Kendall and Logue, 1977), and Hospital Lariboisière in Paris (Djindjian, 1975) – generated a combined effort of independent yet collaborative intellectual investigations that yielded what is referred to as the American/English/French connection classification system. This system separates sVMs into three types, with a fourth subtype later described. Type I lesions, or the "single coiled vessel" type, are dural AVFs and consist of a radicular artery draining into an engorged spinal vein on the dorsal aspect of the dural sheath of a nerve root (Klopper et al., 2009). Kendall and Logue first published a description of this type of lesion in 1977 with the novel recognition that these were actually not intradural sAVMs, but rather dural-based lesions with drainage into the previously not yet described coronal plexus of veins (Kendall and Logue, 1977). In total, this type comprises 80% of sVMs.

Borden et al. proposed a classification system in 1995 that combined cranial and spinal dural-based lesions in a unified and organized system oriented towards treatment paradigms. This is thus a subclassification for type I sVMs, and the authors accordingly use the term dural arteriovenous fistulous malformations (AVFMs). Type I lesions (spinal) have anterograde flow and involve drainage into Batson's epidural venous plexus, type II have retrograde drainage into spinal perimedullary veins, and type III lesions are supplied by meningeal branches of a radicular artery with drainage into the dorsal spinal coronal venous plexus through subarachnoid veins. The authors further classify subtype "a" lesions as having a single arterial feeder, whereas "b" lesions have two or more. Importantly, these lesions can be associated with other AVMs in 15–20% of cases; thus a full angiographic evaluation of dural feeders throughout the central nervous system is warranted, including the cranial circulation, cervical cord, selective thoracic and lumbar intercostal injections, the iliac arteries, aorta, and the lateral sacral arteries (Larsen et al., 1995).

In contrast to the slow flow observed with dural AVFs, an intradural sAVM is a high-flow entity and often presents with hemorrhage and abrupt onset of clinical symptoms. Whereas type I lesions tend to cause neurologic morbidity via venous hypertension, type II lesions – also referred to as spinal glomus AVMs – are rarer lesions that cause damage via hemorrhage, mass effect, vascular steal, and venous hypertension. They comprise 15–20% of sVMs and demonstrate many similarities to the cerebral AVM counterpart, with a medullary artery feeding into a nidus, and then into a venous drainage system. Gross and Du (2013) performed a pooled analysis to demonstrate a 4% annual hemorrhage rate, increased to 10% with prior hemorrhage, similar to cerebral AVMs.

Type III lesions are also intramedullary lesions and are referred to as juvenile spinal AVMs. These are larger type II lesions that occupy the entire cross-sectional area of the spinal cord and actually invade the nearby vertebral body, sometimes producing early spondylolysis and scoliosis. They are supplied by pial, dural, and paraspinal arteries with an expansive and diffuse nidus – in contrast to the compact nidus often observed in type II lesions – often producing neurologic morbidity secondary to both mass effect and venous hypertension. Given their size, type III lesions tend to present at the youngest age, with hemorrhage in 33% of cases, and an associated aneurysm in nearly half at presentation (Gross and Du, 2014). Gross and Du (2014) performed a separate pooled analysis for this subgroup of sVMs and found an annual hemorrhage rate of 2.1%, with associated arterial aneurysms as a risk factor for hemorrhage. The authors further argue that the particularly distinct and unique attribute within this type III category is the intradural and extradural extension of the lesion; thus they refer to the category as "juvenile extradural-intradural AVMs."

Bao and Ling (1997) proposed a classification scheme in 1997 separating lesions into intradural AVFs, dural AVFs, intramedullary AVMs (including both "glomus" type II and "juvenile" type III lesions), paravertebral AVMs, and Cobb's syndrome. Intradural AVFs are defined as direct AV shunts without a nidus, usually located on the surface of the spinal cord or just beneath the pia. They are further subdivided into type I, a small, low-flow single-hole fistula without obvious dilatation of the feeding artery and draining vein; type II, a high-flow single-hole fistula with apparent dilatation of the feeding artery and draining vein; and type III, a giant high-flow fistula with multiple feeders and a huge dilated vein. Dural AVFs in this scheme are similar to the classic type I lesion, with a small fistulous connection in the

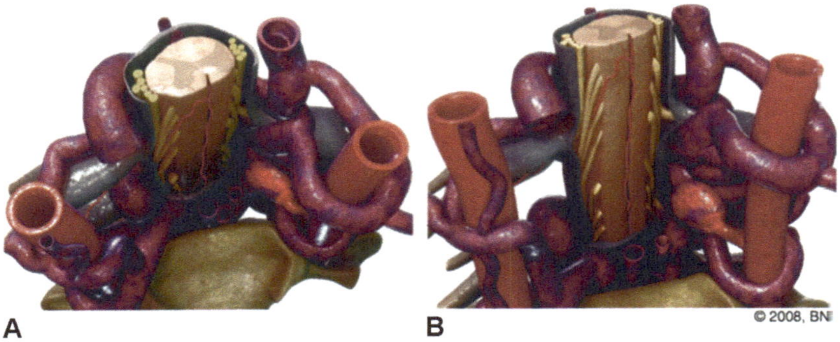

Fig. 13.1. Three-dimensional computer-generated images of extradural spinal arteriovenous fistula with (**A**) axial and (**B**) coronal views. (Used with permission from the Barrow Neurological Institute, Phoenix, Arizona (Dashti et al., 2011).)

spinal dura near the nerve root with medullary venous drainage.

Finally, type IV lesions – described later on by both Djindjian et al. (1977) and Heros et al. (1986) – are intradural and extra/perimedullary. These consist of a direct fistula between a spinal artery (usually artery of Adamkiewicz or ASA) draining into a dilated perimedullary venous plexus without an intervening nidus. Most present with a progressive myelopathy, though over one-third can present acutely with or without associated subarachnoid hemorrhage (Antonietti et al., 2010). Merland et al. (1980) further described three subtypes of type IV lesions (Mourier et al., 1993). In the first type, there is a single arterial feeder from the ASA, mild venous dilatation, and a single, small, and slow fistula. Type IV.II and III lesions have multiple arterial feeders (including both ASAs and at least one PSA) with a dilated ASA; IV.II lesions have multiple AVFs of medium size, whereas the latter group has a single but giant AVF. Both have high fistulous flow. Additionally, type IV.III lesions have giant venous ectasia with rapid venous drainage. Generally, subtype I lesions are best treated surgically due to difficulties catheterizing the feeding artery, type II fistulas are best treated via a combination of embolization and surgery, and type III fistulas are most commonly treated via endovascular intervention (Antonietti et al., 2010).

The Spetzler et al. classification system

Based on 155 surgically treated sVMs, a group of clinicians from the Hôpital Bicêtre, France, proposed a classification system in 2002 whereby lesions were subdivided into AVMs, AVFs, and separate entities, with a separate category allocated for lesions with a genetic background (Rodesch et al., 2002). This same year, Spetzler and his colleagues expanded upon the prior American/English/French system by further delineating the specific anatomic location of sAVFs and sAVMs, adding a separate category for neoplastic vascular lesions (including hemangioblastomas and cavernous malformations), and spinal aneurysms, and introduced AVMs of the conus medullaris for the first time. This was refined by Kim and Spetzler in 2006, whereby AVFs and AVMs were specifically separated (Kim et al., 2013; Osbun et al., 2015). Within this most recent system, sAVFs are classified as extradural, intradural-dorsal, or intradural-ventral. sAVMs are classified similarly based on location as extradural-intradural, intramedullary (with either a diffuse or compact nidus), or conus medullaris. Genders are affected equally, though sAVMs seem to be slightly more common (60%) than sAVFs (40%), with lesions occurring at an increased frequency in the thoracic spine (61%) compared to the cervical (23%), lumbar (14%), and sacral (2%) spine (Rangel-Castilla et al., 2014).

Extradural sAVFs

Extradural sAVFs reside in the epidural space of the spinal canal and typically comprise the minority of representation amongst all sAVFs (Fig. 13.1). Prior to the updated system proposed by Spetzler, these lesions were commonly mistakenly considered to be dural-based lesions (type I). These fistulas arise from an anomalous connection between the epidural segment of a radicular artery and an epidural venous plexus, with the fistula located totally epidurally (not in the nerve root sleeve as with type I lesions), and any intradural venous drainage arising from a venous pouch (not direct drainage). Due to the plethora of venous channels in the epidural space, these lesions tend to present with extensive venous engorgement and local mass effect on to the nerve roots and thecal sac, though venous hypertension and vascular steal are also observed (Kim and Spetzler, 2006). Accordingly, an asymptomatic presentation is less common in this subgroup of sAVFs (Rangel-Castilla et al., 2014). Rangel-Castilla et al. (2011) further subclassified these as: type A, fistula that has both an epidural

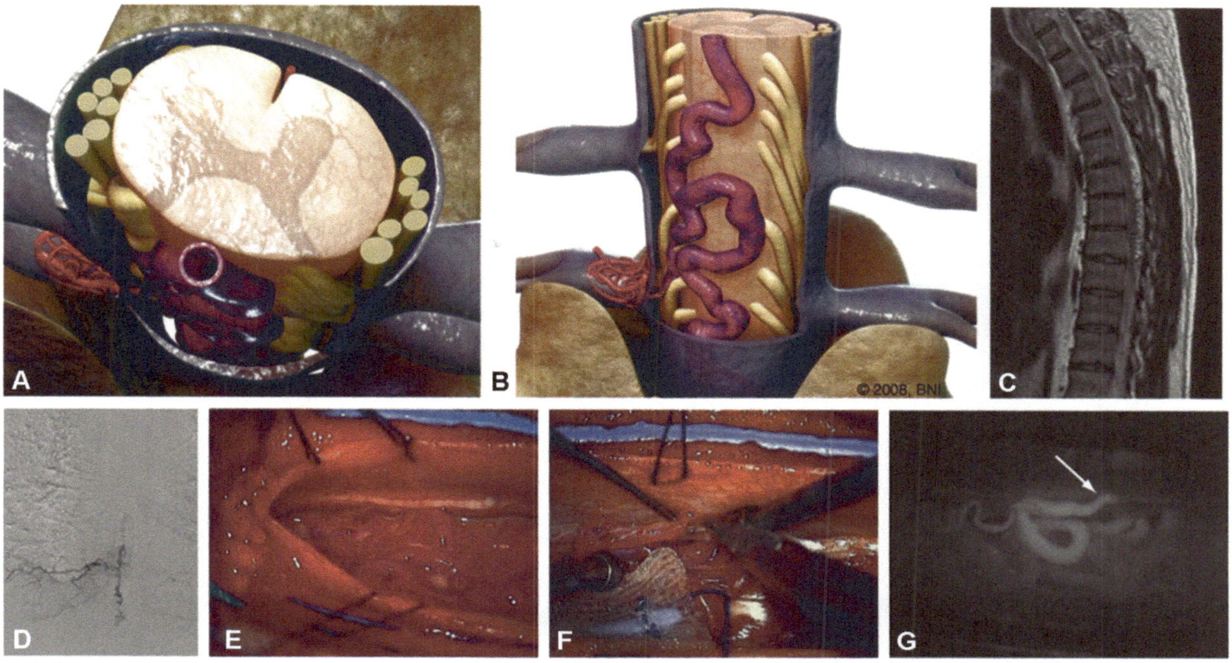

Fig. 13.2. Intradural dorsal spinal arteriovenous fistula. (**A**) Axial and (**B**) coronal computer-generated images. (**C**) Sagittal view of T2-weighted magnetic resonance imaging showing vascular flow voids intradurally on the dorsal aspect of spinal cord. (**D**) Spinal angiogram demonstrating the fistula. (**E**) Intraoperative photograph demonstrating arterialized midline veins on the dorsal spinal cord; (**F**) intraoperative photograph during ligation of the veins; and (**G**) intraoperative indocyanine green angiography with white arrow indicating fistulous connection and beginning of arterialized veins. (Courtesy of the Barrow Neurological Institute, Phoenix, Arizona (Dashti et al., 2011).)

venous plexus and a vein that pierces the dura, travels intradurally, and drains into the venous plexus surrounding the spinal cord; type B, fistula that exclusively drains into Batson's plexus, with B1 lesions causing compression of the thecal sac and/or myelopathy and B2 type without radiographic or clinical sequelae.

Intradural dorsal sAVFs

Intradural dorsal sAVFs loosely correlate to the classic type I lesions, as they entail a midline fistulous connection between a radicular artery after it enters the dural sleeve of a nerve root, and a medullary or radicular vein (Fig. 13.2). This group of sAVF is thought to originate from venous outflow obstruction (e.g., spontaneous thrombosis, trauma) and subsequent arterialization of the coronal venous plexus.

Intradural ventral sAVFs

Intradural ventral sAVFs are composed of an abnormal connection between the ASA and a medullary or radicular vein, which typically drains into a fragile coronal venous plexus that is subject to rupture with venous hypertension (Fig. 13.3). Thus, mass effect, venous hypertension, and hemorrhage are modes of pathophysiology. High flow can also yield vascular steal. Anson and Spetzler (1992) further subdivide these lesions into type A as small with a single feeder, type B as intermediate size with a major ASA feeder and smaller minor feeders, and type C fistula as giant, multipediculated lesions with massively dilated venous channels (Kim and Spetzler, 2006).

Extradural-intradural sAVMs

Extradural-intradural sAVMs correspond to the type III "juvenile" or metameric lesions (Fig. 13.4). There is extensive involvement of the nearby structures, including skin, muscle, bone, spinal cord, and nerve roots. If there is complete involvement of the somite, this is referred to as Cobb's syndrome. These AVMs present with venous engorgement, mass effect, and arterial steal.

Intramedullary sAVMs

Intramedullary sAVMs correspond to the type II "glomus"-type lesions and are considered the true AVM of the spine (Figs 13.5 and 13.6). As discussed above, there is a tremendous amount of similarity to the cerebral AVM counterpart, with an annual

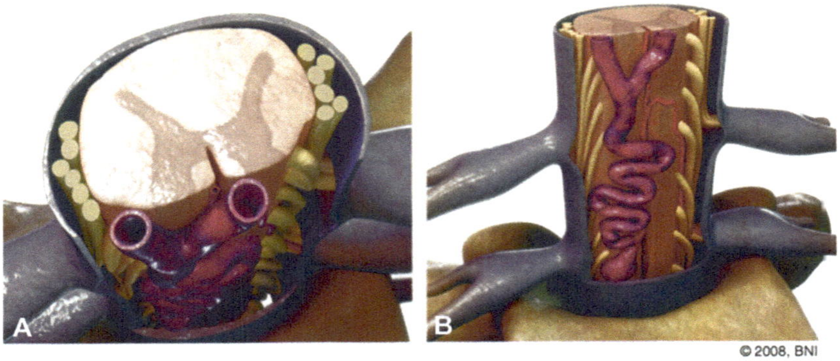

Fig. 13.3. Intradural ventral spinal arteriovenous fistula. (**A**) Axial and (**B**) coronal computer-generated images. (Courtesy of the Barrow Neurological Institute, Phoenix, Arizona (Dashti et al., 2011).)

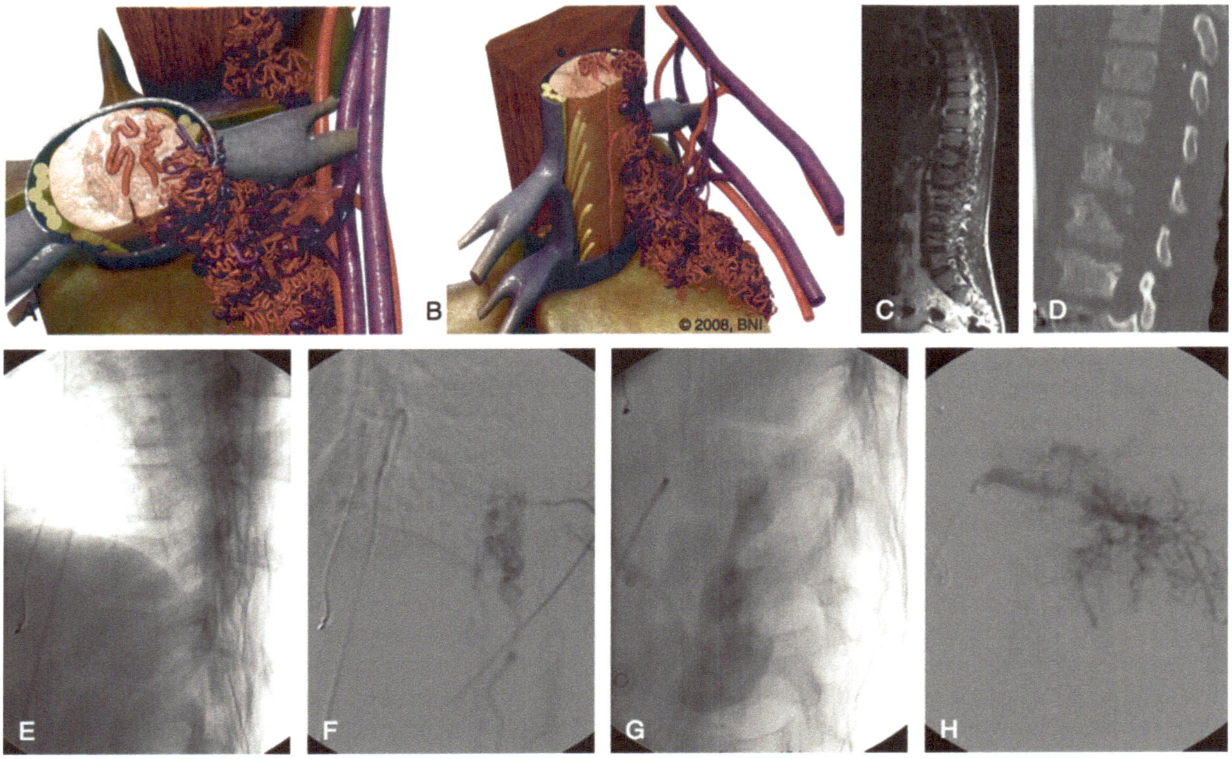

Fig. 13.4. Extradural-intradural spinal arteriovenous malformation (sAVM). (**A**) Axial and (**B**) coronal computer-generated images. (**C**) Sagittal view of T2-weighted magnetic reosonance imaging showing diffuse involvement extending from the intradural space, with vascular flow voids, into the paraspinal musculature and soft tissue posteriorly, with additional involvement anteriorly into the vertebral bodies and intervertebral disc space. (**D**) Sagittal view of computed tomography scan showing erosion of vertebral body secondary to involvement of sAVM. (**E**) Unsubtracted and (**F**) subtracted spinal segmental artery angiogram in the sagittal view of an extradural intradural sAVM encompassing multiple spinal elements, including bone of the vertebral body and posterior elements, paraspinal musculature, dura, and spinal cord parenchyma. (**G**) Unsubtracted and (**H**) subtracted spinal radicular artery angiogram in the magnified sagittal view demonstrating the lesion's extensive involvement of an entire somite. (Courtesy of the Barrow Neurological Institute, Phoenix, Arizona (Dashti et al., 2011).)

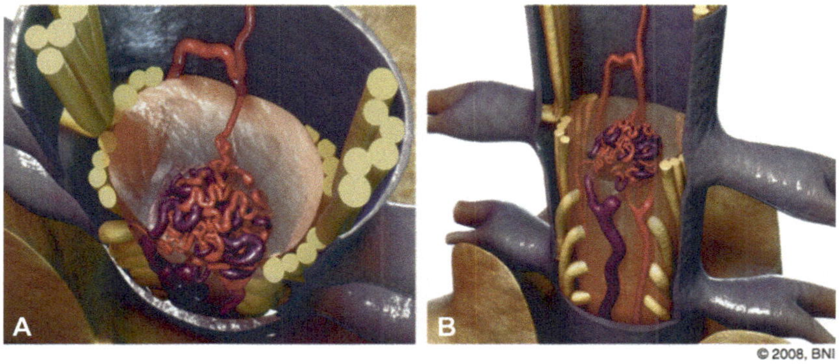

Fig. 13.5. Compact intramedullary spinal arteriovenous malformation. (**A**) Axial and (**B**) coronal computer-generated images. (Courtesy of the Barrow Neurological Institute, Phoenix, Arizona (Dashti et al., 2011).)

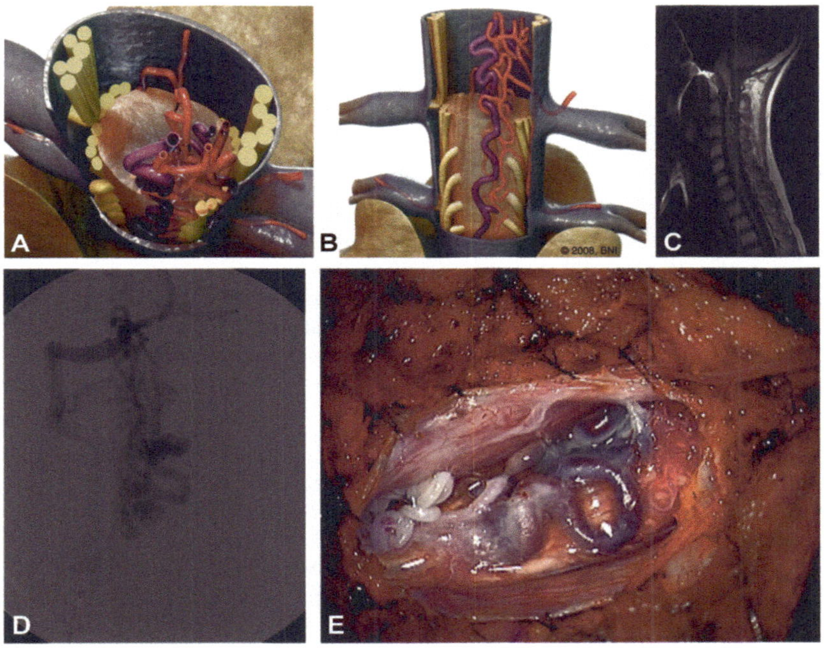

Fig. 13.6. Diffuse intramedullary spinal arteriovenous malformation. (**A**) Axial and (**B**) coronal computer-generated images. (**C**) Sagittal view of T1-weighted magnetic resonance imaging of cervical spine with gadolinium contrast demonstrating vascular malformation on the dorsal aspect of spinal cord at the C2–3 level. (**D**) Spinal angiography demonstrating a diffuse nidus with early venous drainage and a dilated venous system. (**E**) Intraoperative photograph demonstrating large, tortuous, and dilated draining veins on the pial surface, mixed with feeding arteries. (Courtesy of the Barrow Neurological Institute, Phoenix, Arizona (Dashti et al., 2011).)

hemorrhage rate of 4% per year. Intramedullary sAVMs are supplied by one or multiple feeding arteries from the ASA and/or PSA, with a compact or diffuse nidus, and venous drainage into the coronal venous plexus.

Conus medullaris sAVMs

Conus medullaris sAVMs were first described in the Spetzler system (Spetzler et al., 2002) as a separate category (Fig. 13.7). These sAVM are supplied by multiple feeders from both the ASA and PSA, with a nidus centered around the conus medullaris, and marked venous dilatation owing to the amount of open space in the area. Venous hypertension, ischemia, and mass effect all play a role in the pathophysiology of these lesions, and it is not uncommon for cauda equina syndrome to occur.

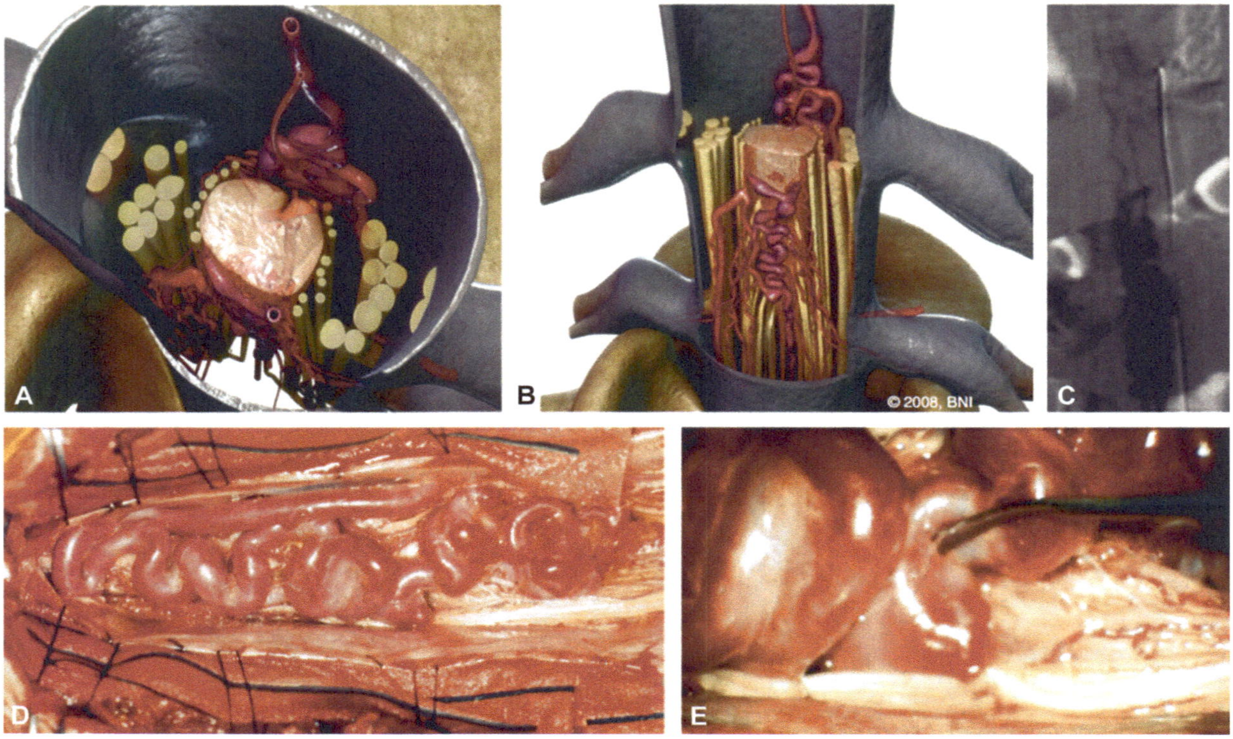

Fig. 13.7. Conus arteriovenous malformation (AVM). (**A**) Axial and (**B**) coronal views of computer-generated images of a conus AVM. (**C**) Spinal angiogram demonstrating the fistula. (**D**) Intraoperative photograph demonstrating arterialized midline veins in the conus region and (**E**) identification and ligation of major feeding arteries. (Courtesy of the Barrow Neurological Institute, Phoenix, Arizona (Dashti et al., 2011).)

References

Aminoff MJ, Barnard RO, Logue V (1974). The pathophysiology of spinal vascular malformations. J Neurol Sci 23: 255–263.

Anson JA, Spetzler RF (1992). Classification of spinal arteriovenous malformations and implications of treatment. BNIQ 8: 2–8.

Antonietti L, Sheth SA, Halbach VV et al. (2010). Long-term outcome in the repair of spinal cord perimedullary arteriovenous fistulas. AJNR Am J Neuroradiol 31: 1824–1830.

Bandyopadhyay S, Sheth RD (1999). Acute spinal cord infarction: vascular steal in arteriovenous malformation. J Child Neurol 14: 685–687.

Bao YH, Ling F (1997). Classification and therapeutic modalities of spinal vascular malformations in 80 patients. Neurosurgery 40: 75–81.

Berenbruch K (1890). Ein fall von multiplen Angio-Lipomen Kombiniert mit einem angiom des Ruckenmarks, Doctoral thesis, Universität Tubingen.

Bergstrand H, Olivecrona H, Tonnis W (1936). Gefassmissbildungen und Gefassgeschwulste des Gehirns, Georg Thieme, Leipzig.

Black P (2006). Spinal vascular malformations: an historical perspective. Neurosurg Focus 21. E11.

Borden JA, Wu JK, Shucart WA (1995). A proposed classification for spinal and cranial dural arteriovenous fistulous malformations and implications for treatment. J Neurosurg 82: 166–179.

Causin F, Gabrielli J, Orru E (2013). Classification and treatment of vascular malformations of the spinal cord. In: M Muto (Ed.), Interventional Neuroradiology of the Spine: Clinical Features, Diagnosis and Therapy, Springer Science & Business Media, Milan.

Cushing H, Bailey P (1928). Tumors Arising from the Blood-Vessels of the Brain Angiomatous Malformation and Hemangioblastomas, Baillière, Tindall & Cox, Springfield, IL.

Dashti SD, Toledo K, Kim LJ et al. (2011). Classification of Spinal Arteriovenous Lesions: Arteriovenous Fistulas and Arteriovenous Malformations. In: W Hr (Ed.), Youmans Textbook of Neurological Surgery, 6th ed. Elsevier, Philadelphia.

Di Chiro G, Wener L (1973). Angiography of the spinal cord. A review of contemporary techniques and applications. J Neurosurg 39: 1–29.

Djindjian R (1975). Embolization of angiomas of the spinal cord. Surg Neurol 4: 411–420.

Djindjian M, Djindjian R, Rey A et al. (1977). Intradural extramedullary spinal arterio-venous malformations fed by the anterior spinal artery. Surg Neurol 8: 85–93.

Djindjian M, Djindjian R, Hurth M et al. (1978). Steal phenomena in spinal arteriovenous malformations. J Neuroradiol 5: 187–201.

Elsberg CA (1916). Diagnosis and Treatment of Surgical Diseases of the Spinal Cord and Its Membranes, WB Saunders, Philadelphia.

Foix C, Alajounanine T (1926). La myélite central angiohypertrophique évolution progressive: Paraplégie amyotrophique lentement ascendan d'abord spasmodique, puis flasque. Rev Neuro 2: 1–42.

Gross BA, Du R (2013). Spinal glomus (type II) arteriovenous malformations: a pooled analysis of hemorrhage risk and results of intervention. Neurosurgery 72: 25–32; discussion 32.

Gross BA, Du R (2014). Spinal juvenile (type III) extradural-intradural arteriovenous malformations. J Neurosurg Spine 20: 452–458.

Hassler W, Thron A (1994). Flow velocity and pressure measurements in spinal dural arteriovenous fistulas. Neurosurg Rev 17: 29–36.

Hassler W, Thron A, Grote EH (1989). Hemodynamics of spinal dural arteriovenous fistulas. An intraoperative study. J Neurosurg 70: 360–370.

Heros RC (2009). Foix-Alajouanine syndrome: what is it? J Neurosurg 111: 900–901.

Heros RC, Debrun GM, Ojemann RG et al. (1986). Direct spinal arteriovenous fistula: a new type of spinal AVM. Case report. J Neurosurg 64: 134–139.

Kataoka H, Miyamoto S, Nagata I et al. (2001). Venous congestion is a major cause of neurological deterioration in spinal arteriovenous malformations. Neurosurgery 48: 1224–1229; discussion 1229–1230.

Kendall BE, Logue V (1977). Spinal epidural angiomatous malformations draining into intrathecal veins. Neuroradiology 13: 181–189.

Kim LJ, Spetzler RF (2006). Classification and surgical management of spinal arteriovenous lesions: arteriovenous fistulae and arteriovenous malformations. Neurosurgery 59: S195–S201; discussion S193–S213.

Kim LJ, Osbun JW, Thompson BG et al. (2013). Arteriovenous Malformations of the Craniovertebral Junction: Spinal and Posterior Fossa AVMs. In: NC Bambakidis, CA Dickman, RF Spetzler et al. (Eds.), Surgery of the Craniovertebral Junction, 2nd edn. Thieme, New York.

Klopper HB, Surdell DL, Thorell WE (2009). Type I spinal dural arteriovenous fistulas: historical review and illustrative case. Neurosurg Focus 26. E3.

Krause F (1911). Chirurgie des Gehirns und Ruckenmarks nach eigenen Erfahrungen. Berlin: Urban and Schwartzenberg 1: 13.

Krings T, Geibprasert S (2009). Spinal dural arteriovenous fistulas. AJNR Am J Neuroradiol 30: 639–648.

Larsen DW, Halbach VV, Teitelbaum GP et al. (1995). Spinal dural arteriovenous fistulas supplied by branches of the internal iliac arteries. Surg Neurol 43: 35–40; discussion 40–41.

Merland JJ, Riche MC, Chiras J (1980). Intraspinal extramedullary arteriovenous fistulae draining into the medullary veins. J Neuroradiol 7: 271–320.

Michon P (1928). Le coup de poignard rachidien. Symptome initial de certaines hemorragies sous-arachnoidiennes. Essai sur les hemorragies meningees spinales. Presse Med 964–966.

Mourier KL, Gobin YP, George B et al. (1993). Intradural perimedullary arteriovenous fistulae: results of surgical and endovascular treatment in a series of 35 cases. Neurosurgery 32: 885–891; discussion 891.

Oldfield EH, Di Chiro G, Quindlen EA et al. (1983). Successful treatment of a group of spinal cord arteriovenous malformations by interruption of dural fistula. J Neurosurg 59: 1019–1030.

Osbun JW, Nerva JD, Kim LJ (2015). Classification of Spinal Arteriovenous Lesions: Arterivenous Fistulas and Arteriovenous Malformations. In: HR Winn (Ed.), Youmans Neurological Surgery, 7th edn. Saunders, Philadelphia, PA.

Rangel-Castilla L, Holman PJ, Krishna C et al. (2011). Spinal extradural arteriovenous fistulas: a clinical and radiological description of different types and their novel treatment with Onyx. J Neurosurg Spine 15: 541–549.

Rangel-Castilla L, Russin JJ, Zaidi HA et al. (2014). Contemporary management of spinal AVFs and AVMs: lessons learned from 110 cases. Neurosurg Focus 37. E14.

Rodesch G, Hurth M, Alvarez H et al. (2002). Classification of spinal cord arteriovenous shunts: proposal for a reappraisal – the Bicetre experience with 155 consecutive patients treated between 1981 and 1999. Neurosurgery 51: 374–379; discussion 379–380.

Spetzler RF, Detwiler PW, Riina HA et al. (2002). Modified classification of spinal cord vascular lesions. J Neurosurg 96: 145–156.

Taveras JM, Wood EH (1976). Diagnostic Neuroradiology, 2nd ed. Williams and Wilkins, Baltimore.

Virchow R (1864). Die Krankhaften Geschwulste, A Hirschwald, Berlin.

Wyburn-Mason R (1943). The vascular abnormalities and tumours of the spinal cord and its membranes, H. Kimpton, London, vii, 196 p.

Zhang HQ, Chen T, Wu SS et al. (2013). The pathophysiology of venous hypertensive myelopathy – study of an animal model: laboratory investigation. J Neurosurg Spine 19: 485–491.

Handbook of Clinical Neurology, Vol. 143 (3rd series)
Arteriovenous and Cavernous Malformations
R.F. Spetzler, K. Moon, and R.O. Almefty, Editors
http://dx.doi.org/10.1016/B978-0-444-63640-9.00014-X

Chapter 14

Epidemiology, clinical presentation, diagnostic evaluation, and prognosis of spinal arteriovenous malformations

ALP OZPINAR, GREGORY M. WEINER, AND ANDREW F. DUCRUET*
Department of Neurological Surgery, University of Pittsburgh Medical Center, Pittsburgh, PA, USA

Abstract

Spinal arteriovenous malformations (sAVM) are rare vascular pathologies whose natural history remains incompletely understood. Advances in diagnostic imaging, coupled with the evolution of endovascular and microsurgical techniques have led to the description of a number of classification schemes for these lesions. An updated method has changed AVM classification from five categories of lesion based on source and location of feeder vessels to three categories based on pathophysiology. These categories include extradural arteriovenous fistulae (AVFs), intradural AVFs, extradural-intradural AVFs, intramedullary AVMs, and conus medullaris AVM each with individual subclassifications. Treatment outcomes have been shown to differ based on classification criteria. The increased use of advanced imaging prior to surgical intervention has facilitated the treatment of AVFs. Definitive diagnosis and characterization have traditionally required digital subtraction angiography, which is now being supplemented with other forms of non-invasive imaging such as computed tomography angiography (CTA) and magnetic resonance angiography (MRA). Epidemiologically, intradural dorsal AVFs account for 80% of all sAVMs, and are characterized by low-pressure shunts located in the sleeve of the dorsal nerve root. Microsurgical treatment has been shown to be highly effective in cases of intradural dorsal AVFs, although many cases are also amenable to durable occlusion using liquid embolics. Conus medullaris AVMs, which has only been recently characterized as a separate category of sAVM, is best treated using a combination of embolization and microsurgery. Successful treatment of sAVM mandates a thorough understanding of the anatomy and classification of these lesions. The purpose of this chapter is to review and summarize the classification, natural history, and prognosis of sAVMs.

INTRODUCTION

Spinal arteriovenous malformations (sAVMs) are rare vascular lesions located within the spinal canal characterized by shunting of arterial blood directly to the venous system. These lesions encompass both AVMs, characterized by a nidus of abnormal vessels interposed between the draining veins and feeding arteries, as well as direct fistulous connections between arteries and veins. The resulting arterial pressure transmitted to the venous system can result in venous hypertension, venous dilation causing compression of surrounding structures, or hemorrhage (Aminoff et al., 1974). Symptoms on presentation include sudden-onset pain accompanied by myelopathy or radiculopathy in the case of hemorrhage, or more gradual or stepwise myelopathy in non-hemorrhagic lesions.

sAVMs were first described in 1888 by Gaupp, who famously characterized them as pia mater "hemorrhoids." His original account of the malformation and its clinical manifestation was most likely consistent with a dural arteriovenous fistula (dAVF), one of the classic subcategories of sAVMs. Fedor Krause was the first known surgeon to identify a spinal lesion intraoperatively as a vascular abnormality (Krause, 1912). He did

*Correspondence to: Andrew F. Ducruet, MD, Department of Neurosurgery, Barrow Neurological Institute, 2910 N 3rd Avenue, Phoenix AZ 85018, USA. Tel: 001-412-864-1839, E-mail: andrew.ducruet@barrowbrainandspine.com

not resect the lesion, but the first steps were made towards *in vivo* recognition of these malformations. By 1943, strides were made to formulate a classification system for sAVMs. Based on postmortem autopsy findings, Wyburn-Mason (1943) catalogued sAVMs histologically into arteriovenous angiomas and purely venous angiomas. This histologic system remained the standard until the 1960s, when spinal angiography allowed for more accurate classification. Despite advances in noninvasive imaging, digital subtraction angiography (DSA) continues to be the gold standard for diagnosing and characterizing the detailed anatomic localization, arteriovenous transit, and venous drainage patterns of sAVMs.

SPINAL VASCULAR ANATOMY

In order to understand the pathophysiology and classification of the different types of sAVMs, it is imperative to understand the vascular anatomy of the spinal cord and adjacent structures. The vascular supply of the spinal cord is provided by three separate anastomotic channels – one anterior spinal artery (ASA) and two posterior spinal arteries (PSA), and numerous radicular arteries. Of the 31 pairs of radicular arteries, 7–8 provide segmental contributions to the cord (Barrow and Awad, 1999). The ASA runs continuously across the ventral surface of the cord, while the paired PSAs run on the posterolateral surface, and form anastomotic junctions at each level. These arteries receive supply from radicular branches of the brachiocephalic, intercostal, vertebral, and lumbar arteries. Of note, the thoracolumbar region is generally supplied by a single radiculomedullary artery, the artery of Adamkiewicz, usually arising from the left side at the level of the 9th–12th intercostal artery (Barrow and Awad, 1999). Thus, the thoracic region represents a watershed zone for the spinal cord, rendering it more susceptible to ischemia.

Anterior spinal artery

The ASA forms from vertebral arterial contributions above the foramen magnum, and courses down the anterior median sulcus, receiving additional input from medullary arteries that branch off the dorsal ramus of the intercostal artery in the thoracic spine. The ASA provides blood flow to the anterior two-thirds to four-fifths of the spinal cord, which encompasses the anterior horn cells, and anterolateral corticospinal and spinothalamic tracts (Barrow and Awad, 1999). The rim of anterior white matter is also supplied by a pial radial plexus branching off the ASA. Circumferential branches of the ASA anastomose with the PSA, forming a vasocorona.

Posterior spinal arteries

These paired arteries often arise from the posterior inferior cerebellar arteries and course on the dorsolateral aspect of the spinal cord and to the level of the conus medullaris, where they anastomose with the distal ASA. They receive input from the vertebral, cervical, intercostal, and lumbar arteries. The PSAs provide blood flow to the dorsal columns and portions of the corticospinal tracts.

EPIDEMIOLOGY

sAVFs and sAVMs account for 3–4% of all intradural spinal cord mass lesions. Little has been published regarding the epidemiology of these lesions. One study out of Stanford analyzed National Inpatient Sample data from 1995 to 2006, screening admissions with a primary diagnostic code for spinal vascular malformation (Lad et al., 2009). In this 11-year period, 3291 patients were admitted with a diagnosis of spinal vascular malformation, with an average of 299 annual admissions with new diagnosis of spinal AVM (Lad et al., 2009). Of note, the final year of the study, 2006, was selected for more detailed analysis of demographics and showed the majority of admissions corresponded to male patients (57%) in the 45–64-year age range (36% of admissions).

DIAGNOSTIC EVALUATION

The definitive diagnosis and characterization of spinal AVMs are made with DSA. However, multiple noninvasive imaging modalities are now being applied to image spinal vascular lesions. Computed tomography angiography (CTA) and magnetic resonance angiography (MRA) have been employed in screening as well as in guiding DSA in the context of spinal AVM (Si-jia et al., 2009; Yano and Hida, 2009). Green and Parker (2003) illustrated the utility of MRA techniques in a small case series, demonstrating that MRA enables localization of the level of feeders prior to using conventional angiography, potentially shortening time of catheter angiography. CTA has been shown beneficial in the evaluation of spinal AVMs since it can allow simultaneous visualization of arteries, lesion nidus, and veins, on a three-dimensional reconstruction. This is especially helpful in cases where an AVM or fistula has multiple arterial feeders (Terae et al., 2004).

NATURAL HISTORY AND PROGNOSIS

To describe the natural history and prognosis of spinal vascular malformation, it is essential to understand the classification of the various pathologic subtypes of sAVMs. Historically, the most commonly used classification of sAVM defines four main categories of lesions based on source of arterial feeder and anatomic relationship to the spinal cord. However, often confusing nomenclature coupled with the description of several new types of spinal vascular lesions spurred the creation of an updated classification system. In 2002, Spetzler and collaborators sought to formulate a classification system that simplified the language of spinal vascular lesions. They defined three broad categories that incorporated all spinal vascular abnormalities: neoplasm, aneurysm, and arteriovenous lesion. Within this scheme, they proposed a new classification system for sAVMs that they believed was more anatomically and pathophysiologically driven (Spetzler et al., 2002). The authors delineated categories of extradural, extradural-intradural, and intradural sAVMs, with further subclassification of the intradural category into ventral or dorsal fistulae, or intramedullary AVMs (compact and diffuse intramedullary lesions) (Spetzler et al., 2002; Kim and Spetzler, 2006). This new system also introduced a new category known as the conus medullaris AVM (Spetzler et al., 2002; Kim and Spetzler, 2006). As seen in Tables 14.1 and 14.2, the modified system can be summarized based on characteristic lesion pathophysiology and presentation, and can be related back to the classic model, with the exceptions of formal definitions of extradural AVFs or conus medullaris lesions.

Extradural AVFs

Extradural AVFs, defined as epidural fistulae in older naming schemes, are lesions defined by fistulous communication between an extradural arterial branch, normally of a radicular artery, and the epidural venous plexus (Fig. 14.1) (Kim and Spetzler, 2006). This high-flow connection engenders venous engorgement, which often produces mass effect on nearby nerve roots and potentially the adjacent spinal cord. It has been proposed that both venous hypertension and vascular steal contribute to symptoms of progressive myelopathy in patients with extradural AVFs (Spetzler et al., 2002; Kim and Spetzler, 2006).

Rangel-Castilla and collaborators (2011) published a study that describes further subclassification of extradural AVFs which were treated with endovascular embolization. They define each subcategory based on venous drainage. Type A lesions may have drainage into both the epidural venous plexus as well as a vein that penetrates the dura and flows into the perimedullary venous plexus.

Table 14.1

Clinical characteristics of spinal arteriovenous fistulae (AVFs)

Characteristic	Extradural	Dorsal intradural	Ventral intradural
Pathophysiology	Cord compression, venous congestion, vascular steal	Venous congestion, rare hemorrhage	Compression (venous aneurysm), hemorrhage, vascular steal
Presentation	Progressive myelopathy	Progressive myelopathy	Progressive myelopathy
Diagnostic modality	MRI, angiography	MRI, angiography	MRI, angiography
Previous nomenclature	Epidural	Dural AVF, long dorsal, type I, others	Types IVa, IVb, IVc; perimedullary

Adapted from Spetzler et al. (2002).
MRI, magnetic resonance imaging.

Table 14.2

Clinical characteristics of spinal arteriovenous malformations (AVMs)

Characteristic	Extradural-intradural	Intramedullary	Conus medullaris
Pathophysiology	Compression, vascular steal, hemorrhage	Hemorrhage, compression, vascular steal	Venous hypertension, compression, hemorrhage
Presentation	Pain, progressive myelopathy	Acute myelopathy, pain, progressive myelopathy	Progressive myelopathy, radiculopathy
Diagnostic modality	MRI, angiography,	MRI, angiography	MRI, angiography
Previous nomenclature	Juvenile or metameric AVM, type III	Classic AVM, glomus type II	None

Adapted from Spetzler et al. (2002).
MRI, magnetic resonance imaging.

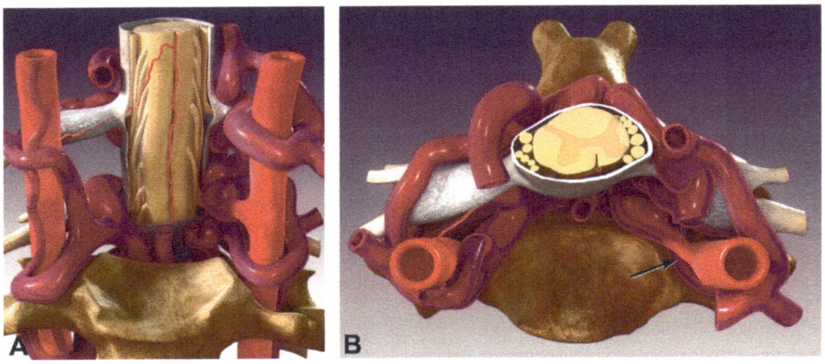

Fig. 14.1. **(A)** Posterior view reveals engorgement of the epidural venous plexus, producing nerve root and spinal cord compression. **(B)** Illustration depicting an extradural arteriovenous fistula arising from a vertebral artery branch (arrow). (Used with permission from Barrow Neurological Institute, Phoenix, Arizona (Ducruet et al., 2013).)

Type B, on the other hand, drain exclusively into Batson's plexus. Further specificity is achieved by dividing type B lesions into B1 and B2 categories, where B1 AVFs represent lesions with engorged venous structures compressing the thecal sac, while B2 AVFs do not exert this mass effect. Rangel-Castilla's study reports 7 patients with extradural spinal AVFs who underwent Onyx embolization. Four of the patients experienced an excellent recovery, while 3 of 7 complained of persistent bowel and/or bladder dysfunction (Rangel-Castilla et al., 2011). Though small, this study provides greater insight into the spectrum of extradural AVFs and the possibilities for therapy.

A more recent comparative analysis from Japan (Takai and Taniguchi, 2012) retrospectively evaluated 45 cases of spinal extradural AVFs from English-language case reports and series. The authors found that patients with type A lesions were typically older (mean age, 63.5 years) than those with type B disease (mean age, 34.3 years). Differences were also noted on T2-weighted magnetic resonance imaging as type A lesions generally demonstrated diffuse high signal intensity of the spinal cord, and no mass effect, while type B exhibited normal signal intensity with significant mass effect secondary to dilated, engorged extradural venous plexus. Location is also a distinguishing factor as type A malformations appear more commonly in the thoracolumbar and lumbar spine, and type B extradural AVFs have a predilection for the cervical and upper thoracic region (Takai and Taniguchi, 2012).

Intradural dorsal AVFs

Intradural dorsal AVFs correlate with type I AVFs in the older classification and are the most common type of sAVM, accounting for 80% of all sAVMs (Grote and Bien, 1996). For historic perspective, these lesions were initially characterized as angioma racemosum venosum by Wyburn-Mason (1943), applying Virchow's nomenclature for vascular abnormalities. Di Chiro et al. (1967) explored preoperative visualization of these lesions with arteriography and popularized the term type I spinal AVMs. In these lesions, the arterial supply arises from a radicular artery (Fig. 14.2). Therefore, the fistulous connection in dorsal dAVFs occurs between one or more radicular artery branches and the perimedullary leptomeningeal venous plexus (Barrow and Awad, 1999). These lesions are extramedullary, low-pressure, and low-flow shunts often located within the dural sleeve of a nerve root. Venous drainage occurs through vessels that penetrate the dura around the dorsal roots. The pathophysiology is thought to be due to progressive myelopathy secondary to venous hypertension. Given the low-flow and low-pressure state of these lesions, they are less prone to rupture.

Multiple case series have evaluated the patient demographics and outcomes following treatment of intradural dorsal AVFs (type I).The series by Saladino et al. (2010) reported a mean age of diagnosis of 63.6 years with a 3.5:1 male/female ratio, and a predilection for the thoracic location. Ropper et al. (2012) reviewed their own case series of 214 craniospinal AVMs and identified 15 spinal dAVFs within this cohort. The mean age for these 15 patients was 60 years, with a male-to-female ratio of 2.8:1. Most spinal dAVF patients in Ropper's study presented with myelopathy: 87% of patients with weakness, 80% presenting with numbness, 60% presenting with paresthesias, and 47% demonstrating bowel and bladder symptoms. Of the 15 type I sAVM patients in the study, 11 underwent microsurgical treatment, and 10 of 11 showed improvement in symptoms following surgical intervention (Ropper et al., 2012).

Intradural ventral AVFs

Intradural ventral AVFs, historically referred to as type IV spinal AVMs, were first described by Heros et al. (1986).

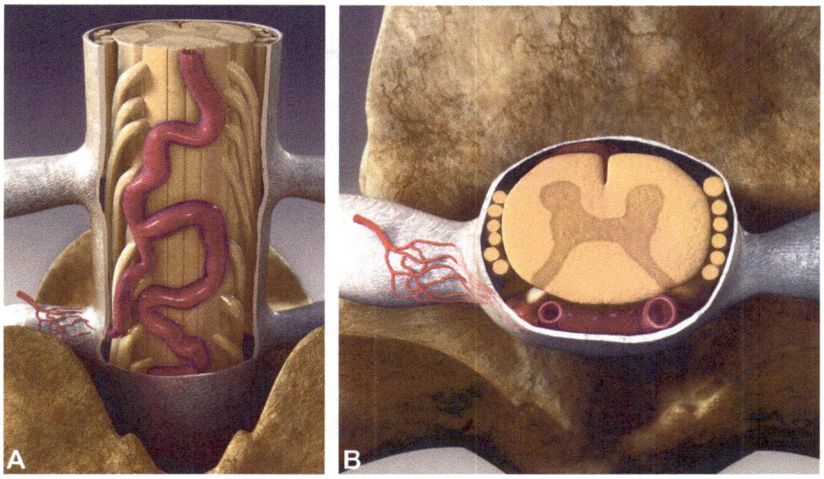

Fig. 14.2. Illustration of an intradural dorsal arteriovenous fistula. **(A)** The fistula is observed along the nerve root sleeve. **(B)** Posterior view demonstrating dilation of the coronal venous plexus, resulting in venous hypertension. (Used with permission from Barrow Neurological Institute, Phoenix, Arizona (Ducruet et al., 2013).)

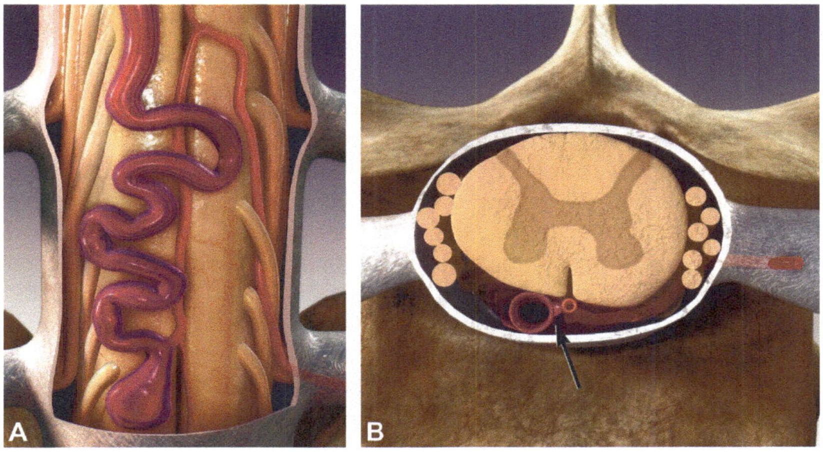

Fig. 14.3. **(A)** Anterior coronal view depicting the fistula along the anterior spinal cord. **(B)** Illustration of an intradural ventral arteriovenous fistula with a fistulous connection (arrow) between the anterior spinal artery and the coronal venous plexus. (Used with permission from Barrow Neurological Institute, Phoenix, Arizona (Ducruet et al., 2013).)

Table 14.3

Anson–Spetzler subclassification of intradural ventral arteriovenous fistulae

Type	Description
A	Single feeder, slow flow
B	Multiple feeders, high flow
C	Giant, multiple feeders, distended venous network

These lesions have been referred to as direct spinal AVFs, perimedullary AVMs, and spinal pial arteriovenous fistulas. The name is derived from the fact that the lesion harbors an intradural, extramedullary connection between a medullary artery and vein (Fig. 14.3). The commonly implicated feeding arteries are the ASA and PSAs, in either case with direct fistulous connection to the spinal coronal venous plexus. Heros et al. (1986) initially postulated that the pathophysiologic mechanism of injury associated with these lesions was cord compression secondary to distended veins, venous hypertension, and a steal phenomenon. Later authors determined that these lesions encompass a heterogeneous group of spinal AVFs and subclassification systems were consequently formulated. Gueguen et al. (1987) made the first attempt to subclassify these lesions, with Anson and Spetzler (1992a) following 5 years later with a modified version of the subclassification system, stratifying subtypes A–C based on increasing size and flow (Table 14.3).

Type A fistulas are single-feeder lesions with slow flow and mild venous hypertension. Type B fistulas are progressively high-flow with multiple feeding arteries, and type C lesions are characterized by a giant fistula with a markedly distended venous network. A recent pooled analysis by Gross and Du (2013b) determined that, based on a cohort of 213 patients, 93% of patients with type IV AVMs presented with neurologic deficits, and 36% presented with hemorrhage. The overall annual hemorrhage rate in this cohort was 2.5%, with a rate of 5.6% for hemorrhagic fistulae. Application of the Anson–Spetzler subclassification revealed that individuals with type IVa lesions were older (mean age, 46.9 years) with a male sex predilection (68%), and patients in the type IVc category were significantly younger (mean age, 18.7 years), with no sex predilection and a higher association with syndromes such as hemorrhagic telangiectasia, Cobb syndrome, and Klippel–Trenaunay–Weber syndrome (Gross and Du, 2013b).

Treatment of intradural ventral AVF differs depending on the particular subtype. In type A fistulas, the ASA is minimally dilated and slow arteriovenous transit is observed. Effective embolisate penetration is difficult without compromising the ASA, thus surgical treatment is generally favored. In type B fistulas, supraselective catheterization is frequently possible. If a stable catheter position is achieved, fistula obliteration may be accomplished using a liquid embolic. Although complete fistula obliteration is necessary to achieve a durable occlusion, partial embolization may facilitate subsequent surgical resection. In type C lesions, the fistulous pouch may be accessed either transarterially or transvenously. Fistula occlusion is then performed using liquid embolics with or without coil embolization. Surgical obliteration of type C lesions is considered hazardous due to the large size and high flow of the fistula. Relatively high rates of obliteration were observed with both surgical and endovascular treatment in this pooled analysis, 88% and 74%, respectively, with 68% of surgical patients showing symptomatic improvement and 75% of endovascular patients demonstrating improvement.

Extradural-intradural AVMs

Extradural-intradural AVMs, also known as juvenile or metameric malformations, are often intramedullary but also have extramedullary and paraspinal extension. Arterial supply is quite variable and may consist of ASAs or PSAs, as well as branches of paraspinal arteries supplying soft tissues (Heros, 1995). Venous drainage is also variable and may include intradural and extradural veins. These lesions can involve most of the spinal cord circumferentially and on cross-section often demonstrate intruding neural parenchyma within the nidus of the AVM (Rosenblum et al., 1987). This extensive and complex distribution underlies the early onset of symptoms and presentation characteristic of these "juvenile" lesions. Given their rarity, the natural history of these lesions is poorly understood. For this reason, Gross and Du (2014) performed another pooled analysis to better ascertain the demographics, natural history, and treatment outcomes related to these type III (extradural-intradural) spinal AVMs. The data revealed a slight male predilection (1.7:1), with a mean age at presentation of 15 years. Thirty-five percentof patients presented with progressive deficits, while 31% presented with hemorrhage and 22% presented with an acute decline not attributable to hemorrhage. The overall annual hemorrhage rate of these lesions was calculated at 2.1% (Gross and Du, 2014). Despite the difficulty associated with treatment of these lesions, 73% of treated patients displayed symptomatic improvement and a statistically significant protective effect against hemorrhage following treatment (Gross and Du, 2014).

Intramedullary AVMs

Intramedullary AVMs, also known as glomus AVMs, are characterized by a compact intramedullary nidus, with pial arterial supply and high-flow AV shunting. They are most commonly fed by the ASA, although some lesions have PSA feeders. These lesions demonstrate AV shunting into the coronal venous plexus around the cord (Fig. 14.4) (Barrow and Awad, 1999). These lesions have been further classified into compact and diffuse, as this relates to surgical resectability (Kim and Spetzler, 2006). Gross and Du (2013a) performed a pooled analysis of 33 spinal glomus AVM studies and found that these lesions tend to occur in younger patients with no sex predilection. Prior studies have shown that most of these lesions are found at either the thoracic or cervical level, although they can occur at any spinal level. Half of these lesions presented with hemorrhage, and 29% had an associated aneurysm (Gross and Du, 2013a). In addition to the high frequency of hemorrhage, these lesions can also present with cord compression due to aneurysmal varices. Gross and Du's (2013a) analysis of the pooled data calculated an overall annual hemorrhage rate of 4%, with elevated higher risk of 10% in cases of prior hemorrhage.

Many authors believe that surgical resection remains the mainstay of treatment for these lesions (Velat et al., 2012). In many cases, there is a role for presurgical embolization, and in select cases embolization may represent definitive treatment. Rates of complete obliteration of lesion were 78% with surgery and 33% with embolization, respectively, and no hemorrhages were noted after complete obliteration was achieved

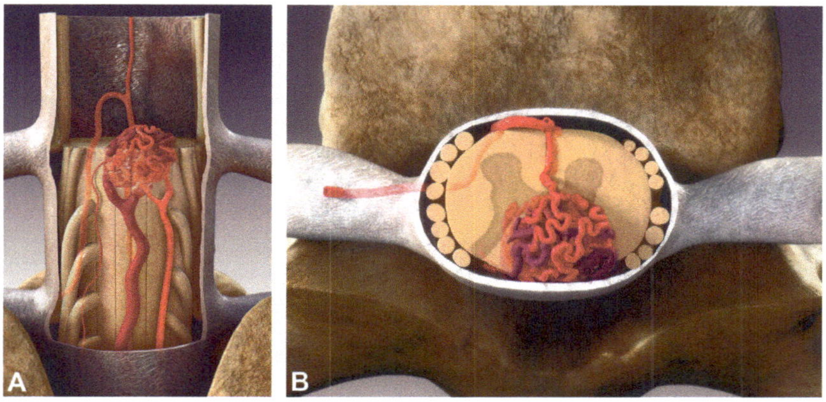

Fig. 14.4. **(A)** Axial illustration of an intramedullary arteriovenous malformation with a compact nidus supplied by an anterior spinal artery feeder. **(B)** Posterior coronal view demonstrating additional supply from the posterior spinal artery. (Used with permission from Barrow Neurological Institute, Phoenix, Arizona (Ducruet et al., 2013).)

(Gross and Du, 2012). While the study demonstrated a decrease in risk of hemorrhage following either surgery or endovascular treatment, one of the inherent challenges of treating intramedullary AVMs is that the nidus and the adjacent segment of the normal spinal cord may share an arterial blood supply (Barrow and Awad, 1999). The first-line treatment for intramedullary sAVMs is embolization due to high intraprocedural risk. In a recent surgical series of 20 type II intramedullary AVM 20% of patients suffered from worsening postoperative neurologic symptoms, thus surgical obliteration has been indicated in a few select cases. With the advances in endovascular treatments, a recent series of 17 patients utilizing Onyx demonstrated complete obliteration of 37.5% of patients, and 82% of patients experienced clinical improvement (Corkill et al., 2007). A meta-analysis of intramedullary AVMs demonstrated that with endovascular treatment the risk for future hemorrhage has been significantly reduced (Gross and Du, 2013b).

Conus medullaris AVMs

Conus medullaris AVMs were first described in Spetzler and colleagues' (2002) modified classification of spinal vascular lesions. These lesions are specific to the conus medullaris and cauda equina and may extend down the entire filum terminale. They are characterized by numerous feeders from the ASA and PSAs, with direct arteriovenous shunts and dilated veins (Kim and Spetzler, 2006). The symptomatology associated with these lesions is attributed to venous hypertension, compressive mass effect from dilated venous structures, and ischemia. Due to the unique location and anatomy of these lesions, patients may present with both upper and motor neuron symptoms (Kim and Spetzler, 2006).

In 2012, Wilson and collaborators published the first series of conus AVMs, retrospectively reviewing 20 years of data on endovascular and microsurgical treatment of these lesions. Sixteen patients were examined in the study. Ten of 16 patients presented with myeloradiculopathy, 4 with radiculopathy alone, 1 with seizure secondary to subarachnoid hemorrhage, and 1 with previous hemorrhage whose symptoms had resolved. Nine of 16 had bowel and/or bladder dysfunction, and 12 of 16 were ambulatory when they presented. Half (8 patients) of the patients underwent microsurgical treatment alone, and the other half had embolization of the lesion followed by microsurgical resection. In terms of discharge outcomes, 10 of 16 patients were neurologically stable (from a motor and bowel/bladder perspective), while 6 exhibited symptoms of new motor weakness or bladder dysfunction. A mean clinical follow-up of 70 months was achieved in 14 of the 16 patients. Six patients experienced neurologic improvement, 6 were neurologically stable, and 2 demonstrated neurologic decline relative to pretreatment motor and bowel/bladder function (Wilson et al., 2012). Of note, for the cohort of 6 patients with worse symptoms at discharge, 3 improved to pretreatment baseline neurologic status, 2 did not improve, and 1 patient was lost to follow-up. Finally, 3 patients were noted to have angiographically verified recurrence of lesion following complete obliteration (Wilson et al., 2012). Though this case series is small, it illustrates the value of combined endovascular and microsurgical approaches to conus AVMs and the need for close follow-up given the possibility of recurrence and long-term neurologic decline.

CONCLUSION

Advances in diagnostic imaging and a greater understanding of pathophysiologic characteristics have allowed for novel classification schemes of spinal AVMs.

The evolution of both endovascular and microsurgical techniques has led to advances in the successful treatment of these pathologies. This chapter has reviewed the epidemiology, imaging characteristics, and prognosis associated with the various subtypes of sAVMs.

References

Aminoff MJ, Barnard RO, Logue V (1974). The pathophysiology of spinal vascular malformations. J Neurol Sci 23: 255–263.

Anson J, Spetzler R (1992a). Classification of spinal arteriovenous malformations and implications for treatment. BNI Q 8: 2–8.

Barrow DL, Awad IA (Eds.), (1999). Spinal vascular malformations. American Association of Neurological Surgeons, Park Ridge, IL.

Corkill RA, Mitsos AP, Molyneux AJ (2007). Embolization of spinal intramedullary arteriovenous malformations using the liquid embolic agent, Onyx: a single-center experience in a series of 17 patients. J Neurosurg Spine 7: 478–485.

Di Chiro G, Doppman J, Ommaya AK (1967). Selective arteriography of arteriovenous aneurysms of spinal cord. Radiology 88: 1065–1077.

Ducruet AF, Crowley RW, McDougall CG, Albuquerque FC (2013). Endovascular management of spinal arteriovenous malformations. Journal of Neurointerventional Surgery 5 (6): 605–611.

Gaupp J (1888). Hamorrhoiden der pia mater spinalis im Gebiet des Lendenmarks. Beitr Pathol 2: 516–518.

Green D, Parker D (2003). CTA and MRA: visualization without catheterization. Semin Ultrasound CT MR 24 (4): 185–191.

Gross BA, Du R (2013a). Spinal glomus (type II) arteriovenous malformations: a pooled analysis of hemorrhage risk and results of intervention. Neurosurgery 72: 25–32; discussion 32.

Gross BA, Du R (2013b). Spinal pial (type IV) arteriovenous fistulae: a systematic pooled analysis of demographics, hemorrhage risk, and treatment results. Neurosurgery 73: 141–151; discussion 151.

Gross BA, Du R (2014). Spinal juvenile (type III) extradural-intradural arteriovenous malformations. J Neurosurg Spine 20: 452–458.

Grote E, Bien S (1996). Arteriovenous malformations of the spinal cord. In: J Youmans (Ed.), Neurological surgery: a comprehensive reference guide to the diagnosis and management of neurosurgical problems, 4th edn, vol. 2. WB Saunders, Philadelphia, PA, pp. 1511–1530.

Gueguen B, Merland JJ, Riche MC et al. (1987). Vascular malformations of the spinal cord: intrathecal perimedullary arteriovenous fistulas fed by medullary arteries. Neurology 37: 969–979.

Heros R (1995). Arteriovenous malformations of the spinal cord. In: R Ojemann, R Heros, R Crowell (Eds.), Surgical management of neurovascular disease. Williams & Wilkins, Baltimore, MD, pp. 519–537.

Heros RC, Debrun GM, Ojemann RG et al. (1986). Direct spinal arteriovenous fistula: a new type of spinal AVM. Case report. J Neurosurg 64: 134–139.

Kim LJ, Spetzler RF (2006). Classification and surgical management of spinal arteriovenous lesions: arteriovenous fistulae and arteriovenous malformations. Neurosurgery 59: S195–S201; discussion S3–13.

Krause F (1912). Surgery of the brain and spinal cord: based on person experiences, Rebman, New York.

Lad SP, Santarelli JG, Patil CG et al. (2009). National trends in spinal arteriovenous malformations. Neurosurg Focus 26: 1–5.

Rangel-Castilla L, Holman PJ, Krishna C et al. (2011). Spinal extradural arteriovenous fistulas: a clinical and radiological description of different types and their novel treatment with Onyx. J Neurosurg Spine 15: 541–549.

Ropper AE, Gross BA, Du R (2012). Surgical treatment of type I spinal dural arteriovenous fistulas. Neurosurg Focus 32: E3.

Rosenblum B, Oldfield EH, Doppman JL et al. (1987). Spinal arteriovenous malformations: a comparison of dural arteriovenous fistulas and intradural AVMs in 81 patients. J Neurosurg 67: 795–802.

Saladino A, Atkinson JLD, Rabinstein AA et al. (2010). Surgical treatment of spinal dural arteriovenous fistulae: a consecutive series of 154 patients. Neurosurgery 67: 1350–1357; discussion 1357–1358.

Si-jia G, Meng-wei Z, Xi-ping L et al. (2009). The clinical application studies of CT spinal angiography with 64-detector row spiral CT in diagnosing spinal vascular malformations. Eur J Radiol 71: 22–28.

Spetzler RF, Detwiler PW, Riina HA et al. (2002). Modified classification of spinal cord vascular lesions. J Neurosurg 96: 145–156.

Takai K, Taniguchi M (2012). Comparative analysis of spinal extradural arteriovenous fistulas with or without intradural venous drainage: a systematic literature review. Neurosurg Focus 32. E8.

Terae S, Kudo K, Asano T et al. (2004). CT angiography with multidetector-row helical CT in spinal arteriovenous malformation. Clin Imaging 28 (1): 23–27.

Velat GJ, Chang SW, Abla AA et al. (2012). Microsurgical management of glomus spinal arteriovenous malformations: pial resection technique: Clinical article. J Neurosurg Spine 16 (6): 523–531.

Wilson DA, Abla AA, Uschold TD et al. (2012). Multimodality treatment of conus medullaris arteriovenous malformations: 2 decades of experience with combined endovascular and microsurgical treatments. Neurosurgery 71: 100–108.

Wyburn-Mason R (1943). The vascular abnormalities and tumours of the spinal cord and its membranes, CV Mosby, St. Louis.

Yano S, Hida K (2009). Current advances in spinal vascular disease. Brain Nerve 61: 645–654.

Handbook of Clinical Neurology, Vol. 143 (3rd series)
Arteriovenous and Cavernous Malformations
R.F. Spetzler, K. Moon, and R.O. Almefty, Editors
http://dx.doi.org/10.1016/B978-0-444-63640-9.00015-1

Chapter 15

Spinal arteriovenous malformations: surgical management

DAVID S. XU, HAI SUN, AND ROBERT F. SPETZLER*
Department of Neurosurgery, Barrow Neurological Institute, St. Joseph's Hospital and Medical Center, Phoenix, AZ, USA

Abstract

Spinal arteriovenous malformations (AVMs) are a diverse population of vascular lesions associated with significant long-term morbidity and neurologic impairment. Spinal AVMs can be classified into three categories: intramedullary, extradural-intradural, and conus, depending on their angioarchitecture and anatomy. Extradural-intradural lesions are typically incurable due to their diffuse nature, but patients may benefit from targeted therapy of symptomatic components. Intramedullary and conus AVMs are amenable to safe surgical resection when combined with endovascular embolization and use of a pial dissection technique that spares entry into the neural parenchyma.

INTRODUCTION

Spinal arteriovenous malformations (AVMs) are a rare and extremely heterogeneous group of spinal vascular lesions (Cogen and Stein, 1983; Gross and Du, 2013, 2014). Patients may demonstrate distinct clinical presentations dependent on the AVM anatomy and its associated pathophysiologic mechanism that will heavily guide the optimal surgical management strategy. To account for this variability, the senior author (RFS) previously created a spinal AVM classification system that takes into account the unique anatomic characteristics of these disparate vascular pathologies (Spetzler et al., 2002; Kim and Spetzler, 2006). As shown in Figure 15.1, true spinal AVMs are divided into three groups depending on location: intramedullary, intradural-extradural, and conus medullaris lesions. This chapter reviews the optimal selection, planning, and treatment of patients based on this classification system.

NATURAL HISTORY

The natural history of patients with spinal AVMs is defined by progressive clinical deterioration due to recurrent hemorrhages or venous hypertension (Yasargil et al., 1984; Rosenblum et al., 1987). However, unlike their counterparts within the brain, spinal AVMs may cause significantly greater morbidity with each bleed due to the tight eloquent confines within the thecal sac. Due to the overall rarity of spinal vascular malformations, the annual risk of rupture of spinal AVMs is poorly defined. Currently, there are no prospective large population studies of patients with these lesions. However, a recent systematic analysis of pooled retrospective studies of type II spinal vascular malformations, which include intramedullary and conus AVMs, revealed an annual rupture risk of 4%; lesions that have already ruptured have a significantly higher annual rupture risk of 10% within the first 10 years after the initial bleed (Gross and Du, 2013). For extradural-intradural AVMs, commonly referred to as type III or metameric AVMs, a systematic review of the literature yielded a population of 51 patients with an annual hemorrhage rate of 2.1% (Gross and Du, 2014).

TREATMENT FUNDAMENTALS

Endovascular collaboration

Optimal contemporary neurosurgical management of spinal AVM patients should occur in a team-based collaborative environment between open-vascular and

*Correspondence to: Robert F. Spetzler, MD, c/o Neuroscience Publications, Barrow Neurological Institute, St. Joseph's Hospital and Medical Center, 350 W. Thomas Rd., Phoenix AZ 85013, USA. Tel: +1-602-406-3593, Fax: +1-602-406-4104, E-mail: Neuropub@dignityhealth.org

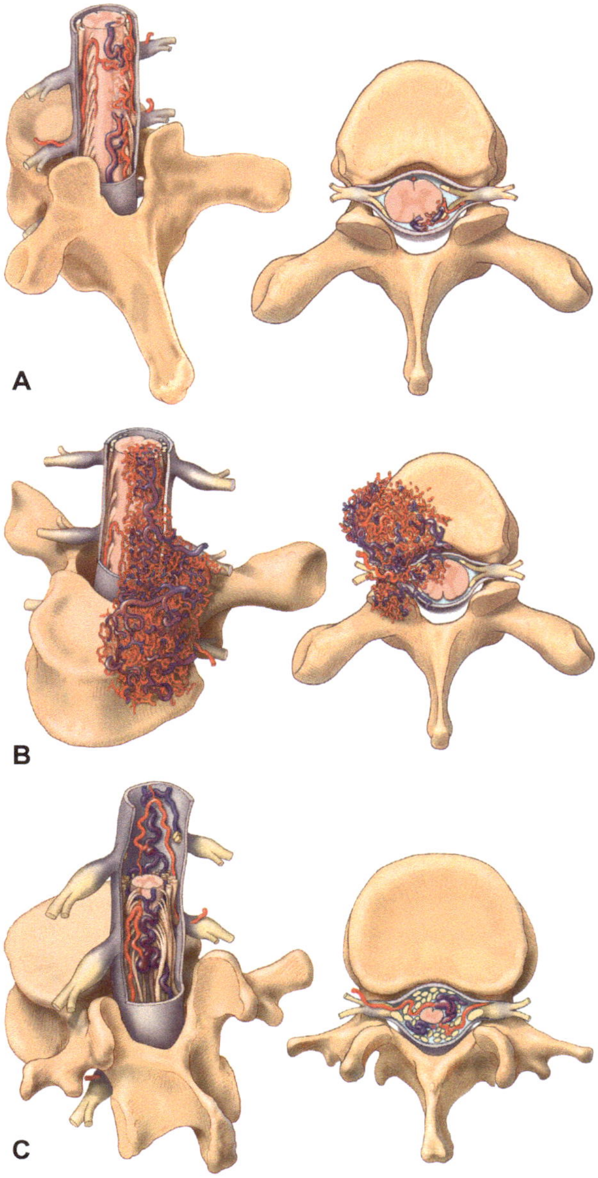

Fig. 15.1. Types of spinal arteriovenous malformations (AVMs). (**A**) Posterior and transverse illustrations showing an intramedullary AVM with a compact nidus that has a posterior portion exposed through the pial surface of the spinal cord. (**B**) Anterior and transverse illustrations of an intradural-extradural AVM showing involvement of all regional tissues around the spinal cord, including bone. (**C**) Posterior and transverse illustrations of a conus AVM showing multiple dilated venous varices compressing adjacent nerves of the cauda equine. (Reproduced from Spetzler et al., 2002, with permission from the *Journal of Neurosurgery*: Spine.)

endovascular neurosurgeons, with participation of each team member driven by the characteristics of the lesion treated. At our institution, we advocate initial endovascular embolization before microsurgical resection of AVMs, to eliminate portions of the lesion that are treacherous to access openly, such as deep intramedullary components and portions located ventral to the spinal cord. Due to the highly eloquent location of spinal AVMs and their complex angioarchitecture, embolization may not always be possible and when used alone rarely results in complete angiographic obliteration of the lesion (Hall et al., 1989; Corkill et al., 2007). Out of 66 consecutively treated patients with AVMs at our institution from 2000 to 2013, endovascular therapy was able to be safely utilized in 56% of cases and was used as the sole therapy only 9.1% of the time, and then only as a means for palliation, not cure (Rangel-Castilla et al., 2014).

Perioperative monitoring and imaging

At our institution, we routinely monitor somatosensory and motor evoked potentials in all patients undergoing AVM resection. For patients with conus lesions, additional monitoring with electromyography can aid identification and preservation of motor nerves that travel to the cauda equina during dissection. Intraoperatively, our vascular imaging modality of choice is indocyanine green fluorescent angiography with formal intraoperative catheter-based angiography performed if there is suspicion of occult residual AVM.

Exposure and approach

We prefer a posterior or posterolateral approach for the majority of cases. Typically, multilevel laminoplasty is performed, extending one level above and below the AVM nidus, and the dura is opened in a paramedian fashion eccentric to the side of the nidus. If significant anterolateral exposure is needed, the bony opening can be extended more laterally through facetectomy in the cervical spine and costotransversectomy in the thoracic spine, with routine sectioning and gentle retraction of the intradural dentate ligament to allow mild rotational mobilization of the spinal cord (Martin et al., 1995). In rare instances where the AVM is contained superficially in the ventral midline, an anterior approach through a corpectomy may be an option (Williams et al., 1991; MacFarlane et al., 2005).

Microsurgical technique and pial resection

There is significant variation with regard to microsurgical techniques used to treat spinal AVMs. Within the past 20 years, the core strategy adopted at our institution and developed by the senior author (RFS) is the previously described pial resection technique, where feeding vessels and the nidus along the surface of the spinal cord are truncated while dissection into the parenchyma is avoided

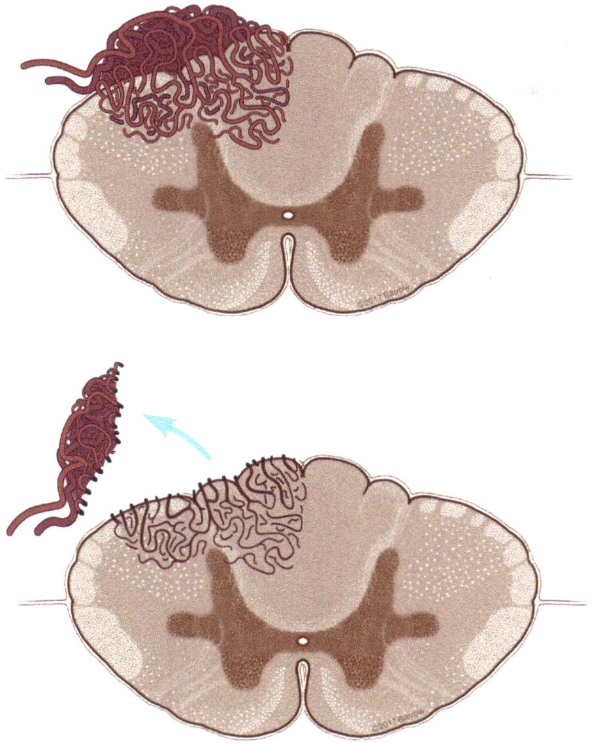

Fig. 15.2. Pial resection technique. Resection of an intramedullary arteriovenous malformation nidus is performed through systematic truncation of blood vessels at the interface of the spinal cord pia. Care is taken not to follow vessels into the parenchyma. (Courtesy of Barrow Neurological Institute, Phoenix, Arizona.)

(Velat et al., 2012). Dissection of the extramedullary angioarchitecture of spinal AVMs is well tolerated and can result in significant interruption of flow to intramedullary components and decrease the burden of venous hypertension in diffuse lesions. In contrast, partial resection of the nidus in brain AVMs frequently results in uncontrollable bleeding. A schematic representation of the strategy is seen in Figure 15.2 and demonstrated in subsequent figures. In uncommon instances, surgical myelotomy may be needed to evacuate a clot, decompress a syrinx, or treat compact purely intramedullary lesions. The four myelotomies we use in these circumstances are the midline dorsal, dorsal root entry zone, lateral, and anterior midline approaches.

INTRAMEDULLARY ARTERIOVENOUS MALFORMATIONS

Presentation

Commonly referred to as type II or glomus AVMs, intramedullary spinal AVMs are similar to their classic intracranial counterparts and are fed by multiple branches of the anterior and posterior spinal arteries. The most common presentation is acute hemorrhage with resultant paralysis (Spetzler et al., 2002). Patients can also present with subacute myelopathy or paresthesias and pain due to mass effect or vascular steal from the spinal cord and adjacent nerve roots (Connolly et al., 1998). With significant hemorrhage, blood may also track into the intracranial subarachnoid space, causing associated medical complications such as vasospasm and hydrocephalus (Caroscio et al., 1980).

Patient selection

Due to the progressive cumulative risk of hemorrhage for intramedullary AVMs, we recommend surgical therapy for young patients and those facing progressive neurologic symptoms. The focus of surgery is devascularization of vessels presenting superficial to the pial surface of the spinal cord. In a small subset of AVMs, the nidus may be contained purely in the intramedullary space. For these rare lesions, we recommend surgery only if the nidus is compact and the patient has experienced either a previous bleed or disabling neurologic deficits.

Surgical technique

After preoperative embolization and exposure of the AVM, the first step is careful identification of the associated AVM vasculature. Figure 15.3 outlines the subsequent steps taken. Intraoperative indocyanine green videoangiography is an invaluable tool that should be used at the start of exposure and repeated throughout the surgery to identify arterial pedicles. Additional embolic material present from preoperative endovascular therapy can also serve as clues to the angioarchitecture of the AVM. Once feeding vessels and the AVM nidus are defined, dissection proceeds with systematic coagulation and truncation of AVM vessels above the pial surface of the spinal cord. Arterial feeders are resected first, followed by the venous components. Care must be taken not to chase after portions of the AVM that course into the spinal cord in order to avoid unintended neurologic injury. Once all visible AVM is resected, formal catheter angiography should be performed immediately to verify the extent of resection.

Outcomes

Using the described combined endovascular and pial dissection strategy, we previously published data from our institution for 20 treated intramedullary AVMs over a 17-year period, revealing a cure rate of 75% (Velat et al., 2012). Of the patients with immediate angiographic cure, one developed a recurrence 7 years later with a small new feeder that has not required additional therapy. Most importantly, no patient experienced

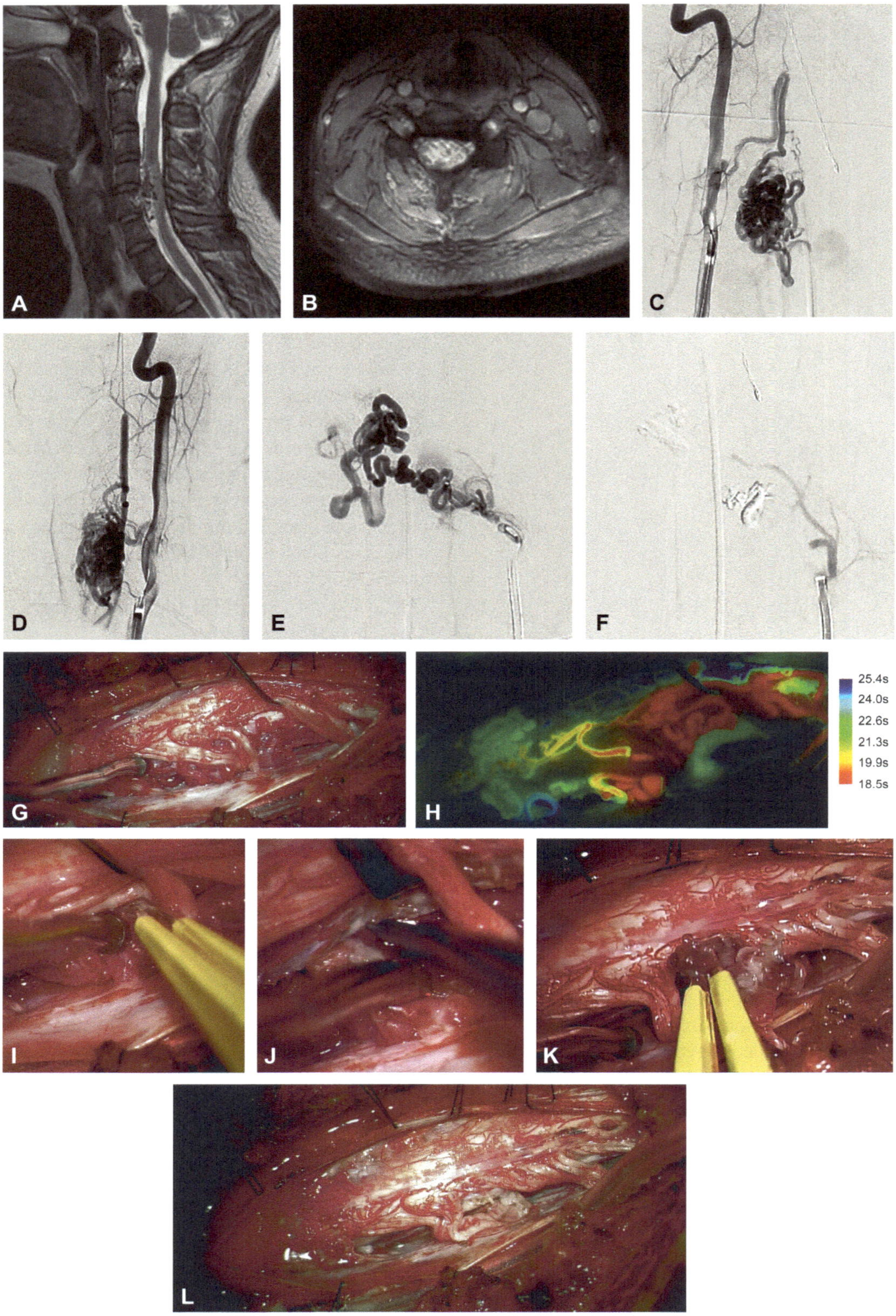

Fig. 15.3. Resection of an intramedullary arteriovenous malformation (AVM) in a 26-year-old woman with sudden onset of severe bilateral arm radiculopathy and right-hand weakness. T2-weighted magnetic resonance images in the (**A**) sagittal and (**B**) axial planes demonstrate a cervical intramedullary AVM filling the spinal canal causing spinal cord edema. Digital subtraction angiography through a right vertebral artery injection in (**C**) anteroposterior and (**D**) lateral projections demonstrate a compact

(Continued)

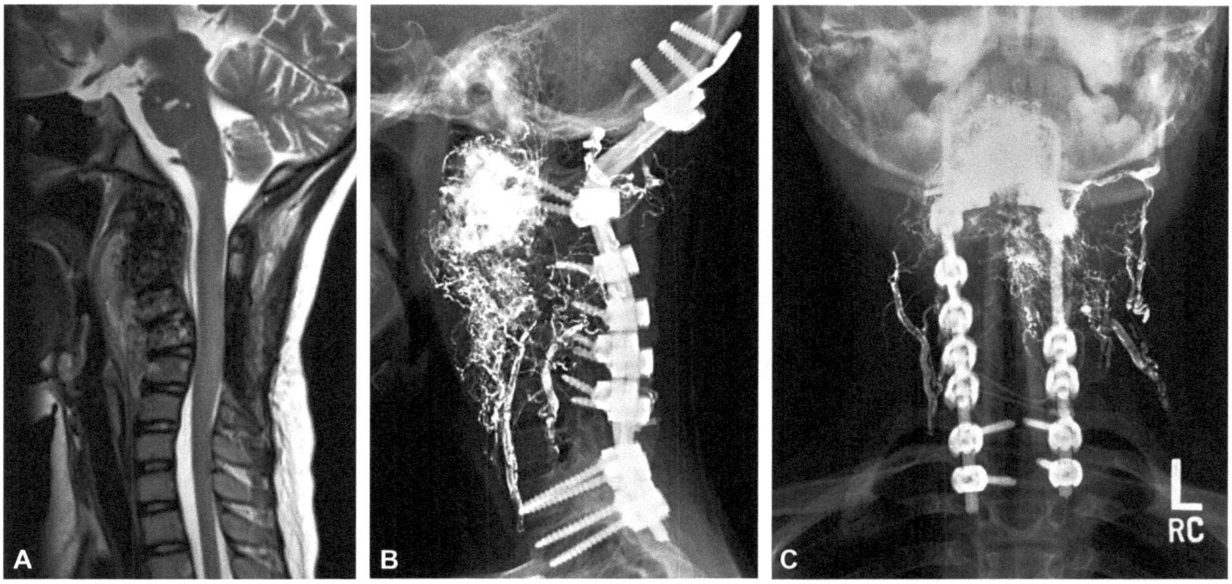

Fig. 15.4. Spinal instability from an extradural-intradural arteriovenous malformation (AVM). A 10-year-old girl with a history of an anterior upper cervical extradural-intradural AVM treated with laminectomy and previous embolizations presented with severe neck pain and progressive myelopathy. (**A**) Sagittal T2-weighted magnetic resonance image of the cervical spine shows the AVM involving the vertebral body and retropharyngeal tissues from C1 to C4 with progressive kyphotic deformity and intramedullary edema thought to be due to venous hypertension. The patient underwent additional embolization and posterior occipital-to-T2 fusion with resolution of her symptoms. (**B**) Lateral and (**C**) anteroposterior radiographs of the cervical spine after therapy reveal the extent of instrumentation in the background of embolic material marking the previously treated AVM. (Courtesy of Barrow Neurological Institute, Phoenix, Arizona.)

clinical deterioration after intervention. Late-term spinal cord tethering is a concern, and this occurred in 3 (15%) patients after a mean follow-up of 5 years. All 3 patients were treated with open detethering and have had no further issues.

EXTRADURAL-INTRADURAL ARTERIOVENOUS MALFORMATIONS

Otherwise known as metameric, juvenile, or type III spinal vascular malformations, extradural-intradural AVMs are diffuse, formidable lesions that cross multiple anatomic planes. These malformations are thought to arise and involve tissues at the somite level and frequently infuse within skin, muscle, bone, dura, spinal cord, and associated nerve roots (Spetzler et al., 1989). Similar to those with intramedullary AVMs, patients can present with acute hemorrhage and progressive myelopathy, but other clinical sequelae from diffuse involvement of the AVM in adjacent tissues can also manifest, such as spinal instability due to bony degradation (Fig. 15.4).

Due to the diffuse nature of extradural-intradural AVMs, treatment is centered on partial control of symptomatic components or high-risk features such as associated arterial aneurysms, which are found in approximately half of lesions (Gross and Du, 2014). Embolization is the mainstay of therapy, specifically targeting symptomatic portions that cause vascular steal or venous hypertension with subsequent surgical resection to relieve mass effect. Aside from isolated case reports, gross surgical obliteration is unlikely, with cumulative cure rates nearing 32% in 51 treated patients (Touho et al., 1991; Gross and Du, 2014). However, partial treatment alone can still result in neurologic improvement more than 70% of the time and reduces the future risk of hemorrhage.

Fig. 15.3 (continued) AVM nidus. This pedicle was unable to be embolized due to loss of motor evoked potential monitoring during test injection of lidocaine. Additional selective angiography of the left thyrocervical trunk through an anteroposterior projection reveals (**E**) additional vascular supply to the AVM that was successfully eliminated (**F**) after embolization. Surgery proceeded through a three-level cervical laminoplasty with (**G**) the initial exposure of the AVM. (**H**) Indocyanine green videoangiography with colorimetric analysis of time to onset of flow reveals arterial components in red. Sequential pial dissection around the AVM nidus then proceeds by (**I**) coagulating then (**J**) cutting vessels as they pierce the pial surface of the spinal cord allowing gross total (**K**) mobilization and (**L**) removal of the nidus. Repeat formal catheter angiography revealed no residual AVM (not shown). (Courtesy of Barrow Neurological Institute, Phoenix, Arizona.)

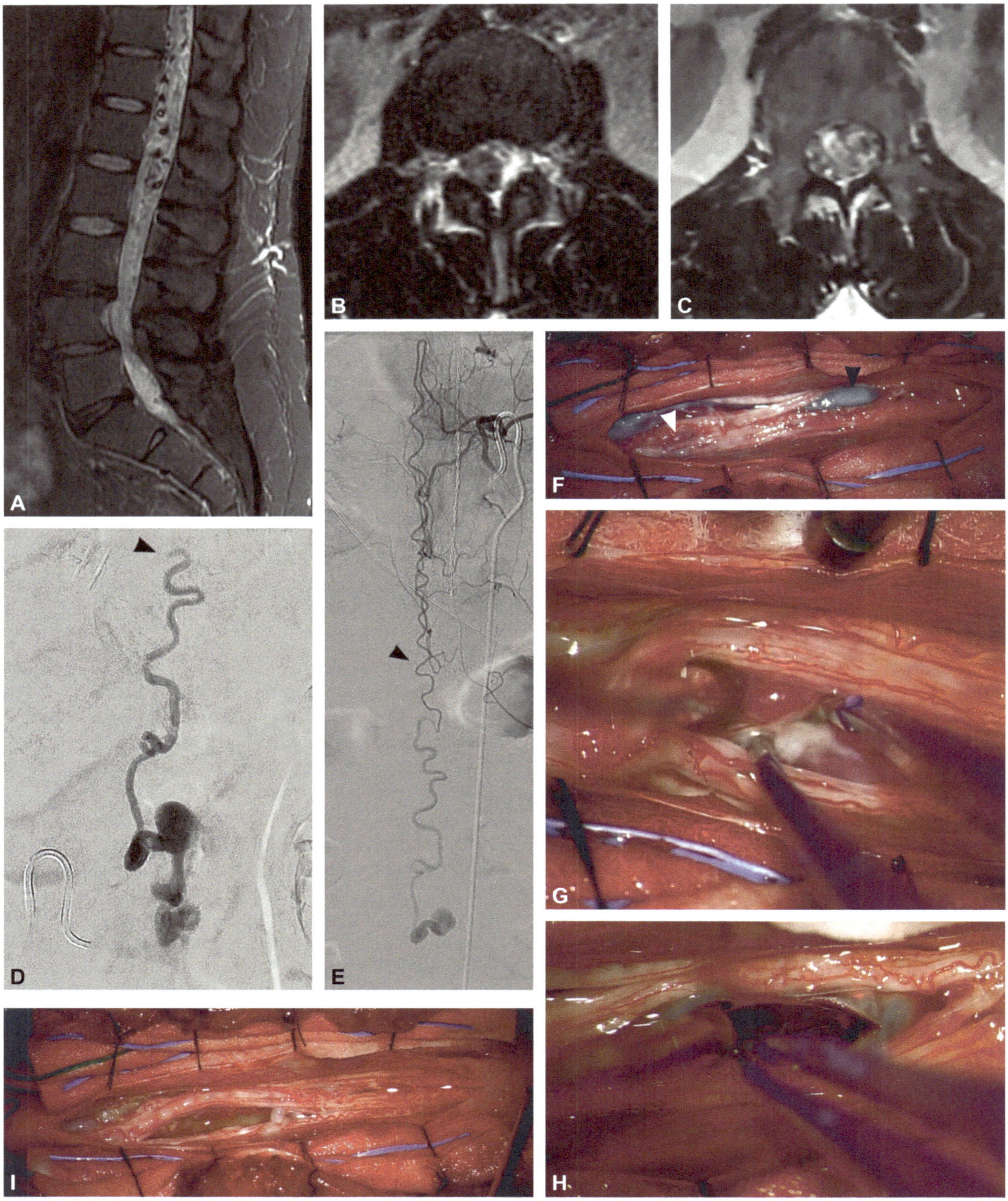

Fig. 15.5. Treatment of a conus arteriovenous malformation (AVM). A 43-year-old man presented to the emergency room with severe progressive lower-extremity radiculopathy and urinary retention. (**A**) Sagittal T2-weighted magnetic resonance imaging (MRI) sequence reveals multiple flow voids within the upper lumbar spinal canal. Axial T2-weighted MRI slices reveal compression of intrathecal contents by (**B**) large varices and (**C**) clumps of smaller vessels. Selective spinal angiography reveals a conus AVM with multiple venous varices fed by several radicular artery contributors. The largest feeder and venous varix are shown (**D**) through a right L2 radicular artery injection in an anteroposterior projection (arrowhead designates proximal

(Continued)

CONUS ARTERIOVENOUS MALFORMATIONS

Presentation

Due to their unique angioarchitecture, conus AVMs represent a distinct category of intradural vascular malformations. These lesions exhibit features similar to those of intramedullary AVMs, such as multiple arterial feeders from the anterior and posterior spinal arteries and discrete nidi concurrent with multiple direct arteriovenous fistulas and complex venous varices. Occurring at the conus medullaris and cauda equina, conus AVMs frequently cause myeloradiculopathy as well as bowel and bladder dysfunction through focal compression, often by dilated venous varices, or through vascular steal and venous hypertension. In the largest consecutive series of conus AVM cases consisting of 16 patients treated over two decades, 63% presented with myeloradiculopathy, 56% had bowel or bladder dysfunction, and 44% of patients had a hemorrhagic event (Wilson et al., 2012).

Patient selection

Previous retrospective natural-history studies evaluating conus AVMs typically pool them together with other intramedullary AVMs due to some overlap of vascular features. Although there are no data on patient selection criteria for conus AVMs, the patient selection criteria for intramedullary AVMs can reasonably be applied to them.

Surgical technique

Surgical resection of a conus AVM begins with careful delineation of anterior and posterior spinal artery branches not involved with the lesion. This often involves extensive arachnoid dissection around the conus medullaris and between the emanating cauda equina. Indocyanine green videoangiography again plays an important role in visualizing the superficial vessel angioarchitecture within the exposure. However, due to crowding of the operative field from nerve roots and dilated venous varices, intraoperative catheter-based angiography may be necessary to provide a fully penetrant view of the vascular anatomy.

Resection of the AVM proceeds with three main objectives, highlighted in Figure 15.5. The first is to perform pial resection of the AVM nidus along the conus and nerve roots. It is possible for the nidus to extend along the entire length of the filum terminale, underscoring the need for full visualization of the vascular anatomy with thorough preoperative or intraoperative angiography. Secondly, direct arteriovenous fistulas that are not amenable to endovascular therapy must be eliminated by ligation. Lastly, dilated venous varices should be decompressed if possible, even if they were previously embolized, to reduce the mass effect on adjacent nerve roots within the thecal sac.

Outcomes

The largest available case series of 16 treated conus AVMs from our institution revealed an angiographic cure rate of 88% (Wilson et al., 2012). Neurologic worsening occurred in 14% of patients, but did not significantly affect motor or ambulation ability. Importantly, in 3 of 4 patients who presented in a nonambulatory state, surgical therapy significantly improved motor function, leading to restoration of ambulation postoperatively. Despite these successes, late-term recurrence did occur in 2 patients at 3 and 16 years after therapy, both with recurrent neurologic deterioration. Therefore, we recommend all treated patients undergo close long-term follow-up with a high index of suspicion if recurrent symptoms arise.

CONCLUSION

Despite their complex diversity, the majority of spinal AVMs can be treated safely and successfully through a combination of endovascular and open surgical therapy with excellent results. These success rates reflect an evolution in the treatment paradigm toward a multimodality approach combining endovascular therapy with an open surgical technique that emphasizes pial resection and preservation of eloquent neurologic tissue.

Fig. 15.5 (continued) site of flow). (**E**) A left T7 radicular artery injection through an anteroposterior projection reveals additional feeders off the anterior spinal artery (arrowhead). The patient's L2 feeder was successfully embolized and he underwent laminoplasty of T12–L2 for resection of the AVM. (**F**) The initial intraoperative exposure reveals the previously seen large venous varix (black arrowhead) as well as focal areas of AVM nidus (white arrowhead) incorporated around nerve roots. Surgical resection proceeded with (**G**) pial dissection of nidus off surrounding nerve roots and (**H**) debulking of the venous varices when flow occlusion was verified. (**I**) At the end of surgery, the spinal canal was fully decompressed with angiographic cure of the AVM. The patient was symptom-free and made a full neurologic recovery. (Courtesy of Barrow Neurological Institute, Phoenix, Arizona.)

References

Caroscio JT, Brannan T, Budabin M et al. (1980). Subarachnoid hemorrhage secondary to spinal arteriovenous malformation and aneurysm: report of a case and review of the literature. Arch Neurol 37: 101–103.

Cogen P, Stein BM (1983). Spinal cord arteriovenous malformations with significant intramedullary components. J Neurosurg 59: 471–478.

Connolly Jr ES, Zubay GP, McCormick PC et al. (1998). The posterior approach to a series of glomus (type II) intramedullary spinal cord arteriovenous malformations. Neurosurgery 42: 774–785; discussion 785–786.

Corkill RA, Mitsos AP, Molyneux AJ (2007). Embolization of spinal intramedullary arteriovenous malformations using the liquid embolic agent, Onyx: a single-center experience in a series of 17 patients. J Neurosurg Spine 7: 478–485.

Gross BA, Du R (2013). Spinal glomus (type II) arteriovenous malformations: a pooled analysis of hemorrhage risk and results of intervention. Neurosurgery 72: 25–32; discussion 32.

Gross BA, Du R (2014). Spinal juvenile (type III) extradural-intradural arteriovenous malformations. J Neurosurg Spine 20: 452–458.

Hall WA, Oldfield EH, Doppman JL (1989). Recanalization of spinal arteriovenous malformations following embolization. J Neurosurg 70: 714–720.

Kim LJ, Spetzler RF (2006). Classification and surgical management of spinal arteriovenous lesions: arteriovenous fistulae and arteriovenous malformations. Neurosurgery 59 (Suppl 5): S3-195–S3-201.

MacFarlane MR, Burn PJ, Evison J (2005). Excision of high and mid cervical spinal cord arteriovenous malformations by anterior operation. J Clin Neurosci 12: 71–79.

Martin NA, Khanna RK, Batzdorf U (1995). Posterolateral cervical or thoracic approach with spinal cord rotation for vascular malformations or tumors of the ventrolateral spinal cord. J Neurosurg 83: 254–261.

Rangel-Castilla L, Russin JJ, Zaidi HA et al. (2014). Contemporary management of spinal AVFs and AVMs: lessons learned from 110 cases. Neurosurg Focus 37: E14.

Rosenblum B, Oldfield EH, Doppman JL et al. (1987). Spinal arteriovenous malformations: a comparison of dural arteriovenous fistulas and intradural AVMs in 81 patients. J Neurosurg 67: 795–802.

Spetzler RF, Zabramski JM, Flom RA (1989). Management of juvenile spinal AVM's by embolization and operative excision. Case report. J Neurosurg 70 628–632.

Spetzler RF, Detwiler PW, Riina HA et al. (2002). Modified classification of spinal cord vascular lesions. J Neurosurg 96: 145–156.

Touho H, Karasawa J, Shishido H et al. (1991). Successful excision of a juvenile-type spinal arteriovenous malformation following intraoperative embolization. Case report. J Neurosurg 75: 647–651.

Velat GJ, Chang SW, Abla AA et al. (2012). Microsurgical management of glomus spinal arteriovenous malformations: pial resection technique. Clinical article. J Neurosurg Spine 16: 523–531.

Williams FC, Zabramski JM, Spetzler RF et al. (1991). Anterolateral transthoracic transvertebral resection of an intramedullary spinal arteriovenous malformation. Case report. J Neurosurg 74: 1004–1008.

Wilson DA, Abla AA, Uschold TD et al. (2012). Multimodality treatment of conus medullaris arteriovenous malformations: 2 decades of experience with combined endovascular and microsurgical treatments. Neurosurgery 71: 100–108.

Yasargil MG, Symon L, Teddy PJ (1984). Arteriovenous malformations of the spinal cord. Adv Tech Stand Neurosurg 11: 61–102.

Handbook of Clinical Neurology, Vol. 143 (3rd series)
Arteriovenous and Cavernous Malformations
R.F. Spetzler, K. Moon, and R.O. Almefty, Editors
http://dx.doi.org/10.1016/B978-0-444-63640-9.00016-3

Chapter 16

Endovascular treatment of spinal arteriovenous malformations

WALEED BRINJIKJI[1] AND GIUSEPPE LANZINO[2]*

[1]*Department of Radiology, Mayo Clinic, Rochester, MN, USA*

[2]*Department of Neurosurgery, Mayo Clinic, Rochester, MN, USA*

Abstract

Spinal arteriovenous malformations (AVMs) are a rare but treatable cause of myelopathy and spinal cord hemorrhage. The clinical presentation and ideal treatment of these lesions vary widely, primarily due to differences in anatomic and angioarchitectural features. Six well-defined types of fistula have been defined: (1) intradural dorsal AVM; (2) intramedullary AVM; (3) extradural-intradural AVM; (4) intradural ventral AVF; (5) extradural/epidural AVF; and (6) conus medullaris AVM. Each of these lesions have unique clinical presentations and variable natural histories and endovascular treatment options.

The principles of endovascular treatment include: (1) a complete understanding of the angioarchitecture and anatomic features of the lesion, including identification of radiculomedullary arteries; (2) careful spinal angiography; (3) identification of treatment goals prior to the procedure (i.e., complete occlusion versus partial occlusion aimed at altering natural history); (4) selection of proper embolic agent; and (5) careful follow-up in both the immediate postoperative setting and long term.

INTRODUCTION

Spinal arteriovenous malformations (AVMs) are a rare but treatable cause of myelopathy. The clinical presentation and ideal treatment of these lesions vary widely, primarily due to differences in anatomic and angioarchitectural features. Over the past several decades, our understanding of these lesions has substantially improved, and as a result, many different classification schemes targeted at better identifying the natural history, clinical presentation, and ideal treatment of these lesions have been proposed. With advancements in knowledge regarding the anatomic and physiologic features of these lesions, as well as improvements in endovascular technologies and techniques, more of these lesions have been approached with endovascular only or combined endovascular and microsurgical treatments. In this chapter we provide an overview of spinal arteriovenous fistulas (AVFs) and AVMs with an emphasis on the endovascular treatment of these lesions.

CLASSIFICATION SCHEMES

Due to improved understanding of the vascular anatomy and pathophysiology of spinal AVMs, numerous classification schemes have been proposed. The most commonly employed classification scheme categorizes these lesions into four types. Type I lesions are spinal dural AVFs (SDAVFs), which are direct fistulous connections between the artery and vein and lack a nidus. These are also described as intradural dorsal AVFs according to the Spetzler classification and are described elsewhere. Type II lesions are intramedullary AVMs and are also called glomus malformations. These lesions have a true nidus. Type III lesions are referred to as juvenile or metameric lesions and often span multiple vertebral levels extending into the intradural and extradural space. According to the Spetzler classification, these are known as extradural-intradural AVMs. Type IV lesions are perimedullary lesions and are located along the ventral surface of the spinal cord and are

*Correspondence to: Giuseppe Lanzino, M.D., Mayo Clinic, Department of Neurosurgery, 200 1st Street SW, Rochester MN 55905, USA. Tel: +1-507-284-2511, E-mail: lanzino.giuseppe@mayo.edu

divided into three types based on number of arterial feeders, rapidity of blood flow, and presence of venous hypertension. According to the Spetzler classification, these are known as intradural ventral AVFs (Kim and Spetzler, 2006).

In addition to the classic four-type classification system, Spetzler and colleagues described two other types of fistula. Extradural AVFs have a direct connection between the radicular artery branch and the epidural venous plexus. These lesions are divided into three types based on the type of venous drainage. Lastly, there are conus medullaris AVMs which are characterized by multiple direct shunts between the spinal and radicular arteries and the pial veins. These lesions have a true nidus and are located along the pial surface at the conus medullaris, cauda equina, or filum terminale. For the purposes of this chapter we will describe these lesions according to the classification scheme proposed by Spetzler and colleagues (Kim and Spetzler, 2006).

BASIC SPINAL VASCULAR ANATOMY

Numerous volumes have been written about spinal vascular anatomy (Thron and Rossberg, 1988; Berenstein and Lasjaunias, 1992). A deep understanding of spinal vascular anatomy is essential to developing an understanding of the natural history of these lesions, endovascular treatment options, and avoiding complications. Here we provide a brief overview of the salient aspects of spinal vascular anatomy for the endovascular treatment of these lesions.

Segmental arteries supplying the spinal cord, vertebral bodies, and meninges arise from the vertebral arteries, deep and ascending cervical arteries, median and lateral sacral arteries, intercostal and lumbar arteries, and internal iliac and iliolumbar arteries (Thron and Rossberg, 1988; Berenstein and Lasjaunias, 1992). At each segmental level, there is a radicular/radiculomeningeal artery which supplies the dura. At multiple levels there are also radiculopial arteries which supply the two parasagittal posterior spinal arteries as well as radiculomedullary arteries which supply the anterior spinal artery.

The most common configuration of spinal arterial supply is two parasagittal posterior spinal arteries and one dominant anterior spinal artery. The posterior spinal arteries are generally smaller in caliber than the anterior spinal artery as the anterior spinal artery supplies about 75% of the spinal cord. There are generally 5–10 radiculomedullary arteries which supply the anterior spinal artery, most of which are below the resolution of conventional angiography (Thron and Rossberg, 1988; Berenstein and Lasjaunias, 1992). There are typically two angiographically visible arteries: the artery of the cervical enlargement, which arises from the cervical vertebral artery, and the artery of Adamkiewicz, which typically arises between T9 and T12, more commonly on the left (Murthy et al., 2010).

While classic anatomic diagrams demonstrate the posterior spinal arteries as two small paired vessels along the posterior surface of the cord, in reality, the posterior spinal arteries are a discontinuous system supplied by multiple radiculopial vessels. These can be recognized by their straight, off-midline appearance and by the fact that these arteries rarely span more than a few levels. There is some communication between the anterior and posterior spinal arteries through penetrating vessels within the spinal cord parenchyma (rami perforantes) and along the circumference of the cord (vasa corona). These vessels are rarely visible on normal angiograms but can be apparent in patients with pial AVF nidus-type AVMs (Thron and Rossberg, 1988; Berenstein and Lasjaunias, 1992; Martirosyan et al., 2015).

The spinal cord venous system has three compartments: intradural veins, including intramedullary and extramedullary veins, epidural veins, and intraosseous paravertebral veins. The intradural veins, including the coronal venous plexus, are connected to the epidural venous plexus through radiculomedullary bridging veins. Venous drainage from the cord goes from the intradural venous plexus into the epidural veins through the radiculomedullary bridging veins and then into the inferior vena cava, azygous, and hemiazygous veins (Thron et al., 2015). These bridging veins have a reflux-impeding mechanism which under normal physiologic circumstances inhibits reflux of blood from the epidural venous plexus into the perimedullary veins due to narrowing and zigzagging of the bridging vein when it crosses the dura (Thron et al., 2015).

ANGIOARCHITECTURE BY LESION TYPE

Intramedullary spinal AVMs

Intramedullary AVMs have a similar angioarchitecture to cerebral parenchymal AVMs in that they consist of an intramedullary nidus. These lesions generally have multiple arterial feeders and drain into the coronal venous plexus. Arterial feeders can either be direct or indirect (Rangel-Castilla et al., 2014). Direct arterial feeders are the angiographically larger branches entering the AVM. These arteries typically are only supplying the AVM and provide little to no supply to the cord itself. In contrast, indirect feeders are smaller-caliber vessels which also feed normal territories. These generally form as a result of a sump effect and local collateral circulation recruitment. About one-third of these lesions have associated aneurysms.

Extradural-intradural AVMs

Extradural-intradural AVMs are also known as metameric AVMs and can involve the spinal cord, vertebral body, epidural space, paraspinal soft tissue, skin, and subcutaneous tissues involving one or more of the 31 spinal segments. These lesions typically have both intradural and extradural components (Linfante et al., 2012). The intradural lesion is most commonly an intramedullary nidus-type spinal AVM. In addition, these patients can get nerve root AVMs which are frequently missed on imaging. Arterial supply is typically from multiple segmental branches and venous drainage is intradural, epidural, and extraspinal. Associated aneurysms are seen in about 50% of cases (Niimi et al., 2013).

Intradural ventral AVFs

Intradural ventral AVFs consist of a direct fistulous connection between an anterior or posterior spinal artery and a pial vein. These lesions form more often on the ventral surface of the spinal cord and typically have a midline location. There are three typical angioarchitectural features of these lesions: (1) single-feeder lesions with slow blood flow and mild venous hypertension; (2) high-flow lesions with multiple feeding arteries; and (3) giant fistulae with a markedly distended venous network. The giant fistulae can have multiple large venous varices which can result in compressive symptoms (Gross and Du, 2013b).

Spinal epidural AVFs

Spinal epidural AVFs (SEDAVFs) consist of a fistulous connection between a segmental artery and an epidural vein. These lesions typically have multiple feeding arteries. Unlike the above-described fistulae, these lesions are not typically supplied by radiculomeningeal and radiculomedullary arteries. However, close inspection of angiographic images is necessary to evaluate for the presence of a radiculomedullary artery arising from the feeding segmental artery. The feeding arteries typically drain into a shunted ventral epidural venous pouch (Choi and Berenstein, 1988; Lasjaunias and Berenstein, 1990).

These lesions are divided into three types based on pattern of venous drainage (Rubin and Rabinstein, 2013). Type A have drainage into the perimedullary venous plexus, type B1 lesions have only epidural venous drainage with thecal sac compression due to engorgement of the epidural venous plexus, and type B2 lesions have epidural venous drainage without compression. In type A lesions, the venous pouch drains through a bridging vein to a radiculomedullary and/or perimedullary vein. In type B1 and B2 lesions, the pouch drains via radiculoemissary veins into the paravertebral veins. Interestingly, cervical SEDAVFs generally demonstrate only paravertebral/extradural drainage (90% of cases) while lumbosacral SEDAVFs generally demonstrate intradural venous dranage (60%) (Huang et al., 2013b).

Conus medullaris AVMs

Conus medullaris AVMs are characterized by multiple feeding arteries, multiple nidi, and complex venous drainage. These lesions typically have multiple direct arteriovenous shunts from the anterior and posterior spinal arteries as well as nidal-type arteriovenous connections that are usually extramedullary. These lesions are always located in the conus medullaris and cauda equina and can extend along the entire filum terminale (Rangel-Castilla et al., 2014).

CLINICAL PRESENTATION

Intramedullary spinal AVMs

Intramedullary spinal AVMs present with hematomyelia or subarachnoid hemorrhage, with devastating neurologic deficits in about 50% of cases. Spinal hemorrhage typically results in severe pain in the back, neck, and legs. Subarachnoid extension of the hemorrhage can result in headaches, loss of consciousness, and seizures. About half of patients with intramedullary spinal AVMs present with myelopathy (Gross and Du, 2013a).

Extradural-intradural AVMs

Extradural-intradural AVMs present with hemorrhage in about one-third of cases, and progressive neurologic deficit in a third of cases, and acute neurologic deterioration in 20% of cases (Linfante et al., 2012; Niimi et al., 2013). Acute thrombosis of the draining vein is thought to be the cause of acute neurologic deterioration in the absence of hemorrhage. Another rarer type of presentation of these lesions is painful radiculopathy secondary to nerve root AVMs. These lesions are rarely diagnosed as incidental findings.

Intradural ventral AVFs

The presentation of intradural AVFs is fairly constant by angioarchitecture. About one-third of lesions present with hematomyelia or subarachnoid hemorrhage. Another third present with acute deficits presumably due to thrombosis of the draining veins. Lastly, one-third present with progressive myelopathy. Because of the large venous varices associated with some of these lesions, these patients can also present with a compressive myelopathy (Gross and Du, 2013b).

Spinal epidural AVFs

Because of their extradural location, hematomyelia and subarachnoid hemorrhage are exceedingly rare in patients with SEDAVFs. Epidural hematoma occurs in about 5% of patients. In patients with type A fistulas, the presentation is often due to a congestive myelopathy due to venous drainage along the coronal venous plexus. In type B1 fistulas, patients can present with a compressive myelopathy due to an enlarged epidural venous plexus or a compressive radiculopathy due to mass effect on the nerve root. Type B2 fistulas, which lack mass effect, can be seen as incidental findings in patients being imaged for nonspecific symptoms such as back pain (Huang et al., 2013a).

Conus medullaris AVMs

Conus medullaris AVMs typically present with progressive myelopathy, radiculopathy, or myeloradiculopathy secondary to venous hypertension or mass effect. Hemorrhage can occur in about 10% of these lesions due to the fragile nature of the multiple nidal arteriovenous connections (Rangel-Castilla et al., 2014).

INDICATIONS FOR TREATMENT

In general, spinal vascular malformations of any type should undergo treatment with endovascular therapy, microsurgery, or multimodality treatment. The goal of treatment should be to achieve complete obliteration of the AVM or AVF without subjecting the patient to additional morbidity (Rodesch et al., 2002, 2003, 2005; Rodesch and Lasjaunias, 2003; Lee et al., 2014; Rangel-Castilla et al., 2014). In cases where a complete cure cannot be achieved without risk of substantial morbidity, partial treatment focused on obliterating weak points such as intranidal aneurysms or pseudoaneurysms or reducing the flow through the malformation should be performed. Indications and aggressiveness of treatment are dictated by the angioarchitecture of the lesion as well as the age and clinical presentation of the patient (Rodesch et al., 2002, 2003, 2005; Rodesch and Lasjaunias, 2003; Lee et al., 2014; Rangel-Castilla et al., 2014).

The timing of treatment in patients with spinal hemorrhage is a subject of debate. Gross and Du (2013a) demonstrated a 10% annual rate of hemorrhage in lesions that previously bled, suggesting that urgent intervention should be considered when deemed safe. Immediately following the hemorrhage, patients may present with severe loss of function at and below the level of the hematoma. However, many patients do regain some function following the initial insult and some authors suggest that waiting for natural recovery to occur is essential (Berenstein and Lasjaunias, 1992; Gross and Du, 2013a). An exception is made in the setting of a pseudoaneurysm as these can result in early rehemorrhage (Rodesch et al., 2004; Yahiro et al., 2004; Alnaami et al., 2013).

In patients with a fixed deficit following a hemorrhage or infarct, treatment is indicated to reduce the risk of rehemorrhage or progression of the lesion. In these cases, the decision to treat should be based on whether or not it is feasible to treat without increasing the neurologic deficit. This is especially true for cervical or thoracic lesions where rehemorrhage or reinfarction can result in upper-extremity symptoms, respiratory distress, or loss of consciousness. In patients with lower thoracic or lumbar AVMs who have fixed deficits, treatment is appropriate if complete occlusion can result in preservation of sphincter control or sexual function (Berenstein and Lasjaunias, 1992).

ANESTHESIA CONSIDERATIONS

Unless there is a major contraindication, general endotracheal anesthesia is strongly recommended when performing endovascular intervention for treatment of spinal vascular malformations. It is widely accepted that there is no substitute for high-resolution spinal angiography during an embolization procedure and performing these procedures under conscious sedation or local anesthesia restricts operator ability to perform a high-quality exam. In addition, for lumbar and sacral vascular malformations, bowel movements associated with normal peristalsis can be an issue limiting the resolution of angiography. To overcome this limitation, administration of glucagon (which diminishes peristalsis) is a valid adjunct to increase quality of the images.

Spinal cord monitoring during the procedure is a topic of debate (Sala et al., 1999, 2000, 2001; Niimi et al., 2000, 2004). No study has systematically compared outcomes of patients with and without spinal cord monitoring, and any such study would be limited by the fact that operative monitoring is likely performed primarily for more complex lesions. The decision as to whether or not to perform intraoperative spinal cord monitoring should be based on operator and anesthesiologist preference.

Spinal cord monitoring should be considered primarily for treatment of intramedullary AVMs, intradural-extradural lesions, intradural ventral AVFs, and conus medullaris AVMs. It can also be considered for SEDAVFs or SDAVFs in the vicinity of the anterior spinal artery. Electrophysiologic monitoring with somatosensory evoked potential (SEP) and motor evoked potential (MEP) with or without provocative testing with pharmacologic agents can be performed (Sala et al., 1999, 2000, 2001; Niimi et al., 2000, 2004).

SEPs and MEPs should be recorded at the beginning of the procedure immediately after induction of general anesthesia. Monitoring of upper and lower extremities as well as the bulbocavernosus reflexes can be performed. SEP and MEP monitoring can be performed throughout the procedure for early detection of ischemia due to blocked flow from thrombus, vasospasm, the microcatheter, or embolic agents. MEPs can be obtained even in patients with moderate deficits, while SEPs cannot be obtained in patients who have lost proprioception. One thing to keep in mind is that MEPs and SEPs are sensitive to the depth of anesthesia (Sala et al., 1999, 2000, 2001; Niimi et al., 2000, 2004).

According to Berenstein and colleagues, provocative testing should be performed immediately prior to injection of the liquid embolic agent with the tip of the microcatheter located somewhat remotely from the nidus or fistula site (Niimi et al., 2004). Injection under a roadmap can be performed to determine the optimal rate of injection to distribute the anesthetic without creating reflux. Following this, 50 mg of sodium amytal can be injected, with monitoring of SEPs or MEPs. If SEPs and MEPs do not change, then 20–40 mg of lidocaine can be injected. If SEPs and MEPs do not change following this injection, then embolization can safely be performed. If SEPs and MEPs do change with either amytal or lidocaine injection, then repositioning of the microcatheter and/or careful risk–benefit assessment should be performed. In cases where the amytal injection is positive, lidocaine injection need not be performed. According to their experience, Berenstein and colleagues reported that the negative predictive value of provocative testing was high (97.6%) (Niimi et al., 2004). Furthermore, positive provocative testing resulted in change in management in some cases so as to prevent embolization of arterial feeders supplying the normal spinal cord. Other authors have reported similar findings (Sala et al., 2007).

GENERAL TECHNIQUES OF ENDOVASCULAR TREATMENT

Embolic agent selection

The choice of embolic agent depends largely on the angioarchitecture of the lesion as well as the goals of treatment and operator preference. Below is a description of the various embolic agents used for the treatment of spinal vascular malformations as well as potential pitfalls.

Coils

Detachable coils have a role as both adjunctive and primary embolic agents in the treatment of spinal vascular malformations. The advantage of detachable as opposed to nondetachable coils is the fact that the operator can select the location and configuration of the detachable coil as it can be easily repositioned and retrieved. In addition, a control angiographic run can be performed immediately prior to detachment so as to ensure that coil deployment results in the proper hemodynamic effect (Berenstein et al., 2004a, b; Berenstein and Lasjaunias, 2012).

In the treatment of spinal vascular malformations, detachable coils are primarily used for giant intradural ventral AVFs, epidural/paraspinal AVFs, and or extradural-intradural lesions as they can be used to obliterate large fistulous connections as well as venous and arterial aneurysms. The use of coils has been advocated in the treatment of giant AVFs (Fig. 16.1) as they can be used as a frame for liquid embolic materials in high-flow lesions, thus preventing migration of glue or Onyx to the venous side. Coil embolization of large arterial feeders can also be performed in an attempt to alter the hemodynamics of some lesions for both palliative purposes as well as prior to definitive embolization with liquid embolic agents or microsurgery.

Unlike liquid embolic agents or particles, detachable coils require positioning of the microcatheter directly at the site of coil deployment. Selection of proper coil size is important. Coils should be large enough to allow the coil to remain stable within the treated lesion so that a scaffold can be created for additional coiling. It is important that operators select a microcatheter with an inner diameter that is large enough to allow for the deployment of coils. These larger microcatheters are more rigid, thus limiting the accessibility of some lesions (Berenstein et al., 2004a, b; Berenstein and Lasjaunias, 2012).

Particles

Particle embolization is used primarily for palliative or presurgical embolization. Particles can also be used to alter the hemodynamics of a lesion in preparation for definitive treatment with liquid embolic agents (Berenstein et al., 2004a, b; Berenstein and Lasjaunias, 2012). Because of their high coefficient of friction, polyvinyl alcohol (PVA) particles adhere to the vessel wall, resulting in stagnation of blood flow and eventual clotting. However, recanalization rates following PVA embolization of spinal AVMs are as high as 80%. Despite this, in certain cases where the goal is to alter natural history or in cases where other treatment options are not feasible, particle embolization is a good alternative. One thing to keep in mind is that surveillance angiography with re-embolization is the rule, not the exception, so ensuring good clinical follow-up is key. Lesions with large arteriovenous connections or high flow may not be amenable to particle embolization due to the risk of

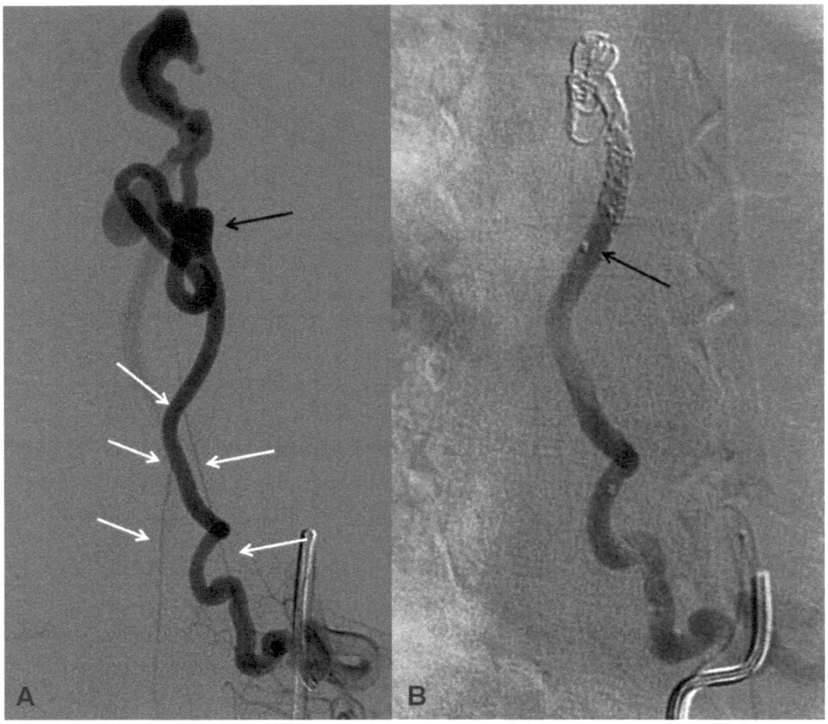

Fig. 16.1. (**A**) Pre-embolization angiogram demonstrated the feeding artery/posterolateral spinal artery arising off T9 with the fistulous connection at the site of the ectatic vein (black arrow). The artery of Adamkiewicz also is fed by the left T9 intercostal artery but is not supplying the fistula (white arrows). (**B**) The microcatheter tip was placed in the distal aspect of the feeding artery during coil embolization (black arrow). Coil embolization of the fistulous connection and distal feeding artery was performed with complete angiographic occlusion.

shunting of the particles to the pulmonary circulation (Berenstein et al., 2004a, b; Berenstein and Lasjaunias, 2012). This is especially true in patients with right-to-left shunts as this can result in systemic embolization as well.

PVA embolic particles come in a range of sizes (100—1000 μm). For the purposes of spinal vascular malformation embolization, larger particles are primarily used. Particles are mixed with contrast so as to allow for visualization of the particles during embolization. Care must be taken to not create too high a concentration of particles as this can cause plugging of the microcatheter. Particles are generally injected by connecting a 10-cc syringe full of a contrast-particle mixture and a 1-cc syringe to a three-way stopcock. The mixture is aspirated into the 1-cc syringe and then slowly injected into the microcatheter. The particles should be injected under a blank roadmap image so as to allow for visualization of slowing of blood flow as well as reflux of particles. Prior to performing control angiograms through the microcatheter or prior to repositioning of the catheter, the microcatheter should be flushed with 1 cc of saline under the roadmap so as to rid the system of any left-over particles.

Other type of particles, including silicon sphere, hydrogel-based spheres, and Embospheres, have been used for embolization of spinal vascular malformations but are mostly of historic interest. The rules of embolization with these agents are similar to that of particles.

NBCA or "glue"

Prior to the advent of Onyx, embolization with *N*-butyl cyanoacrylate (NBCA) was the preferred endovascular treatment for most spinal vascular malformations. NBCA is an acrylic glue that polymerizes when it comes in contact with any ionic solution, including contrast or blood. Prior to injection of NBCA, the microcatheter must be prepared with an injection of 5% dextrose to prevent the glue from hardening within the microcatheter (Berenstein et al., 2004a, b; Berenstein and Lasjaunias, 2012). In order to make the glue radiopaque, it is mixed with either tantalum powder or lipiodol. Tantalum powder is the preferred mixing agent in high-flow fistulas. In order to control the embolization of glue, a slow injection is required, preferably through a marked 1-cc syringe. Similar to PVA particle embolization, a blank roadmap can be used to examine the flow of the glue as well as the presence of any reflux.

The goal of a controlled glue injection is to create a complete cast of the fistulous connection and draining

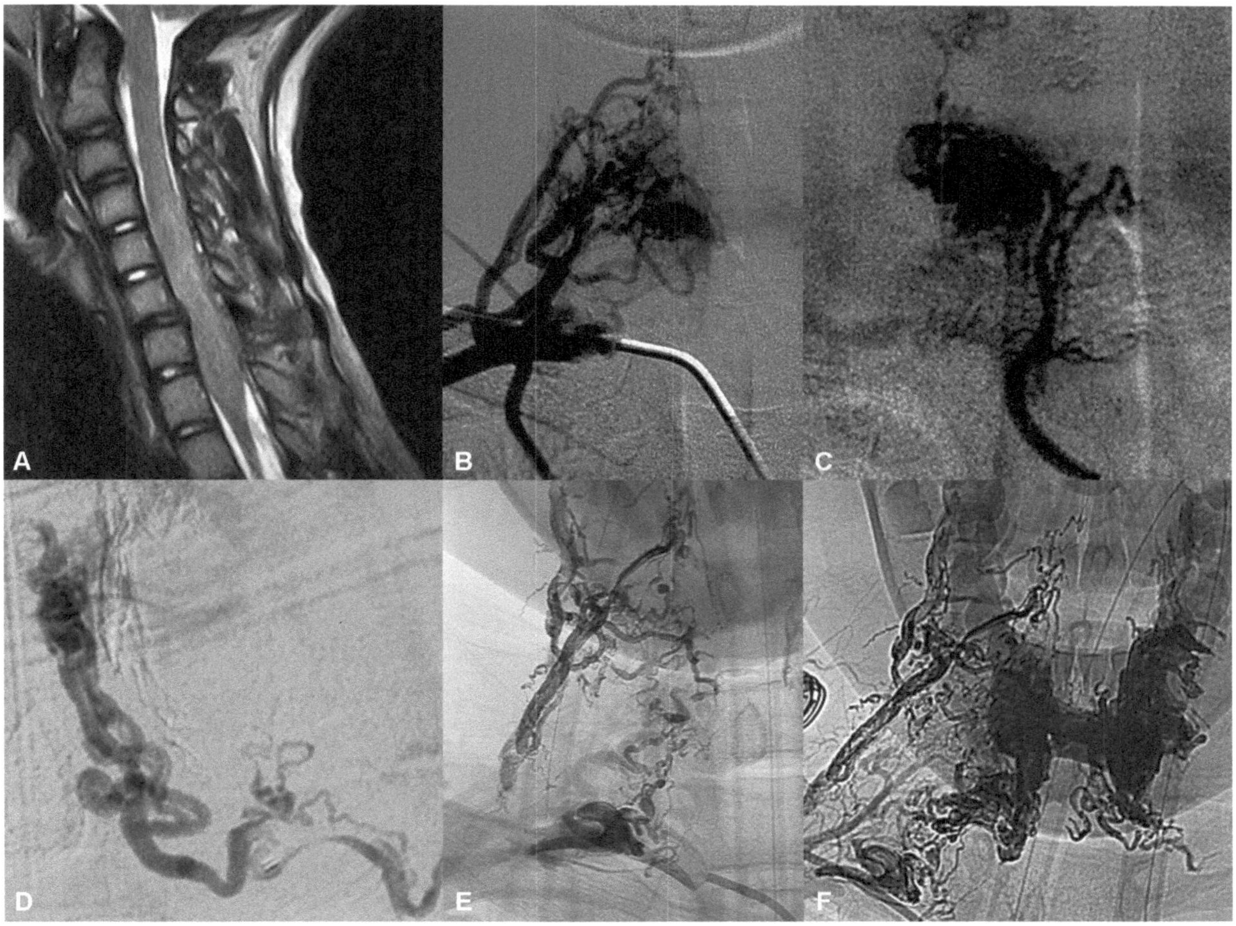

Fig. 16.2. 12 F with a cervical-spine epidural arteriovenous malformation (AVM). Patient presented with sudden onset of weakness and sensory loss in the hands and legs. (**A**) Cervical spinal cord magnetic resonance imaging shows marked enlargement and high T2 signal in the cord. Spinal angiogram demonstrated a complex extradural-intradural AVM which involved the spinal elements at C6–C7 with filling from multiple arterial branches, including the right thyrocervical trunk (**B**), a large branch from the aortic arch (**C**), and the left deep cervical artery (**D**). The patient was treated with multiple sessions of staged Onyx and particle embolization (**E** and **F**). The patient made a near-complete recovery with near-complete resolution of high T2 cord signal on follow-up imaging and was able to walk normally 2 years following treatment.

vein(s), otherwise recanalization can occur. The use of adjunctive coil embolization during the embolization of large AVFs has been suggested as a means to reduce distal migration of glue. There are a few key technical points to glue embolization that must be discussed. First, if superselective catheterization of an arterial feeder is achieved and there is no risk of embolization of normal spinal cord, NBCA can be injected as a continuous column to create a cast of the vessel. In some cases, the catheter can be wedged into the arterial feeder causing flow arrest (Berenstein et al., 2004a, b; Berenstein and Lasjaunias, 2012). In this case, glue can be injected into the nidus and reflux into territories of the nidus fed by other arteries. Care must be taken when removing the microcatheter following embolization as the microcatheter may became "glued" to the vessel and in some cases may be left behind to avoid extreme traction. However, with the advent of detachable microcatheters and hydrophilic coating this has become less of a problem.

Onyx

Onyx has become the preferred embolic agent in the treatment of spinal vascular malformations (Figs 16.2 and 16.3). This is largely due to higher rates of complete angiographic obliteration as well as reduced recanalization rates. Onyx is composed of ethylene vinyl alcohol, which is a permanent nonadhesive liquid embolic agent. Onyx comes as a prepared radiopaque solution and does not require any additional preparation. When planning an Onyx embolization, preparation is key. First, Onyx vials should be agitated prior to use and the vials should be

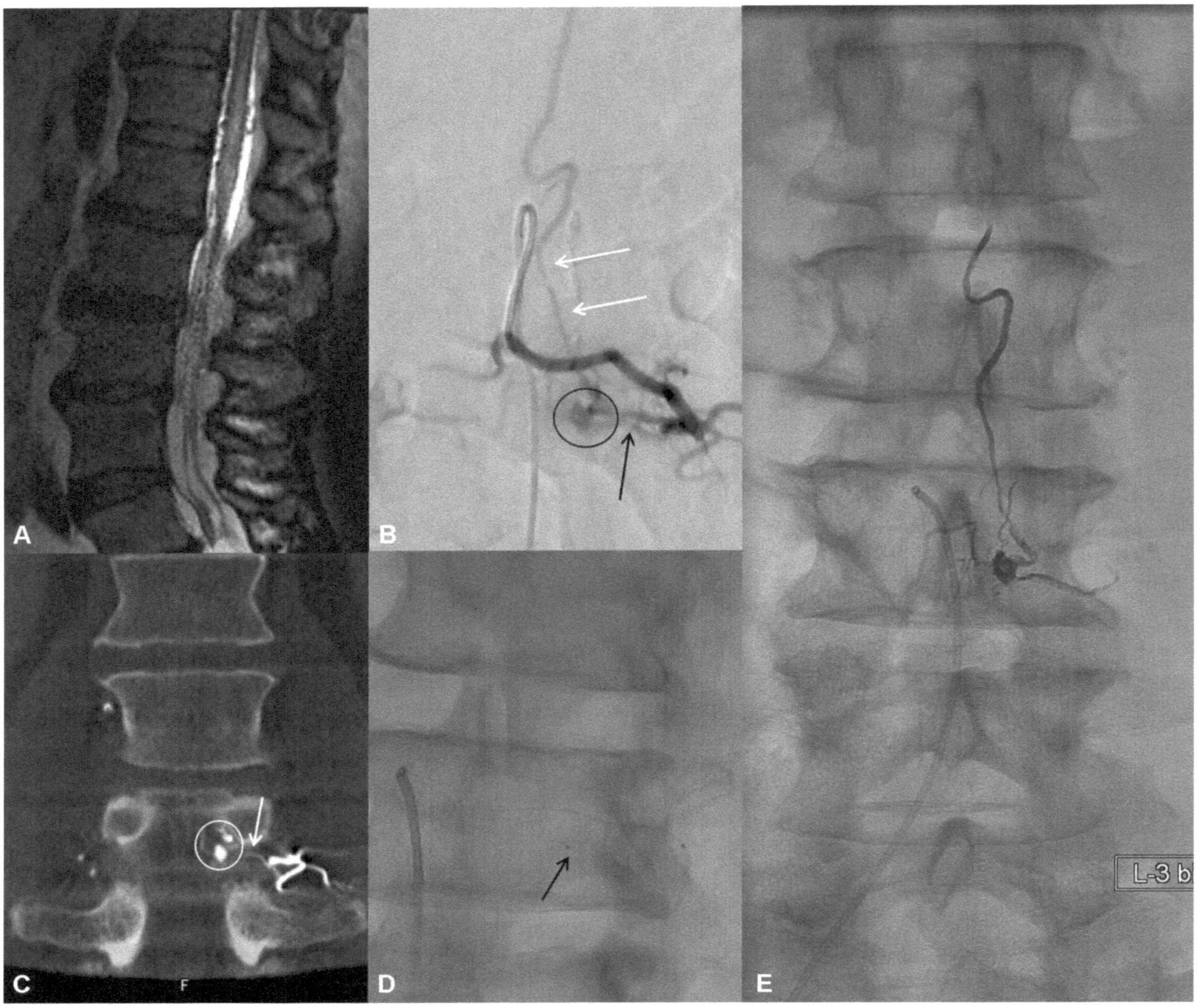

Fig. 16.3. Type A spinal epidural arteriovenous fistula in a patient status postlaminectomy. Patient had a prior history of L3 laminectomy. He presented with new myelopathic symptoms about 1 year following his surgery. (**A**) Magnetic resonance imaging demonstrates multiple intradural flow voids and high T2 signal in the conus. (**B**) Conventional spinal angiographic images following injection of the left L3 segmental artery in anteroposterior and lateral views demonstrate a medially coursing artery (black arrow) connecting with a venous pouch in the epidural space (black circle). The intradural draining vein is seen coursing up the spinal canal (white arrows). (**C**) Flat-panel computed tomography images demonstrate the medially coursing artery (white arrow) with the venous pouch in the epidural space representing the site of the fistula (white circle). (**D**) The tip of the microcatheter is located just proximal to the fistulous pouch (black arrow). (**E**) Following injection of Onyx, there is Onyx filling of the fistulous pouch as well as the draining intradural vein and feeding artery.

on the "shaker" before the operator enters the room. In addition, it is important to make sure that the microcatheter that is being used is dimethyl sulfoxide (DMSO)-compatible as DMSO is required to prepare the microcatheter so that Onyx does not precipitate within the microcatheter itself. An adapter must be screwed on to the hub of the microcatheter prior to DMSO infusion. It is important to inject enough DMSO so that the entire microcatheter volume is prepared. Following this, Onyx is injected, through a 1-cc syringe, through the microcatheter. A proximal plug can be created around the distal tip of the microcatheter to produce sufficient proximal flow arrest and allow distal penetration of the embolic agent to the nidus or fistulous pouch. The advantage of Onyx is the associated retrograde occlusion of other arterial feeders. One disadvantage of Onyx is the propensity to reflux along the course of the microcatheter.

Perioperative anticoagulation

No specific preoperative medication or anticoagulation is required prior to embolization of spinal vascular malformations. Systemic heparinization during the procedure is recommended (Berenstein et al., 2004a, b; Berenstein and Lasjaunias, 2012). Typically, a loading dose of 3000 units of heparin is given once the sheath is inserted with adjustments made during the procedure according to activated clotting time measurements. Flush bags are typically heparinized as well with 2000 U/L in children and 4000 U/L in adults.

Postoperatively, anticoagulation is continued in some cases. For lesions with large venous varices such as SEDAVFs or intradural ventral AVFs with large venous varices, thrombosis of these varices can result in a compressive myelopathy or even induce ischemia. For this reason, many advocate the use of systemic heparinization for 24–48 hours following the procedure followed by aspirin.

Equipment choices

Equipment choices are extremely important when preparing to treat a spinal vascular malformation. It is important that the microcatheter that is selected is compatible with the embolic agent that is being used, and in the case of Onyx embolization, the catheter must be DMSO-compatible. When using coils, a coil-compatible, rigid microcatheter is needed. If detachable coils are used the presence of a double marker on the catheter is another useful prerequisite. For particle embolization, the inner diameter of the microcatheter must be large enough to not be plugged up by the particles.

When choosing a microcatheter, operators must choose between using over-the-guidewire-type microcatheters or flow-directed microcatheters. Flow-directed microcatheters are the preferred tool for embolization of spinal vascular malformations with glue; however, most flow-directed microcatheters are not DMSO-compatible, precluding their use in Onyx embolization. These catheters are designed with a gradually changing flexibility and diameter between the proximal and distal ends of the microcatheter. The distal tip is more flexible, which allows the microcatheter to be directed by the flow of blood, while the proximal end is stronger, allowing for more torquability. This design allows for navigation of tortuous anatomy which is generally present in all types of spinal vascular malformations. In addition, these catheters are less traumatic, resulting in lower rates of dissection and vasospasm.

Embolizations are usually performed using a standard 5 F access system. A 5 F catheter such as a Cobra catheter or Mikaelson catheter is typically used to select the main arterial feeder and provide mechanical support for the microcatheter. All microcatheters are compatible with 5 F systems.

Principles of endovascular treatment

Before going into depth regarding the goals and techniques of embolization of specific types of spinal vascular malformations, a brief overview of the general principles of embolization is important. The key to achieving a good treatment result is to develop an understanding of the angioarchitecture of the lesion, the goals of treatment, and the location of the anterior and posterior spinal arteries. While the posterior spinal arteries generally have sufficient collaterals, the anterior spinal artery does not. Thus, errant embolization of a normal anterior spinal artery can have devastating consequences. Prior to embolization, the levels above and below the vascular malformation should be injected to determine if there is any risk of embolizing the anterior spinal artery through arterial–arterial anastomoses. In addition, following injection of embolic material, careful control angiograms should be performed to determine if changes in the hemodynamic effects of embolization resulted in redirection of flow to a normal anterior spinal artery (Berenstein et al., 2004a, b; Berenstein and Lasjaunias, 2012).

In some cases, embolization of individual branches of the anterior or posterior spinal arteries is needed to obliterate the fistula. Identification of pathologic vessels that are supplying the fistula and not the cord itself is important. Such vessels can usually be embolized without incident (Fig. 16.1). However, great care must be taken to avoid reflux of embolic material into the normal anterior spinal artery.

Once the vascular anatomy has been mapped out and the target vessel has been identified, the radicular artery from which the feeding vessel arises can be selectively catheterized with the 5 F catheter. Achieving stability of the 5 F catheter is essential. If the catheter is not in stable position, one can easily lose access following selective catheterization of a distal feeder, which can prolong operating time and result in vascular injury or vasospasm. Once the 5 catheter is in good position, slow introduction of the microcatheter can be performed. The target of embolization should be the nidus or site of the AVF itself. Thus, the operator should attempt to have the microcatheter reach as far distally as possible so that the site of the AVM/AVF can be targeted. If a catheter or wire is not working, operators should have a low threshold for changing equipment. Once the microcatheter is in place a superselective angiogram should be performed to ensure proper position and lack of any unsuspected arterial anastomoses.

When embolizing with particles or liquid embolic material, care should be taken to try to avoid embolization of a wedged microcatheter. When the microcatheter is wedged, there can be changes in the flow patterns and the embolic material might reach arteries that were not opacified on the pre-embolization angiogram. With liquid embolic materials it is also preferable to have some of the agent reach the draining vein(s) or reflux into other feeding arteries from the nidus/venous pouch in order to achieve a better angiographic result and reduce the risk of recanalization. Obliteration of the venous drainage in the setting of a persistent fragile nidus or fistulous connection can result in hemorrhage of the lesion due to hypertension in the nidus. Also, embolization of the feeding artery proximal to the nidus or fistulous connection without successful targeting of the nidus is usually ineffective as other arterial anastomoses will inevitably be recruited to the shunt.

LESION-SPECIFIC ENDOVASCULAR TREATMENT

Intramedullary spinal AVMs

Embolization is the preferred initial treatment for intramedullary AVMs. The main contraindication to endovascular treatment is inability to access the nidus due to either a long route or unstable catheter positioning. Features that make a lesion more favorable to surgery are superficial location or location in the cervical spine or filum terminale (conus medullaris-type fistula). While complete embolization of the lesion is preferable, partial embolization of angiographic weak points is generally sufficient to alter the natural history of these lesions. When treating these lesions, selective catheterization of the anterior or posterior spinal arteries feeding the AVM is performed. A flexible, flow-directed microcatheter may be needed to navigate this potentially tortuous course. The microcatheter should be positioned inside the nidus or as close to the nidus as possible. When injecting, especially in the case of feeding vessels arising from the anterior spinal artery, the key is to avoid any reflux of embolic material into the normal anterior spinal artery.

Extradural-intradural AVMs

Treatment decisions for extradural-intradural AVMs are generally made on a case-by-case basis. These lesions are extremely rare and fewer than 100 cases have been reported in the literature. A majority of these lesions are treated with endovascular therapy alone or combined endovascular and surgical treatment, while only a minority are treated with surgery alone. Complete cure of these lesions is extremely difficult and thus treatment is considered to be palliative, targeting at relieving venous hypertension, mass effect, or reducing risk of bleeding (Fig. 16.2). Treatment is targeting at the spinal component rather than the soft-tissue component of these lesions. Specific targeting of nidal aneurysms or pseudoaneurysms is recommended to improve the natural history of these lesions (Linfante et al., 2012; Niimi et al., 2013; Rangel-Castilla et al., 2014).

Intradural ventral AVFs

Choice of therapy for spinal intradural ventral AVFs depends largely on angioarchitecture. In general, single-feeder lesions with slow blood flow and mild venous hypertension are primarily treated with surgical treatment while high-flow lesions with multiple feeding arteries are treated with either surgery or combined surgical and endovascular therapy. When these two lesions are located on the posterior or lateral surface of the cord they are easily treated with surgery; however, if located on the anterior surface of the cord, endovascular therapy is the treatment of choice. Giant fistulae with a markedly distended venous network are predominantly treated with endovascular therapy alone (Fig. 16.1). As with other classes of spinal vascular malformations, the target for embolization is the site of the arteriovenous connection. Liquid embolic materials are the agents of choice for these lesions and PVA particles are generally reserved for presurgical embolization. In giant fistulae with a markedly distended venous network, coils can be placed in the fistula to prevent migration of liquid embolic material or particles to the venous side of the fistula. This requires the use of a larger, coil-compatible microcatheter. Following this, the operator can decide whether or not to proceed with additional embolization with PVA particles or liquid embolic agents.

Similarly to intramedullary AVMs, the microcatheter should be positioned inside the nidus or as close to the venous connection as possible. When injecting, especially in the case of feeding vessels arising from the anterior spinal artery, the key is to avoid any reflux and flow of embolic material into the normal anterior spinal artery. In cases of both anterior and posterior spinal artery feeders, the posterior spinal artery should be embolized first. Unlike intramedullary AVMs, there is little benefit in partial embolization of these lesions (Berenstein et al., 2004a, b; Berenstein and Lasjaunias, 2012).

Spinal epidural AVFs

Currently, endovascular treatment is the favored method of treatment, with about two-thirds of studies and case reports reporting the use of endovascular treatment. Transarterial embolization was the primary means of treatment. Endovascular treatment of type A SEDAVFs is focused on occlusion of the venous pouch along with the proximal intradural draining vein (Fig. 16.3). Failure to obliterate the pouch and draining vein can result in recanalization due to *de novo* recruitment of additional arterial feeders (Silva et al., 2007). Transarterial embolization is often difficult to perform when the feeding arteries are small or tortuous. The primary complication of transarterial embolization is inadvertent embolization of the radiculomedullary artery and careful examination of the arterial distribution at a few levels above and below the actual connection is paramount to avoid this disastrous complication (Ramanathan et al., 2014a).

In the setting of large venous pouches, embolization can be difficult due to the large amount of embolic material required to achieve complete obliteration. This is a particular problem in the treatment of type B lesions, where symptom onset often occurs once the venous pouch has become large enough to produce mass effect. The primary embolic agents used to treat SEDAVFs are Onyx and NBCA glue (Rangel-Castilla et al., 2011; Wang et al., 2011; Lanzino et al., 2012; Ramanathan et al., 2014b; D'Souza, 2015). Onyx is nonthrombogenic and cohesive and allows for better packing of the epidural venous pouch without the risk of thrombus formation and embolization. Coil embolization has also been reported in the treatment of SEDAVFs.

Transvenous embolization of SEDAVFs has also been reported and, according to the review by Huang et al. (2013b), has been used in about 10% of reported cases. Transvenous embolization has been reported to be more effective than transarterial embolization in the setting of multiple arterial feeders. One unique approach to transvenous embolization was reported by Ramanathan et al. (2014a), who performed direct percutaneous embolization of an epidural venous pouch with NBCA glue.

Conus medullaris AVMs

Conus medullaris AVMs are best treated with a multimodality approach using superselective embolization followed by surgical resection. The reason for this is the presence of multiple arterial feeders, many of which are angiographically occult. One factor complicating endovascular management of these lesions is the presence of anterior and posterior spinal artery vessels supplying the conus, thus complete obliteration with endovascular management alone is difficult to achieve. Techniques employed for management of these fistulas are similar to those of pial AVFs and glomus-type AVMs (Hsu et al., 2002; Ramanathan et al., 2014a).

POSTOPERATIVE CARE

Patients undergoing embolization of spinal vascular malformations should be monitored in a neurointensive care unit for at least 24 hours following the procedure. Systemic heparin administration is recommended for patients treated for fistulas with an intradural or intramedullary component with a target activated partial thromboplastin time of 50–60 seconds in order to prevent thrombosis of normal vessels. Regular neurologic exams are recommended in order to detect any acute deterioration. Such a deterioration could be secondary to thrombosis of venous varices or a large feeding intradural artery or from hemorrhage of the lesion due to hemodynamic alterations.

ANGIOGRAPHIC AND CLINICAL OUTCOMES OF ENDOVASCULAR THERAPY

Intramedullary spinal AVMs

There are over 100 cases of intramedullary AVMs treated with endovascular embolization reported in the literature, compared to over 200 cases of surgically treated lesions. Complete obliteration rates with endovascular therapy are low (33%); however, long-term clinical outcome among these patients is improved in about 66% of cases and worsening has only been reported in about 10% of cases. Among patients treated with endovascular therapy, no postembolization AVM hemorrhages have been reported, even for lesions with only partial embolization. In their meta-analysis, Gross and Du (2013a) found that the rate of rehemorrhage for nontreated lesions was significantly higher than that of partially or completely treated lesions, thus demonstrating how even partial embolization can alter the natural history of these lesions. Interestingly, recanalization rates with PVA are about 20% per year while recanalization rates with liquid embolic materials are less than 5% per year.

Extradural-intradural AVMs

Extradural-intradural AVMs are notoriously difficult to treat and are exceedingly rare, with fewer than 100 cases reported in the literature. As mentioned previously, the goal of treatment is not complete obliteration of the fistula, but rather improvement in symptoms and the natural history of these lesions. In their review of 35 treated cases, Gross and Du (2014) reported that complete obliteration with endovascular embolization alone was about 30% and over 70% of patients reported improvement of symptoms with endovascular therapy. About 20% of patients however reported worsening of symptoms. No cases of hemorrhage on follow-up were described in the series by Gross and Du (2014). The largest single series to date on the treatment of these lesions was reported by Niimi et al. (2013) and included 28 patients. These authors reported a follow-up hemorrhage rate of 22%, with only about 50% of patients reporting improvement following endovascular treatment. Follow-up angiography was performed in 20 patients in this series and progression of the fistula with formation of new aneurysms was reported in over a third of cases.

Intradural ventral AVFs

In select intradural ventral AVFs, endovascular embolization can be performed with good result. In one systematic review performed by Gross and Du (2013b), complete obliteration rates were 90% for single-feeder lesions with slow blood flow and mild venous hypertension and 80% for high-flow lesions with multiple feeding arteries. Giant fistulae with a markedly distended venous network are primarily treated with endovascular therapy alone. Complete obliteration rates for these lesions are about 70%. Symptom improvement is reported in 75% of embolized patients, with only 10% of patients reporting worsening of their findings. Rehemorrhage rates following complete angiographic occlusion are exceedingly low – no cases have been reported. One case of rehemorrhage following partial obliteration has been reported (Gross and Du, 2013b).

Spinal epidural AVFs

Huang et al. (2013a) recently reported a systematic review of surgical and endovascular treatment of epidural AVFs. As mentioned previously, endovascular therapy is the preferred approach to these lesions, with nearly 70% of lesions treated in this fashion. Complete obliteration rates are over 90% with endovascular therapy and improvement in symptoms is reported in over 90% of cases. Eighty-five percent of patients report improvement in motor function and pain, 90% of patients report improvement in sensory function, and 70% of patients report improvement in bladder function following endovascular treatment. Recanalization rates with endovascular therapy are exceedingly low.

Conus medullaris AVMs

The largest series of conus medullaris AVMs reported to date is that of Wilson et al. (2012) from Barrow Neurological Institute. Of the 16 treated patients, 8 underwent embolization followed by microsurgical resection and no patients underwent endovascular treatment alone. Endovsacular embolization is largely an adjunct to microsurgical resection as achieving a complete angiographic cure is difficult due to the high potential for embolizing the vascular supply to the conus medullaris. Complete angiographic obliteration rates were about 90% and 86% of patients were improved or stable following treatment (Wilson et al., 2012).

CONCLUSION

In conclusion, spinal AVMs represent a variety of complex lesions. Treatment decisions and natural history vary widely based on the angioarchitecture of the lesion, clinical presentation, and age of the patient. A variety of endovascular tools are available for treatment of these lesions and, in a majority of cases, endovascular treatment can be considered a first-line treatment modality. The principles of endovascular treatment include: (1) a complete understanding of the angioarchitecture and anatomic features of the lesion, including identification of radiculomedullary arteries; (2) spinal angiography; (3) identification of treatment goals prior to the procedure (i.e., complete occlusion versus partial occlusion aimed at altering natural history); (4) selection of proper embolic agent; and (5) careful follow-up in both the immediate postoperative setting and long-term.

References

Alnaami I, Lam FC, Steel G et al. (2013). Arteriovenous fistula and pseudoaneurysm of the anterior spinal artery caused by an epidural needle in a 5-year-old patient. J Neurosurg Pediatr 11: 340–345.

Berenstein A, Lasjaunias P (1992). Spine and spinal cord vascular lesions. Surgical neuroangiography. Springer, Berlin.

Berenstein A, Lasjaunias P (2012). Surgical neuroangiography: Vol. 2. Clinical and endovascular treatment aspects in adults. Springer Science & Business Media, Berlin.

Berenstein A, Lasjaunias P, ter Brugge KG (2004a). Spinal arteriovenous malformations. Surgical neuroangiography. Springer, Berlin.

Berenstein A, Lasjaunias P, ter Brugge KG (2004b). Spinal dural arteriovenous fistulae. Surgical neuroangiography. Springer, Berlin.

Choi I, Berenstein A (1988). Surgical neuroangiography of the spine and spinal cord. Radiol Clin North Am 26: 1131–1141.

D'Souza S (2015). Aneurysmal subarachnoid hemorrhage. J Neurosurg Anesthesiol 27: 222–240.

Gross BA, Du R (2013a). Spinal glomus (type II) arteriovenous malformations: a pooled analysis of hemorrhage risk and results of intervention. Neurosurgery 72: 25–32; discussion 32.

Gross BA, Du R (2013b). Spinal pial (type IV) arteriovenous fistulae: a systematic pooled analysis of demographics, hemorrhage risk, and treatment results. Neurosurgery 73: 141–151; discussion 151.

Gross BA, Du R (2014). Spinal juvenile (type III) extradural-intradural arteriovenous malformations. Journal of neurosurgery. Spine 20: 452–458.

Hsu SW, Rodesch G, Luo CB et al. (2002). Concomitant conus medullaris arteriovenous malformation and sacral dural arteriovenous fistula of the filum terminale. Interventional neuroradiology : journal of peritherapeutic neuroradiology, surgical procedures and related neurosciences 8: 47–53.

Huang W, Gross BA, Du R (2013a). Spinal extradural arteriovenous fistulas: clinical article. Journal of neurosurgery. Spine 19: 582–590.

Huang W, Gross BA, Du R (2013b). Spinal extradural arteriovenous fistulas: clinical article. J Neurosurg Spine 19: 582–590.

Kim LJ, Spetzler RF (2006). Classification and surgical management of spinal arteriovenous lesions: arteriovenous fistulae and arteriovenous malformations. Neurosurgery 59: S195–S201; discussion S193–S213.

Lanzino G, D'Urso PI, Kallmes DF et al. (2012). Onyx embolization of extradural spinal arteriovenous malformations with intradural venous drainage. Neurosurgery 70: 329–333.

Lasjaunias P, Berenstein A (1990). Surgical neuroangiography. Part 3. Functional vascular anatomy of brain, spinal cord and spine, Springer-Verlag, Berlin.

Lee YJ, Terbrugge KG, Saliou G et al. (2014). Clinical features and outcomes of spinal cord arteriovenous malformations: comparison between nidus and fistulous types. Stroke; a journal of cerebral circulation 45: 2606–2612.

Linfante I, Tari Capone F, Dabus G et al. (2012). Spinal arteriovenous malformation associated with spinal metameric syndrome: a treatable cause of long-term paraplegia? Journal of neurosurgery. Spine 16: 408–413.

Martirosyan NL, Kalani MY, Lemole Jr GM et al. (2015). Microsurgical anatomy of the arterial basket of the conus medullaris. Spine 22: 672–676.

Murthy NS, Maus TP, Behrns CL (2010). Intraforaminal location of the great anterior radiculomedullary artery (artery of Adamkiewicz): a retrospective review. Pain Med 11: 1756–1764.

Niimi Y, Sala F, Deletis V et al. (2000). Provocative testing for embolization of spinal cord AVMs. Interv Neuroradiol 6 (Suppl 1): 191–194.

Niimi Y, Sala F, Deletis V et al. (2004). Neurophysiologic monitoring and pharmacologic provocative testing for embolization of spinal cord arteriovenous malformations. AJNR Am J Neuroradiol 25: 1131–1138.

Niimi Y, Uchiyama N, Elijovich L et al. (2013). Spinal arteriovenous metameric syndrome: clinical manifestations and endovascular management. Am J Neuroradiol 34: 457–463.

Ramanathan D, Levitt MR, Sekhar LN et al. (2014a). Management of spinal epidural arteriovenous fistulas: interventional techniques and results. Journal of neurointerventional surgery 6: 144–149.

Ramanathan D, Levitt MR, Sekhar LN et al. (2014b). Management of spinal epidural arteriovenous fistulas: interventional techniques and results. J Neurointerv Surg 6: 144–149.

Rangel-Castilla L, Holman PJ, Krishna C et al. (2011). Spinal extradural arteriovenous fistulas: a clinical and radiological description of different types and their novel treatment with Onyx. J Neurosurg Spine 15: 541–549.

Rangel-Castilla L, Russin JJ, Zaidi HA et al. (2014). Contemporary management of spinal AVFs and AVMs: lessons learned from 110 cases. Neurosurg Focus 37E14.

Rodesch G, Lasjaunias P (2003). Spinal cord arteriovenous shunts: from imaging to management. Eur J Radiol 46: 221–232.

Rodesch G, Hurth M, Alvarez H et al. (2002). Classification of spinal cord arteriovenous shunts: proposal for a reappraisal – the Bicetre experience with 155 consecutive patients treated between 1981 and 1999. Neurosurgery 51: 374–379; discussion 379–380.

Rodesch G, Hurth M, Alvarez H et al. (2003). Embolization of spinal cord arteriovenous shunts: morphological and clinical follow-up and results – review of 69 consecutive cases. Neurosurgery 53: 40–49; discussion 49–50.

Rodesch G, Hurth M, Alvarez H et al. (2004). Angioarchitecture of spinal cord arteriovenous shunts at presentation. Clinical correlations in adults and children. The Bicetre experience on 155 consecutive patients seen between 1981–1999. Acta Neurochir (Wien) 146: 217–226; discussion 226–227.

Rodesch G, Hurth M, Alvarez H et al. (2005). Spinal cord intradural arteriovenous fistulae: anatomic, clinical, and therapeutic considerations in a series of 32 consecutive patients seen between 1981 and 2000 with emphasis on endovascular therapy. Neurosurgery 57: 973–983; discussion 973–983.

Rubin MN, Rabinstein AA (2013). Vascular diseases of the spinal cord. Neurol Clin 31: 153–181.

Sala F, Niimi Y, Krzan MJ et al. (1999). Embolization of a spinal arteriovenous malformation: correlation between motor evoked potentials and angiographic findings: technical case report. Neurosurgery 45: 932–937; discussion 937–938.

Sala F, Niimi Y, Berenstein A et al. (2000). Role of multimodality intraoperative neurophysiological monitoring during embolisation of a spinal cord arteriovenous malformation. A paradigmatic case. Interv Neuroradiol 6: 223–234.

Sala F, Niimi Y, Berenstein A et al. (2001). Neuroprotective role of neurophysiological monitoring during endovascular procedures in the spinal cord. Ann N Y Acad Sci 939: 126–136.

Sala F, Beltramello A, Gerosa M (2007). Neuroprotective role of neurophysiological monitoring during endovascular procedures in the brain and spinal cord. Neurophysiologie clinique = Clinical neurophysiology 37: 415–421.

Silva Jr N, Januel AC, Tall P et al. (2007). Spinal epidural arteriovenous fistulas associated with progressive myelopathy. Report of four cases. Spine 6: 552–558.

Thron AK, Rossberg C (1988). Vascular anatomy of the spinal cord: neuroradiological investigations and clinical syndromes, Springer, Berlin.

Thron A, Krings T, Otto J et al. (2015). The transdural course of radicular spinal cord veins – a microangiographical and microscopical study. Clin Neuroradiol 25 (4): 361–369.

Wang Q, Song D, Chen G (2011). Endovascular treatment of high-flow cervical direct vertebro-vertebral arteriovenous fistula with detachable coils and Onyx liquid embolic agent. Acta Neurochir (Wien) 153: 347–352.

Wilson DA, Abla AA, Uschold TD et al. (2012). Multimodality treatment of conus medullaris arteriovenous malformations: 2 decades of experience with combined endovascular and microsurgical treatments. Neurosurgery 71: 100–108.

Yahiro T, Hirakawa K, Iwaasa M et al. (2004). Pseudoaneurysm of the thoracic radiculomedullary artery with subarachnoid hemorrhage. Case report. J Neurosurg 100: 312–315.

Handbook of Clinical Neurology, Vol. 143 (3rd series)
Arteriovenous and Cavernous Malformations
R.F. Spetzler, K. Moon, and R.O. Almefty, Editors
http://dx.doi.org/10.1016/B978-0-444-63640-9.00017-5

Chapter 17

Radiosurgical ablation of spinal cord arteriovenous malformations

ERIC S. SUSSMAN, JOHN R. ADLER*, AND ROBERT L. DODD
Department of Neurosurgery, Stanford University, Stanford, CA, USA

Abstract

Spinal cord arteriovenous malformations (SCAVMs) are rare entities that account for less than 20% of spinal masses. These lesions represent a unique clinical challenge, in that surgical or endovascular treatment is often associated with devastating functional consequences. Over the past few decades, radiosurgery has become a well-established treatment modality for SCAVMs, and may be a first-line treatment in many patients afflicted with this devastating disease. This chapter summarizes the data available regarding radiosurgical treatment of SCAVMs.

INTRODUCTION

Spinal cord arteriovenous malformations (SCAVMs) are rare, congenital lesions that account for approximately 1% of spinal cord masses and approximately 20–30% of spinal vascular lesions (Rosenblum et al., 1987; Oldfield and Doppman, 1988). Although they were described in the literature as early as 1885 by Hebold and Gaupp (Aminoff and Logue, 1974a), our understanding of the pathophysiology of these lesions remained limited until the advent of angiography in the 20th century. In 1972, Djindjian defined spinal vascular malformations as an abnormal connection between an artery and vein that bypasses the normal capillary plexus (Riché et al., 1982). There have since been multiple attempts to further classify these lesions and to distinguish SCAVMs from other spinal vascular malformations. The most widely accepted classification system subdivides spinal vascular malformations into four distinct subtypes based on the precise anatomy of the anomalous arteriovenous connection (Doppman et al., 1985). Type I lesions are dural arteriovenous fistulas consisting of a fistulous connection between a dural artery and a medullary vein. Type II malformations, also known as glomus arteriovenous malformations (AVMs), represent the most common type of spinal vascular malformation. These lesions typically involve a compact intramedullary nidus fed by multiple arterial feeders off the anterior and posterior spinal arteries, with nidal drainage into the coronal venous plexus. Type III, or juvenile AVMs, are rare, large malformations composed of a diffuse intramedullary nidus as well as an extramedullary and/or paraspinal component. Finally, type IV lesions are similar to type I lesions in that they consist of a fistulous connection between an artery and vein without an intervening nidus; however they are distinct in that type IV lesions are typically fed by a medullary artery and the fistula is often intraspinal, whereas type I lesions are fed by a radiculomedullary artery and the fistula is intradural but extramedullary (Rosenblum et al., 1987; Sinclair et al., 2006).

This chapter will review the radiosurgical treatment of true SCAVMs – specifically type II and type III lesions. Importantly, the large and diffuse nature of type III lesions often precludes the option of radiosurgical treatment, as the risk of damage to the normal spinal cord would be unacceptably high with such a large treatment target. Thus, the vast majority of SCAVMs that have been treated with radiosurgery are type II lesions.

*Correspondence to: John R. Adler, M.D., Stanford University, Department of Neurosurgery, 300 Pasteur Drive, MC5327, Stanford CA 94305, USA. Tel: +1-650-725-5562, E-mail: jra@stanford.edu

EPIDEMIOLOGY

Until recently, our epidemiologic understanding of SCAVMs was derived predominantly from case reports and single-center case series. Over the past decade, the development of multiple large national databases has allowed for a more thorough analysis of the epidemiology of these rare lesions. A 2009 review of the National Inpatient Sample estimated that around 300 patients a year are admitted with a diagnosis of spinal vascular malformation. Type II SCAVMs are estimated to occur with an incidence of 0.02 per 100 000 person-years and a prevalence of 0.08 per 100 000 persons (Lad et al., 2009). These lesions are thought to be congenital, and some authors have reported that SCAVMs arise during formation of the spinal vasculature in the second to third week of embryogenesis. They typically present clinically by the third decade of life. A slight male preponderance has been reported by some authors; however, interestingly, this gender predilection has not been observed among the subset of patients diagnosed prior to age 40 (Aminoff and Logue, 1974a; Rosenblum et al., 1987; Lad et al., 2009).

PATHOPHYSIOLOGY AND CLINICAL PRESENTATION

As noted above, the earliest known clinical description of SCAVMs was in the late 19th century by Hebold and Gaupp. Almost a century later, Aminoff and Logue published a seminal series of papers detailing the clinical presentation and pathophysiology of these lesions (Aminoff and Logue, 1974a; Aminoff et al., 1974).

From a pathophysiologic perspective, there are four major mechanisms by which SCAVMs may present (Aminoff et al., 1974; Brown et al., 2005). One mechanism involves direct mass effect of the lesion itself: increased blood flow in the involved vessels leads to vascular dilatation, which may directly compress the adjacent spinal cord. Alternatively, the vascular malformation may result in ischemic signs and symptoms as a result of a vascular steal phenomenon. SCAVMs may also lead to venous congestion, wherein the spinal venous network is overflowed by the high flow of shunted blood from arterial feeders, thereby blocking venous drainage of the nearby normal spinal cord tissue and leading to a progressive myelopathy. Finally, the normally low-flow venous system may not be able to adapt to the high-flow arterial shunting, which may lead to rupture of the malformation and subsequent hemorrhage. Rarely, these lesions may be diagnosed incidentally during radiographic or clinical workup of unrelated signs and symptoms.

In clinical practice, the precise mechanism underlying the presenting signs and symptoms often cannot be ascertained, and so clinical presentations are more practically dichotomized into hemorrhagic and ischemic episodes. Hemorrhagic signs and symptoms tend to be acute in onset and dramatic in scope, with the specific signs and symptoms dependent on the location of the lesion. Ischemic presentations may also be acute and catastrophic in onset; however, many patients with ischemia secondary to an underlying SCAVM will present with more subtle, recurrent symptoms specific to the affected spinal cord levels.

With regard to specific clinical signs and symptoms, pain is the most frequent initial symptom, occurring in almost 50% of patients, according to one large observational series (Aminoff and Logue, 1974a). The specific nature of this pain may be spinal nerve root pain, local back pain, or referred pain to a site remote from the malformation. Weakness and sensory disturbance are also common presenting signs and symptoms, whereas bowel, bladder, and sexual dysfunction are more rarely identified at the time of initial presentation (Aminoff and Logue, 1974a). Aminoff and Logue also reported hemorrhage as a rare finding at initial presentation (5% in their series of 60 patients with SCAVM); however, this is significantly lower than has been reported in more recent series, and may be a reflection of the antiquated imaging technology available at the time of their publication. In a recent meta-analysis of 11 studies including 135 patients with type II SCAVMs, hemorrhage rate was reported to be 4% a year, increasing to as high as 10% per year for those with a history of prior hemorrhage. A multivariate analysis of risk factors for hemorrhage in this patient cohort identified prior hemorrhage as the only clinical variable independently associated with an increased risk of subsequent hemorrhage (Gross and Du, 2013).

TREATMENT MODALITIES

Prior to the 1960s, there are only intermittent reports of surgical treatment of such lesions, including Krause's description of surgical treatment of an SCAVM in 1911 and the first major series by Wyburn-Mason in 1943 (Jellema et al., 2006). As radiographic imaging technology continued to improve throughout the second half of the 20th century, these lesions were identified with increasing frequency, and this led to a growing demand for new and effective treatment options for this often incapacitating disease. Microsurgical advances certainly facilitated the treatment of some SCAVMs; however, the intimate relationship of these lesions with normal-functioning spinal cord tissue and the overlap of SCAVM vessels with indispensable spinal vasculature exclude many of these lesions from surgical management. Nonsurgical treatment modalities, including endovascular embolization and radiosurgical techniques, have therefore become integral in the management of SCAVMs.

There are currently three primary modalities used for the treatment of SCAVMs: microsurgical techniques, endovascular embolization, and radiosurgery. Microsurgery remains the consensus gold-standard therapy for definitive and permanent obliteration; however, the morbidity of microsurgery may be prohibitively high for certain malformations. Endovascular embolization allows for a more minimally invasive approach to SCAVM treatment, and provides the unique advantage over microsurgery of allowing for the precise identification of involved vessels and the selective treatment of those vessels that selectively feed or drain the malformation without affecting the physiology of the normal spinal cord. However, particularly in the case of type II SCAVMs, which are often intrinsic to the cord itself and therefore intimately intertwined with essential spinal cord vasculature, even endovascular treatment may be prohibitively morbid. In Berenstein et al.'s (2004) series of 44 patients with SCAVMs treated with endovascular embolization, morbidity occurred in >10% of patients, despite good or excellent outcomes in 77%.

Due to the tremendous advances made in radiotherapy technology over the past three decades, radiosurgery has become a viable treatment option for many patients with SCAVMs. The availability of high-resolution and three-dimensional imaging modalities for treatment planning combined with the precision of radiosurgical equipment has made it possible to deliver very high doses of radiation to the lesion itself, while sparing the adjacent spinal cord. In fact, due to the high morbidity associated with the microsurgical and endovascular treatment of SCAVMs, radiosurgery is now the favored treatment modality for many SCAVM patients.

RADIOSURGICAL EXPERIENCES

The published experience with treating SCAVMs using any method of radiation therapy is mostly limited to a handful of case reports and case series (Wolkov and Bagshaw, 1988; Hida et al., 2003; Kishi et al., 2005). Wolkov and Bagshaw (1988) were the first to report their experience using conventional radiotherapy to treat AVMs of the central nervous system in 1988. Two patients in this series had SCAVMs, which were treated with 45.0–50.0 Gy in 1.8–2.0-Gy fractions with subsequent resolution of weakness. Neither patient experienced clinical evidence of neurologic deterioration or radiation injury throughout 5 and 12 years of follow-up, respectively. No posttreatment imaging was obtained to assess the radiographic response of these lesions to radiotherapy.

Fifteen years later, Hida et al. (2003) published a case series of 10 patients who underwent fractionated Linac stereotactic radiotherapy for SCAVMs. The first 7 patients in this series received 32–40 Gy in 1.8–2.0-Gy fractions, whereas the last 3 patients were treated with a hypofractionated schedule of 30 Gy over eight sessions or 20 Gy over four sessions. The median follow-up was 49 months (range 26–124), during which no hemorrhages or radiation-related adverse events occurred. Seven patients in this series underwent follow-up angiogram, which revealed decreased nidus size in 5 patients, but no evidence of complete obliteration in any patient. In the largest series to date, Adler et al. (2010) described the Stanford series of 30 patients with SCAVMs treated with CyberKnife radiosurgery (CKRS). The description of this series is updated in the following section.

THE STANFORD EXPERIENCE

Patient population

Stanford University Medical Center has the largest known series of radiosurgically treated SCAVMs. To date, 45 CKRS treatments have been performed on 40 patients. All patients undergoing CKRS for SCAVMs at Stanford between 1997 and 2014 were enrolled in a prospective study. Males comprise 45% ($n=18$) of this series, which is consistent with the lack of a clear gender predilection that has been previously cited in the literature. Among patients in this series, the median age at the time of diagnosis was 32.7 years (range 9.4–57.2 years), and the median time between diagnosis and CKRS treatment was 31.7 months (range 1–270 months). The median SCAVM volume treated was 1.57 cm^3 (range 0.14–15.01 cm^3).

Twenty-three patients (57.5%) presented initially with hemorrhage, whereas 14 patients (35%) presented with symptoms of ischemia, and 1 (2.5%) presented with local back pain. Two patients (5%) were diagnosed incidentally during workup for unrelated medical issues. While the initial presentation was hemorrhagic in 23 patients, 5 additional patients presented initially with ischemic symptoms and subsequently experienced hemorrhage prior to CKRS treatment. With regard to presenting signs and symptoms, 30 patients (75%) presented initially with pain, while 26 (65%) presented with motor deficit, 21 (52.5%) with sensory disturbance, 10 (25%) with gait difficulty, and 7 (17.5%) with bowel/bladder incontinence (Table 17.1). As mentioned above, 2 patients were diagnosed incidentally, and were asymptomatic at the time of presentation. Radiosurgery was the first treatment administered in 21 (52.5%) of the patients in this series. Three patients had previously been treated subtotally with microsurgery, 12 with embolization, and 4 with both microsurgery and endovascular embolization.

Radiosurgical technique

A detailed technical description of CKRS has been previously reported and is beyond the scope of this chapter (Sinclair et al., 2006). In short, each patient underwent high-resolution magnetic resonance imaging (MRI), computed tomography angiogram, and conventional spinal angiogram immediately prior to radiosurgical treatment. These scans were fused and analyzed using proprietary semiautomated image fusion software, and treatment planning was performed using this combined data set. The radiosurgical technique was not uniform among all patients in the series, as the CKRS technique evolved throughout the course of the study period. For instance, some of the earliest patients in the series required percutaneous fiducial placement prior to placement, whereas the need for such fiducial placement was eliminated in 2004 with a technologic development that allowed for the localization of target lesions relative to spinal anatomic landmarks (Ho et al., 2007). Three-dimensional rotational angiography became a routine part of pre-procedural planning thereafter.

The radiation dose and treatment regimen also evolved over the course of the study period. In the early years of the study, there was no experience with radiosurgical treatment of SCAVMs on which to base a study treatment protocol. The radiation doses were therefore intentionally conservative in the first few patients, and the radiation dose was gradually increased only as the group's experience with CKRS for SCAVMs continued to grow (Fig. 17.1). In the 45 CKRS treatments

Table 17.1

Initial presenting signs and symptoms ($n=40$)

Sign/symptom	Number (%)
Asymptomatic	2 (5%)
Pain	30 (75%)
Motor deficit	26 (65%)
Sensory disturbance	21 (52.5%)
Gait difficulty	10 (25%)
Bowel/bladder incontinence	7 (17.5%)

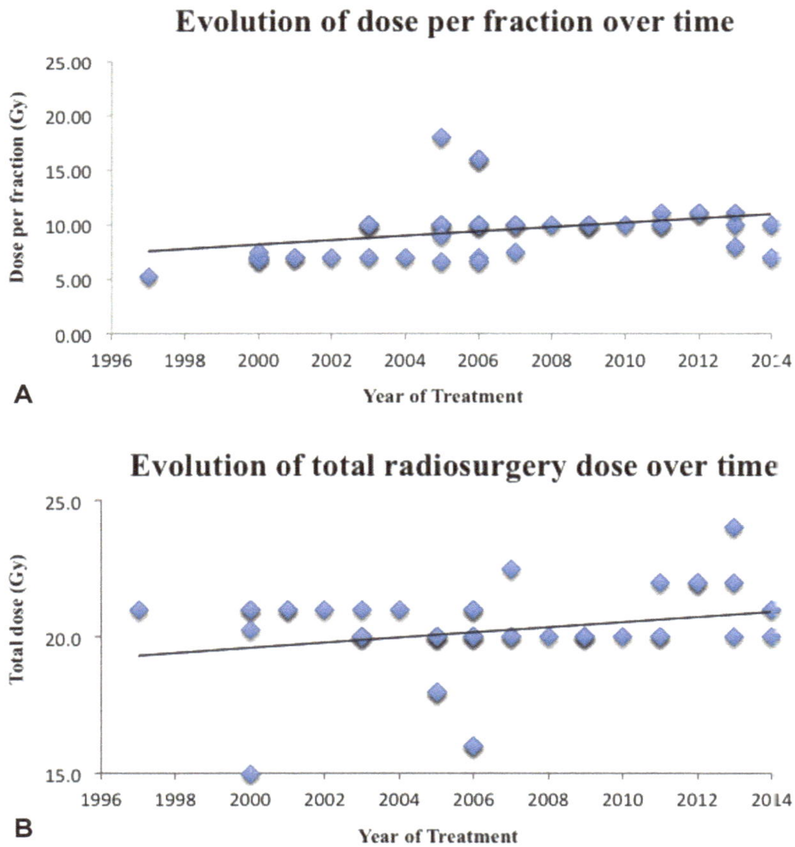

Fig. 17.1. Graphical depiction of the upward trend in dose per fraction (**A**) and total radiation dose (**B**) throughout the course of the study period from 1997 to 2014.

administered to patients with SCAVMs at Stanford, the median number of fractions administered was 2 (range 1–4). However, in the cohort of patients that were treated since 2006, all treatments were administered in either two or three fractions, with the majority of patients (72%) receiving two fractions. Radiation dose per fraction was also lower at the early stages of the study, with an average dose per fraction of 525–750 cGy in the first 9 patients. In the subsequent 36 treatments, dosages of at least 1000 cGy per fraction were routine. Overall, the median dose per fraction was 10.0 Gy (range 5.25–18.0), and the median maximum dose (D_{max}) was 26.0 Gy (range 17.7–30.0). Importantly, despite these trends, the particular treatment plan for each patient was developed as part of an iterative process that took into account each AVM's unique volume, location, and anatomic morphology (Fig. 17.2).

Follow-up evaluations

Routine clinical and radiographic follow-up was pre-established as part of the study protocol. Postradiosurgery serial MRI scans were obtained at 6 months, 1, 2, and 3 years following CKRS treatment. A spinal angiogram was also obtained after 3 years, or sooner if an earlier MRI scan suggested obliteration of the lesion. Clinical assessment occurred with each interval radiographic follow-up. Additional clinical and radiographic follow-up was obtained only as needed depending on each individual patient's response to radiosurgery.

Among the 45 CKRS treatments administered in this trial, 39 patients (86.7%) had clinical follow-up of at least 6 months after the completion of radiosurgery. The median duration of clinical follow-up was 41.5 months (range 7.4–122 months). Similarly, 37 patients (82.2%) had radiographic follow-up at least 6 months after CKRS completion, with a median duration of radiographic follow-up of 37 months (range 6–105 months). Four patients were lost to follow-up, which was defined as no clinical or radiographic follow-up beyond 6 months after the completion of CKRS treatment. As CKRS for SCAVMs is a novel therapy for which patients are referred to Stanford from around the world, this is a relatively noteworthy follow-up rate.

Clinical outcomes

Clinical outcome was evaluated in the 39 patients with at least 6 months of clinical follow-up after completion of CKRS treatment. Among this subgroup, 7 patients (17.9%) reported subjective clinical improvement at the time of last follow-up, whereas 27 patients (69.2%) noted no change in clinical symptoms, and 5 patients (12.8%) reported clinical worsening. With regard to specific signs and symptoms, pain and gait difficulty were the most responsive to CKRS, improving in 61.5% and 60% respectively. Sensory disturbance was the least responsive to CKRS, improving in only 45% and worsening in 20% of patients. Importantly, pain did not worsen in any patients undergoing CKRS for SCAVM (Fig. 17.3A). As noted above, the clinical presentation of patients with SCAVMs is rather heterogeneous. In light of this, the outcome analysis discussed here is qualitative in nature, as the number of patients presenting with any one specific sign or symptom is too small to justify a quantitative statistical analysis.

Twenty-three patients (57.5%) presented initially with hemorrhage. Among this subgroup, the median time between diagnosis and CKRS was 30 months (range 1–351). Two of these patients (8.7%) clinically improved after CKRS, 3 (12%) worsened, and 15 (65.2%) remained unchanged. Three of the patients in this subgroup were lost to clinical follow-up. Pain was the most common presenting symptom among patients presenting initially with hemorrhage, occurring in 21/23 patients (91%). A motor deficit was identified in 17/23 patients (74%), and sensory disturbance was endorsed in 11/23 (47.8%). Less commonly, patients with hemorrhagic

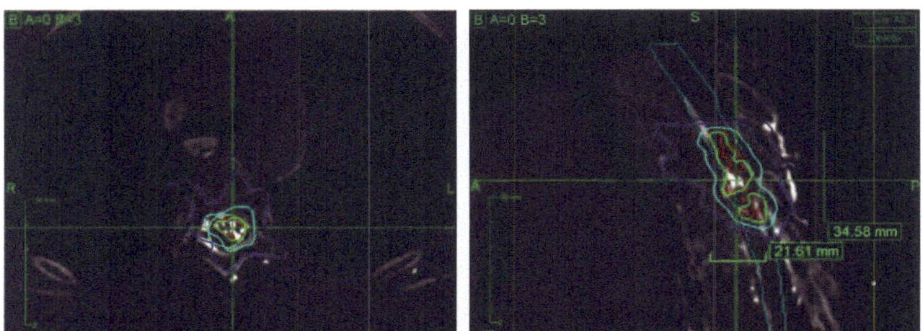

A B

Fig. 17.2. Axial (**A**) and sagittal (**B**) images superimposed with the CyberKnife radiosurgery treatment plan for a patient with a T4–T6 spinal cord arteriovenous malformation.

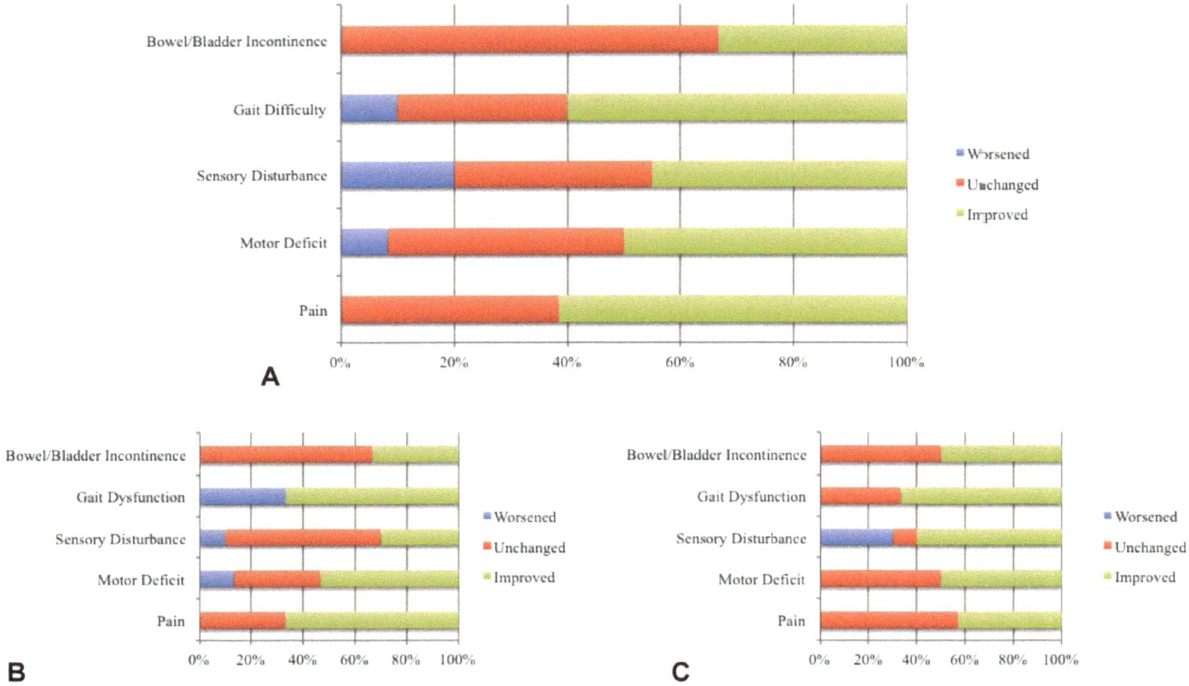

Fig. 17.3. Graphical depiction of the effect of CyberKnife radiosurgery on each presenting sign and symptom in (**A**) all patients in the Stanford series; (**B**) in the cohort of patients who presented with hemorrhage; and (**C**) in the cohort of patients who presented without evidence of hemorrhage.

presentations suffered from gait difficulty (13%) and/or bowel/bladder incontinence (17.4%). The response of these symptoms was assessed in patients with at least 6 months of clinical follow-up. Pain, motor deficit, and gait dysfunction were the most responsive signs/symptoms, improving in 66.7%, 53.3%, and 66.7% of patients in this subgroup, respectively. Sensory disturbance and bowel/bladder incontinence, on the other hand, were less responsive to radiosurgery, remaining unchanged in 60% and 66.7%, respectively, of patients who initially presented with hemorrhage (Fig. 17.3B).

The natural history of hemorrhagic injury involves gradual and progressive clinical improvement as hemorrhagic blood products reabsorb, secondary inflammatory injury abates, and spinal cord physiology and vascular perfusion are reestablished (Aminoff and Logue, 1974b). As such, in patients presenting initially with hemorrhage and subsequently undergoing CKRS, it is difficult to ascertain what component of clinical outcome is the result of normal posthemorrhage disease progression, versus what may be directly attributable to radiosurgical intervention. The clinical outcome of patients presenting with nonhemorrhagic signs and symptoms, on the other hand, may be a better barometer of the direct impact of CKRS on the natural history of SCAVMs, as nonhemorrhagic signs and symptoms tend to be more constant and persistent. Fifteen patients (37.5%) in the Stanford series had nonhemorrhagic presentations, including signs/symptoms secondary to vascular steal, venous congestion, and/or direct mass effect. Within this cohort, 4 patients (26.7%) clinically improved after CKRS, 2 (13.3%) worsened, and 8 (53.3%) remained unchanged. One patient was lost to follow-up. In contrast to the hemorrhagic cohort, pain was less common among patients with nonhemorrhagic presentations, occurring in only 9/15 patients (60%). It is also notable that gait difficulty and sensory disturbance were significantly more common in the nonhemorrhagic cohort, occurring in 53.8% and 73.3%, respectively. Motor deficits were identified in 53.3% and bowel/bladder incontinence occurred in 20% of patients in this subgroup. Among the patients in the nonhemorrhagic subgroup with at least 6 months of clinical follow-up, the response of specific signs/symptoms to radiosurgery was assessed (Fig. 17.3C).

It is also worth mentioning that 2 patients (5%) were diagnosed incidentally amidst workup for other medical conditions. These patients were asymptomatic at the time of presentation, with the exception of intermittent back pain or other nonspecific symptoms noted in retrospect, and all of these patients remained clinically stable throughout the entirety of the post-CKRS follow-up period.

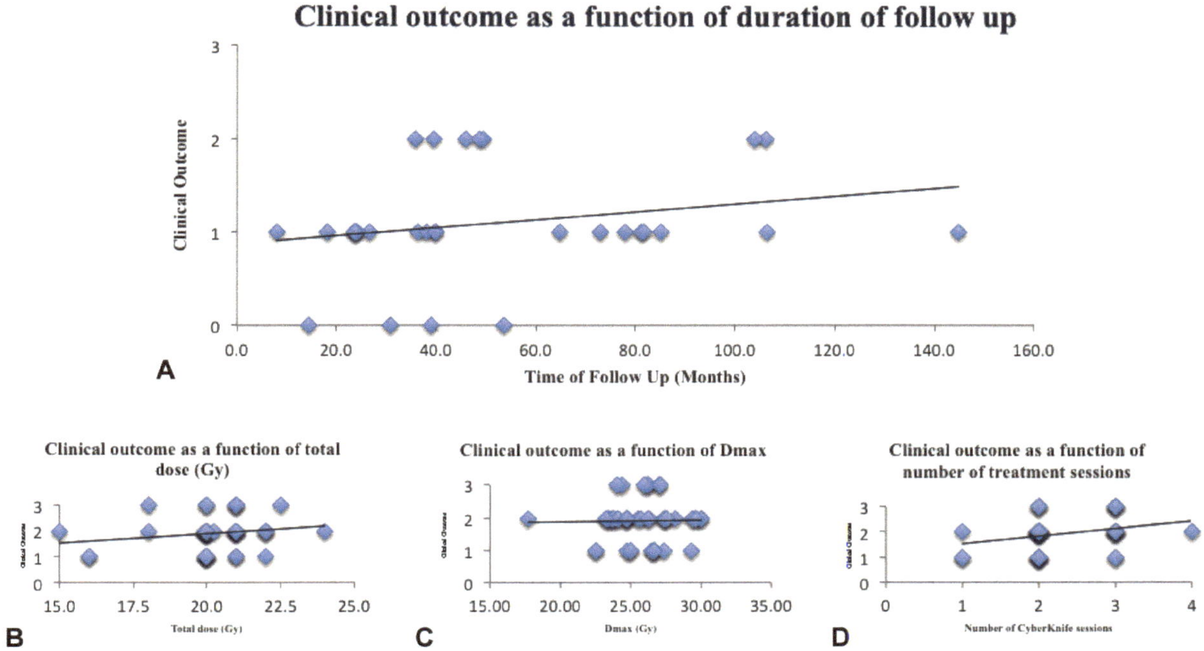

Fig. 17.4. Graphical depiction demonstrating a trend towards better clinical outcomes with (**A**) longer duration of follow-up; (**B**) higher total radiation dose; (**C**) maximum dose; and (**D**) increased number of treatment sessions. Clinical outcomes are grossly stratified as worse (=1), unchanged (=2), or improved (=3). D_{max} – maximum dose.

A slight positive correlation was identified between clinical outcome and the duration of follow-up, dose, D_{max}, and number of treatment sessions (Fig. 17.4). Additional experience with radiosurgical treatment of SCAVMs is necessary to identify the optimal dose and treatment regimen for these lesions.

Effect of CyberKnife radiosurgery on hemorrhage risk

As described previously, 28 patients in the Stanford series had SCAVM rupture and subsequent hemorrhage prior to CKRS treatment, 23 of which occurred as the initial presentation of the lesion, and an additional 5 that occurred some time after the lesion had already been diagnosed. Among all patients who experienced SCAVM hemorrhage, the median number of hemorrhages prior to CKRS treatment was one (range 1–4). Eighteen patients (64.3%) experienced only one hemorrhage, whereas 6 (21.4%), 2 (7.1%), and 2 (7.1%) suffered two, three, and four hemorrhages, respectively.

Throughout the duration of the greater than 1700 patient-months of clinical follow-up in this series – 1170 months of which were specifically in patients who suffered from hemorrhage prior to radiosurgical treatment – no patients experienced hemorrhage after CKRS (Fig. 17.5).

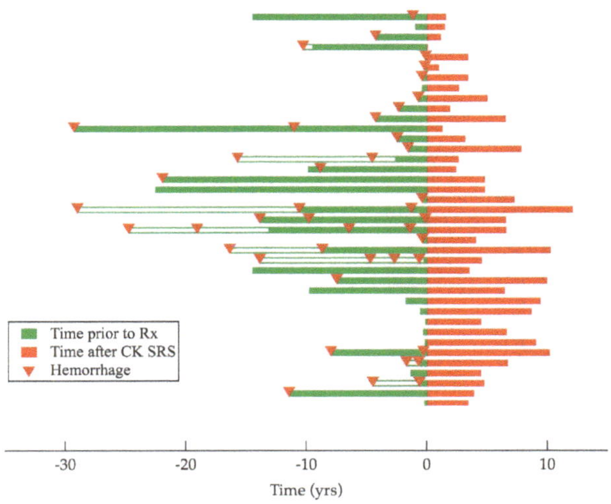

Fig. 17.5. Chart illustrating the timing of each individual patient's hemorrhage(s) (red arrows) relative to CyberKnife radiosurgery (CK SRS) treatment (midline intersection between green and red bar). The open part of the green line represents the period of time prior to definitive diagnosis of spinal cord arteriovenous malformation. It is important to note that there were no hemorrhages after radiosurgery treatment in this series.

Radiographic outcomes

Radiographic outcome was evaluated in the 37 patients in whom follow-up imaging was obtained at least 6 months after the completion of a course of radiosurgery. For the purposes of this study, patients were stratified into four subgroups depending on the percent reduction of the lesion on the last available follow-up scan, as compared to the pre-radiosurgery scan. Five patients (24%) were considered to be nonresponders, defined as those patients with a less than 25% reduction in SCAVM volume. Twelve patients (32.4%) had a reduction of SCAVM volume between 25 and 75%, and were considered to be partial responders. Ten patients (27.0%) demonstrated a near-complete response of >75% but <100% reduction in SCAVM volume, while 10 patients (27%) achieved complete obliteration.

Interestingly, there was a slight negative correlation between degree of SCAVM obliteration and the duration of follow-up, total dose, D_{max}, and number of treatment sessions (Fig. 17.6). This may be attributable to the more aggressive treatment strategies (i.e., high marginal dose and D_{max}, and higher number of treatment sessions) employed for larger, more aggressive lesions that were inherently less likely to completely obliterate.

Repeat radiosurgery

Five patients in the series underwent a second course of CKRS treatment due to incomplete response to the first treatment. These cases warrant special consideration, since repeat CKRS treatment for residual SCAVM after an initial course of radiosurgery has not previously been reported in the literature.

Among the 5 patients in the Stanford series who underwent a second course of CKRS, 2 patients were female, and the median age at the time of first CKRS treatment was 35.4 years (range 18.8–41.5 years). Four of the five cases were located in the cervical cord, and one was located at the thoracolumbar junction. Also of note is that four of the five cases presented initially with SCAVM hemorrhage. The median time between CKRS treatments was 4.5 years (range 3–6 years). The median percent residual SCAVM after the first treatment was 50% (range 20–75%), whereas after the second treatment, the median percent residual was 10% (range 0–75%). Of note, the individual in the study who was reported to have had 75% residual on radiographic follow-up after a second course of CKRS treatment was lost to follow-up 1 year after undergoing repeat CKRS, and so long-term clinical and radiographic outcome data are not available for this patient. Among the remaining 4 patients with long-term (at least 3-year) follow-up after the second course of treatment, 2 patients achieved complete obliteration, while the other 2 achieved near-total obliteration (20% and 10% radiographic residual, respectively). See representative case #2, below, for a detailed discussion of a patient who underwent repeat CKRS for residual SCAVM.

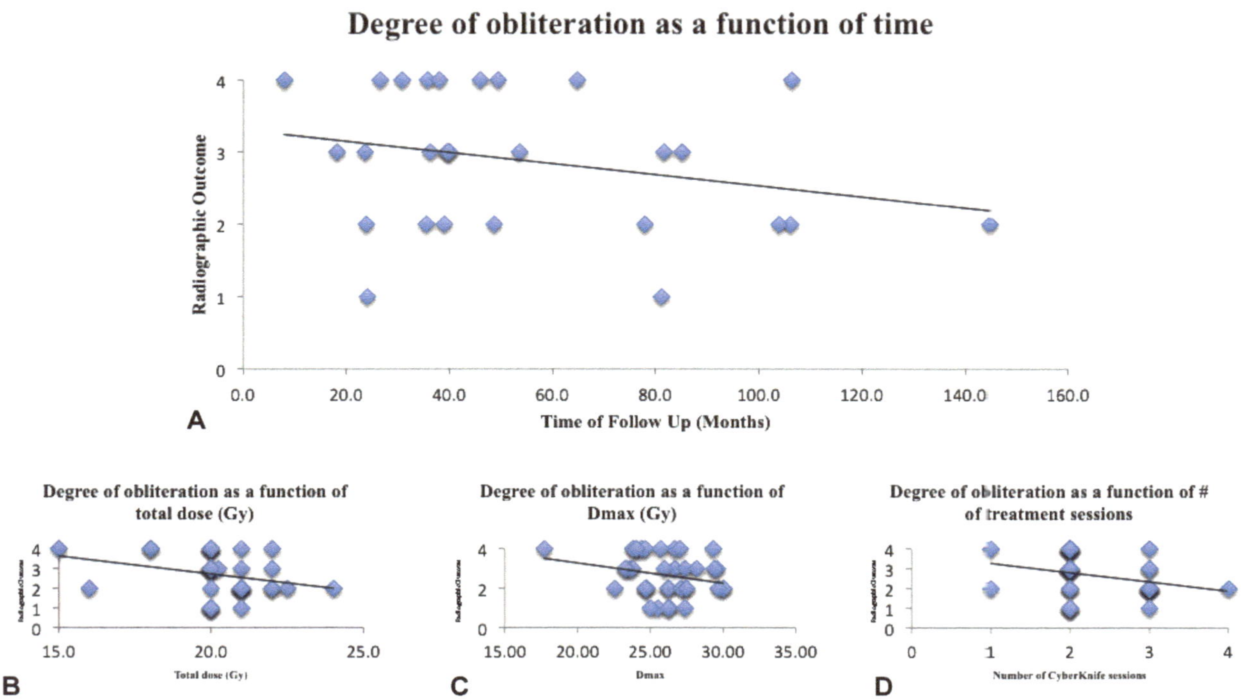

Fig. 17.6. Graphical depiction demonstrating a trend towards worse radiographic outcome with (**A**) longer duration of follow-up; (**B**) higher total radiation dose; (**C**) maximum dose; and (**D**) increased number of treatment sessions. Radiographic outcomes are stratified into four cohorts: <25% reduction in nidus volume (=1), 25–75% reduction (=2), >75% but <100% reduction (=3), and complete obliteration (=4). D_{max}, maximum dose.

Adverse outcomes

Five patients (12.5%) experienced clinical worsening of symptoms after CKRS. In other words, 87.5% of patients in this series were clinically unchanged or improved at follow-up after CKRS treatment. This suggests a significant improvement in disease course, as compared with the 50% rate of progression to disability among untreated patients, which was noted in Aminoff and Logue's (1974b) series.

Two patients who experienced clinical worsening of signs/symptoms after CKRS had radiographic findings consistent with radiation injury. A 34-year-old female who presented with 2 years of intermittent bilateral hand weakness and numbness as well as back pain was subsequently found to have a C5–C7 SCAVM without radiographic or clinical evidence of hemorrhage. Due to the complexity of the lesion's vascular anatomy, endovascular and microsurgical intervention were deferred. CKRS was administered over two sessions of 10 Gy each to a 2.24-cm^3 lesion using a maximum dose of 26.67 Gy. Following treatment, she experienced worsening of right-hand weakness and numbness, as well as new-onset weakness and numbness of her right lower extremity. Her left-upper-extremity symptoms improved, as did her back pain. Follow-up MRI at 6 months (Fig. 17.7A–C) demonstrated a dramatic increase in cord signal abnormality from C2 to T3, which was significantly improved at 1 year and nearly completely resolved on the 3-year follow-up MRI (Fig. 17.7D and E). Digital subtraction angiogram 3 years after radiosurgery treatment demonstrated 80% reduction in SCAVM volume; however, she continued to have right upper- and lower-extremity weakness, numbness, and spasticity that was worse than prior to CKRS.

Similarly, a 19-year-old female who had suffered four prior hemorrhages from a C7–T1 SCAVM was also deemed to not be a candidate for surgical or endovascular treatment due to the location and morphology of her lesion. She was treated with CKRS over two sessions of 10 Gy each to a 1.57-cm^3 lesion, using a maximum dose of 26.67 Gy. Nine months after treatment, she developed worsening paraplegia and spasticity, and MRI at that time revealed T2 hyperintensity within the spinal cord, consistent with radiation-induced myelopathy. Three years after treatment, angiogram demonstrated complete obliteration of the SCAVM nidus.

Three other patients in this series experienced clinical worsening following CKRS without radiographic

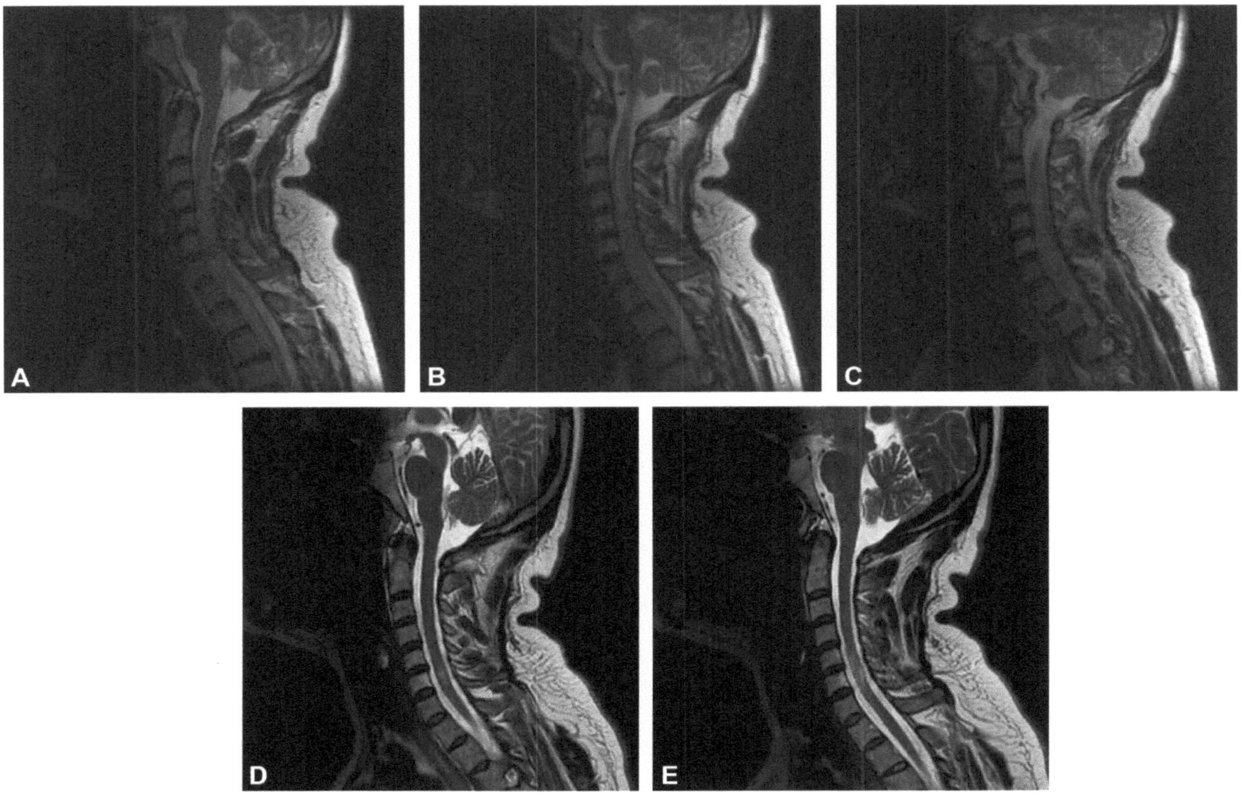

Fig. 17.7. Six-month follow-up magnetic resonance imaging (MRI) (**A–C**) after CyberKnife radiosurgery demonstrating residual C5–C7 arteriovenous malformation and significant cord edema from C2–T3. Three-year follow-up MRI (**D, E**) revealed nearly complete resolution of cord edema.

evidence of radiation injury. It is difficult to say with certainty which of these patients truly experienced an adverse effect from radiosurgical intervention, as opposed to treatment-refractory disease progression. However, as described above, only 2 patients (5%) in this series demonstrated both clinical and radiographic evidence of radiation-induced myelopathy. The radiation dose and treatment strategy employed in these patients were wholly typical as compared with other treated patients, emphasizing the need for further investigation to better predict those patients at risk for radiosurgery-related complications.

REPRESENTATIVE CASES

Case #1

The patient was a 21-year-old male who initially presented with 6 weeks of right-upper-extremity pain progressing to right-hand weakness and numbness. Workup, including an MRI (Fig. 17.8A and B), identified a focus of hemorrhage within the spinal cord, and further assessment with spinal cord angiogram revealed a 1.62-cm^3 AVM at C4–C6 (Fig. 17.8C). The patient underwent endovascular embolization of a nidal aneurysm; however, an attempt to embolize the remaining SCAVM was compromised by the fact that intraprocedural physiologic (amytal) testing demonstrated significant neurologic dysfunction. Hence, 30 months after diagnosis of the SCAVM, the patient underwent CKRS over two sessions of 10 Gy each, for a D_{max} of 25.7 Gy. He experienced gradual and progressive clinical improvement of his pain, weakness, and numbness; however, he endorsed mild residual weakness and numbness at 3-year follow-up. Angiogram at that time revealed complete obliteration of the SCAVM nidus (Fig. 17.8D).

Case #2

This patient was a 39-year-old male who presented with 6 months of sensorimotor symptoms, including right-lower-extremity weakness, gait difficulty, and paresthesias in the right upper and lower extremities. Imaging revealed a 1.56-cm^3 C5–C6 AVM (Fig. 17.9A and B), and attempted endovascular embolization at that time was aborted due to concern for potentially occluding the anterior vertebral artery. CKRS was administered as three sessions of 7.0 Gy, for a total dose of 21.0 Gy

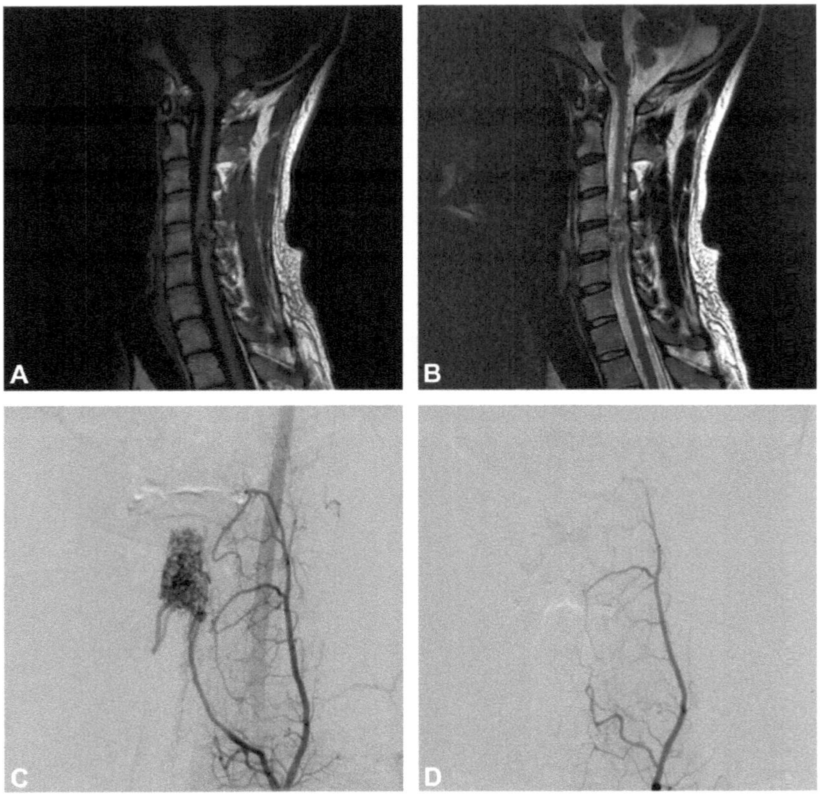

Fig. 17.8. Pre-radiosurgery (**A**) T1- and (**B**) T2-weighted magnetic resonance imaging demonstrating an intramedullary spinal cord arteriovenous malformation extending from C4 to C6. Pre-radiosurgery angiogram (**C**) reveals the nidus being fed exclusively by the left thyrocervical trunk. Three-year follow-up angiogram (**D**) demonstrates complete obliteration of the arteriovenous malformation.

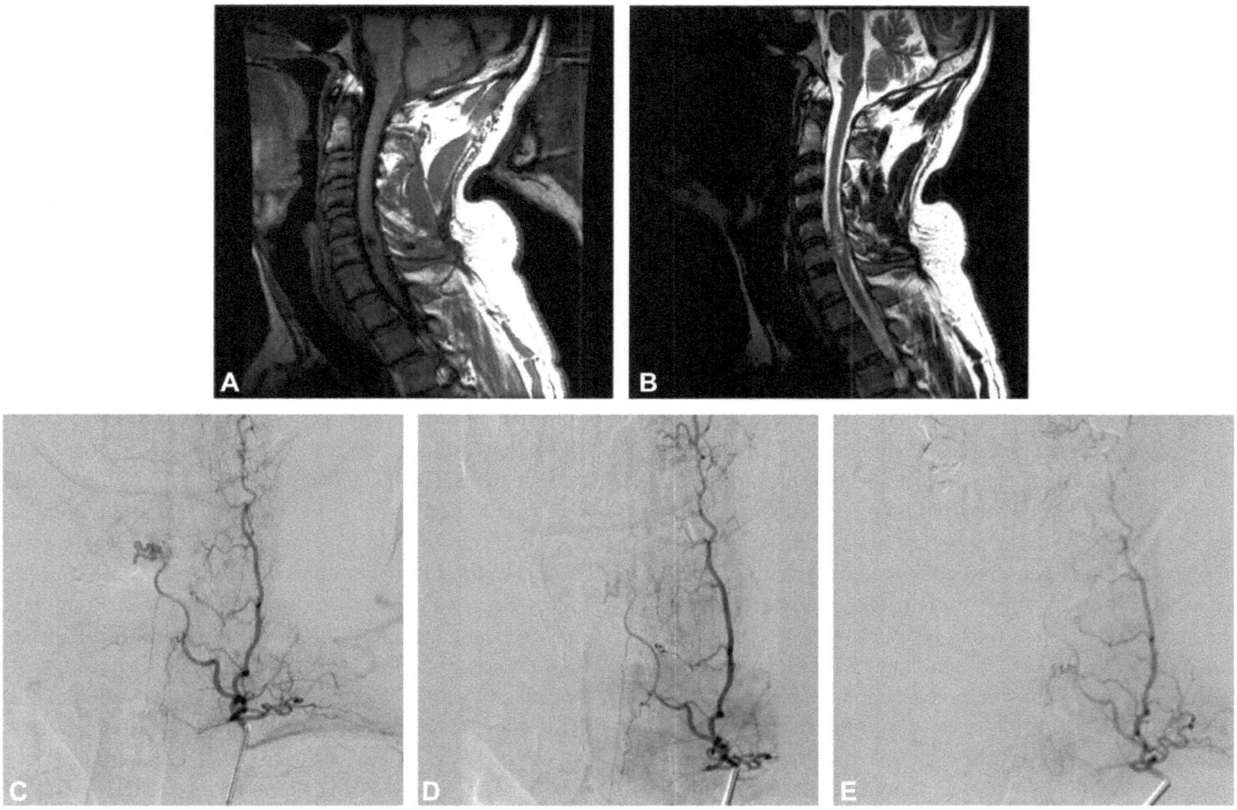

Fig. 17.9. Pre-radiosurgery (**A**) T1- and (**B**) T2-weighted magnetic resonance imaging images demonstrating an intramedullary spinal cord arteriovenous malformation at C5–C6. These images are prior to the first CyberKnife treatment. Follow-up angiogram (**C**) 4.5 years after the first treatment reveals a 50% reduction in the size of the nidus, fed primarily by the left thyrocervical trunk. After a second CyberKnife treatment, 3-year follow-up angiogram (**D**) reveals approximately 90% reduction in nidus size, which progressed to complete obliteration on the 4-year follow-up angiogram (**E**).

and a D_{max} of 26.3 Gy. Follow-up angiogram was performed 54 months after this treatment, and revealed a 50% reduction in nidus volume (Fig. 17.9C). The decision was made to proceed with a repeat CyberKnife treatment, which was performed 4.5 years after the first treatment as two sessions of 9.0 Gy, for a total dose of 18 Gy and a D_{max} of 24 Gy to a 0.14-cm^3 target lesion. Three-year follow-up angiogram after the second treatment revealed approximately 10% residual nidus (Fig. 17.9D), and repeat angiogram 1 year later revealed complete obliteration of the SCAVM nidus (Fig. 17.9E). At clinical follow-up, the patient reported mild right-lower-extremity weakness that was significantly improved compared with his pretreatment assessment.

DISCUSSION

Radiation therapy has been an evolving treatment modality for a variety of central nervous system lesions for many years, and its efficacy has been well described in the literature. In the setting of AVMs, radiation therapy causes ionizing injury to blood vessels, thereby producing hyperplasia of the endothelium and subsequent intraluminal thrombosis and hemostasis (Steiner et al., 1974). When treating spinal cord pathology, attention must be focused on preventing injury to the adjacent spinal cord and preserving the function of normal tissue. This is particularly relevant in the setting of radiation therapy, wherein the dose and target planning must be carefully selected so as to minimize the amount of radiation that the normal spinal cord tissue is exposed to.

Tolerance to radiosurgery

Despite tremendous advances in radiosurgical technology, and the ability to precisely deliver ionizing radiation to a specific target using stereotactic-based systems, the normal tissue surrounding a treatment target will inevitably be exposed to some ionizing radiation. Therefore, it is critical to understand the tolerance of the normal spinal cord tissue to radiation.

This question was partially addressed in 1991 by a task force commissioned by the National Cancer Institute, which was intended to investigate the tolerance of normal tissue organs to radiation. In this assessment, it was suggested that the tolerance dose of the spinal cord that within 5 years will result in a 5% rate of radiation injury (TD 5/5) is 50 Gy using conventional fractionation of 1.8–2.0 Gy/fraction, and it was suggested that only "sporadic, idiosyncratic" radiation-induced complications would occur up to this dose (Emami et al., 1991). In 2010, an updated comprehensive review suggested that, with conventional fractionated radiotherapy, the estimated risk of radiation-induced spinal cord injury is <1% at 54 Gy and <10% at 61 Gy, and further suggested that the risk of injury is strongly dependent on dose per fraction (Kirkpatrick et al., 2010). This is particularly relevant in the setting of stereotactic radiosurgery, as is typically used to treat SCAVM patients. That is, radiosurgery implies less than or equal to five high-dose fractions, as opposed to radiotherapy, which involves a higher number of lower-dose fractions. With regard to stereotactic radiosurgery, the published experience available for review is quite limited, but the data suggest that the risk of radiation-induced injury is <1% when the maximum dose to the spinal cord is less than 13 Gy in a single fraction or less than 20 Gy in three fractions (Kirkpatrick et al., 2010). In the Stanford series, the routine dosing regimen was at the upper limit or above these recommended doses, and only 2 patients (5%) experienced radiation-induced spinal cord injury. Of note, these 2 patients did not receive unusually high radiation doses, nor did they have unusually large SCAVMs, as compared with the other patients in the series.

Of course, radiosurgery is not a categorically effective treatment in all patients, and so the safety of repeat treatment is an important consideration. Given the concern for radiation-induced injury, it is important to consider not only the cumulative radiation dose in patients treated multiple times, but also the impact of the time that lapses between treatments. Preclinical and clinical data suggest that subclinical radiation damage is partially repaired over time, which becomes relevant by 6 months following radiation treatment (Kirkpatrick et al., 2010). Among the 5 patients in the Stanford series who were treated with repeat radiosurgery for residual SCAVM after an initial CKRS treatment, none were found to suffer from radiation-induced injury. The interval between treatments in this cohort of patients was significantly greater than the 6-month threshold cited here.

Effective dose of radiosurgery

The goal in treating SCAVMs is to produce the maximum benefit at a minimum cost to the patient. Therefore, in addition to understanding the tolerance of the spinal cord to radiation, it is also important to understand the effective dose in the setting of the particular pathology being treated. As discussed previously, Wolkov and Bagshaw (1988) published the first reported experience using radiation therapy to treat SCAVMs. Two patients were treated with conventional fractionation of 1.8–2.0 Gy for a total dose of 45 and 50 Gy, respectively, which led to clinical improvement. Unfortunately, no follow-up imaging was obtained to assess the radiographic efficacy of these treatment regimens. Hida et al. (2003) treated 7 patients with 32–40 Gy in 1.8–2.0-Gy fractions, and an additional 3 patients were treated with 20–30 Gy over four to eight sessions. Among the subgroup of these patients who underwent follow-up angiogram, 71.4% achieved reduction in nidus size without evidence of complete obliteration. In the Stanford series, 45 treatments were administered to 40 patients, with an overall median dose of 20 Gy (range 15–24 Gy) administered over one to four treatments (median 2), and a median dose per fraction of 10 Gy (range 5 25–18.0 Gy). Among the patients with adequate clinical and radiographic follow-up, 54% had complete or near-complete (>75%) reduction in SCAVM nidus volume, and almost 90% demonstrated at least 25% reduction in nidus volume. Additionally, 72% had clinical improvement of at least one presenting sign or symptom. Perhaps most importantly, no patient across any of these three series experienced an SCAVM hemorrhage after radiation therapy.

There are insufficient data currently to determine a single effective radiosurgery dose and treatment regimen for SCAVMs, and certainly there are many variables that must be considered in addressing this clinical question, including, but not limited to, SCAVM size and morphology, radiation dose, and fractionation schedule, as well as the individual patient's susceptibility to radiation treatment. Furthermore, the goal of treatment (i.e., symptomatic improvement vs. prevention of future hemorrhage vs. radiographic obliteration) is an essential consideration when attempting to determine treatment efficacy.

Observed hemorrhage before and after radiosurgery

Twenty-eight patients (70%) presented with a prior history of AVM bleed, and in over 35% there were multiple hemorrhages. Five of the 28 patients presented with other symptoms before suffering a bleed. Despite a history of hemorrhage, intranidal aneurysms were observed on angiography in only a minority of cases. Among the 28 patients who bled, 44 pre-radiosurgical hemorrhages were recorded; 6 patients had at least two bleeds, 2 had at least three, and 2 had four episodes of bleeding. It is widely assumed that persistence of any AVM nidus on angiography is associated with a risk of future hemorrhage that is little changed from the natural history of an untreated lesion. In other words, only complete obliteration or resection is protective. Despite this general belief, no rebleeding was observed among any of the partially obliterated spinal cord AVMs in current study. While it is impossible to statistically estimate a treatment effect of CKRS on future hemorrhage rate when none were following radiosurgery, the current results stand as a powerful observation.

CONCLUSIONS

Radiosurgery is a well-established treatment option for patients with SCAVM, and should be the first-line modality for patients in whom the size, location, or anatomy of the lesion precludes the safe performance of microsurgery or endovascular embolization. In addition to radiographically reduced nidus volume and improved clinical signs/symptoms, radiosurgery is this series also appears associated with a dramatic decrease in the incidence of SCAVM hemorrhage. Future investigation of this promising treatment modality should focus on optimizing radiosurgery treatment doses and regimens and minimizing radiation injury.

References

Adler JR, Gupta G, Chang SD et al. (2010). CyberKnife ablation for intramedullary spinal cord arteriovenous malformations (AVMs): a new promising new therapeutic approach. Cureus 2. http://dx.doi.org/10.7759/cureus.14.

Aminoff MJ, Logue V (1974a). Clinical features of spinal vascular malformations. Brain 97: 197–210.

Aminoff MJ, Logue V (1974b). The prognosis of patients with spinal vascular malformations. Brain 97: 211–218.

Aminoff MJ, Barnard RO, Logue V (1974). The pathophysiology of spinal vascular malformations. J Neurol Sci 23: 255–263.

Berenstein A, Lasjaunias P, Bruggeter KG (2004). Spinal arteriovenous malformations, Springer, Berlin, pp. 737–847.

Brown RD, Flemming KD, Meyer FB et al. (2005). Natural history, evaluation, and management of intracranial vascular malformations. Mayo Clin Proc 80: 269–281.

Doppman JL, Di Chiro G, Oldfield EH (1985). Origin of spinal arteriovenous malformation and normal cord vasculature from a common segmental artery: angiographic and therapeutic considerations. Radiology 154: 687–689.

Emami B, Lyman J, Brown A et al. (1991). Tolerance of normal tissue to therapeutic irradiation. Int J Radiat Oncol Biol Phys 21: 109–122.

Gross BA, Du R (2013). Spinal glomus (type II) arteriovenous malformations: a pooled analysis of hemorrhage risk and results of intervention. Neurosurgery 72: 25–32.

Hida K, Shirato H, Isu T et al. (2003). Focal fractionated radiotherapy for intramedullary spinal arteriovenous malformations: 10-year experience. J Neurosurg 99: 34–38.

Ho AK, Fu D, Cotrutz C et al. (2007). A study of the accuracy of cyberknife spinal radiosurgery using skeletal structure tracking. Neurosurgery 60: ONS147–ONS156.

Jellema K, Tijssen CC, Gijn JV (2006). Spinal dural arteriovenous fistulas: a congestive myelopathy that initially mimics a peripheral nerve disorder. Brain 129: 3150–3164.

Kirkpatrick JP, van der Kogel AJ, Schultheiss TE (2010). Radiation dose-volume effects in the spinal cord. Int J Radiat Oncol Biol Phys 76: S42–S49.

Kishi K, Shirai S, Sonomura T et al. (2005). Selective conformal radiotherapy for arteriovenous malformation involving the spinal cord. Br J Radiol 78: 252–254.

Lad SP, Santarelli JG, Patil CG et al. (2009). National trends in spinal arteriovenous malformations. Neurosurg Focus 26: 1–5.

Oldfield EH, Doppman JL (1988). Spinal arteriovenous malformations. Clin Neurosurg 34: 161–183.

Riché MC, Modenesi-Freitas J, Djindjian M et al. (1982). Arteriovenous malformations (AVM) of the spinal cord in children. A review of 38 cases. Neuroradiology 22: 171–180.

Rosenblum B, Oldfield EH, Doppman JL et al. (1987). Spinal arteriovenous malformations: a comparison of dural arteriovenous fistulas and intradural AVM's in 81 patients. J Neurosurg 67: 795–802.

Sinclair J, Chang SD, Gibbs IC et al. (2006). Multisession CyberKnife radiosurgery for intramedullary spinal cord arteriovenous malformations. Neurosurgery 58: 1081–1089.

Steiner L, Leksell L, Forster DM et al. (1974). Stereotactic radiosurgery in intracranial arterio-venous malformations. Acta Neurochir (Wien) (Suppl 21): 195–209.

Wolkov HB, Bagshaw M (1988). Conventional radiation therapy in the management of arteriovenous malformations of the central nervous system. Int J Radiat Oncol Biol Phys 15: 1461–1464.

Handbook of Clinical Neurology, Vol. 143 (3rd series)
Arteriovenous and Cavernous Malformations
R.F. Spetzler, K. Moon, and R.O. Almefty, Editors
http://dx.doi.org/10.1016/B978-0-444-63640-9.00018-7

Chapter 18

Spinal arteriovenous fistulae: surgical management

ARTHUR L. DAY[1]*, ALI HASSOUN TURKMANI[1], AND P. ROC CHEN[2]
[1]*Department of Neurosurgery, University of Texas Medical School at Houston, Houston, TX, USA*
[2]*Department of Neurosurgery, University of Texas Health Science Center, Houston, Texas, USA*

Abstract

Spinal vascular malformations include several types of pathologic entities with multiple variations in their neurologic presentation and risks. Spinal dural arteriovenous fistulae (SDAVF) are the most frequent type, and account for 70% of all lesions affecting the spinal cord. The nonspecific early neurologic symptoms of SDAVF often make early diagnosis of this clinical entity challenging, and a delay in diagnosis or appropriate treatment often has devastating consequences on spinal cord function. The lesion can invariably be obliterated with surgical occlusion of the intradural arterialized venous reflux, and progression of neurologic deficits is usually arrested and sometimes reversed.

INTRODUCTION

Spinal cord vascular lesions represent a heterogeneous group of disorders that affect the spinal cord parenchyma either directly or indirectly. Lesion types include arteriovenous malformations (AVMs), arteriovenous fistula (AVFs), cavernous malformations, aneurysms, vascular spinal cord tumors (hemangioblastomas), and paraspinal arteriovenous AVMs or AVFs whose venous drainage enters the spinal canal through the intervertebral foramina (Table 18.1). Clinical presentations include intraspinal hemorrhage (subarachnoid, subdural, or intraparenchymal), infarction, mass effect with cord compression, venous congestion with progressive myelopathy, or combinations.

This chapter deals with the surgical management of spinal AVFs. Early recognition of spinal cord AVFs is important, as these entities often yield devastatingly severe neurologic deficits if untreated over time. Symptoms from spinal AVFs invariably arise from spinal cord edema caused by shunting of arterial blood directly into the venous system, leading to congestion of the cord's microcirculation. Spinal AVFs are almost always curable with direct open microsurgical intervention. This chapter discusses the planning, judgment, and operative techniques useful in obtaining excellent outcomes for this patient group.

ANATOMY OF SPINAL VASCULATURE

Adamkiewicz's monograph, published in 1881, was the first comprehensive treatise on the subject of spinal cord vascularization, and provided the definitive description of the artery that came to bear his name. In 1889, Kadyi published a superbly accurate manuscript on the arterial and venous vertebrospinal anatomy. Following the development of spinal angiography, important contributions were made by Doppman, Djindjian and Lazorthes (Lazorthes et al., 1973). In 1988, Thron summarized a lifetime of observations into the anatomy and pathophysiology of the spinal vasculature. The landmark Surgical Neuroangiography series by Lasjaunias et al. (2001) crystallized our present-day knowledge of spinal vasculature and the role of endovascular treatment in the management of spinal vascular pathology.

Arterial anatomy

Segmental arteries arise from the vertebral arteries, aorta, and iliac arteries at each vertebral level to supply the

*Correspondence to: Arthur L. Day, MD, 6400 Fannin, Suite 2800, Houston TX 77030, USA. Tel: +1-713-704-7375, E-mail: Arthur.L.Day@uth.tmc.edu

Table 18.1

Types of lesions that may affect the spinal cord and/or dural vasculature

Spinal cord arteriovenous malformation (AVM)
Arteriovenous fistula (AVF)
Dural AVF
Perimedullary AVF
Paraspinal AVM or AVF whose venous drainage enters the spinal canal through an intervertebral foramen
Cavernous malformation
Aneurysm of the anterior spinal artery
Vascular spinal cord tumor (i.e., hemangioblastoma)

bone, paraspinous musculature, spinal cord, and conus. From its origin, each segmental artery moves dorsally towards the neural foramina on its respective side of the vertebral body, where it divides into three major trunks: (1) a radicular artery, which supplies the dura and the nerve roots and cord at every segmental level (Lasjaunias et al., 2001; Krings 2010); (2) a dorsal branch that supplies the posterior spinal musculature and the facet regions; and (3) lateral branches – a continuation of the segmental artery as the intercostal or lumbar artery that circumferentially wraps around the chest or abdominal wall from dorsal to ventral (Fig. 18.1).

The radicular artery and its branches can be classified based on their region of supply (i.e., radicular, radiculopial, and radiculomedullary (Lasjaunias et al., 2001)). The radicular artery is a small branch present at every segmental level whose supply is restricted to the dura and the nerve root. The radiculopial branch supplies blood to the dorsal nerve root and posterior superficial pial system, including the posterior spinal arteries. The radiculomedullary branch supplies the ventral nerve root, anterior superficial pial system, and the anterior spinal artery (ASA) (Fig. 18.2).

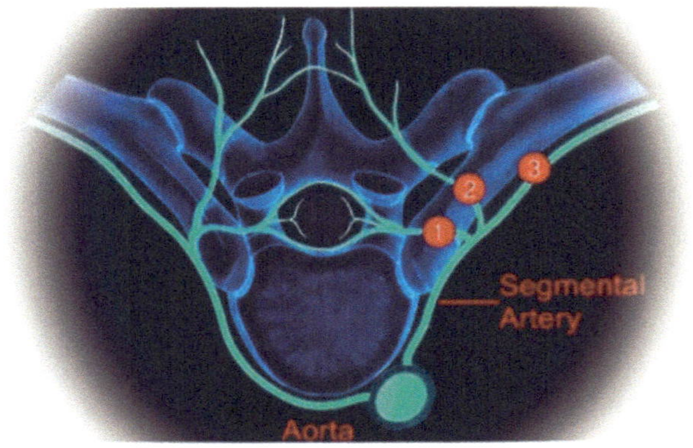

Fig. 18.1. The segmental arteries and their branches. The segmental arteries have three major trunks: 1, radicular (radiculomedullary or radiculopial artery); 2, dorsal (facets, dorsal muscular branches); and 3, lateral (continuation of the intercostal or lumbar artery).

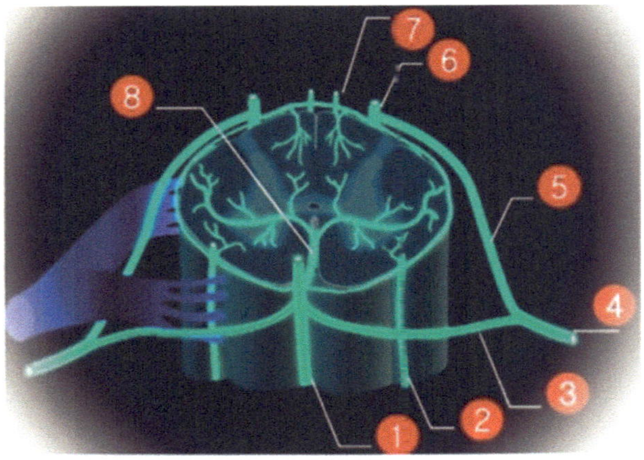

Fig. 18.2. The branches of the radicular arteries. 1, anterior spinal artery; 2, circumferential arterial plexus; 3, radiculomedullary branch (anterior); 4, radicular artery – the source of branches to the spinal cord and dura; 5, radiculopial branch (posterior); 6, posterior spinal artery; 7, circumferential arterial plexus; 8, sulcocommissural artery

The ASA typically originates superiorly from the union of two small branches off the terminal portion of the intracranial vertebral arteries just before they merge to form the basilar artery. The ASA then descends caudally along and within the anterior median sulcus, coursing from the cervical spine to the conus medullaris. Along its way, it is reinforced by radicular branches off the segmental arteries (Lasjaunias et al., 2001). Blood flow derived from these segmental arteries may travel in caudal or cranial directions to maintain cord perfusion. The most well-known radicular arteries are the artery of Adamkiewicz or the arteria radicularis magna, which arises close to the thoracolumbar enlargement (usually on the left side between T8 and T11), and Lazorthe's artery, which supplies the cervical enlargement (usually around the C5 level). In the thoracic region, however, the ASA is joined by only a few terminal vessels descending from the subclavian and vertebral arteries or those ascending from the abdominal aorta. This watershed region is particularly vulnerable to vascular insufficiency, and spinal cord infarction is most likely to occur at T4 to T9.

The posterior spinal arteries originate from the pre-atlantal part of the vertebral artery or from the posterior inferior cerebellar artery. In contrast to the anterior spinal system, the posterior spinal arterial arcade is rather discontinuous beyond several adjacent levels. Supported by multiple radicular artery branches, the posterior spinal arteries are always located off midline and show no connection to the ASA (Nelson and Shapiro, 2015) through the cervical and thoracic levels. Both systems (anterior and posterior) anastomose with adjacent levels via an extensive pial network. At the level of the conus medullaris, the posterior spinal arteries connect to the ASA through anastomotic semicircles (Krings, 2010).

Venous anatomy

The spinal venous drainage system can be subdivided into: (1) an intradural portion, which consists of an intrinsic venous network draining the spinal cord parenchyma; (2) an extrinsic system, including surface veins of the spinal cord linking the intradural venous system with (3) the extradural venous plexus (Batson's plexus), which consists of veins both surrounding and within the vertebrae and the paraspinal musculature (Batson, 1967). The extradural system includes a ventral and dorsal epidural and a foraminal venous plexus (Batson, 1967; Thron, 1988).

Despite great variability, most anatomic specimens demonstrate a single anterior median spinal vein within the ventral median sulcus and a single posterior median spinal vein along the dorsal median fissure. These two veins, connected by multiple transmedullary as well as circumferential veins that follow the curvature of the cord, represent an effective mechanism for pressure equalization throughout the surface spinal channels. The anterior and posterior venous segments collect blood over an equal territory, unlike the arterial system that exhibits more anterior dominance. The exception is the thoracolumbar region, where ventral sulcal veins are more prominent compared to their dorsal counterpart (Lasjaunias and Berenstein, 1990). Between T11 and L3, a large anterior radiculomedullary vein often travels with either the anterior or posterior nerve roots and it can be mistaken for the artery of Adamkiewicz because of its size and location.

CLASSIFICATION OF SPINAL ARTERIOVENOUS FISTULAE

Spinal cord vascular malformations that contain at least one direct arteriovenous shunt that affects the spinal cord and/or dural vasculature include AVMs and AVFs (Table 18.2). Each type affects the spinal cord parenchyma either directly or indirectly. Rosenblum et al. (1987) classified spinal vascular lesions with arteriovenous shunts into four types: type 1 (dural AVF), type 2 (glomus AVM), type 3 (juvenile AVM), and type 4 (perimedullary or pial AVF). In this chapter, we will consider the features and surgical management of AVFs types 1 and 4 as defined by this system (Fig. 18.3). Pial AVFs have also been termed dorsal and ventral intradural AVFs by Spetzler et al. (Spetzler et al., 2002; Kim and Spetzler, 2006).

Table 18.2

Type of vascular malformations that contain at least one direct arteriovenous shunt that affects the spinal cord and/or dural vasculature either directly or indirectly

Arteriovenous malformation (AVM)
Glomus AVM (type 2 vascular malformation)
Juvenile AVM (type 3 vascular malformation)
Arteriovenous fistula (AVF)
Dural AVF (lateral dural or type 1 vascular malformation)
Perimedullary AVF (ventral intradural or type 4 vascular malformation)
Dorsal pial AVF

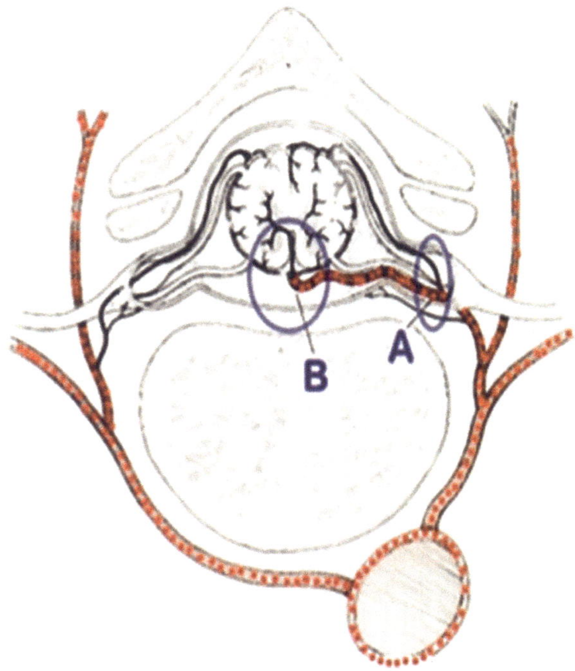

Fig. 18.3. The location of the typical dural and perimedullary arteriovenous fistula. A, Type I dural arteriovenous fistula – this type is most often a slow-flow fistula in the dural root sleeve in the thoracolumbar region. B, Type IV perimedullary fistula – this type most often represents an aberrant ventral midline anastomosis between branches of the anterior spinal artery and the coronal venous plexus.

Spinal dural AVFs (type 1)

Spinal dural AVFs account for 70% of all spinal arteriovenous shunting lesions (AVFs and AVMs). This lesion is approximately five times more common in men than women. Spinal dural AVFs are not congenital, but are acquired lesions that usually become symptomatic in older individuals between the fifth and eighth decades of life. Similar to cranial dural AVFs, obstruction of a venous outlet as a result of thrombosis or fibrosis related to aging is likely associated with their generation.

These lesions are most common in the thoracolumbar region above the level of the conus. The fistula is generally a slow-flowing connection located on the external surface of the dura mater adjacent to a spinal nerve root. The AVF is supplied by a single radicular artery, although some can have multiple feeders into the fistula. At the fistulous site lies a small tuft of tiny arteriovenous channels, from which exits a draining radicular vein that enters the subarachnoid space alongside the adjacent nerve root.

Because the arterialized venous drainage is directed into the subarachnoid space, the fistula creates an increase in spinal venous pressure, secondarily impairing drainage of normal spinal veins, with subsequent venous congestion and intramedullary edema within the spinal cord. The diminished arteriovenous pressure gradient leads to chronic hypoxia and progressive myelopathy. The radicular draining vein in the well-developed dural AVF will be thickened and resemble an artery, and it will connect to several dilated and serpiginous veins on the dorsal surface of the spinal cord. The venous congestion affecting the spinal cord can be transmitted above and/or below the level of the fistula to create a falsely localizing clinical level.

The initial symptoms of congestive myelopathy are unspecific but relentless. Usually, the deficit progresses gradually, but acute deficits and/or interruptions and remissions are also possible. Without therapy, the prognosis of this condition is grim. The typical patient experiences progressive and irreversible motor weakness, sensory disturbance, and bladder and bowel dysfunction (Krings and Geibprasert, 2009a; Fugate et al., 2012), and will be wheelchair-bound within 2–3 years of symptom onset. Focal mass effect and hemorrhagic changes are rare as presenting symptoms, although the cord will be somewhat swollen by the increased venous outflow resistance.

Spinal perimedullary fistulae (type 4)

Approximately 20–30% of all spinal AV shunts are spinal cord pial AVMs fed by arteries of the spinal cord and drained by spinal cord veins (Krings et al., 2005; da Costa et al., 2009). Some of these spinal cord lesions are fistulae, and consist of aberrant direct anastomoses between branches of the ASA and the adjacent ventral venous plexus, hence the terms perimedullary or ventral intradural AVFs.

Unlike dural AVFs, most perimedullary fistulae are midline lesions located within the ventral subarachnoid space, in the thoracolumbar and sometimes cervical regions. These lesions also appear to be acquired lesions, and their origins have been linked by some authors to trauma. Unlike the dural variety, they are not associated with a dural vein thrombosis in their genesis, and they most often become symptomatic in young adults (Van Dijk et al., 2002). Because they are located in the subarachnoid space, arteries feeding the lesion may generate aneurysms, and hemorrhage presentations can occur. Similar mechanisms of cord ischemia are operant in these lesions as for the dural AVF variety, but the distension of dorsally located surface pial veins is often not as prominent.

PREOPERATIVE EVALUATION

Preoperative evaluation consists of a detailed history and neurologic examination. Further workup includes baseline urodynamic status and appropriate imaging studies that confirm the diagnosis and location of a spinal fistula (Fig. 18.4). A thoracolumbar (or total axis) spinal

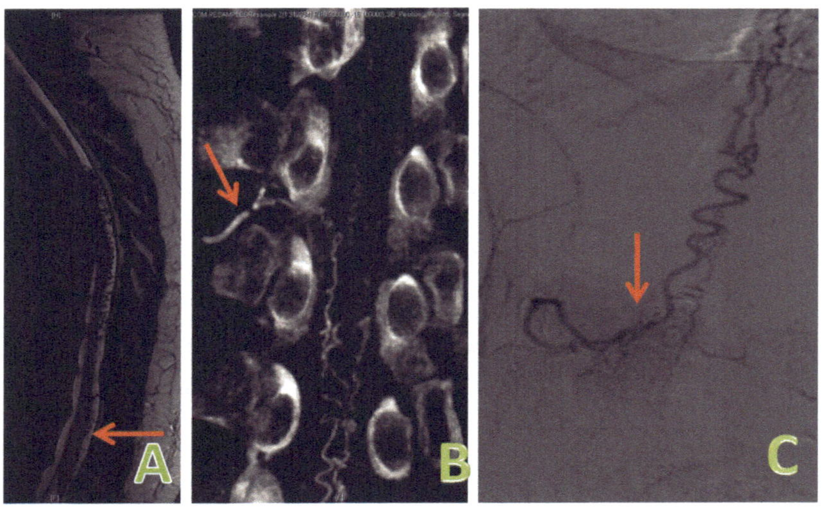

Fig. 18.4. The preoperative imaging characteristics of a typical spinal dural arteriovenous fistula (AVF). (**A**) Sagittal T2-weighted images demonstrate signal voids within the thoracic spinal canal with evidence of venous congestion and signal changes within the cord below the fistula (red arrow). (**B**) Multiphase intra-arterial (intra-aortic) contrast injection computed tomography angiography (IA-CTA) demonstrating segmental artery feeding dural AVF (red arrow). (**C**) Selective spinal digital subtraction angiography showing radicular artery feeding radicular vein through a fistula on the dural root sleeve (red arrow). The arterialized vein then joins the spinal cord venous system, usually dorsally.

magnetic resonance imaging (MRI) is the first diagnostic procedure of choice, and it usually demonstrates the combination of intramedullary signal alterations and perimedullary dilated vessels (Fugate et al., 2012). On T2-weighted images, the cord shows a central hyperintensity over multiple segments. On T1-weighted images, the swollen cord is slightly hypointense and enlarged. Following contrast administration, diffuse intramedullary enhancement may be seen as a sign of chronic venous congestion with a breakdown of the blood–spinal cord barrier. In the late stages of the disease, the spinal cord may be atrophic. For dural AVFs, the dilated and tortuous perimedullary vessels are mainly on the dorsal part of the cord and can be seen on the T2-weighted images as flow void signals. Small dural AVFs or perimedullary fistulae, however, may have venous congestion seen only after contrast enhancement (Krings and Geibprasert, 2009a).

MRI can identify the level and location of the nidus in relation to the spinal cord and dura, particularly when combined with a spinal contrast-enhanced dynamic magnetic resonance angiogram (MRA) (Zampakis et al., 2006; Ali et al., 2007). The technique of first-pass gadolinium-enhanced MRA can demonstrate early venous filling which may indicate the level of the shunt. Multidetector-row helical computed tomography (CT) angiography with intravenous contrast injection (IV-CTA) can provide spinal vessel images showing the surrounding bony structure (Lai et al., 2005; Si-jia et al., 2009). MRA TRICKS (time-resolved imaging of contrast kinetics) is a method of three-dimensional contrast-enhanced MRA that allows for an ultra-resolved three-dimensional contrast-enhanced MRA without venous contamination (Albini Riccioli et al., 2007).

Although MRI, MRA, and CTA are less invasive, their findings often do not provide all of the hemodynamic information needed to plan surgical therapy. Presently, selective spinal digital subtraction angiography (DSA) is the definitive diagnostic test for diagnosis, treatment, and follow-up examination (Zampakis et al., 2006; Krings and Geibprasert, 2009b). Complete spinal DSA includes selective injection of all vessels that may feed the spinal cord, i.e., all intercostal and lumbar arteries, both vertebral arteries, the thyrocervical trunk, the deep cervical artery, and the internal iliac arteries.

SURGICAL TREATMENT

Treatment planning for spinal AVFs is based on accurate location of the fistula in both the axial and longitudinal planes. Endovascular treatment of spinal DAVFs has significantly improved over the last several decades, largely because of the use of N-butyl cyanoacrylate or other liquid embolic agents. Although systemically less invasive, endovascular therapy is associated with lower initial success rates (70–90%) and higher rates of recurrence compared to microsurgical therapy (Van Dijk et al., 2002; Sivakumar et al., 2009).

Spinal dural AVFs (type 1) (Figs 18.5 and 18.6)

Surgery for spinal AVFs is almost always definitive and curative. Best results are obtained in patients treated before advanced clinical deterioration occurs, as recovery of preexisting deficits is often limited. Operative approaches are generally no more systemically stressful than that required for a microdiscectomy or lumbar laminectomy for stenosis. When the procedure is done by experienced surgeons, operative risks are very low, obliteration of the fistula certain, and easing of the venous congestion begins immediately.

The procedure is carried out under general anesthesia. For the typical thoracolumbar dural AVF, the patient is placed in the prone position on a radiolucent frame, with the thoracolumbar spine segment centrally located in a neutral or slightly kyphotic position. After positioning, the appropriate procedure level is confirmed with fluoroscopy. A midline vertical skin incision is made down to the fascia, following which the paraspinal muscles are dissected from the midline until the articular facets on each side are identified. A two-level laminectomy is carried out, centered on either side of the feeding radicular artery. The opening is slightly wider on the side of the lesion to identify the site of the fistula and some of the rostral and caudal dura above and below the tuft of the arteriovenous shunt.

Once the exposure is complete, the fistula must be clarified. Most dural AVFs are fed by a single radicular artery, but other arterial feeders from adjacent levels above and below the actual fistula may reach the lesion through transdural communications. The fistula itself is seen as a small tuft of tiny arteriovenous channels on the dura alongside the adjacent nerve root sleeve. The local epidural veins are usually congested and arterialized. The problematic drainage is not seen externally, and generally consists of a single radicular vein that enters the subarachnoid space along the nerve root to join the dorsal veins of the spinal cord. Transdural venous communications can allow the arterialized venous blood to enter the intradural space at adjacent segments.

After the appropriate level of the fistula is defined, the dura is opened longitudinally and retracted with stay sutures to visualize the spinal cord and the nerve root accompanied by the arterialized vein. The draining vein is invariably thickened by the chronic effects of arterialization and is usually easily distinguishable

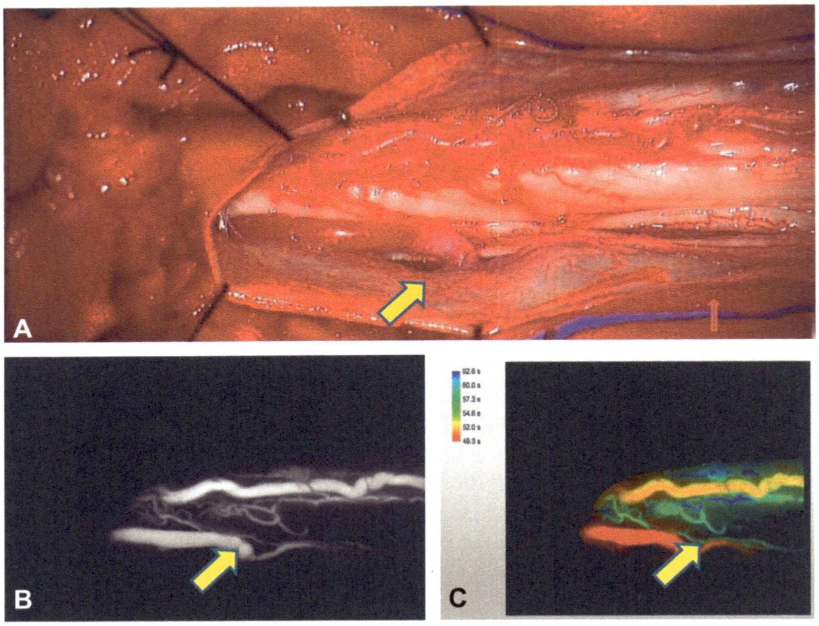

Fig. 18.5. Operative view and pretreatment intraoperative imaging of dural arteriovenous fistula. (**A**) Intraoperative intradural view of arterialized vein (yellow arrow) associated with left-sided nerve root; note engorged dorsal venous plexus. The draining vein is invariably thickened by the chronic effects of arterialization and is usually easily distinguishable from other normal vascular channels. (**B**) Standard indocyanine green injection demonstrating shunting into the spinal venous system (yellow arrow indicates beginning of arterialized radicular vein into the dorsal venous plexus). (**C**) Variation in flow between arterialized radicular vein (red), the distended dorsal veins over the cord (yellow), and the parenchymal veins (green).

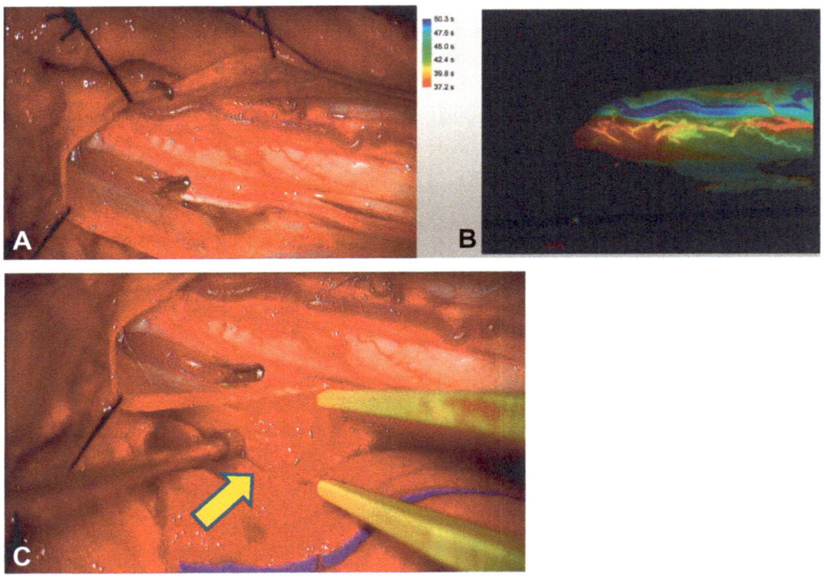

Fig. 18.6. Operative view and posttreatment intraoperative imaging of dural arteriovenous fistula. (**A**) A titanium hemoclip is placed on the subarachnoid vein draining the fistula. (**B**) Postligation indocyanine green injection showed normalization of venous flow; the dilated dorsal veins (blue) now appear as ghost-like remnants of their former arterialized state. (**C**) The dura is reflected, and the fistulous site observed as a small tuft of tiny arteriovenous channels (yellow arrow), from which exits the draining vein that enters the subarachnoid space alongside the adjacent nerve root. The tuft and the dura surrounding it are coagulated to ensure that the fistula does not remain and transmit its flow to an adjacent level. The intradural vein is then coagulated and sectioned just within the dura, leaving the hemoclip as a radiographic marker for future level reference if needed.

from other normal vascular channels. Intravenous videoangiography is then done using indocyanine green (ICG), and the fistulous drainage into the intradural vessel is confirmed (Iacopino et al., 2001; Colby et al., 2009; Hanel et al., 2010; Murakami et al., 2011).

A temporary clip is then placed on this vein, and the ICG injection repeated. Since the fistula is often slow-flow, it often takes a few minutes for the initial contrast to wash through the system. If there is only one vein draining the fistula, no early contrast filling will appear within the distended spinal cord veins on the second injection. After confirmation that the fistula is interrupted, the offending vein is coagulated intradurally and divided between two titanium hemoclips. We then prefer to bipolar cauterize the fistula itself to obliterate both its extra- and intradural surfaces, along with some of the adjacent dura surrounding it, so as to eliminate the possibility that the fistula might survive and transmit its arterialized blood to adjacent venous outlets. Arteries that supply the cord must be left in place (Doppman et al., 1985).

If any question remains, a third ICG injection is then done to validate cure. Before this injection is done, the stump of the sectioned arterialized vein can be gently aspirated to remove any residual contrast within the now stagnant dorsal veins. Thereafter, if abnormal arteriovenous shunting still appears, then an adjacent level communication should be suspected and sought. ICG's value is limited to those vessels seen by direct visualization. By having two levels exposed, a greater panoramic view of the cord's circulation is appreciated. More longitudinal exposures can be quite valuable if more than one feeder or draining vein is suspected.

Intraoperative motor and sensory-evoked potential monitoring is often used until the end of the procedure. The mean arterial pressure is kept between 60 and 80 mmHg to prevent hypotension and cord hypoperfusion before the fistula is obliterated.

Spinal perimedullary fistulae (type 4) (Figs 18.7 and 18.8)

The surgery for perimedullary fistulae begins with a similar midline posterior exposure, followed by a ventrolateral approach to the lesion. The incision is longer, and the number of segments opened is greater (at least three to four segments) than that required for dural AVFs. The exposure is also carried out more laterally, and the skin and paraspinous musculature retracted with fish hooks on the chosen side of entry so as to allow a more lateral view of the ventral spinal canal. In the thoracic region, the ipsilateral facet and pedicle may be removed to enhance the lateral exposure. The dura is opened longitudinally, and the dentate ligaments on both sides sectioned for three to four segments to allow the cord to be mobilized. Cotton balls placed ventral to the cord at the rostral and caudal exposure limits help to maintain a wider working space and view. Sutures placed in the

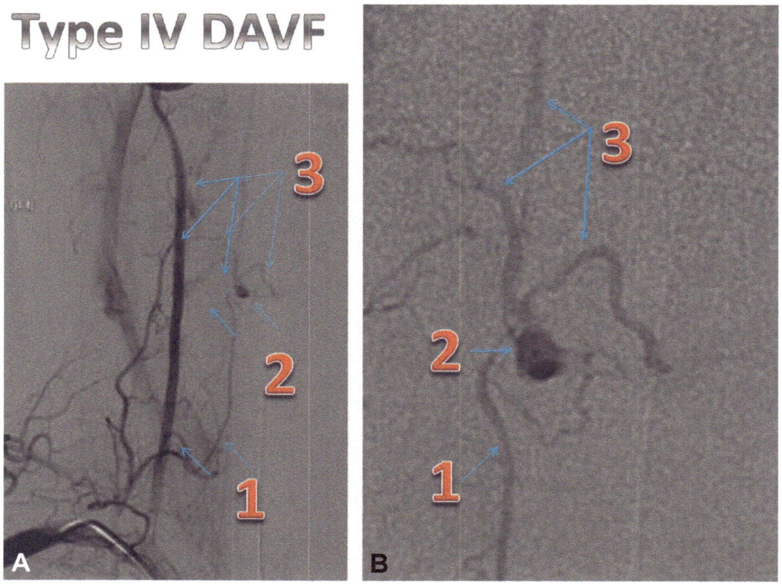

Fig. 18.7. The preoperative imaging characteristics of a typical spinal perimedullary arteriovenous fistula. (**A**) Subclavian artery injection showing thyrocervical and vertebral arteries: note the branch ascending from the thyrocervical trunk to reinforce the anterior spinal artery (1). Note the fistula (2) (arrow) and the adjacent venous aneurysm on the ventral surface of the cervical spinal cord. Also note the multiple draining veins (blue arrows) (3). (**B**) Magnified images showing the anterior spinal artery (1), the fistula and venous aneurysm (2), and the draining veins (3). DAVF, dural arteriovenous fistula.

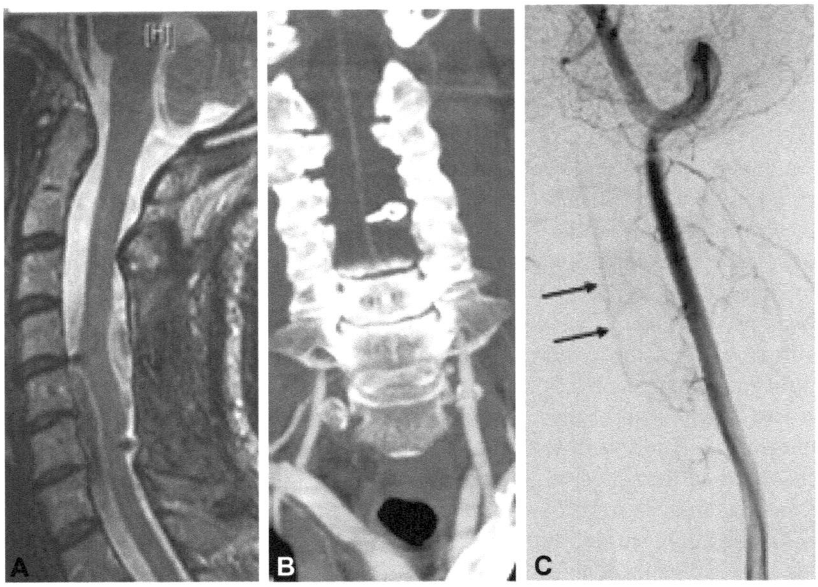

Fig. 18.8. Postoperative images of status post obliteration of type IV cervical spinal dural arteriovenous fistula. (**A**) Magnetic resonance imaging demonstrating multilevel laminectomy and residual anterior inflammation from rupture of aneurysm associated with the fistula. (**B**) Computed tomography angiogram showing complete aneurysm (clip) and fistula obliteration with patency of the anterior spinal artery. (**C**) Vertebral artery selective injection with no evidence of arteriovenous shunting and complete fistula obliteration, anterior spinal artery preservation (arrows).

dentate remnant help the cord to be rotated up and away from the ipsilateral ventral spinal canal surface.

These lesions are usually ventrally located, and are fed by a direct branch of the ASA. Aneurysms may form along the fistula (from an arterial branch or a vein nearby draining the fistula), and be the source of subarachnoid hemorrhage. As with dural AVFs, ICG is very useful in clarifying the fistula and/or aneurysm location and in assessing the completeness of their obliteration. Once the fistula is identified, the arteriovenous connection is interrupted with a thin aneurysm clip, followed by another ICG injection. Obviously, the main channel of the ASA must be preserved.

POSTOPERATIVE MANAGEMENT

The dura and deep fascial structures are closed tightly, so that patients can be mobilized shortly after surgery. Prophylaxis for deep venous thrombosis is essential, particularly so for those with preexisting deficits and reduced lower-extremity function. We use steroids in the perioperative period in the hopes of reducing the effects of the venous congestion and its associated edema. Aspirin is also used as a deterrent to unwanted spinal cord venous thrombosis from stagnation following treatment.

Long-term postsurgical outcome is closely related to preoperative neurologic function. With early diagnosis and surgical treatment, at least 90% of patients retain their pretreatment neurologic function or improve (Cecchi et al., 2008; Wilson et al., 2012).

A follow-up MRI should be done 3–6 months after surgery, and the cord edema and distended veins should no longer be visible. Any worrisome areas or clinical events should prompt strong considerations for another arteriogram.

SURGICAL OUTCOME

Successful fistula obliteration is higher for surgery than with embolization. In one single-institution series and metaanalysis, 98% of patients treated surgically had their fistulas obliterated, versus 46% achieving obliteration with endovascular embolization (Steinmetz et al., 2004).

Surgical treatment of spinal dural AVFs is also highly effective in stabilizing or improving neurologic symptoms (Tacconi et al., 1997; Behrens and Thron, 1999). Tacconi et al. (1997) reported clinical stabilization or improvement following surgical intervention in 84% of patients. In a report by Behrens and Thron (1999), the greatest improvement was found in muscle strength (67%), followed by walking distance (57%), while deteriorations were most often noted in male potency (28%) and pain (24%). Neurologic outcomes is closely related to the preoperative neurologic function of the patient, the duration of the symptoms, and the diffuse longitudinal and axial distributions of the intramedullary high-intensity signal in T2-weighted MR images. Patients with poor preoperative neurologic function generally show less benefit from surgery, but their symptoms are more likely to stabilize (Cecchi et al., 2008).

CONCLUSION

Despite the considerable advances that have been made in imaging and endovascular technology, spinal AVFs continue to pose significant therapeutic challenges. Surgery for spinal AVFs is almost always definitive and curative. Operative approaches are generally no more systemically stressful than those required for a microdiscectomy or lumbar laminectomy for stenosis. When done by experienced surgeons, operative risks are very low, obliteration of the fistula is certain, and easing of the venous congestion begins immediately.

References

Adamkiewicz A (1881). Die blutgefäße des menschlichen Rückenmarkes. I theil. Sitzungsber. Kaiserl. Akad. Wiss., Wien. Math-Naturwiss Cl 84 (3): 469–502.

Albini Riccioli L, Marliani AF, Ghedin P et al. (2007). CE-MR angiography at 3.0 T magnetic field in the study of spinal dural arteriovenous fistula. Interv Neuroradiol 13: 13–18.

Ali S, Cashen TA, Carroll TJ et al. (2007). Time-resolved spinal MR angiography: initial clinical experience in the evaluation of spinal arteriovenous shunts. Am J Neuroradiol 28: 1806–1810.

Batson OV (1967). The vertebral system of veins as a means for cancer dissemination. Prog Clin Cancer 3: 1–18.

Behrens S, Thron A (1999). Long-term follow-up and outcome in patients treated for arteriovenous fistula. J Neurol 241: 181–185.

Cecchi PC, Musumeci A, Facciolo F et al. (2008). Surgical treatment of spinal dural arteriovenous fistulae: long-term results and analysis of prognostic factors. Acta Neurochir (Wien) 150: 563–570.

Colby GP, Coon AL, Sciubba DM et al. (2009). Intraoperative indocyanine green angiography for obliteration of a spinal dural arteriovenous fistula. J Neurosurg Spine 11: 705–709.

da Costa L, Dehdashti AR, terBrugge KG (2009). Spinal cord vascular shunts: spinal cord vascular malformations and dural arteriovenous fistulas. Neurosurg Focus 26: E6.

Doppman JL, Di Chiro G, Oldfield EH (1985). Origin of spinal arteriovenous malformation and normal cord vasculature from a common segmental artery: angiographic and therapeutic considerations. Radiology 154: 687–689.

Fugate JE, Lanzino G, Rabinstein AA (2012). Clinical presentation and prognostic factors of spinal dural arteriovenous fistulas: an overview. Neurosurg Focus 32: E17.

Hanel RA, Nakaji P, Spetzler RF (2010). Use of microscope-integrated near-infrared indocyanine green videoangiography in the surgical treatment of spinal dural arteriovenous fistulae. Neurosurgery 66: 978–984; discussion 984–985.

Iacopino DG, Giusa M, Conti A et al. (2001). Intraoperative microvascular Doppler monitoring of blood flow within a spinal dural arteriovenous fistula: a precious surgical tool. Case report. Neurosurg Focus 10: ECP1.

Kadyi H (1889). Uber die Blutgefasse des menschlichen Ruckenmarkes, Gubrynowicz and Schmidt, Lemberg, Poland.

Kim LJ, Spetzler RF (2006). Classification and surgical management of spinal arteriovenous lesions: arteriovenous fistulae and arteriovenous malformations. Neurosurgery 59 (5 Suppl 3): S195–S201; discussion S3–S13.

Krings T (2010). vascular malformations of the spine and spinal cord: anatomy, classification, treatment. Clin Neuroradiol 20: 5–24.

Krings T, Geibprasert S (2009a). Spinal dural arteriovenous fistulas. Am J Neuroradiol 30: 639–648.

Krings T, Geibprasert S (2009b). Spinal dural arteriovenous fistulas. Am J Neuroradiol 30: 639–648.

Krings T, Mull M, Gilsbach JM et al. (2005). Spinal vascular malformations. Eur Radiol 15: 267–278.

Lai PH, Pan HB, Yang CF et al. (2005). Multi-detector row computed tomography angiography in diagnosing spinal dural arteriovenous fistula: initial experience. Stroke 36: 1562–1564.

Lasjaunias P, Berenstein A (1990). Spinal and spinal cord arteries and veins. In: Surgical Neuroangiography: Functional Vascular Anatomy of Brain, Spinal Cord, and Spine, Springer-Verlag, New York.

Lasjaunias PL, Berenstein A, terBrugge K (2001). Surgical neuroangiography. Clinical vascular anatomy and variations, vol 1. Springer, Berlin.

Lazorthes G, Gouaze A, Djindjian R (1973). Vascularisation et circulation de la moelle épinière, Masson, Paris.

Murakami T, Koyanagi I, Kaneko T et al. (2011). Intraoperative indocyanine green videoangiography for spinal vascular lesions: case report. Neurosurgery 68 (1 Suppl Operative): 241–245; discussion 245.

Nelson PK, Shapiro M (2015). Spinal vascular anatomy. In: RF Spetzler et al. (Eds.), Neurovascular Surgery, 2nd edn; http://dx.doi.org/10.1055/b-0035-122319 THIEME.

Rosenblum B, Oldfield EH, Doppman JL et al. (1987). Spinal arteriovenous malformations: a comparison of dural arteriovenous fistulas and intradural AVMs in 81 patients. J Neurosurg 67: 795–802.

Si-jia G, Meng-wei Z, Xi-ping L et al. (2009). The clinical application studies of CT spinal angiography with 64-detector row spiral CT in diagnosing spinal vascular malformations. Eur J Radiol 71: 22–28.

Sivakumar W, Zada G, Yashar P et al. (2009). Endovascular management of spinal dural arteriovenous fistulas. A review. Neurosurg Focus 26: E15.

Spetzler RF, Detwiler PW, Riina HA et al. (2002). Modified classification of spinal cord vascular lesions. J Neurosurg 96 (2 Suppl): 145–156.

Steinmetz MP, Chow MM, Krishnaney AA et al. (2004). Outcome after the treatment of spinal dural arteriovenous fistulae: a contemporary single-institution series and meta-analysis. Neurosurgery 55: 77–87.

Tacconi L, Lopez Izquierdo BC, Symon L (1997). Outcome and prognostic factors in the surgical treatment of spinal dural arteriovenous fistulas. Br J Neurosurg 11: 298–305.

Thron A (1988). Vascular anatomy of the spinal cord: neuroradiological investigations and clinical syndromes, Springer, Berlin.

Van Dijk JM, terBrugge KG, Willinsky RA et al. (2002). Multidisciplinary management of spinal dural arteriovenous fistulas: clinical presentation and long term follow-up in 49 patients. Stroke 33: 1578–1583.

Wilson DA, Ablaa AA, Uschold TD et al. (2012). Multimodality treatment of conus medullaris arteriovenous malformations :two decades of experience with combined endovascular and microsurgical treatments. Neurosurgery 71: 100–108.

Zampakis P, Santosh C, Taylor W et al. (2006). The role of non-invasive computed tomography in patients with suspected dural fistulas with spinal drainage. Neurosurgery 58: 686–694; discussion 694.

Handbook of Clinical Neurology, Vol. 143 (3rd series)
Arteriovenous and Cavernous Malformations
R.F. Spetzler, K. Moon, and R.O. Almefty, Editors
http://dx.doi.org/10.1016/B978-0-444-63640-9.00019-9

Chapter 19

Endovascular management of spinal dural arteriovenous fistulas

PATRICK A. BROWN[1], ALI R. ZOMORODI[2], AND L. FERNANDO GONZALEZ[2]*
[1]*Department of Radiology, Duke University, Durham, NC, USA*
[2]*Department of Neurosurgery, Duke University, Durham, NC, USA*

Abstract

Spinal dural arteriovenous fistulas (dAVF) are the most common subset of the larger group of spinal vascular malformations. In this chapter, we discuss the definition, epidemiology, clinical presentation, diagnosis, treatment, and outcomes of spinal dAVF. Special attention is given to clinical approach, imaging, and diagnosis with discussion of newer spinal magnetic resonance angiographic techniques. Endovascular management techniques are discussed, including technical and safety considerations for the performance of spinal angiography and various methods of embolization. Embolization media, including liquid embolics, particles, and coils, are introduced. Finally, clinical and imaging outcomes are discussed with attention to the clinical and imaging findings of dAVF recurrence.

INTRODUCTION

Definition

Dural arteriovenous fistulas (dAVFs) represent a subset of the larger group of spinal vascular malformations. While classification schemes and pathophysiology have been discussed at length elsewhere in this volume, a brief discussion of underlying mechanisms is warranted. The pathophysiology of dAVF was first proposed by Aminoff et al. in 1974. In brief, the key principle to understanding dAVF is the presence of high-pressure arterial blood flow entering into the valveless coronal venous system without the presence of an intervening capillary bed. As a result, high pressure is transmitted to the spinal cord parenchyma, resulting in the most common presentation, progressive myelopathy (Anson and Spetzler, 1993).

Epidemiology

dAVFs are uncommon lesions, representing 3–4% of all spinal cord lesions (Veznedaroglu et al., 2006). Despite the rarity of their presentation, accurate and timely diagnosis and treatment are of the utmost importance, as they carry with them significant clinical consequences, including radicular pain, extremity weakness and paresis, sensory deficits, and loss of bowel and bladder function, if not managed appropriately (Aminoff and Logue, 1974). These lesions are thought to be acquired lesions, most commonly affecting male patients after the fourth decade, most commonly in the fifth to sixth decades of life. They are most commonly found in the lower thoracic or upper lumbar vertebrae, usually comprised of a single fistulous connection, though they may occasionally consist of two or more fistulae (Barrow, 1997).

Anatomy and pathophysiology

Classically, spinal dAVFs have been considered as type I of the four broad types of spinal dural vascular malformations. They are the most common of the spinal dural vascular malformations and represent 70–80% of all lesions (Kendall and Logue, 1977). More recent revisions of the varying classifications schemes have further broken down the classification of these lesions into extradural and intradural types, which may be further classified into dorsal and ventral lesions (Spetzler et al., 2002). A detailed examination of these classification schemes may be found elsewhere in this volume.

*Correspondence to: L. Fernando Gonzalez, MD, Duke University, Department of Neurosurgery, 200 Trent Drive DUMC 3807, Durham NC 27710, USA. Tel: +1-919-668-0650, E-mail: fernando.gonzalez@duke.edu

DIAGNOSIS

Clinical presentation

The clinical presentation of spinal dAVF may be quite broad and nonspecific, but the most common presentation is related to venous hypertension and congestion. These symptoms most commonly present in males in the sixth decade of life and frequently consist of vague lower-extremity weakness and sensory disturbance that may manifest in gait abnormalities, sensory loss, ill-defined lower-back pain, loss of bowel and/or bladder control, as well as sexual dysfunction. As this constellation of clinical findings mimics that of spinal stenosis and neurogenic claudication, further clinical and imaging evaluation is directed at that of compressive myelopathy. However, the possible presence of spinal dAVF should be considered as a diagnostic possibility, especially in the middle-aged male patient (Jellema et al., 2006). Rarely, spinal dAVF may present with spinal subarachnoid hemorrhage or spinal cord hemorrhage. In these patients, the clinical presentation is much more acute in onset. Additionally, spinal dAVF may present as a rare cause of intracranial subarachnoid hemorrhage when located in the cervical spine (Kai et al., 2005). While this is a rare cause of subarachnoid hemorrhage, at the authors' institution, care is taken to rule out this entity in cases of subarachnoid hemorrhage without evidence of intracranial causative vascular lesion.

Imaging

Magnetic resonance imaging (MRI) and spinal MR angiography

Conventional MRI

The most common initial imaging study obtained in evaluation of spinal dAVF is MRI of the thoracic or lumbar spine, with or without the use of intravenous gadolinium-based contrast. The most common initial imaging protocol at the authors' institution consists of T1- and T2-weighted imaging in the axial and sagittal planes as well as short tau inversion recovery (STIR) imaging in the sagittal plane. The addition of T1-weighted imaging and contrast-enhanced MR angiography (discussed below) after intravenous gadolinium-based contrast administration improves the sensitivity of diagnosis and enables improved lesion localization (Saraf-Lavi et al., 2002).

The classic conventional MRI appearance of spinal dAVF is that of T2-hyperintense, T1-hypointense enlargement at the distal spinal cord and conus with associated pial vascular engorgement, typified by the presence of tortuous, T2-hypointense vascular "flow voids," typically along the dorsal surface of the cord (Fig. 19.1). Additionally, diffuse parenchymal enhancement is frequently seen on contrast-enhanced T1-weighted images in the region of T2-weighted signal abnormality (Farb et al., 2002). This constellation of findings on conventional MRI is highly suggestive of the presence of a dAVF. However, enlarged pial vascular flow voids, the finding considered to be most specific for the presence of dAVF, is only seen in approximately one-half of confirmed cases (Hurst and Grossman, 2000). As such, one must be careful to remember other potential causes of T2-hyperintense expansion and enhancement of the spinal cord when confronted with such cases. The most common of these are intramedullary neoplasms, infection, inflammation, and demyelination.

Spinal MR angiography

While classic conventional MRI findings of spinal dAVF are highly sensitive for their detection, conventional MRI is poorly sensitive at localization of the level of fistula formation. While the gold-standard technique for localization is conventional catheter angiography (discussed below), noninvasive methods for localization are frequently useful, especially when catheter angiography is being contemplated for definitive diagnosis or therapy. Spinal MR angiography has been seen in several forms over the past decade (Farb et al., 2002), but as MR technology has improved with increased imaging speed and decreased imaging times, technology has arisen that is capable of performing noninvasive, dynamic, time-resolved angiographic examinations (Saindane et al., 2015). These techniques are known under various trade names depending upon the scanner manufacturer (i.e., TWIST: Siemens Healthcare Erlangen, Germany; TRICKS: GE Healthcare, Buckinghamshire, UK), but all involve multiphasic image acquisition after an initial intravenous bolus of gadolinium-based contrast agent (Fig. 19.2). Typical imaging parameters include acquisition of a nonenhanced image followed by several acquisitions during arterial and venous phases. The nonenhanced image serves as a "mask" which is then digitally subtracted from the contrast-enhanced images to isolate the areas that are contrast-enhanced. These data can then be displayed as a projection of vascular enhancement as a function of time elapsed during the imaging period. This "time-resolved" component is critical to securing the diagnosis of a spinal dAVF and to localization of the precise level of abnormality. The imaging data may be displayed in any plane or projection, most frequently in sagittal, coronal, and axial projections. These projection data may be used for localization of the fistula and to focus catheter angiography to the level of interest.

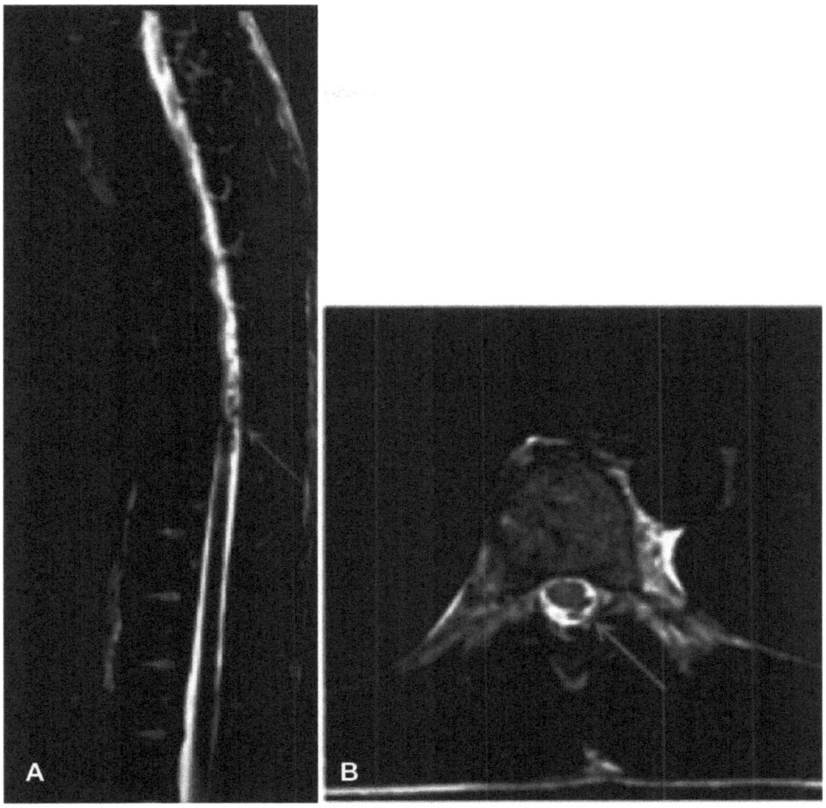

Fig. 19.1. Sagittal (**A**) and axial (**B**) T2-weighted magnetic resonance images of a middle-aged patient presenting with lower thoracic pain and progressive lower-extremity weakness demonstrate multiple tortuous flow voids along the left lateral aspect of the cord (arrows) with ill-defined T2 hyperintense cord signal. Findings raised suspicion for presence of spinal dural arteriovenous fistula.

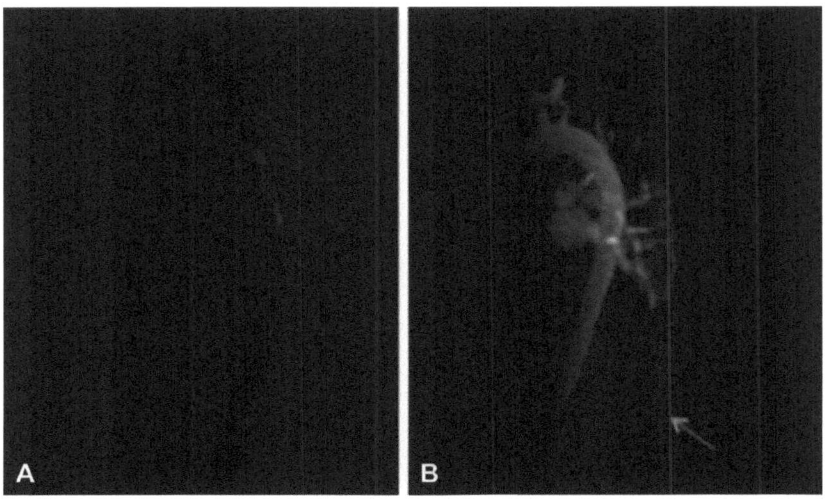

Fig. 19.2. Sagittal short tau inversion recovery (STIR) (**A**) and dynamic TWIST spinal magnetic resonance (MR) angiography (**B**) images. Conventional STIR MR image demonstrates diffuse expansile T2 hyperintense cord signal with irregular dorsal flow voids while the MR angiographic images demonstrate abnormal early venous drainage (arrow) correlating with the region of signal abnormality.

The hallmark finding of spinal dAVF on dynamic contrast-enhanced MR angiography is the simultaneous opacification of large perimedullary veins in phase with arterial opacification of spinal radicular artery branches. Recent publications indicate that contrast-enhanced MR angiography is capable of detecting spinal dAVF in at least 80% of suspected patients who later went on to undergo catheter angiography demonstrating the presence of a dAVF (Lindenholz et al., 2014). While dynamic contrast-enhanced MR angiography is highly effective in identifying and predicting the level of dAVF, it is not currently intended to replace diagnostic spinal angiography. Its current role is that of an adjunct to catheter angiography for purposes of procedural planning by narrowing the number of levels that must be catheterized to the region of interest during the definitive diagnostic procedure, thereby reducing radiation dose to both patient and operator as well as decreasing volume of contrast agent administered.

Computed tomography (CT)

Conventional CT and CT angiography

CT is a workhorse technique for diagnosis of a multitude of conditions, neurologic or otherwise, though in the evaluation of nonosseous spinal pathology, CT tends to be less useful than MR due to the decreased contrast resolution in soft-tissue evaluation. Conventional CT findings of spinal dAVF include diffuse enlargement of the distal spinal cord with diffuse pial vascular enhancement on contrast-enhanced images. CT angiographic findings parallel those of MR angiography, which includes simultaneous opacification of arterial and venous systems via a fistulous connection, usually within a low thoracic or lumbar neuroforamen (Fig. 19.3). While CT angiography is less technically challenging to perform than MR angiography and provides moderate agreement with spinal catheter angiography for localization of dAVF, it appears to be somewhat less reliable overall for localization than contrast-enhanced MR angiography (Oda et al., 2014). However, with increased detector rows in newer scanners, there appears to be improved sensitivity for detection and localization of dAVF (Cao et al., 2014). While newer multiphasic dynamic CT angiographic imaging techniques are becoming more common for imaging of the head and neck (Bi et al., 2015) and may find utility in diagnosis of cervical dAVF, they require special equipment and may result in relatively high doses of ionizing radiation compared to that of conventional CT and CT angiography. Whether these techniques will find widespread use is yet to be seen (Yamaguchi et al., 2013).

CT myelography

CT tends to be most often utilized in the evaluation of spinal dAVF in the setting of a contraindication to MRI, i.e., patients with retained metallic foreign bodies or implanted medical devices such as cardiac pacemakers/defibrillators, medication pumps, or spinal cord stimulators, among others. In these patients, the instillation of myelographic iodinated contrast followed by CT imaging may be useful for detection of subtle pial engorgement when MR is inconclusive or contraindicated (Gilbertson et al., 1995).

Angiography

The gold-standard technique for diagnosis of spinal dAVF is catheter angiography. Catheter angiography

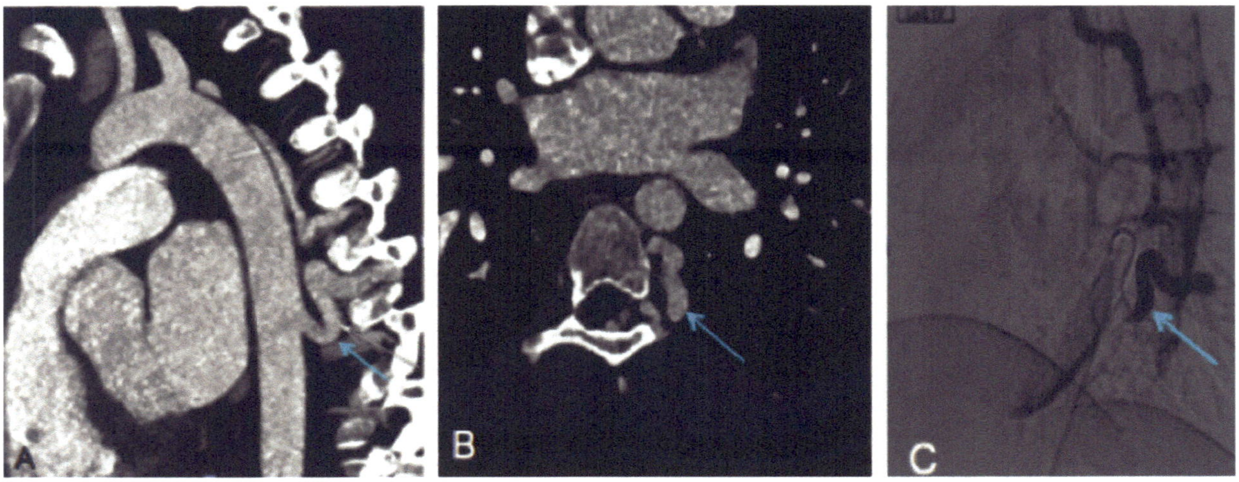

Fig. 19.3. Sagittal oblique (**A**) and axial (**B**) computed tomography angiographic images demonstrate incidental finding of tortuous venous varix (arrows) in left dorsal spinal canal on arterial-phase imaging, suggesting spinal dural arteriovenous fistula (dAVF). Catheter angiogram (**C**) confirmed presence of spinal dAVF at the left T7 level with multiple associated enlarged venous channels (arrow).

provides localization of the dAVF connection, the arterial inflow vessels, and venous outflow, and also provides a means for access for various endovascular treatment techniques. Additionally, spinal angiography provides critical information regarding normal structures that may be at risk during the treatment of these lesions. One of the most important findings on spinal angiography is localization of the anterior spinal artery, or the artery of Adamkiewicz. The artery of Adamkiewicz is variable in its origin but most commonly arises from the T9–T12 intercostal arteries, most frequently from the left; however, anatomy may be quite variable, and it may arise from lumbar arteries at the upper lumbar levels, from upper thoracic intercostal arteries, or even the intercostobronchial trunk, which supplies bronchial arteries. The artery of Adamkiewicz is the dominant arterial inflow to the thoracic and upper lumbar cord, and identification of its origin is mandatory prior to any type of spinal embolization, as inadvertent occlusion can produce anterior cord ischemia and subsequent paresis (Yoshioka et al., 2006).

A full diagnostic spinal angiogram requires systematic catheterization and injection of each of the intercostal arteries bilaterally as well as the lumbar arteries bilaterally (Fig. 19.4). Depending on location of the MR abnormality and level of suspicion for cervical or lower lumbar involvement, angiographic interrogation may be extended to include the subclavian arteries and associated branches, including the costocervical and thyrocervical trunks, the vertebral arteries, internal carotid arteries, external carotid arteries, as well as the internal iliac arteries. This systematic approach can lead to long procedure times, high radiation doses, and large-volume administration of contrast media. Pre-procedure localization with noninvasive imaging such as dynamic MR angiography may help to reduce procedure time, radiation dose, and contrast media utilized. Many authors prefer to perform these procedures under general anesthesia in order to reduce motion artifact that may obscure subtle lesions. In preparing for the procedure, patients should be instructed about risks including bleeding and infection, vascular injury at the site of arterial access, stroke, paralysis, exposure to iodinated contrast media, as well as risks associated with exposure to ionizing radiation.

Spinal dAVFs have a characteristic appearance on catheter angiography, paralleling the appearance on dynamic MR angiography described previously. The classic appearance is described as that of opacification of engorged and tortuous pial veins during the arterial phase of injection, which may extend to multiple levels above and below that of the fistula (Fig. 19.5). Most spinal dAVFs arise from a single arterial feeder from a radicular branch arising from either an intercostal or lumbar artery, though the angiographer should catheterize several levels above and below the fistulous level, as adjacent levels may occasionally provide arterial inflow (Jeng et al., 2015).

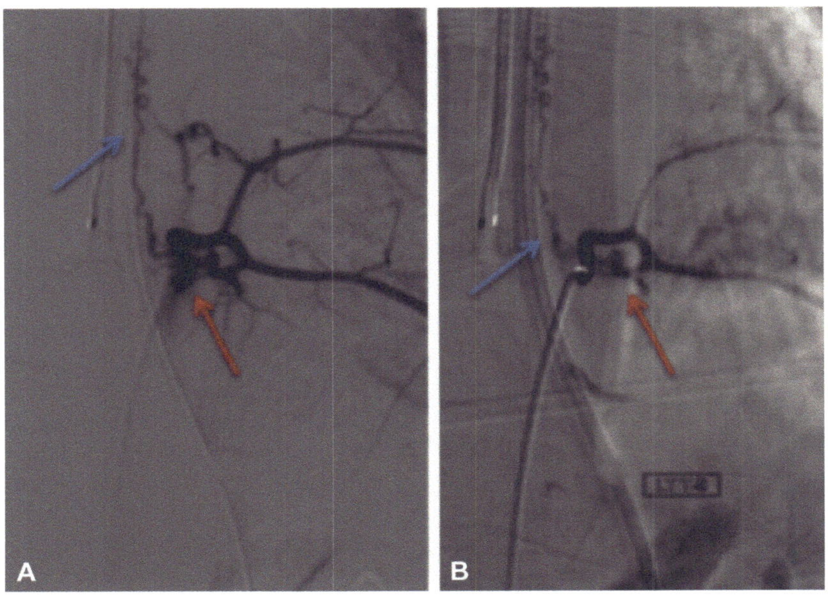

Fig. 19.4. Anteroposterior angiographic images from spinal angiography left T9 intercostal injection in early arterial (**A**) and late arterial (**B**) phases demonstrate simultaneous opacification of a dilated, tortuous dorsal pial vein (blue arrows) in phase with the left T9 and conjoined T8 intercostal arteries. Irregular and tortuous fistulous connection is also seen (red arrows).

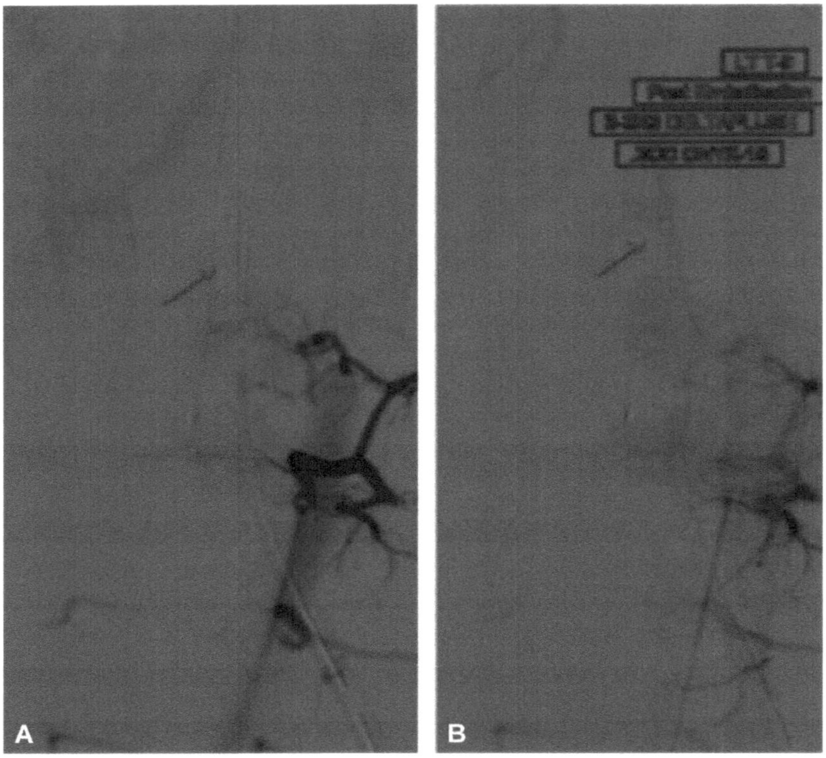

Fig. 19.5. (**A** and **B**) Postembolization anteroposterior angiography after combined coil and Onyx embolization at the dural arteriovenous fistula nidus demonstrates cessation of opacification of the enlarged tortuous pial veins. Incidental note is made of the anterior spinal (artery of Adamkiewicz) arising from the left T8 radicular artery.

TREATMENT

The main objective for treatment of spinal dAVF is obliteration of the fistulous connection with restoration of normal anterograde arterial flow and venous drainage. Surgical obliteration of the fistula involves laminectomy or hemilaminectomy with isolation and ligation of the fistulous connection. It is highly effective, with reported initial success rates of 98% (Steinmetz et al., 2004), with relatively low rates of morbidity. Still, since the majority of patients with suspected spinal dAVF are referred for angiography and with decreased recovery time compared with surgical approaches, angiographic embolization is frequently the first-line approach when a single feeder is identified and the arteriovenous connection is direct, as opposed to a plexal, diffuse one. Multiple techniques for embolization have been utilized, with the most common described in the following paragraphs.

Endovascular management

Technique

Patient preparation, as with any invasive procedure, for endovascular treatment of spinal dAVF is critical for success. Prior to treatment, the patient must be informed of the risks, benefits, and alternatives to therapy. Risks that must be discussed with the patient include bleeding, infection, injury to the femoral artery during access, cord ischemia and infarction, paralysis, exposure to iodinated contrast with possible renal function impairment, and exposure to ionizing radiation. The operator should carry out a thorough review of the patient's past medical history, with special attention to bleeding diatheses, allergies, medications, renal impairment, prior vascular procedures, and history of anesthesia A checklist approach to the pre-procedure evaluation and consent process has been shown to improve preparation and is encouraged (Ripley et al., 2015).

After the preparation and consent process, the patient is taken to the angiography suite and positioned on the table. Biplane angiography capability, while not absolutely necessary, is preferred, especially during the embolization process. The authors recommend the use of intraoperative neurophysiologic monitoring (Niimi et al., 2004; Sala et al., 2007). After standard sterile preparation, and induction of anesthesia, the common femoral artery may be accessed with a standard micropuncture set. Most commonly, a 5 F standard introducer sheath is utilized at the groin access site. For particularly tortuous iliac anatomy, a longer 6 F introducer sheath such as

Cook Shuttle (Cook Medical, Bloomington, IN) or a braided Super Arrow sheath (TeleFlex, Limerick, PA) may be used to aid in distal access, especially for access to upper thoracic segments. Alternatively, a standard 6 F sheath may be placed to allow utilization of various guide and delivery catheters for stabilization of distal access. Typically, any one of a number of reverse-curve catheters is used for catheterization of the intercostal or lumbar arteries. Generally, the catheter is advanced over a guidewire into the proximal aorta and then formed after removal of the guidewire. The catheter is then withdrawn until it engages the targeted intercostal or lumbar artery. Further retraction serves to "seat" the catheter tip within the vessel. Contrast angiography is then performed. This sequence of steps is repeated until the desired vessels have been imaged.

Ideal cases for endovascular management involve direct artery-to-vein connections, as endovascular treatment becomes problematic when arterial contribution takes on a more plexal morphology with multiple feeder channels. After identification of the fistulous connection, if the morphology is amenable for endovascular management, attention may be turned to embolization. A microcatheter and microwire arrangement can be advanced in a coaxial fashion via the access catheter and positioned at the fistulous connection. Roadmap guidance is utilized to aid in precise positioning of the microcatheter. Typically the goal is to position the microcatheter as far as possible, even into the fistula itself. Microangiography via the microcatheter may be used to confirm catheter positioning and better define the anatomy of the fistula. At this point, the embolic material may be administered via the microcatheter.

Various embolic agents are discussed below, but the overall goal of catheter-directed embolization is occlusion of the fistulous connection and the proximal outflow vein. Care must be taken not to begin embolization too far proximally within the inflow artery, since proximal occlusion will not adequately occlude the fistula and is associated with early fistula recurrence. Additionally, proximal occlusion obliterates access to the fistula and precludes further access. It is for these reasons that proximal embolization with embolic coils or particles is not recommended for treatment of spinal dAVF. The most effective endovascular techniques utilize liquid embolic agents and are discussed individually below (Krings and Geibprasert, 2009).

Agents

Liquid

***n*-Butyl cyanoacrylate (*n*-BCA).** Trufill (Cordis, Miami Lakes, FL) is an adhesive suspension of n-BCA, ethiodized oil, and tantalum powder (to provide increased radiopacity during angiographic embolization). It carries a Food and Drug Administration (FDA) indication for use in preoperative embolization of intracranial arteriovenous malformations but is frequently used off-label for extracranial applications. Its mechanism of action is that of rapid polymerization of the *n*-BCA when it comes into contact with ionic fluids such as blood or crystalloid solutions. The rate of polymerization is highly dependent on the ratio of *n*-BCA to the ethiodized oil. This ratio determines the effective viscosity of the material. Decisions regarding viscosity of the material should depend upon the position of the catheter in relation to the fistula and the rate of flow through the fistula.

The Trufill kit comes with separately packaged *n*-BCA, tantalum powder, and ethiodized oil which can be mixed in varying ratios of ethiodized oil:*n*-BCA in order to achieve the desired effect during embolization. In general, the higher the ethiodized oil:*n*-BCA ratio, the slower the polymerization rate, and the longer the polymerization time. More rapid polymerization rates produce hardening and polymerization more closely to the catheter tip and are recommended when the catheter is able to be positioned within the lesion, while slower polymerization rates allow more distal penetration prior to hardening and are recommended when the catheter is farther proximal to the target lesion. Prior to embolization, the embolic delivery catheter is primed with a 5% dextrose solution in water in order to prevent polymerization inside the catheter. After the embolic material comes into contact with blood, it begins to immediately polymerize and hardens over a few seconds. During this hardening process, it is possible for the embolic material to adhere to the delivery catheter, which must be considered during the embolization process.

n-BCA-compatible syringes and microcatheters (Prowler (Cordis, Hialeah, FL), Transit (Cordis, Hialeah, FL), Regatta (Cordis, Hialeah, FL), among others) are mandatory. After satisfactory positioning of the microcatheter into the fistula or within the supplying artery, the microwire is removed, and the microcatheter is primed with 5% dextrose in water. The *n*-BCA mixture is prepared, and then injected in small pulses via the microcatheter under fluoroscopic observation. After injection of embolic material, the microcatheter is aspirated at the hub and quickly withdrawn to prevent nontarget embolization during withdrawal of the microcatheter and to prevent adhesion of the catheter to the embolic material within the targeted lesion.

n-BCA has been shown to be quite an effective method for spinal dAVF embolization. Some authors recommend it as a first-line embolic agent due to its ability to penetrate the lesion and reach the proximal draining vein, suggestive of improved durability of treatment compared with other embolic devices and agents (Blackburn et al., 2014). Potential drawbacks to the use of *n*-BCA for dAVF embolization are related to the relatively fast polymerization rate. The required rapid injection rate increases the possibility for nontarget embolization. Additionally, the rapid polymerization rate and adhesive quality of *n*-BCA increase the risk for adhering the catheter to the *n*-BCA material inside the artery. Currently, the authors recommend the use of *n*-BCA for lesions in which intralesional placement of the delivery catheter is not possible and the delivery catheter is located within the proximal inflow artery.

Onyx. Onyx liquid embolic agent (Covidien, Mansfield, MA) is a nonadhesive suspension of ethylene vinyl alcohol (EVOH) copolymer in dimethyl sulfoxide (DMSO) suspended with tantalum powder to provide radiopacity during angiographic procedures. It carries an FDA indication for use in the preoperative embolization of intracranial arteriovenous malformations but, like *n*-BCA, is frequently used off-label for extracranial neurovascular procedures and peripheral interventions. Its mechanism of action is similar to that of *n*-BCA except that the polymerization process occurs in a much slower, more controlled fashion. While suspended in DMSO, the Onyx material remains in a liquid state; however, when it comes into contact with an ionic fluid such as blood or crystalloids, it begins to precipitate into its solid form. The precipitation process occurs in a cohesive, nonadhesive fashion at the leading edge of the flow of embolic material, from peripheral to central, and requires approximately 5 minutes to complete. This, in effect, enables the interventionalist to redirect the flow of embolic material by allowing the leading edge of administered embolic to harden before injecting additional material. It may be injected over long periods of time due to its nonadhesive properties, such that it does not readily adhere to the delivery microcatheter (Guimaraes and Wooster, 2011). Furthermore, delivery microcatheters have recently been developed with detachable distal tips (Apollo, Covidien, Mansfield, MA) in the event of adhesion to the delivery microcatheter.

Onyx is available in varying degrees of viscosity measured in centipoises, branded as Onyx-18 (6% EVOH) and Onyx-34 (8% EVOH). Lower-viscosity embolic material travels more easily distal from the delivery microcatheter and tends to penetrate the targeted lesion more easily, with the disadvantage of relatively less control over distal migration of embolic material. Higher-viscosity embolic material provides greater control over distal migration at the potential expense of diminished penetration of the targeted lesion. Given these characteristics, lower-viscosity embolic material is recommended for lesions with slower flow rates to permit full lesion penetration and obliteration of the proximal outflow vein, while higher-viscosity embolic material is recommended for lesions with faster flow rates to prevent embolization too far distally into outflow vein or possible nontarget embolization.

Any of a number of DMSO-compatible microcatheters (e.g., Headway Duo (Microvention, Tustin, CA), Echelon (Medtronic, Dublin, Ireland), Marathon (Medtronic, Dublin, Ireland), Apollo (Medtronic, Dublin, Ireland)) may be used for Onyx delivery. After confirmation of satisfactory positioning of the microcatheter within the fistula or at the distal aspect of the inflow artery, the microwire is removed, and the microcatheter is primed with a specified volume of DMSO, listed on the catheter packaging. After the microcatheter is primed, the Onyx material may be administered into the lesion under fluoroscopic guidance. The authors recommend making use of blank roadmap fluoroscopy in addition to live fluoroscopy in order to better visualize embolic material as it exits the microcatheter. As the Onyx material exits the microcatheter, it can be intermittently allowed to harden at its leading edge as needed in order to allow redirection into the target lesion. A short segment of reflux along the microcatheter (10–15 mm or less) is allowable. As this segment of reflux hardens along the margins of the catheter, it can serve as a functional "plug" to prevent further reflux and direct additional material distally into the targeted lesion. An alternative to the use of the functional Onyx plug technique is to use a dual-lumen balloon-tipped microcatheter (Scepter-C, Microvention, Tustin, CA) in order to provide proximal occlusion of the inflow vessel, prevent reflux of embolic material, and direct embolic material into the targeted lesion (Dabus et al., 2014).

Onyx embolization of spinal dAVF has been shown to be quite effective, enabling penetration of embolic material into the draining vein without evidence of lesion recurrence by clinical or imaging criteria (Noguiera et al., 2009). However, some recent studies have called into question the long-term effectiveness of Onyx, citing poor venous penetration and higher rates of recurrence versus *n*-BCA and surgical resection (Adamczyk et al., 2012; Blackburn et al., 2014). Currently the authors prefer the use of Onyx when the operator is able to position the microcatheter very distally into the inflow artery or within the fistula. If unable to obtain this distal access, the operators recommend consideration of dual-lumen balloon occlusion or use of *n*-BCA (Figs 19.6–19.11).

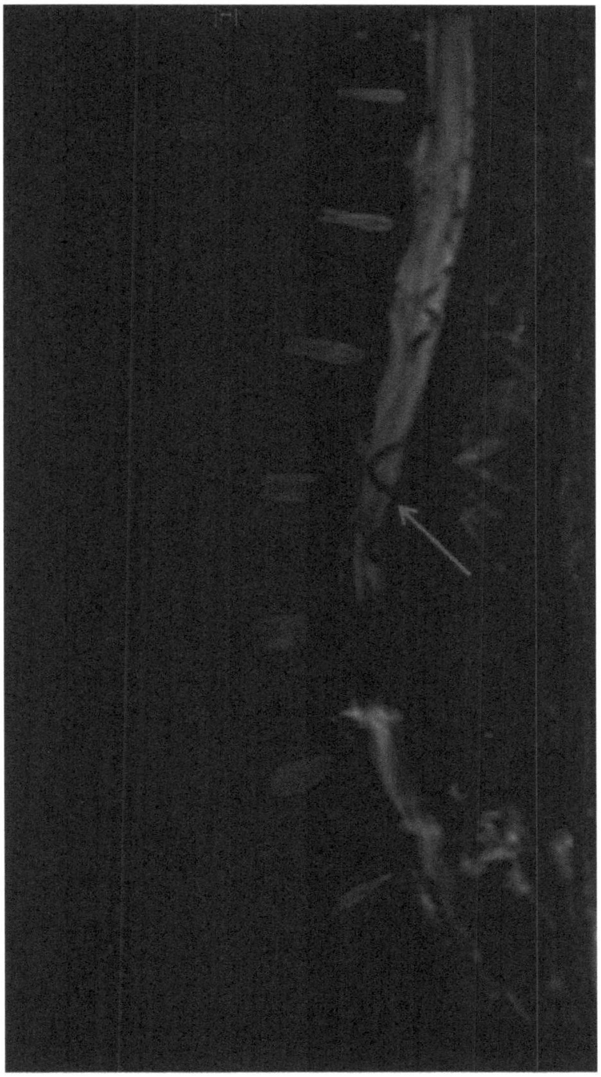

Fig. 19.6. Middle-aged male patient with history of lumbar fusion with new, progressive back pain, ill-defined lower-extremity weakness, and difficulty urinating. T2-weighted sagittal magnetic resonance imaging demonstrates multiple ill-defined flow voids within the lower lumbar spine.

Coils and particles

As discussed above, the authors do not recommend the routine use of coil or particulate embolization for spinal dAVF due to the high recurrence rates related to proximal embolization. As stated above, in order to increase the likelihood of achieving a durable embolization and decrease the likelihood of lesion recurrence, the goal of endovascular management is placement of embolic material within the distal aspect of the inflow artery, within the fistulous connection itself, and into the proximal aspect of the venous outflow. Coil and particulate embolization is not particularly well suited for this goal.

There are, however, a couple of notable exceptions to this concept. The first of these is in the setting of a fistula that is unable to be treated with endovascular techniques alone. In these cases, the authors recommend placement of a coil into the feeding artery near the fistula to aid in surgical localization of the fistula. An additional setting in which coil embolization finds use is in the presence of a large venous varix associated with the AVF; in these cases, coil embolization may be carried out to exclude the varix serving as an adjuvant for liquid embolics (Figs 19.12 and 19.13).

OUTCOMES AND FOLLOW-UP

Clinical outcomes

Successful endovascular occlusion of spinal dAVF lesions is correlated with significant clinical improvement in associated symptoms (Fig. 19.14–19.16). The most profound clinical improvements appear to be those related to muscle weakness and gait disturbance. A 2004 study of 44 patients treated primarily by embolization demonstrated a clinical improvement in gait disturbance in 64% of patients and improvement in muscle strength in 56% of patients (Jellema et al., 2004). Other studies have reported gait improvement in 50–100% of patients. Clinical improvement in bowel and bladder function, pain, and muscle spasm appears to be comparatively less common (Jellema et al., 2006).

An important factor in improving clinical outcome appears to be prompt diagnosis. In a 2016 series of 53 ambulatory spinal dAVF patients with a mean of 6 months delayed diagnosis, more than half (32) of patients had developed worsened Aminoff motor scores or modified Rankin Scale, requiring wheelchair or walker for ambulation. The majority of these patients did not demonstrate improvement in ambulation in 90-day follow-up after endovascular or surgical treatment (Brinjiki et al., 2016). This underscores the need for accurate and timely diagnosis whenever a diagnosis of spinal dAVF is considered.

Follow-up

The most relevant follow-up method for spinal dAVF patients after treatment is that of clinical symptomatology. Very few series exist on posttreatment imaging follow-up of spinal dAVF patients. However, a 2011 imaging follow-up series suggests that patients successfully treated with surgical exclusion of spinal dAVF continue to have persistent cord signal abnormalities on follow-up MR studies. The tendency is for mild

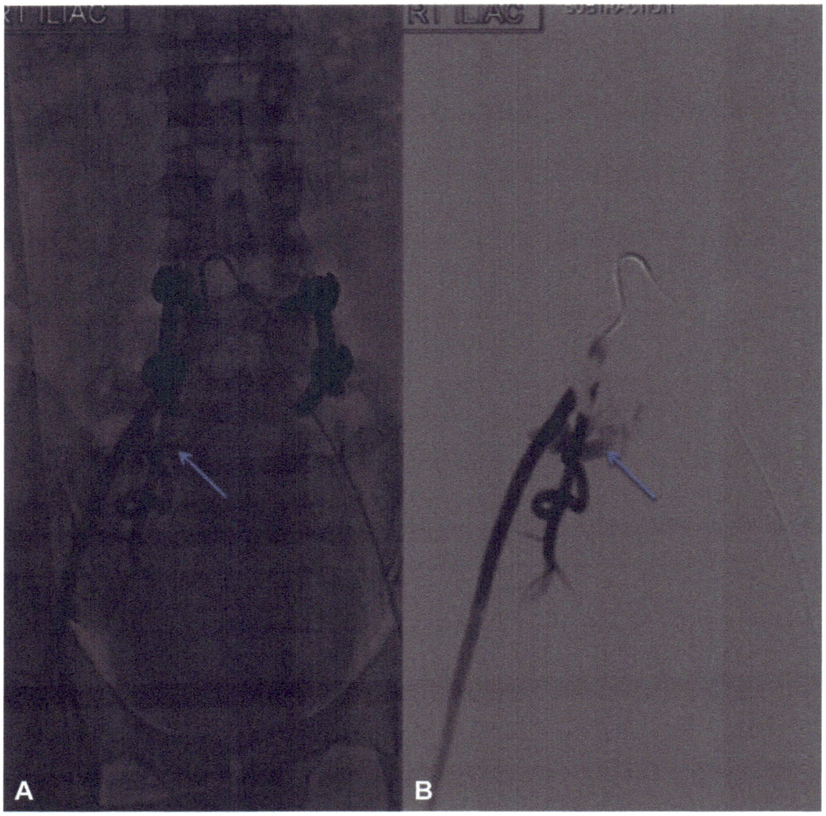

Fig. 19.7. (**A** and **B**) Anteroposterior angiography of the right common iliac artery demonstrates simultaneous arterial and venous opacification (arrows) with irregular tortuous venous return, suspicious for the presence of arteriovenous fistula.

improvement in signal abnormalities over time, though these improvements do not appear to correlate with clinical improvement. Yet, clinical decline in motor function after initial improvement is correlated with fistula recurrence and should prompt further imaging investigation (Kaufmann et al., 2011). If fistula recurrence is suspected, repeat angiography with possible embolization should be considered.

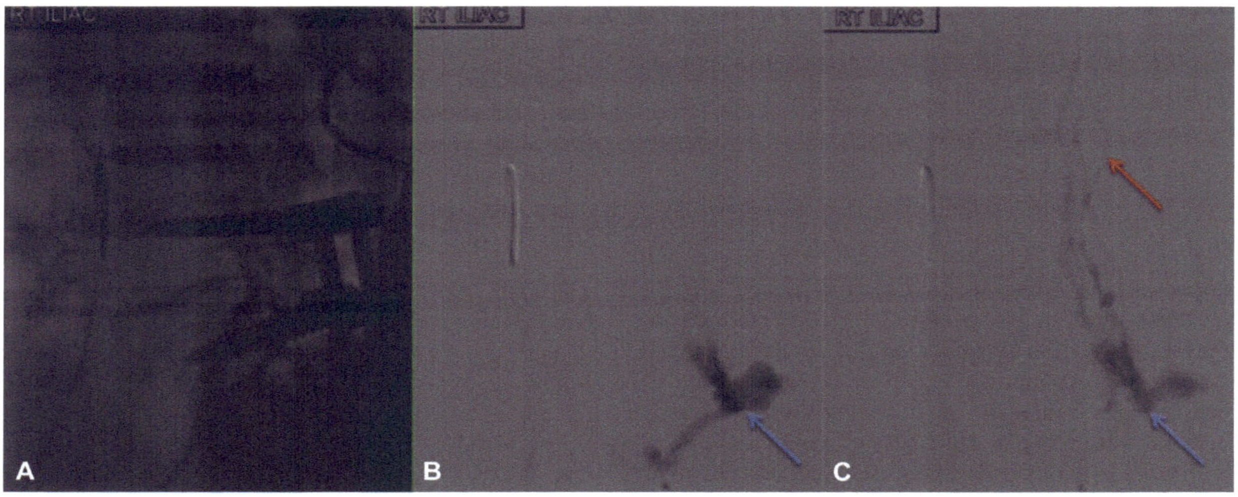

Fig. 19.8. (**A–C**) Lateral right common iliac angiography redemonstrates enlarged and tortuous arteriovenous fistula (blue arrows) with opacification of multiple enlarged pial veins (red arrow).

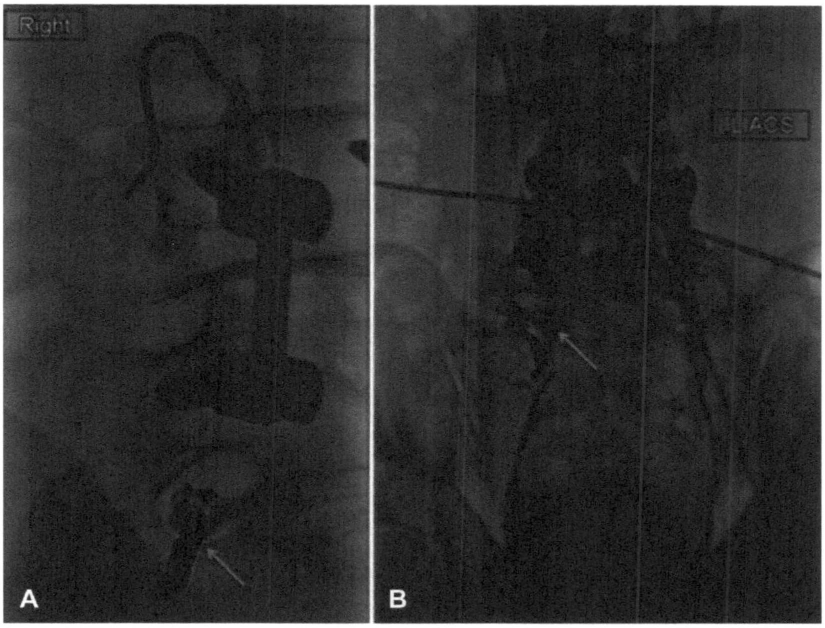

Fig. 19.9. (**A** and **B**) Anteroposterior angiography after Onyx embolization of fistula nidus demonstrates distribution of embolic material at the distal aspect of the arterial inflow, within the fistula nidus, and within the proximal outflow vein (arrow). Iliac artery injection (**B**) demonstrates cessation of opacification of the outflow vein (arrow).

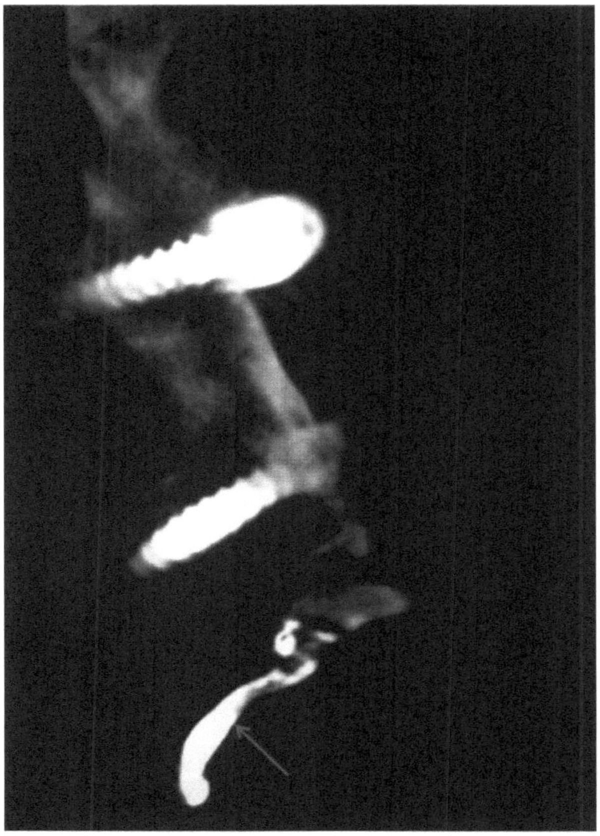

Fig. 19.10. Computed tomography image demonstrating Onyx embolic material (blue arrow) extending to near the site of previously placed fixation screw, raising suspicion for traumatic dural arteriovenous fistula.

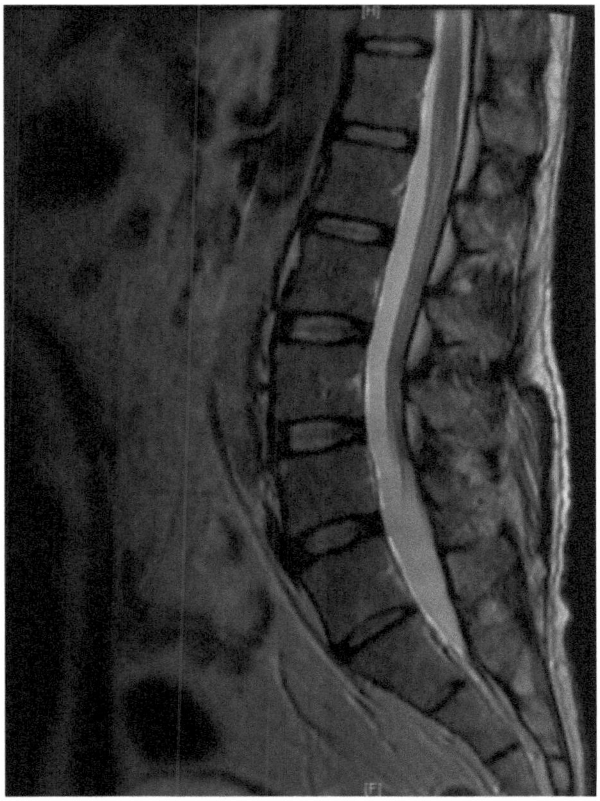

Fig. 19.11. Six-month follow-up sagittal T2-weighted magnetic resonance imaging demonstrating complete resolution of venous flow voids. Patient endorsed complete resolution of symptoms.

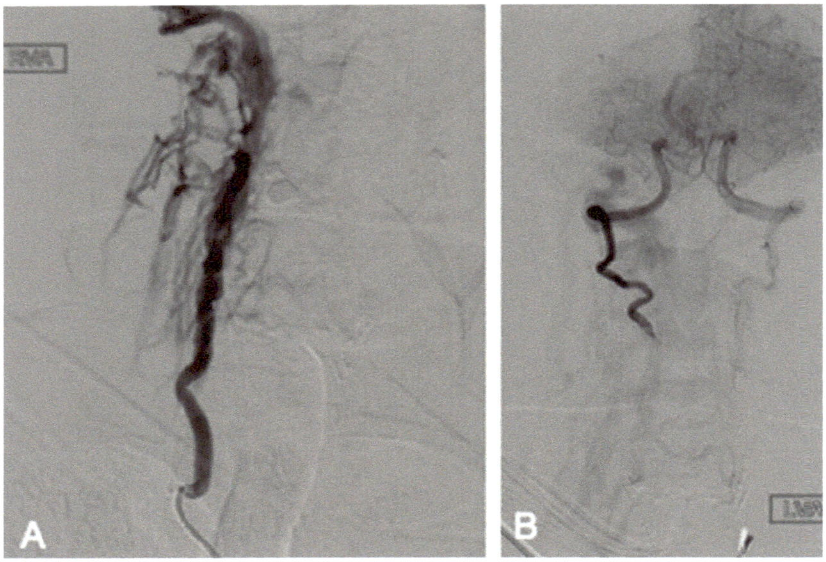

Fig. 19.12. Middle-aged patient with neck pain and complaints of pulsatile tinnitus. Case demonstrates the combined use of coils and Onyx embolization. Right vertebral artery (**A**) and left vertebral artery (**B**) angiography demonstrates large abnormal early venous opacification with multiple enlarged and tortuous intraspinal and extraspinal veins, compatible with dural arteriovenous fistula.

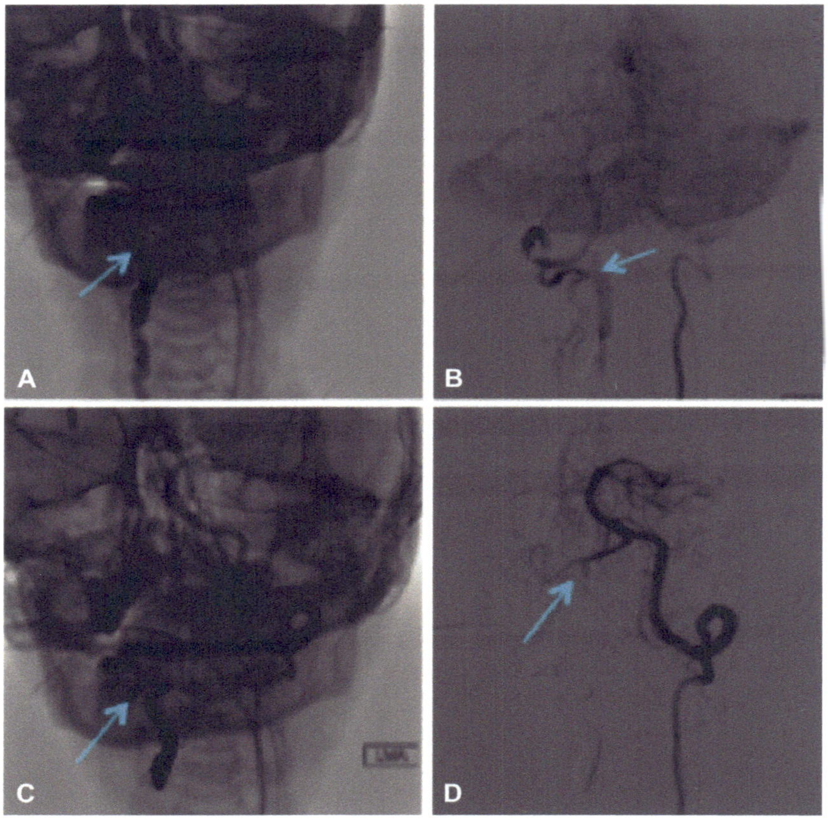

Fig. 19.13. Example of combined coil and liquid embolization. (**A**) Embolic coils (arrow) were utilized to reconstruct the right vertebral artery after Onyx embolization of a small feeder vessel with proximal balloon occlusion. (**B**) Left vertebral artery angiography demonstrates persistent early venous opacification via left vertebral artery injection (arrow). The fistulous connection was accessed via a retrograde approach from the left vertebral artery across the vertebrobasilar junction, with Onyx liquid embolic delivered cranial to the fistulous connection. Angiographic images after retrograde Onyx embolization (**C** and **D**) demonstrate cessation of retrograde arterial fistulous inflow (arrows).

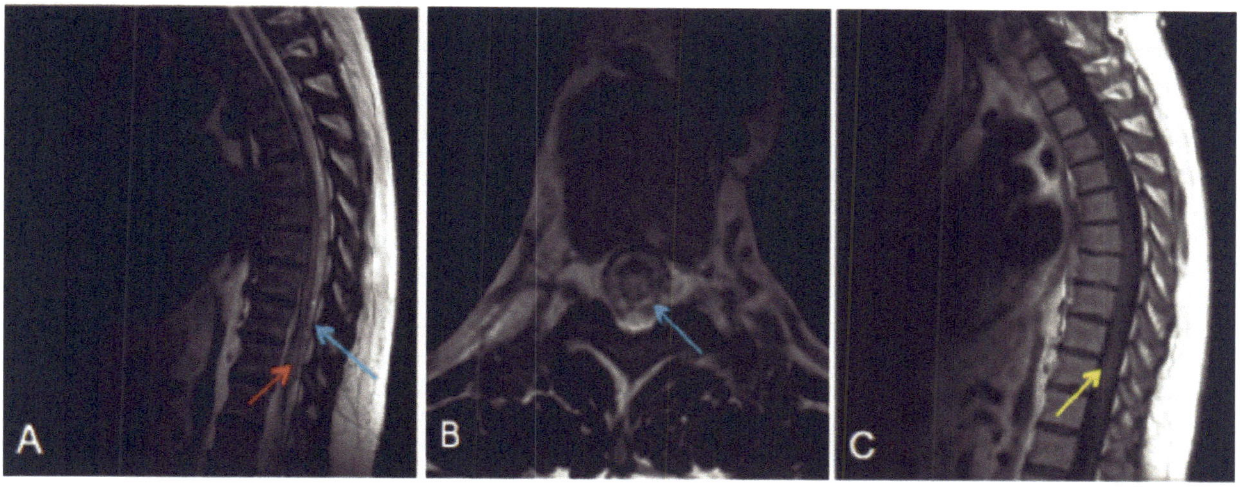

Fig. 19.14. Sagittal T2-weighted (**A**), axial T2-weighted (**B**), and sagittal postcontrast magnetic resonance images in a middle-aged male patient with vague back pain and lower-extremity weakness demonstrate long-segment expansile T2-hyperintense cord signal (red arrow), tortuous dorsal flow voids (blue arrows), and faint enhancement (yellow arrow), raising suspicion for spinal dural arteriovenous fistula.

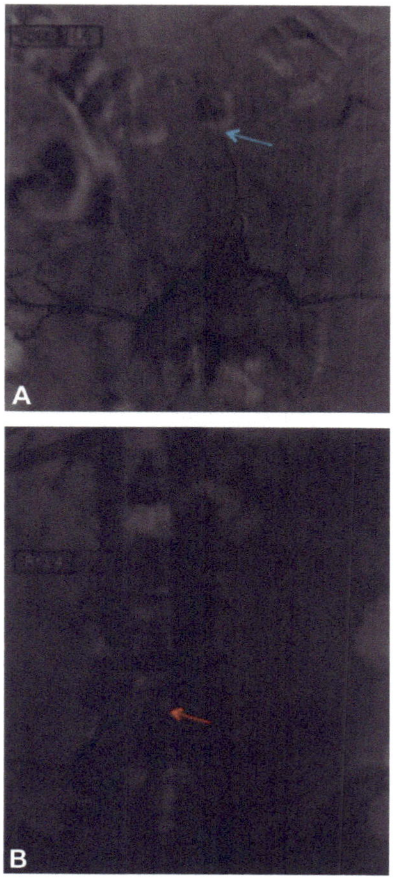

Fig. 19.15. Pre-embolization (**A**) and postembolization (**B**) catheter angiographs demonstrate early venous drainage within a tortuous dorsal medullary vein (blue arrow) compatible with spinal dural arteriovenous fistula. Postembolization image after placement of Onyx embolic material (red arrow) demonstrates cessation of opacification of fistula outflow vein.

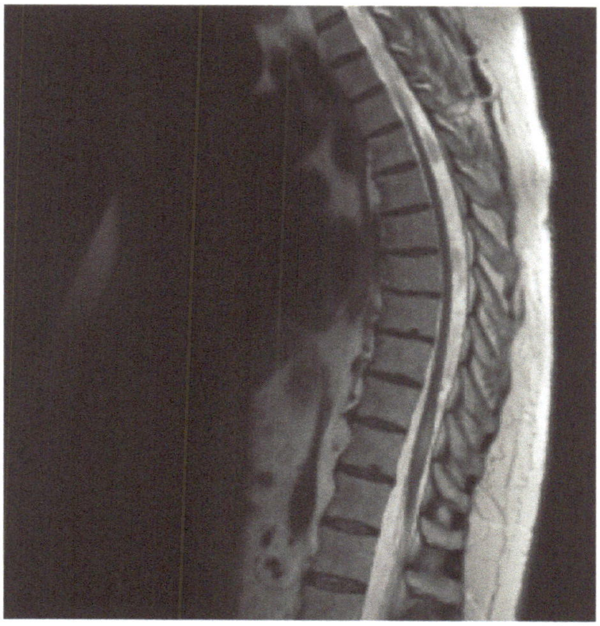

Fig. 19.16. Follow-up sagittal T2-weighted magnetic resonance image demonstrates resolution of dorsal flow voids and near resolution of medullary T2-hyperintense expansion. Patient endorsed significant symptomatic improvement.

SUMMARY AND CONCLUSION

Spinal dAVFs are rare lesions that pose an interesting and problematic clinical scenario. The clinical presentation is often vague, frequently leading to delays in imaging evaluation and diagnosis. The imaging characteristics are quite specific and can guide the treating physician to definitive therapy either by surgical or endovascular management. Clinical outcomes for both endovascular and surgical management are quite good, with surgical management appearing to be the more effective of the two in the long term. Still, due to its relatively low morbidity and short recovery time, angiographic embolization has become first-line therapy at many institutions. Embolic therapy is aimed at embolization of the inflow artery, the fistulous connection, and the proximal outflow vein. Liquid embolic therapies are preferred over coil due to better long-term durability and lower recurrence rates. Clinical improvement is best in motor and gait disturbance with less reliable improvement in sensory, bowel, and bladder disturbances. Clinical improvement is dependent on timely, definitive diagnosis and treatment.

ACKNOWLEDGMENTS

The authors would like to acknowledge Dr. Tony P. Smith and Dr. Mary I.H. Cobb, whose ongoing work continues to contribute to the literature and improve the lives of patients with neurovascular conditions.

REFERENCES

Adamczyk P, Amar AP, Mack WJ et al. (2012). Recurrence of "cured" dural arteriovenous fistulas after Onyx embolization. Neurosurg Focus 32. E12.

Aminoff MJ, Logue V (1974). The prognosis of patients with spinal vascular malformations. Brain 97: 211–218.

Aminoff MJ, Barnard MR, Logue V (1974). The pathophysiology of spinal vascular malformations. J Neurol Sci 23 (2): 255–263.

Anson J, Spetzler R (1993). Spinal dural arteriovenous malformations. In: A Awad, D Barrow (Eds.), Dural arteriovenous malformations. Park Ridge, IL, American Association of Neurological Surgeons.

Barrow DL (1997). Spinal vascular malformations. In: KM Welch (Ed.), Primer on cerebrovascular diseases. Gulf Professional Publishing, Houston, TX.

Bi WL, Brown PA, Abolfotoh M et al. (2015). Utility of dynamic computed tomography angiography in the preoperative evaluation of skull base tumors. J Neurosurg 123: 1–8.

Blackburn SL, Kadkhodayan Y, Ray WZ et al. (2014). Onyx is associated with poor venous penetration in the treatment of spinal dural arteriovenous fistulas. J Neurointerv Surg 6: 536–540.

Brinjiki W, Nasr D, Morris J et al. (2016). Clinical outcomes of patients with delayed diagnosis of spinal dural arteriovenous fistulas. AJNR Am J Neuroradiol 37: 380–386.

Cao JB, Cui LL, Jiang XY et al. (2014). Clinical application and diagnostic value of noninvasive spinal angiography in spinal vascular malformations. J Comput Assist Tomogr 38: 474–479.

Dabus G, Linfante I, Martinez-Galdamez M (2014). Endovascular treatment of dural arteriovenous fistulas using dual lumen balloon microcatheter: technical aspects and results. Clin Neurol Neurosurg 117: 22–27.

Farb RI, Kim JK, Willinsky RA et al. (2002). Spinal dural arteriovenous fistula localization with a technique of first-pass gadolinium-enhanced MR angiography: initial experience. Radiology 222: 843–850.

Gilbertson JR, Miller GM, Goldman MS et al. (1995). Spinal dural arteriovenous fistulas: MR and myelographic findings. AJNR Am J Neuroradiol 16: 2049–2057.

Guimaraes M, Wooster M (2011). Onyx (ethylene-vinyl alcohol copolymer) in peripheral applications. Semin Intervent Radiol 28: 350–356.

Hurst RW, Grossman RI (2000). Peripheral spinal cord hypointensity on T2-weighted MR images: a reliable imaging sign of venous hypertensive myelopathy. AJNR Am J Neuroradiol 21: 781–786.

Jellema K, Tijssen CC, van Rooij WJ et al. (2004). Spinal dural arteriovenous fistulas: long-term follow-up of 44 treated patients. Neurology 62: 1839–1841.

Jellema K, Tijssen CC, van Gijn J (2006). Spinal dural arteriovenous fistulas: a congestive myelopathy that initially mimics a peripheral nerve disorder. Brain 129: 3150–3164.

Jeng Y, Chen DY, Hsu HL et al. (2015). Spinal dural arteriovenous fistula: imaging features and its mimics. Korean J Radiol 16: 1119–1131.

Kai Y, Hamada J, Morioka M et al. (2005). Arteriovenous fistulas at the cervicomedullary junction presenting with subarachnoid hemorrhage: six case reports with special reference to the angiographic pattern of venous drainage. AJNR Am J Neuroradiol 26: 1949–1954.

Kaufmann TJ, Morris JM, Saladino A et al. (2011). Magnetic resonance imaging findings in treated spinal dural arteriovenous fistulas: lack of correlation with clinical outcomes. J Neurosurg Spine 14: 548–554.

Kendall BE, Logue V (1977). Spinal epidural angiomatous malformations draining into intrathecal veins. Neuroradiology 13: 181–189.

Krings T, Geibprasert S (2009). Spinal dural arteriovenous fistulas. AJNR Am J Neuroradiol 30: 639–648.

Lindenholz A, TerBrugge KG, van Dijk JM et al. (2014). The accuracy and utility of contrast-enhanced MR angiography for localization of spinal dural arteriovenous fistulas: the Toronto experience. Eur Radiol 24: 2885–2894.

Niimi Y, Sala F, Deletis V et al. (2004). Neurophysiologic monitoring and pharmacologic provocative testing for embolization of spinal cord arteriovenous malformations. AJNR Am J Neuroradiol 25: 1131–1138.

Nogueira R, Dabus G, Rabinov J et al. (2009). Onyx embolization for the treatment of spinal dural arteriovenous fistulae: initial experience with long-term follow-up. Technical case report. Neurosurgery 64: E197–E198.

Oda S, Utsunomiya D, Hirai T et al. (2014). Comparison of dynamic contrast-enhanced 3 T MR and 64-row multidetector CT angiography for the localization of spinal dural arteriovenous fistulas. AJNR Am J Neuroradiol 35: 407–412.

Ripley BA, Tiffany D, Lehmann LS et al. (2015). Improving the informed consent conversation: a standardized checklist that is patient centered, quality driven, and legally sound. J Vasc Interv Radiol 26: 1639–1646.

Saindane AM, Boddu SR, Tong FC et al. (2015). Contrast-enhanced time-resolved MRA for pre-angiographic evaluation of suspected spinal dural arterial venous fistulas. J Neurointerv Surg 7: 135–140.

Sala F, Beltramello A, Gerosa M (2007). Neuroprotective role of neurophysiological monitoring during endovascular procedures in the brain and spinal cord. Neurophysiol Clin 37: 415–421.

Saraf-Lavi E, Bowen BC, Quencer RM et al. (2002). Detection of spinal dural arteriovenous fistulae with MR imaging and contrast-enhanced MR angiography: sensitivity, specificity, and prediction of vertebral level. AJNR Am J Neuroradiol 23: 858–867.

Spetzler RF, Detwiler PW, Riina HA et al. (2002). Modified classification of spinal cord vascular lesions. J Neurosurg 96: 145–156.

Steinmetz MP, Chow MM, Krishnaney AA et al. (2004). Outcome after the treatment of spinal dural arteriovenous fistulae: a contemporary single-institution series and meta-analysis. Neurosurgery 55: 77–87; discussion 87–88.

Veznedaroglu E, Nelson PK, Jabbour PM et al. (2006). Endovascular treatment of spinal cord arteriovenous malformations. Neurosurgery 59: S202–S209; discussion S203–S213.

Yamaguchi S, Takeda M, Mitsuhara T et al. (2013). Application of 4D-CTA using 320-row area detector computed tomography on spinal arteriovenous fistulae: initial experience. Neurosurg Rev 36: 289–296; discussion 296.

Yoshioka K, Niinuma H, Ehara S et al. (2006). MR angiography and CT angiography of the artery of Adamkie-wicz: state of the art. Radiographics 26 (Suppl 1): S63–S73.

Handbook of Clinical Neurology, Vol. 143 (3rd series)
Arteriovenous and Cavernous Malformations
R.F. Spetzler, K. Moon, and R.O. Almefty, Editors
http://dx.doi.org/10.1016/B978-0-444-63640-9.00020-5

Chapter 20

Spinal subarachnoid hemorrhage and aneurysms

TANMOY K. MAITI, SHYAMAL C. BIR, AND ANIL NANDA*
Department of Neurosurgery, Louisiana State University Health Sciences Center, Shreveport, LA, USA

Abstract

Spinal subarachnoid hemorrhage (SAH) is a rare disease. Spinal aneurysms are even rarer and mostly undetected unless they rupture and become symptomatic. In this chapter we aim to review the available literature about spinal subarachnoid hematoma with special emphasis on spinal aneurysms. As most reports of spinal aneurysms describe a single case or a small case series, the diagnostic algorithm is often lacking. The outcome is also different based on the etiologies; therefore management strategy must be individualized. We addressed these issues in this chapter. The reported incidence of spinal SAH is less than epidural hematoma and more than subdural hematoma. Spinal aneurysms can present as isolated entity or can be associated with other vascular anomalies. Microsurgical clipping and/or resection is possible, especially when they are located dorsally or dorsolaterally. Endovascular approach is also a feasible option unless negotiation of microcatheter becomes difficult in tortuous small-caliber arteries. Successful obliteration leads to good outcome, especially when present in posterior spinal artery. A detailed knowledge of spinal SAH and spinal aneurysms is important to detect them in time. Clinicians must consider several factors to choose an appropriate treatment strategy to ensure the safety of their patients.

INTRODUCTION

Spinal subarachnoid hemorrhage (SAH) constitutes about 0.05–1.5% of all cases of SAH (Komiyama et al., 1997) and 15% of all spinal hematomas (Kreppel et al., 2003).

The relatively low incidence, confusion over nomenclature, wide spectrum of clinical manifestations, and association with various diseases make spinal SAH a unique entity. Although the spinal SAH causing spinal cord compression is relatively low, clinicians must be aware of this possibility, as favorable preoperative clinical status and early intervention are critical for successful recovery (Sunada et al., 1995). Spinal aneurysm as a cause of spinal SAH is even rarer and different from its cranial counterparts in several ways. Most reports in the literature are single case reports. The majority of the reported cases were mostly diagnosed only after they became symptomatic. Therefore the epidemiology, appropriate management strategy, obliteration rate, and outcome are largely unknown.

HISTORY AND NOMENCLATURE

Spinal hematoma with "spinal apoplexy" was first described by Duverney in 1682 in an autopsy study (Plagne, 1961; Kreppel et al., 2003). Jackson is credited with the first clinical description of spinal hematoma in 1869. There is substantial controversy in the literature about who first described spinal SAH or spinal aneurysm. Gaupp (1888), Balck (1900), Harris (1911), Hammes (1920), and Slavin (1937) are often credited for initial descriptions of spinal SAH. However, Plotkin et al. (1966) were first to describe the anatomic features and surgical removal of a spinal SAH in 1966. Berney et al. introduced the term spinal subarachnoid hematoma in 1967. They believed it is extremely rare for an SAH to clot and therefore cause spinal cord compression. Though the two terms spinal SAH and spinal subarachnoid hematoma have been used interchangeably in most reports, few authors believe them to be distinct entities (Komiyama et al., 1997). According to them,

*Correspondence to: Anil Nanda, MD, MPH, FACS, Department of Neurosurgery, Louisiana State University Health Sciences Center – Shreveport, 1501 Kings Hwy, Shreveport LA 71130, USA. Tel: +1-318-675-6404, Fax: +1-318-675-7111, E-mail: ananda@lsuhsc.edu

subarachnoid hematoma forms a localized clot in subarachnoid space and therefore can compress the cord. In contrast, SAH denotes a diffuse hemorrhage in subarachnoid space.

Babonneix and Wediez (M Babonneix, 1930; Rengachary et al., 1993) described one of the first known cases of spinal aneurysm in a syphilis patient. Echols and Holcombe are also credited with one of the initial description of spinal aneurysm in 1941 (Madhugiri et al., 2013). Vincent reported the first angiographically confirmed rupture of an anterior spinal artery aneurysm in 1981.

ETIOLOGY OF SAH

Kreppel et al. (2003) found that spinal SAH has been reported in patients aged 3–83 years with peak incidence between 15 and 20 years. In their analysis, spinal SAH was found to be more common in males (M:F = 2:1).

A wide spectrum of conditions can lead to spinal subarachnoid hematoma/SAH. Tumor-related hemorrhage and vascular malformation are the two most common causes of SAH (Kreppel et al., 2003). Kreppel et al. (2003) suggested idiopathic spinal subarachnoid represents only the fifth common cause, and it is usually the diagnosis of exclusion (Moore et al., 2015). Most of the other causes are listed in Table 20.1 (Nassar and Correll, 1968; Djindjian, 1978b; Kito et al., 1983; Kreppel et al., 2003; Klingler et al., 2009; Nakhla et al., 2016). The major causes are discussed in next section.

Multiple reports suggest that less than 1% of SAHs result from rupture of spinal aneurysms. Spinal aneurysms are slightly more prevalent in female (1.11:1) with mean age of 38.1 ± 21.5 years (Madhugiri et al., 2013). Multiplicity is rare, with an incidence of only 6.25% according to Madhugiri et al. (2013).

Table 20.1

Etiology of spinal subarachnoid hemorrhage

Tumor hemorrhage
- Ependymoma
- Schwann cell tumors (neurofibroma, schwannoma)
- Meningeal tumors (meningioma, meningosarcoma)
- Gliomas (astrocytoma, neurospongioma)
- Hemangioblastoma
- Unspecified

Vascular
- Vascular malformation (arteriovenous malformation, arteriovenous fistula, moyamoya disease)
- Spinal aneurysm
- Coarctation of aorta

Anticoagulant therapy

Trauma, lumbar puncture, spinal anesthesia

Spontaneous with or without trivial trauma or minor exertion (urination, defecation, vomiting, exercise, turning over in bed, coughing, sexual intercourse)

Systemic
- Infective: syphilis, pertussis
- Pregnancy
- Vasculitis: Behçet disease, systemic lupus erythematosus, Sjögren syndrome, polyarteritis nodosa
- Connective tissue disorders: Ehlers–Danlos syndrome, Marfan syndrome, pseudoxanthoma elasticum, fibromuscular dysplasia
- Alcohol abuse
- Associated with spine pathology: spondyloarthrotic processes, arachnoiditis, disk herniation, hypertrophy of the yellow ligaments

Extension of bleeding from subdural space

Spinal tumors

In a report in 1978, Djindjian et al. described approximately 50 cases where spinal SAH was associated with tumor hemorrhage. Ependymomawas found to be the most common cause (Nassar and Correll, 1968; Dijindjian et al., 1978). Ulrich et al. (2008) tabulated 18 cases of spinal ependymoma with SAH. Interestingly, all except one case were located at the lumbar level. The next common causes are neurofibroma, astrocytoma, and meningioma. We reported a case of spinal schwannoma causing cranial and spinal SAH (Kukreja et al., 2014) (Fig. 20.1).

Spinal vascular malformations

Aminoff and Logue (1974a, b) are credited with publishing one of the first large clinical studies with a detailed description of symptoms and established a famous clinical scale for spinal arteriovenous malformation (AVM). Henson and Croft (1956) suggested spinal angiomas are the most common cause of spinal subarachnoid hematomas. Fody et al. (1980) echoed this opinion as well. They found the incidence to be 13%, with a female predominance. Djindjian (1978a) and Swann et al. (1984) reported AVM as the most common cause of SAH. Pia (1978a, b) mentions rupture of spinal angiomas in 50% of arteriovenous forms and 30% of capillary angiomas. Pia (1978a, b) postulated that angiomas lead to repeated small hemorrhages and arachnoiditis causing spinal compression rather than acute hemorrhage. Another report suggests that spinal AVM can lead to spinal SAH in 6% of cases, but the incidence rises to as high as 75–90% when they are associated with aneurysms.

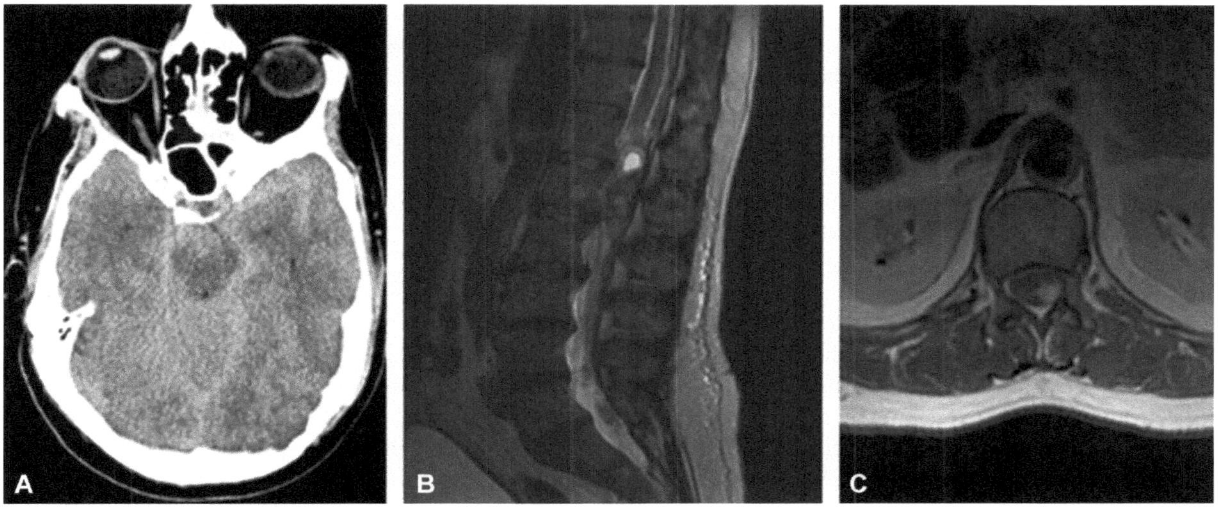

Fig. 20.1. (**A–C**) A 47-year-old patient had presented to us with features suggestive of cranial vascular event. The details of this patient have been reported earlier. To summarize, the initial imaging documented subarachnoid hemorrhage (SAH), but the source of SAH was undetected. Spinal magnetic resonance imaging was sought after the patient developed back and leg pain 3 months after initial symptoms. A spinal tumor with bleed was detected. The patient underwent successful surgical resection of tumor and had an uneventful recovery. The tumor was histopathologically confirmed as schwannoma.

Anticoagulation therapy

Anticoagulation therapy, especially in combination with spinal anesthesia, can cause spinal arachnoid hemorrhage. A time interval of 1–6 hours has been suggested for initiation of spinal anesthesia after discontinuation of anticoagulation (Wille-Jorgensen et al., 1991; Vandermeulen et al., 1994). An interval of 1–12 hours has been suggested for initiation of anticoagulation following spinal anesthesia (Wille-Jorgensen et al., 1991; Vandermeulen et al., 1994; Kreppel et al., 2003).

Spinal aneurysms

Though the spinal aneurysm is not a common cause of spinal SAH, the true incidence may be underestimated. Spinal aneurysms are most prevalent in the anterior spinal artery. The posterior spinal artery (PSA) and artery of Adamkiewz are the next two common locations.

PATHOGENESIS OF SPINAL SAH AND ANEURYSMS

The pathogenesis of spontaneous subarachnoid is unknown, as the name suggests. One proposed mechanism suggests a minor trauma as initial event which causes rapid changes in intrathoracic and intra-abdominal pressure. The resulting increased intraluminal pressure causes subsequent tearing of vessels within subarachnoid space. Mechanical factors like spondylosis, disc herniation, arachnoiditis, or thickening of yellow ligament can also contribute to the pathogenesis (Domenicucci et al., 2005). Cerebrospinal fluid (CSF) may dilute a subarachnoid hematoma. CSF can also act as fibrinolytic agent (Gambacorta et al., 1987). Further, defibrination by pulsation of the spinal cord reduces the chance of formation of a clot. Masdeu et al. (1979) suggested the bleeding is initially from radicular vein or artery. Though it is initially limited to subarachnoid space, it can soon extend to subdural space. Gambacorta et al. (1987) postulated the symptoms can be acute, subacute, or even chronic (Harris, 1911; Khosla et al., 1985; Gambacorta et al., 1987) based on associated pathology and progression.

On the other hand, inflammatory processes, intimal defects, and arterial dissections are described as initiating factors for the development of spinal aneurysms (Kito et al., 1983; Berlis et al., 2005; Sato et al., 2012). They are mostly associated with vascular malformations (AVM, arteriovenous fistula, moyamoya disease (Walz et al., 2006)), coarctation (Jiarakongmun et al., 2002; Aoun et al., 2014), and vasculitis (Massand et al., 2005). Most of the risk factors are associated with intracranial aneurysms as well except Chiari 1 malformation (Pollock et al., 2009). However association with autosomal-dominant polycystic kidney disease is not known. Spinal aneurysms differ from intracranial aneurysms in various ways (Massand et al., 2005). They are mostly flow-related but can be congenital like their cranial counterparts (Rengachary et al., 1993). They occur in the course of an artery rather than at its branching points. They are smaller in caliber than their cranial counterparts and less affected by atherosclerosis (Rengachary et al., 1993). Most of the spinal aneurysms have a diameter of less than 3 mm. Fusiform aneurysms are more common than saccular aneurysms. Dissection aneurysms are common as well. Partial thrombosis is noted in a few cases. Most of the aneurysms occur in the anterior spinal artery. The PSA is the second most common location. Kim and Choi (2012) identified eight PSA aneurysms described in the literature prior to 2012, all dissecting in nature, while another recent report identified 13 PSA aneurysms (including their own case) (Johnson et al., 2015).

Kurita et al. (2009) documented a fourth case of lateral spinal artery (LSA) aneurysm. On the first three occasions there were occlusions or stenosis of posterior inferior cerebellar artery or vertebral artery, from which the LSA arises. This led to increased hemodynamic stress, dilation, and aneurysm formation of LSA, which is a tiny perforating vessel otherwise.

Son et al. (2013) noticed aneurysms in the Adamkiewicz artery; such aneurysms have been described only seven times in the literature (including their own case description).

Nogueira et al. (2010) observed that all 4 cases of spinal lateral sacral artery aneurysms were associated with renal transplantation and occurred behind S1 vertebral body. High flow from transpelvic collaterals between both internal iliac arteries was the probable reason behind their formation.

CLINICAL FEATURES: RELEVANT POINTS

Obviously the clinical features depend on the level of lesion and associated pathology. However, clinicians must be aware of a few key differentiating features which distinguish spinal subarachnoid hematoma from epidural or subdural hematoma. For example, Henson and Croft (1956) considered constellation of sudden back pain, radicular pain, and meningeal irritation as pathognomonic of spinal SAH. Swann et al. (1984) noticed absence of spinal compression and cerebral manifestations with meningitic symptoms, impaired consciousness, and epileptiform seizures for these lesions. Caroscio et al. (1980) documented cranial symptoms in 80% of cases with spinal SAH. Fincher (1951) proposed that, in contrast to epidural or subdural hematoma, spinal SAH results in severe radiating, sciatiform pain which may lead to opisthotonos.

Tumor and vascular malformations are common at lumbar and cervical level respectively (Kreppel et al., 2003). Spontaneous and traumatic hematoma tends to occur at the thoracic and cervicothoracic level (Kreppel et al., 2003). Spinal anesthesia and anticoagulation-related hematoma are more prevalent at the lumbar level (Kreppel et al., 2003). Spinal aneurysms generally occur in the upper cervical area beyond the C3 and thoracolumbar junction (Kim and Choi, 2012; Madhugiri et al., 2013).

The hematoma can be ventral and dorsal as well, depending on its relation to the spinal cord (Komiyama et al., 1997).

Spine aneurysm causing SAH

Spine aneurysms are primarily diagnosed only in symptomatic patients. The natural history is largely unknown, and the available literature may represent only the tip of the iceberg. Most of these aneurysms are only diagnosed after hemorrhage. The hemorrhage can lead to spinal and/or cranial symptoms. The possibilities of cranial symptoms are more common when the pathology is higher along the spinal column. Madhugiri et al. (2013) found the hemorrhage rate is higher in isolated spinal aneurysms than when they are associated with other vascular malformations. This may be due to the fact that isolated spinal aneurysms are only diagnosed when they present with hemorrhage. On the other hand, spinal aneurysm with vascular malformation may present with compressive symptoms as well.

The hemorrhage rate is 92.2% in isolated spinal aneurysmand 79.2% in spinal aneurysm-AVM, whereas it is 70% when there is a spinal AVM without spinal aneurysm (Madhugiri et al., 2013). Intranidal spinal aneurysms were found to bleed more than those with spinal aneurysm AVMs (Madhugiri et al., 2013). Older patients were found to present with isolated spinal aneurysm and hemorrhage than the younger population (Madhugiri et al., 2013).

Madhugiri et al. (2013) noted the presence of cranial SAH in one-third of their cases. Though most of them were from a cervical spinal aneurysm, a thoracic or thoracolumbar aneurysm was also responsible for a few of the cases. An associated cranial aneurysm was rather rare. Banna et al. (1973) described a 40-year-old male with coarctation of the aorta, where a 1-mm diameter unruptured posterior cerebral artery aneurysm was noted along with a spinal aneurysm with craniospinal SAH. Therefore, a patient with angiography-negative cranial SAH must be examined for spinal digital subtraction angiogram, especially when SAH is most prominent in the fourth ventricle or posterior fossa.

The aneurysmal bleed may also present in any plane, such as intramedullary, subdural, or in multiple planes. Hematomyelia, i.e., blood within the cord, results in a dismal prognosis.

The spinal aneurysm may sometimes be detected without bleed as well. Children younger than 5 years may present with persistent bed wetting, failure to thrive, or breath-holding spells (Binder et al., 1982;

Table 20.2

Classification of spinal aneurysm

Type 1	Spinal aneurysm associated with vascular malformation (AVM/AVF)
1A	Intramedullary AVM/AVF (1A fv: feeding vessel, 1A n: within the nidus)
1B	Extramedullary AVM
Type 2	Isolated spinal aneurysm
Type 3	Dissecting/traumatic/pseudoaneurysm

AVM, arteriovenous malformation; AVF, arteriovenous fistula.

Madhugiri et al., 2013). Features suggestive of spinal tethering (such as scoliosis, asymmetric wasting, and patchy sensory loss) may be presenting symptoms for older children (Scarff and Reigel, 1979). Understandably the cervical and thoracic spinal artery AVM presents with compressive myelopathy, and lumbar and lumbosacral lesions present with cauda equina syndrome in most cases (Madhugiri et al., 2013). Identification of isolated aneurysms without hemorrhage was rare (<0.5%) and they were at least 1.5 cm when detected as they produced compression symptoms like backache and radicular pain (Madhugiri et al., 2013). Madhugiri et al. (2013) postulated a classification of spinal aneurysms based on etiopathogenesis (Table 20.2).

Spinal SAH with intracranial extension

Spinal SAH can extend into the cranial compartment. Nam et al. (2014) found 10 such cases in the literature. However, the incidence of vasospasm is less in comparison to cranial SAH. Espinosa-Aguilar et al. (2012) documented vasospasm with transcranial Doppler and cerebral angiography in one of their patients. Nam et al. (2014) shared their experience with a patient whose vasospasm led to multiple cerebral infarctions. Therefore, the spine surgeon should consider this possibility if a patient develops delayed neurodeficit.

DIAGNOSIS

Neuroradiologic diagnosis of spinal subarachnoid hematoma/SAH is extremely difficult. Most of the earlier cases were diagnosed at autopsy or after surgical exploration.

Frager et al. (1982) described a characterized filling defect in myelography, called capping. Gambacorta et al. (1987) described a case of chronic spinal subarachnoid hematoma, where this capping was particularly noticeable. However, Komiyama et al. (1997) doubted the utility of myelography, as it gives a false impression of the epidural mass or intraparenchymal mass. As the hematoma is located within the bony canal, precise sagittal location of hematoma is less likely with computed tomography (CT). CT myelography fails to provide a clear picture as well because the contrast material does not permeate the cleavage between the spinal cord and the hematoma. CT angiography may provide better information depicting the relation between the aneurysm and the spinal canal (Gutierrez Romero et al., 2014).

Based on the above finding, magnetic resonance imaging (MRI) seems to be the initial diagnostic modality of choice. The SAH in acute stage is hyperintense or isointense on T1 and hyperintense or hypointense on T2 MRI. In the subacute stage, T2 MRI reveals a hyperintense signal reflecting the presence of strongly paramagnetic methemoglobin. Towards the end of the subacute phase, the signal intensity is reduced further because of deposition of hemosiderin in the periphery. Domenicucci et al. (2005) described a case where hematoma was surrounded by cerebrospinal fluid, separating it from the internal surface of dura mater and thus distinguishing it from subdural hematoma. However, when the amount of blood is more, the identification of blood–arachnoid interface is difficult. Identification becomes more difficult when the hematoma is located ventral to the spinal cord.

Similarly, blood in the cerebrospinal fluid is also not specific for spinal subarachnoid hematoma. It can occur in approximately 13% of spinal subdural hemorrhages according to Domenicucci et al. (2005).

A spinal angiogram is required to establish the cause of SAH. A cranial angiogram may be required in the presence of cranial symptoms. It is advisable to perform a cranial angiogram for a cervical vascular malformation and a spinal angiogram when the cranial angiogram does not reveal any obvious cause of posterior fossa SAH.

Multiple reports suggest that if surgery is delayed for any reason or if emergency surgery is not required, a repeat angiogram must be performed to rule out any spontaneous resolution (Berlis et al., 2005; van Es et al., 2013).

TREATMENT

Spinal SAH may warrant emergency exploration to prevent irreversible damage. However, ventral SAH may be managed successfully without any surgical intervention (Komiyama et al., 1997). Conversely, dorsal SAH may need exploration, but Ruelle et al. (2001) contradicted this. Close observation and an individual approach are warranted.

Spinal aneurysms have been successfully managed by surgical as well as endovascular approaches. In a report, 2 of the 3 patients who underwent conservative treatment for PSA aneurysm had a fatal rebleed

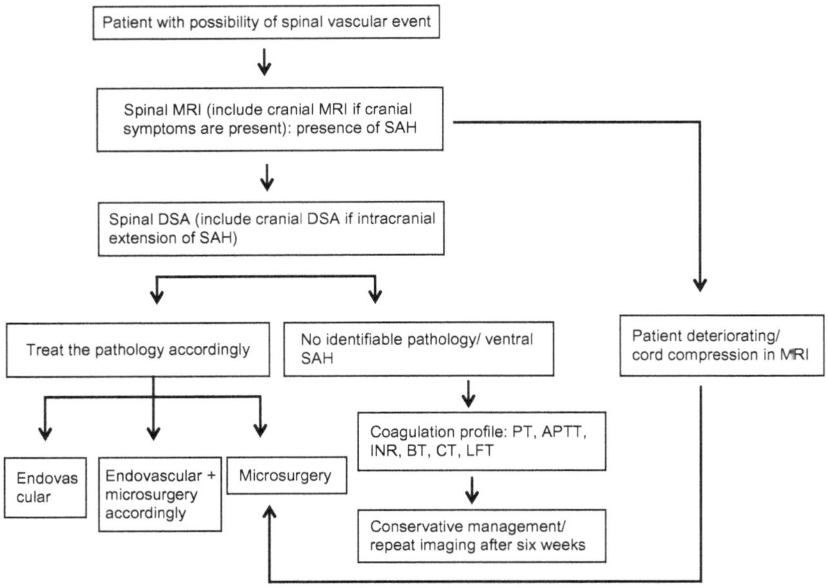

Fig. 20.2. Management algorithm for management of spinal subarachnoid hemorrhage (SAH) and aneurysm. MRI, magnetic resonance imaging; DSA, digital subtraction angiography; PT, prothrombin time; APTT, activated partial thromboplastin time; INR, international normalized ratio; BT, bleeding time; CT, clotting time; LFT, liver function test.

(van Es et al., 2013). However, the obliteration of associated AVM/arteriovenous fistula may lead to obliteration or regression of spinal aneurysm. Resection of spinal aneurysm is often preferred if associated with AVM. Clipping of aneurysm may be performed in isolated spinal aneurysms. Aneurysms present in small radicular and radiculopial branches may be sacrificed without any sequelae. The approach depends on the location of aneurysm.

The endovascular therapy seems to have worse obliteration rate than microsurgery in patients with spinal aneurysm AVM. Therefore surgery only or surgery with endovascular therapy seems to be the treatment of choice for spinal aneurysm AVMs. On the other hand, either modality of treatment may successfully obliterate the isolated aneurysms. The choice of treatment should be based on the level of the lesion and availability of expertise. A management algorithm has been proposed in Figure 20.2.

Endovascular therapy for spinal aneurysms: technical points

Direct coiling is often adequate for spinal aneurysms. However, it may be technically more difficult to treat spinal aneurysms in comparison to cranial aneurysms. Catheterization may be difficult for smaller spinal arteries with lower flow rate in comparison to cranial vessels. A large coil mass can potentially lead to compressive symptoms in the long term, especially at a higher spinal level. Cannulation of femoral artery may be difficult in cases with coarctation and the brachial artery may need to be cannulated.

Surgery for spinal aneurysms: technical points

Surgical approach is more commonly employed in posteriorly localized aneurysms due to their dorsolateral and superficial location (Geibprasert et al., 2010; Gutierrez Romero et al., 2014). The mass effect of the aneurysm or the blood clot may be an indication for microsurgery (Son et al., 2013). If the aneurysm is fusiform, direct clipping may not be possible. As PSA aneurysms are often small and without a neck, preservation of the parent vessel is usually not possible. Typically, a surgical resection is performed with favorable clinical outcome. In contrast, preservation of anterior spinal artery and artery of Adamkiewicz must be ensured (Son et al., 2013), to avoid devastating complications. Again, a small parent vessel may be sacrificed if there is no flow distal to the aneurysm. If there is evidence of distal blood flow at surgery the aneurysm must be wrapped with clip reinforcement when it is not amenable for direct clipping (Vishteh et al., 1997). When the bulk of the aneurysm makes wrapping impossible, microvascular reconstruction/vascular bypass can be attempted.

The position of the aneurysm with respect to the spinal cord is a key determinant of the surgical approach. For aneurysms located anterolateral or posterior to the spinal cord, traditional posterior laminotomy or laminoplasty can be performed (Gonzalez et al., 2005;

Son et al., 2013). Sectioning the dentate ligament improves exposure from a posterior approach and also can be used to retract the spinal cord (Son et al., 2013). If the aneurysm is located in the midline, anterior to the spinal cord in the lower thoracic region, an anterior approach can be used to avoid traction. Intraoperative monitoring such as somatosensory and motor evoked potentials can be a useful tool in surgery (Son et al., 2013).

Medical therapy for spinal aneurysms

Spontaneous regression of dissecting aneurysm is possible. Iihoshi et al. (2011) described one such case in a 60-year-old female. Spinal aneurysms associated with vasculitis or infection may regress with optimum treatment (such as immune modulation, steroid or antibiotic therapy) of the primary cause. A careful nonsurgical treatment is adequate for these patients. Komiyama et al. (1997) postulated that the ventral SAH often leads to a benign course. Therefore, careful observation is enough in most of the ventral but not dorsal SAH cases.

OUTCOME

Most of the reported cases of spinal SAH and aneurysms are discrete case reports without adequate follow-up period. Therefore, the natural history and rate of rebleed for spinal aneurysms are largely unknown. Komiyama et al. (1997) suggested dorsal, but not ventral, SAH has symptomatic compression effects. Domenicucci et al. (2005) observed that patients with combined spinal SAH and subdural hematoma had worst prognosis, whereas patients with isolated dorsal SAH had better outcome. Ruelle et al. (2001) described two cases of spinal SAH, one with a ventral location and the other with a dorsal location. Patients were treated conservatively. They did not experience any neurologic deterioration or recurrence in the next 2 years (Ruelle et al., 2001). Komiyama et al. (1997) suggested that if a hematoma comes through an acute stage, recurrence is less likely. However, in the presence of vascular malformation, multiple repeated bleeding is the norm, and this mandates a definite treatment at earliest.

Madhugiri et al. (2013) found that spinal aneurysms with or without vascular malformation had similar prognosis. However, in spinal aneurysm patients with vascular malformations, the prognosis depends on the type of associated anomaly. As the prognosis for intramedullary AVM is worse in comparison to other vascular malformations, the outcome is worse when spinal aneurysm is associated with them. Patients with poor neurologic status (such as myelopathy, bladder involvement, weakness, or sensory dysfunction) had poorer outcome regardless of type of spinal aneurysm. As spinal aneurysms presenting with hemorrhage are more likely to have neurodeficit at presentation, they are more prone to have worse outcome. Cervical aneurysms have worse prognosis than thoracolumbar aneurysms (Kim and Choi, 2012). Spinal cord infarction constitutes the most dreaded complication after both surgical and endovascular treatments (Gutierrez Romero et al., 2014).

CONCLUSION

Spinal subarachnoid hematoma can be multifactorial. Diagnosis of cause is important in formulating the management plan. Spinal aneurysms are usually not diagnosed until after they rupture. The clinical features, management, and outcome depend on associated pathology and location. There is scarce existing literature on these entities, and a careful individualistic approach should be contemplated when they are diagnosed. The etiology and location of the lesion and distal flow to the lesion should be considered when planning treatment.

References

Aminoff MJ, Logue V (1974a). Clinical features of spinal vascular malformations. Brain 97: 197–210.

Aminoff MJ, Logue V (1974b). The prognosis of patients with spinal vascular malformations. Brain 97: 211–218.

Aoun SG, El Ahmadieh TY, Soltanolkotabi M et al. (2014). Ruptured spinal artery aneurysm associated with coarctation of the aorta. World Neurosurg 81 (441): e417–e422.

Banna MM, Rose PG, Pearce GW (1973). Coarctation of the aorta as a cause of spinal subarachnoid hemorrhage. Case report. J Neurosurg 39: 761–763.

Berlis A, Scheufler KM, Schmahl C et al. (2005). Solitary spinal artery aneurysms as a rare source of spinal subarachnoid hemorrhage: potential etiology and treatment strategy. AJNR Am J Neuroradiol 26: 405–410.

Berney MJ, Gauthier G, Werner A (1967). Acute myeloradicular compression by a spontaneous subarachnoid hematoma. Neurochirurgie 13: 771–773.

Binder B, Eng GD, Milhorat TH et al. (1982). Spinal arteriovenous malformations in an infant: unusual symptomology and pathology. Dev Med Child Neurol 24: 380–385.

Caroscio JT, Brannan T, Budabin M et al. (1980). Subarachnoid hemorrhage secondary to spinal arteriovenous malformation and aneurysm. Report of a case and review of the literature. Arch Neurol 37: 101–103.

Djindjian R (1978a). Angiography in angiomas of the spinal cord. In: RD Hw Pia (Ed.), Spinal angiomas. Advances in diagnosis and therapy. Springer, Berlin.

Djindjian R (1978b). Treatment of spinal angiomas by embolization. In: RD Hw Pia (Ed.), Spinal angiomas. Advances in diagnosis and therapy. Springer, Berlin.

Dijindjian M, Djindjian R, Houdart R et al. (1978). Subarachnoid hemorrhage due to intraspinal tumors. Surg Neurol 9: 223–229.

Domenicucci M, Ramieri A, Paolini S et al. (2005). Spinal subarachnoid hematomas: our experience and literature review. Acta Neurochir (Wien) 147: 741–750; discussion 750.

Echols D, Holcombe RJ (1941). Extramedullary aneurysm of the spinal cord. New Orleans Med Surg J 93: 582–583.

Espinosa-Aguilar M, Hernandez-Palazon J, Fuentes-Garcia D (2012). Intraspinal and intracranial subarachnoid haemorrhage with severe cerebral vasospasm after spinal anaesthesia for assisted delivery. Br J Anaesth 108: 885–886.

Fincher EF (1951). Spontaneous subarachnoid hemorrhage in intradural tumors of the lumbar sac; a clinical syndrome. J Neurosurg 8: 576–584.

Fody EP, Netsky MG, Mrak RE (1980). Subarachnoid spinal hemorrhage in a case of systemic lupus erythematosus. Arch Neurol 37: 173–174.

Frager D, Zimmerman RD, Wisoff HS et al. (1982). Spinal subarachnoid hematoma. AJNR Am J Neuroradiol 3: 77–79.

Gambacorta D, Reale F, De Falco D (1987). Spontaneous chronic spinal subarachnoid hematoma. Report of a case and review of the literature. Spine (Phila Pa 1976) 12: 716–718.

Geibprasert S, Krings T, Apitzsch J et al. (2010). Subarachnoid hemorrhage following posterior spinal artery aneurysm. A case report and review of the literature. Interv Neuroradiol 16: 183–190.

Gonzalez LF, Zabramski JM, Tabrizi P et al. (2005). Spontaneous spinal subarachnoid hemorrhage secondary to spinal aneurysms: diagnosis and treatment paradigm. Neurosurgery 57: 1127–1131; discussion 1127–1131.

Gutierrez Romero D, Batista AL, Gentric JC et al. (2014). Ruptured isolated spinal artery aneurysms. Report of two cases and review of the literature. Interv Neuroradiol 20: 774–780.

Hammes E (1920). Hemorrhage in the cauda secondary to lumbar puncture. Arch Neurol Psychiat 3: 595.

Harris W (1911). Two cases of spontaneous haematorrhachisor intrameningeal spinal haemorrhage-one cured by laminectomy. Proc R Soc Med 5: 115–122.

Henson RA, Croft PB (1956). Spontaneous spinal subarachnoid haemorrhage. Q J Med 25: 53–66.

Iihoshi S, Miyata K, Murakami T et al. (2011). Dissection aneurysm of the radiculomedullary branch of the artery of Adamkiewicz with subarachnoid hemorrhage. Neurol Med Chir (Tokyo) 51: 649–652.

Jackson R (1869). Case of spinal apoplexy. Lancet 2: 5–6.

Jiarakongmun P, Chewit P, Pongpech S (2002). Ruptured anterior spinal artery aneurysm associated with coarctation of aorta. Case report and literature review. Interv Neuroradiol 8: 285–292.

Johnson J, Patel S, Saraf-Lavi E et al. (2015). Posterior spinal artery aneurysm rupture after 'Ecstasy' abuse. J Neurointerv Surg 7. e23.

Khosla VK, Kak VK, Mathuriya SN (1985). Chronic spinal subdural hematomas. Report of two cases. J Neurosurg 63: 636–639.

Kim HJ, Choi IS (2012). Dissecting aneurysm of the posterior spinal artery: case report and review of the literature. Neurosurgery 71: E749–E756; discussion E756.

Kito K, Kobayashi N, Mori N et al. (1983). Ruptured aneurysm of the anterior spinal artery associated with pseudoxanthoma elasticum. Case report. J Neurosurg 58: 126–128.

Klingler JH, Glasker S, Shah MJ et al. (2009). Rupture of a spinal artery aneurysm attributable to exacerbated Sjogren syndrome: case report. Neurosurgery 64: E1010–E1011; discussion E1011.

Komiyama M, Yasui T, Sumimoto T et al. (1997). Spontaneous spinal subarachnoid hematoma of unknown pathogenesis: case reports. Neurosurgery 41: 691–693; discussion 693–694.

Kreppel D, Antoniadis G, Seeling W (2003). Spinal hematoma: a literature survey with meta-analysis of 613 patients. Neurosurg Rev 26: 1–49.

Kukreja S, Ambekar S, Sharma M et al. (2014). Cauda equina schwannoma presenting with intratumoral hemorrhage and intracranial subarachnoid hemorrhage. J Neurosurg Spine 21: 357–360.

Kurita M, Endo M, Kitahara T et al. (2009). Subarachnoid haemorrhage due to a lateral spinal artery aneurysm misdiagnosed as a posterior inferior cerebellar artery aneurysm: case report and literature review. Acta Neurochir (Wien) 151: 165–169.

M Babonneix AW (1930). Sclerose combinee; lesions diffuses et inflammatoires du nevraxe. Anevrysme de l'artere spinale anterieure. Syphilis probable. Rev Neurol 1: 1214–1217.

Madhugiri VS, Ambekar S, Roopesh Kumar VR et al. (2013). Spinal aneurysms: clinicoradiological features and management paradigms. J Neurosurg Spine 19: 34–48.

Masdeu JC, Breuer AC, Schoene WC (1979). Spinal subarachnoid hematomas: clue to a source of bleeding in traumatic lumbar puncture. Neurology 29: 872–876.

Massand MG, Wallace RC, Gonzalez LF et al. (2005). Subarachnoid hemorrhage due to isolated spinal artery aneurysm in four patients. AJNR Am J Neuroradiol 26: 2415–2419.

Moore JM, Jithoo R, Hwang P (2015). Idiopathic spinal subarachnoid hemorrhage: a case report and review of the literature. Global Spine J 5: e59–e64.

Nakhla J, Nasser R, Yassari R et al. (2016). Ruptured anterior spinal artery aneurysm from a herniated cervical disc. A case report and review of the literature. Surg Neurol Int 7: 10.

Nam KH, Lee JI, Choi BK et al. (2014). Intracranial extension of spinal subarachnoid hematoma causing severe cerebral vasospasm. J Korean Neurosurg Soc 56: 527–530.

Nassar SI, Correll JW (1968). Subarachnoid hemorrhage due to spinal cord tumors. Neurology 18: 87–94.

Nogueira RG, Kasper E, Walcott BP et al. (2010). Lateral sacral artery aneurysm of the lumbar spine: hemorrhage resulting in cauda equina syndrome. J Neurointerv Surg 2: 399–401.

Pia H (1978a). Operative treatment of spinal angiomas. In: Hw Pia, R Djindjian (Eds.), Spinal angiomas. Advances in diagnosis and therapy. Springer, Berlin.

Pia H (1978b). Symptomatology of spinal angiomas. In: Hw Pia, R Djindjian (Eds.), Spinal angiomas. Advances in diagnosis and therapy. Springer, Berlin.

Plagne R (1961). L'hématome extra-dural rachidien non traumatique (hématome épidural spontané). Thèse pour le Doctorat en Médecine (Diplôme d'Etat). Université de Clermont, Faculté mixte de Médecine et de Pharmacie.

Plotkin R, Ronthal M, Froman C (1966). Spontaneous spinal subarachnoid haemorrhage. Report of 3 cases. J Neurosurg 25: 443–446.

Pollock JM, Powers AK, Stevens EA et al. (2009). Ruptured anterior spinal artery aneurysm: a case report. J Neuroimaging 19: 277–279.

Rengachary SS, Duke DA, Tsai FY et al. (1993). Spinal arterial aneurysm: case report. Neurosurgery 33: 125–129; discussion 129–130.

Ruelle A, Zerbi D, Andrioli G (2001). Spinal subarachnoid bleeding of unknown etiology. Case reports. J Neurosurg Sci 45: 53–57.

Sato K, Roccatagliata L, Depuydt S et al. (2012). Multiple aneurysms of thoracic spinal cord arteries presenting with spinal infarction and subarachnoid hemorrhage: case report and literature review. Neurosurgery 71: E1053–E1058; discussion E1058.

Scarff TB, Reigel DH (1979). Arteriovenous malformations of the spinal cord in children. Childs Brain 5: 341–351.

Slavin H (1937). Spontaneous intraspinal subarachnoid hemorrhage. J Nerv Ment Dis 86: 425–427.

Son S, Lee SG, Park CW (2013). Solitary ruptured aneurysm of the spinal artery of Adamkiewicz with subarachnoid hemorrhage. J Korean Neurosurg Soc 54: 50–53.

Sunada I, Akano Y, Kidosaki Y et al. (1995). Spontaneous spinal subarachnoid hematoma – case report. Surg Neurol 44: 133–136.

Swann KW, Ropper AH, New PF et al. (1984). Spontaneous spinal subarachnoid hemorrhage and subdural hematoma. Report of two cases. J Neurosurg 61: 975–980.

Ulrich CT, Beck J, Seifert V et al. (2008). Ependymoma of conus medullaris presenting as subarachnoid haemorrhage. Acta Neurochir (Wien) 150: 185–188.

Vandermeulen EP, Van Aken H, Vermylen J (1994). Anticoagulants and spinal-epidural anesthesia. Anesth Analg 79: 1165–1177.

van Es AC, Brouwer PA, Willems PW (2013). Management considerations in ruptured isolated radiculopial artery aneurysms. A report of two cases and literature review. Interv Neuroradiol 19: 60–66.

Vincent FM (1981). Anterior spinal artery aneurysm presenting as a subarachnoid hemorrhage. Stroke 12: 230–232.

Vishteh AG, Brown AP, Spetzler RF (1997). Aneurysm of the intradural artery of Adamkiewicz treated with muslin wrapping: technical case report. Neurosurgery 40: 207–209.

Walz DM, Woldenberg RF, Setton A (2006). Pseudoaneurysm of the anterior spinal artery in a patient with moyamoya: an unusual cause of subarachnoid hemorrhage. AJNR Am J Neuroradiol 27: 1576–1578.

Wille-Jorgensen P, Jorgensen LN, Rasmussen LS (1991). Lumbar regional anaesthesia and prophylactic anticoagulant therapy. Is the combination safe? Anaesthesia 46: 623–627.

PART II

CAVERNOUS MALFORMATIONS

Handbook of Clinical Neurology, Vol. 143 (3rd series)
Arteriovenous and Cavernous Malformations
R.F. Spetzler, K. Moon, and R.O. Almefty, Editors
http://dx.doi.org/10.1016/B978-0-444-63640-9.00021-7

Chapter 21

Natural history of cerebral cavernous malformations

CHIBAWANYE ENE, ANAND KAUL, AND LOUIS KIM*
Department of Neurological Surgery, Harborview Medical Center, University of Washington, Seattle, WA, USA

Abstract

Cerebral cavernous malformations (CCM) are vascular abnormalities of the central nervous system with an incidence of 0.4–0.5% and an annual rate of hemorrhage ranging from 0.7% to 1%. Most lesions are located in the cerebral hemisphere but some occur in deeper locations such as the basal ganglia and pons. The most common symptoms during presentation are headache, seizures, and focal neurologic deficits. Surgery remains the most effective treatment modality for symptomatic CCM, while the management of incidental CCM remains controversial. Factors associated with increased risk of hemorrhage include being female and less than 40 years old. This finding, however, is not consistent in all natural history studies evaluated. During follow-up, the most important and consistent risk factor for rebleed was a prior hemorrhage. Here, we provide an indepth but concise review of the literature regarding the natural history of CCMs.

INTRODUCTION

Cavernous malformations (aka cavernomas, capillary hemangiomas, cavernous hemangiomas, or cavernous angiomas) are sinusoidal vascular channels lined by a single layer of simple endothelium and a thin fibrous adventitia without intervening brain parenchyma or large feeding or draining vessels. Gross specimens appear reddish-blue in color and the cavities of these lesions are often comprised of thrombi at different stages of organization (Giombini and Morello, 1978). Cerebral cavernous malformations (CCMs) are referred to as "cryptic" or "angiographically occult" as they cannot be visualized on routine cerebral angiograms (Wakai et al., 1985; Zabramski et al., 1999). Historically, discerning the incidence of CCMs relied on surgery and postmortem autopsy.

Nowadays, the use of magnetic resonance imaging (MRI) and computed tomography (CT) has allowed for the detection of more CCMs within the population (Rigamonti et al., 1987, 1988; Robinson et al., 1991; Campbell et al., 2010). Specifically, MRI has become the gold standard for diagnosis of CCM. It has improved the sensitivity and specificity of diagnosis of this lesion and invariably led to the increased reported incidence of CCM in the general population (Del Curling et al., 1991; Robinson et al., 1991). Previously, CT scans have shown excellent rate of detection of CCM (Vaquero et al., 1983; Tagle et al., 1986). More recent analysis, however, demonstrates that CT scans are not specific for preoperative diagnosis of CCM, with only 51% of total lesions verified by CT scan in one retrospective review (Rigamonti et al., 1987).

The appropriate management of patients with CCM requires an understanding of the presentation, natural history, and treatment options available. The purpose of this chapter is to provide an indepth but concise review of the literature regarding the natural history of CCMs.

NATURAL HISTORY

The natural history of CCMs has been difficult to elucidate due to the tendency to opt for surgical resection of symptomatic lesions (Aiba et al., 1995). With MRI driving diagnosis of incidental lesions, however, we can now follow the clinical course/natural history of untreated

*Correspondence to: Louis Kim, M.D., Department of Neurological Surgery, Box 359924, University of Washington/Harborview Medical Center, 325 9th Avenue, Seattle WA 98104, USA. Tel: +1-206-744-9330, Fax: +1-206-744-9944, E-mail: ljkim1@u.washington.edu

CCMs. Ultimately, understanding the natural history of CCMs will shed light on the appropriate management options for patients as it allows us to better predict risk of progression and recurrence of symptoms.

Several natural history studies exist for CCMs (Table 21.1). Based on these studies, supratentorial lesions are most common and have been shown to occur in 72–90% of reported cases (Voigt and Yaşargil, 1976; Giombini and Morello, 1978; Del Curling et al., 1991; Robinson et al., 1991; Moriarity et al., 1999). Among these cases, lesions were typically found within the frontal lobe (22%), temporal lobe (26%), parietal lobe (9%), occipital lobe (9%), and insula (4%) (Moriarity et al., 1999). Deep intracerebral lesions identified within the thalamus, basal ganglia, and brainstem have been reported with incidence between 8% and 37% (Kondziolka et al., 1995; Porter et al., 1997; Moriarity et al., 1999). Infratentorial lesions have been reported to occur between 10% and 22% (Voigt and Yaşargil, 1976; Giombini and Morello, 1978; Del Curling et al., 1991; Robinson et al., 1991; Moriarity et al., 1999).

Although, seizures are the most common presenting symptom for patients with CCMs, other common presentations include headaches, focal neurologic deficits, and hemorrhage. Conversely, asymptomatic or incidental CCMs occurred in 1.5–19% of cases reported (Del Curling et al., 1991; Robinson et al., 1991; Kim et al., 1997; Porter et al., 1997; Moriarity et al., 1999).

Seizures

Patients presented with seizures on average of 40.6% (23–50%) (Del Curling et al., 1991; Robinson et al., 1991; Kondziolka et al., 1995; Kim et al., 1997; Porter et al., 1997; Moriarity et al., 1999). Del Curling et al. (1991) performed a retrospective analysis of 32 patients and reported a seizure risk of 1.5% per patient per year. In a prospective analysis of 68 patients by Moriarity et al. (1999), seizures where found to occur at a rate of 2.4% per patient per year. Supratentorial lesions were more likely to present with seizures, whereas brainstem CCMs were much more likely to present with focal neurologic deficits (Kim et al., 1997).

It is believed that the constant leakage and breakdown of blood products within CCMs lead to the buildup and deposition of hemosiderin within cerebral tissue surrounding the lesion (Kim et al., 1997). This iron buildup is known to act as an epileptogenic focus in pre-clinical mouse models (Willmore et al., 1978). Therefore, iron buildup associated with chronic microhemorrhage from CCMs may instigate seizures, especially in a supratentorial location (Baumann et al., 2006; Upchurch et al., 2010).

Hemorrhage

Appropriate comparison of hemorrhage rates (number of hemorrhages identified per person per year) across multiple studies is difficult given the large variance in definition of CCM hemorrhage and calculation of either retrospective or prospective hemorrhage rates. In general, hemorrhage has been defined as symptomatic presence of extralesional blood on MRI scans (Zabramski et al., 1994; Kim et al., 1997), "gross blood on imaging studies" (Kondziolka et al., 1995), and clinically significant or radiographically identifiable hemorrhage. Robinson et al. (1991) defined overt hemorrhage as the presence of extralesional blood on an MRI scan, positive lumbar puncture and fresh clot outside the confines of lesion at surgery.

The calculated risk of hemorrhage has been reported in literature based on both retrospective (McCormic, 1984; Del Curling et al., 1991; Kondziolka et al., 1995; Kim et al., 1997) and prospective analyses (Table 21.2) (Robinson et al., 1991; Zabramski et al., 1994; Kondziolka et al., 1995; Porter et al., 1997).

Table 21.1

Natural history of cavernous malformations

Authors	Year	Patients	M/F	Mean Age (years)	Symptoms (%total)	Location	Study design
Del Curling et al.	1991	32	17/15	37.6	50S, 34HA	86C, 16B/CB	Retrospective
Kim et al.	1997	108	38/24	32.2	40.8S, 8.2FD, 6.1HA	63.3C, 20.4B, 10.2BG	Retrospective
Kondziolka et al.	1995	122	60/62	37.3	23S, 15HA	48C/CB, 17BG	Prospective
Moriarity et al.	1999	68	24/44	34.6	65HA, 49S, 46FD, 13H	22F, 26T, 9P, 9Occ, 14B, 5C	Prospective
Porter et al.	1997	173	88/85	37.5	36S, 25H, 6HA	27.7F, 17.3T, 7P, 3.5Occ	Prospective
Zabramski et al.	1994	31	13/18	25	52HA, 39S, 39AS	37F, 21T, 12P, 11Occ	Familial retrospective

H, hemorrhage; S, seizures; HA, headache; I, incidental; FD, focal deficit; AS, asymptomatic; M, male; F, female; Location (C, cortex; F, frontal; T, temporal; P, parietal; Occ, occipital; B, brainstem; CB, cerebellum; BG, basal ganglia).

Table 21.2

Follow-up and long-term outcomes

Author	Year	Patients	Follow-up (patient-year)	Hemorrhage rate (per patient-year)	Re-hemorrhage rate (per patient-year)	Factor promoting re-bleed
Robinson et al.	1991	57	143*	0.7*	na	Female gender
Kondziolka et al.	1995	122	342	2.6	4.5	History of hemorrhage
Porter et al.	1997	173	427	1.6	na	History of hemorrhage, deep location
Moriarity et al.	1999	68	353	3.1	0	Female gender

*Lesion-year.

Retrospective hemorrhage rates are calculated on the assumption that lesions identified have been present since birth and therefore may underestimate the true hemorrhage rate. Prospective rates are based on patients followed after they have been diagnosed and added to a CCM cohort database; however, these data may inappropriately be skewed by the predominant inclusion of symptomatic cavernous malformations, and failure to include asymptomatic lesions.

Prospective hemorrhage rates have been reported between 0.7% and 6.5% per lesion-year (Robinson et al., 1991; Zabramski et al., 1994; Kondziolka et al., 1995; Porter et al., 1997). Porter and colleagues (1997) prospectively followed 110 patients with a mean follow-up of 3.8 years. They reported hemorrhage in 25% of patients during the follow-up period with a calculated prospective hemorrhage rate of 4.2% per person per year. This study also showed that the location of CCMs within the brain was an important factor in predicting occurrence of hemorrhage. Deep lesions (within the thalamus, basal ganglia, and brainstem) characteristically carried a worse prognosis with respect to annual hemorrhage rates (10.6% per patient per year versus 0.4% per patient per year for superficial cerebral lesions) and subsequent neurologic deterioration.

Kondziolka et al. (1995) prospectively followed 122 patients with radiographically confirmed CCMs. Patients included in this study had a retrospective hemorrhage rate of 1.3% per patient per year. During the follow-up period (mean 34 months) for all patients with and without previous hemorrhage, a prospective hemorrhage rate was calculated to be 2.6%. The authors concluded that prior hemorrhage was the most significant risk factor for subsequent hemorrhage. For patients with no prior history of hemorrhage the prospective hemorrhage rate was calculated to be 0.6%, compared to 4.5% per patient per year in patients with prior history of hemorrhage. Unlike Porter et al. (1997), location had no significant correlation with risk of subsequent hemorrhage.

Moriarity et al. (1999) reported an overall prospective hemorrhage rate of 3.1% per patient per year. Additionally, they reported a statistically significant difference in prospective hemorrhage rates between men and women (0.9% per patient per year versus 4.2% per patient per year respectively; $p=0.04$). Robinson et al. (1991) reported an increased bleeding risk in female gender as well as in pregnancy, with a rate of hemorrhage calculated to be 0.7% per patient per year. These differences based on sex, however, have remained controversial given inconsistencies in study methodology.

Incidental

Moore et al. (2014) prospectively followed a group of 107 patients identified from a previously published database of patients with incidentally discovered CCMs or unclear symptoms. Over a median follow-up of 12.5 years (1311 patient-years), only 1 patient (0.9%) had a radiographically confirmed symptomatic hemorrhage, yielding a prospective symptomatic hemorrhage rate of 0.08% per patient per year. This low rate of hemorrhage is evidence for recommending conservative management for an asymptomatic, incidentally discovered solitary cavernous malformation.

Familial

Unlike sporadic CCMs, familial CCMs are commonly associated with multiple CCMs per patient (Rigamonti et al., 1988; Zabramski et al., 1994; Moriarity et al., 1999). These studies also identified genetic loci that are associated with the development of familial CCM including CCM1 on the long arm of chromosome 7 (7q21 and 7q22), CCM2 on short arm of chromosome 7 (7p12 through 7p15), and CCM3 on the long arm of chromosome

3 (3q25 through 3q27) (Dubovsky et al., 1995; Craig et al., 1998). These genes were found to be inherited in an autosomal-dominant pattern with variable expressivity. In terms of demographics, there is a higher prevalence of familial cases in Hispanic Americans (Revencu and Vikkula, 2006; Labauge et al., 2007; D'Angelo et al., 2011). Asymptomatic familial CCMs have a higher annual hemorrhage rate (4.3% per patient-year) compared to asymptomatic sporadic CCMs (0.6% per patient-year) (Washington et al., 2010). These results suggest that two distinct forms of the disease may exist, with the familial form being more aggressive.

Pediatrics

CCMs represent 2–18% of all vascular malformations in children and 25% of these lesions are observed in children under the age of 18 (Lena et al., 2007). Postmortem studies from the 1980s found that there was a 0.4–0.5% prevalence of CCMs in children (Otten et al., 1989; Robinson et al., 1991). The mean age at presentation was 9.1–10.2 years and there is no sex predominance. CT and MRI scans have altered previous misconceptions that CCMs are lesions predominantly affecting adults (Mazza et al., 1991). Similar to adults, most lesions are found as incidental findings on MRI or with symptoms such as headaches (6–65%) and seizures (40–50%) (Washington et al., 2010). Gross et al. (2015) evaluated 153 pediatric patients (range 0.1–21years old) with 193 CCMs. They found an annual hemorrhage rate of 3.3% per lesion-year. In incidental cases, the annual hemorrhage rate was 0.5% per lesion-year. In familial cases, the annual hemorrhage risk for pediatric CCMs was 1.1% per lesion-year compared with 4.0% per lesion-year in nonfamilial cases (Gross et al., 2015). This suggests that the familial form of the disease may be less aggressive in the pediatric population than in the adult population (see familial section, above) (Washington et al., 2010).

CLINICAL SEQUELAE OF HEMORRHAGE

Multiple natural history studies have assessed the clinical sequelae of patients following CCM hemorrhage (Del Curling Jr et al., 1991; Robinson et al., 1991; Zabramski et al., 1994; Porter et al., 1997; Moriarity et al., 1999; Gross et al., 2011). In the follow-up period, the rate of new-onset seizures was 2.4 per patient-year (Moriarity et al., 1999). Location is the most important factor in determining interval events in follow-up

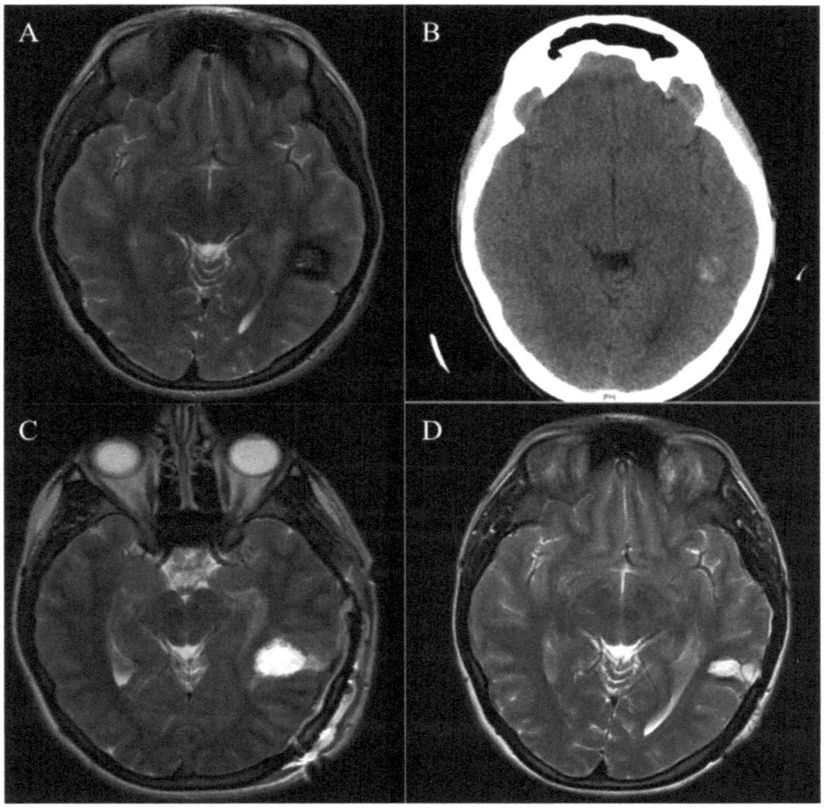

Fig. 21.1. (A–D) Case 1. A 23-year-old woman with left temporal cavernous malformation presenting with seizures.

(Porter et al., 1997). Univariate analysis has shown that female gender, infratentorial location, and prior gross hemorrhage were significantly associated with subsequent neurologic disability. Logistic regression analysis further demonstrated that infratentorial location and previous gross hemorrhage were independent predictors of neurologic disability (Robinson et al., 1991). Age less than 40 was the only significant factor predisposing to functional disability secondary to seizures, while lesion size or multiplicity did not influence disability (Robinson et al., 1991). One retrospective analysis showed that re-hemorrhage from a CCM may lead to increased deficits but rarely results in patient mortality (Kondziolka et al., 1995).

CASE PRESENTATION

This 23-year-old woman presented after a generalized tonic-clonic seizure. She was started on levetiracetam and further workup, including an MRI, demonstrated a posterior left temporal CCM measuring 1.8×1.7 cm (T2-weighted MRI, Fig. 21.1A). A noncontrast head CT showed no signs of hemorrhage (Fig. 21.1B). She was taken to the operating room for stereotactic-guided resection of the CCM via a left frontal temporal parietal craniotomy. Given the location of the lesion, cortical mapping and stimulation were performed awake to preserve any areas associated with language.

Postoperatively she had minor issues with speech and an MRI showed no residual lesion (Fig. 21.1C). Following inpatient and outpatient rehabilitation, she was off seizure medications and had no speech deficits at her 2 year follow-up visit. Furthermore, an MRI obtained as part of routine follow-up showed no evidence of residual disease (Fig. 21.1D).

CONCLUSION

The natural history of CCMs has been extensively studied prospectively and retrospectively. The development of more advanced imaging technologies has increased the reported incidence of CCMs. Patients with CCMs typically present with seizures, hemorrhage, headache, and focal neurologic deficits or in some cases, it is discovered incidentally. Sporadic CCMs are usually found as a solitary lesion, while familial-type CCMs often present in multiple locations throughout the central nervous system. Incidentally discovered/asymptomatic solitary lesions are associated with low risk of hemorrhage, therefore surgical resection may not be necessary for these lesions. Previously hemorrhagic, or symptomatic lesions, however, have an increased risk of recurrent hemorrhage so surgical intervention may be indicated for patients in this group.

REFERENCES

Aiba T, Tanaka R, Koike T et al. (1995). Natural history of intracranial cavernous malformations. J Neurosurg 83: 56–59.

Baumann CR, Schuknecht B, Lo Russo G et al. (2006). Seizure outcome after resection of cavernous malformations is better when surrounding hemosiderin-stained brain also is removed. Epilepsia 47: 563–566.

Campbell PG, Jabbour P, Yadla S et al. (2010). Emerging clinical imaging techniques for cerebral cavernous malformations: a systematic review. Neurosurg Focus 29. E6.

Craig HD, Gunel M, Cepeda O et al. (1998). Multilocus linkage identifies two new loci for a mendelian form of stroke, cerebral cavernous malformation, at 7p15-13 and 3q25.2-27. Hum Mol Genet 7 (12): 1851–1858.

D'Angelo R, Marini V, Rinaldi C et al. (2011). Mutation analysis of CCM1, CCM2 and CCM3 genes in a cohort of Italian patients with cerebral cavernous malformation. Brain Pathol 21: 215–224.

Del Curling Jr O, Kelly Jr DL, Elster AD et al. (1991). An analysis of the natural history of cavernous angiomas. J Neurosurg 75: 702–708.

Dubovsky J, Zabramski JM, Kurth J et al. (1995). A gene responsible for cavernous malformations of the brain maps to chromosome 7q. Hum Mol Genet 4 (3): 453–458.

Giombini S, Morello G (1978). Cavernous angiomas of the brain. Acta Neurochir 40: 61–82.

Gross BA, Lin N, Du R et al. (2011). The natural history of intracranial cavernous malformations. Neurosurg Focus 30. E24.

Gross BA, Du R, Orbach DB et al. (2015). The natural history of cerebral cavernous malformations in children. J Neurosurg Pediatr: 1–6.

Kim D-S, Park Y-G, Choi J-U et al. (1997). An analysis of the natural history of cavernous malformations. Surg Neurol 48: 9–17.

Kondziolka D, Lunsford LD, Kestle JR (1995). The natural history of cerebral cavernous malformations. J Neurosurg 83: 820–824.

Labauge P, Denier C, Bergametti F et al. (2007). Genetics of cavernous angiomas. Lancet Neurol 6: 237–244.

Lena G, Ternier J, Paz-Paredes A et al. (2007). Central nervous system cavernomas in children. Neurochirurgie 53: 223–237.

Mazza C, Scienza R, Beltramello A et al. (1991). Cerebral cavernous malformations (cavernomas) in the pediatric age-group. Childs Nerv Syst 7: 139–146.

McCormic W (1984). Pathology of vascular malformations of the brain. Intracranial Vascular Malformation. William & Wilkins, Baltimore, pp. 44–63.

Moore SA, Brown Jr RD, Christianson TJ et al. (2014). Long-term natural history of incidentally discovered cavernous malformations in a single-center cohort. J Neurosurg 120: 1188–1192.

Moriarity JL, Wetzel M, Clatterbuck RE et al. (1999). The natural history of cavernous malformations: a prospective study of 68 patients. Neurosurgery 44: 1166–1171.

Otten P, Pizzolato GP, Rilliet B et al. (1989). 131 cases of cavernous angioma (cavernomas) of the CNS, discovered by retrospective analysis of 24,535 autopsies. Neurochirurgie 35 (82–83): 128–131.

Porter PJ, Willinsky RA, Harper W et al. (1997). Cerebral cavernous malformations: natural history and prognosis after clinical deterioration with or without hemorrhage. J Neurosurg 87: 190–197.

Revencu N, Vikkula M (2006). Cerebral cavernous malformation: new molecular and clinical insights. J Med Genet 43: 716–721.

Rigamonti D, Drayer BP, Johnson PC et al. (1987). The MRI appearance of cavernous malformations (angiomas). J Neurosurg 67: 518–524.

Rigamonti D, Spetzler RF, Drayer BP et al. (1988). Appearance of venous malformations on magnetic resonance imaging. J Neurosurg 69: 535–539.

Robinson JR, Awad IA, Little JR (1991). Natural history of the cavernous angioma. J Neurosurg 75: 709–714.

Tagle P, Huete I, Méndez J et al. (1986). Intracranial cavernous angioma: presentation and management. J Neurosurg 64: 720–723.

Upchurch K, Stern JM, Salamon N et al. (2010). Epileptogenic temporal cavernous malformations: operative strategies and postoperative seizure outcomes. Seizure 19: 120–128.

Vaquero J, Leunda G, Martínez R et al. (1983). Cavernomas of the brain. Neurosurgery 12: 208–210.

Voigt K, Yaşargil M (1976). Cerebral cavernous haemangiomas or cavernomas. Incidence, pathology, localization, diagnosis, clinical features and treatment. Review of the literature and report of an unusual case. Neurochirurgia 19: 59–68.

Wakai S, Ueda Y, Inoh S et al. (1985). Angiographically occult angiomas: a report of thirteen cases with analysis of the cases documented in the literature. Neurosurgery 17: 549–556.

Washington CW, McCoy KE, Zipfel GJ (2010). Update on the natural history of cavernous malformations and factors predicting aggressive clinical presentation. Neurosurg Focus 29. E7.

Willmore LJ, Sypert GW, Munson JB (1978). Recurrent seizures induced by cortical iron injection: a model of posttraumatic epilepsy. Ann Neurol 4: 329–336.

Zabramski JM, Wascher TM, Spetzler RF et al. (1994). The natural history of familial cavernous malformations: results of an ongoing study. J Neurosurg 80: 422–432.

Zabramski J, Henn J, Coons S (1999). Pathology of cerebral vascular malformations. Neurosurg Clin N Am 10: 395–410.

Handbook of Clinical Neurology, Vol. 143 (3rd series)
Arteriovenous and Cavernous Malformations
R.F. Spetzler, K. Moon, and R.O. Almefty, Editors
http://dx.doi.org/10.1016/B978-0-444-63640-9.00022-9

Chapter 22

Natural history of spinal cavernous malformations

MARC OTTEN AND PAUL McCORMICK*
Department of Neurological Surgery, Columbia University Medical Center, Neurological Institute of New York, New York, NY, USA

Abstract

Spinal cavernous malformations are intramedullary vascular lesions. They have low pressure and flow, so they may take many years to present with clinical symptoms. Because of their relatively benign nature, surgical intervention is not always indicated. An understanding of the natural history of cavernous malformations helps make decisions about when and if to intervene. In patients who do not have surgery, 88.7% have stable or improved neurologic function, whereas 89.3% have these outcomes in the surgical group. Of note, 51.5% of patients were found to improve in the surgical group, compared to 30.2% in the nonsurgical group. Characteristics that correlated with better neurologic outcome were: resection within 3 months of the onset of symptoms, gross total resection, presentation with motor symptoms, and an acute course. Sensory symptoms correlated with worse outcome. Given the natural history of spinal cavernous malformations, surgery may be considered for symptomatic patients, when general medical health and lesion location permit safe resection. The severity of symptoms must also be considered, as the natural history of the disease can be benign.

INTRODUCTION

Cavernous malformations are low-pressure, vascular abnormalities, which most commonly affect the brain and spinal cord. They have also been called cavernous angiomas, cavernomas, or cryptic vascular malformations (Batra et al., 2009). Because of their relatively benign nature when compared to other vascular anomalies, it is possible that observation is an appropriate approach for some patients with spinal cavernous malformations.

Some projected that the increasing availability of magnetic resonance imaging (MRI) would reveal more asymptomatic lesions, although there are no data to support that this has happened (McCormick et al., 1988; Kharkar et al., 2007; Steiger et al., 2010). The increasing availability of MRI has, however, allowed for the diagnosis of spinal cavernous malformations without tissue analysis. This diagnosis without surgery has increased our understanding of their natural history. A clinician must understand the natural history of cavernous malformations to present the risks and benefits of surgical resection, and how surgery can alter outcomes.

The purpose of this chapter is to review the literature, and present what is known about the natural history of spinal cavernous malformations. Because spinal cavernous malformations are a relatively rare pathologic entity, the literature is composed of retrospective case studies. Furthermore, because early literature indicated that surgical resection was safe and effective, there is a selection bias against observation of patients who present with spinal cavernous malformations. In a meta-analysis of patients with intramedullary spinal cavernous malformations, just 10.1% (64 of 631 patients) did not have operative resection (Badhiwala et al., 2014). Our understanding of the natural history of this disease must, therefore, depend on a limited data set of nonoperative patients, and on the preoperative courses of surgical patients.

*Correspondence to: Paul McCormick, Department of Neurological Surgery, Columbia University Medical Center, Neurological Institute of New York, 710 West 168th Street, New York NY 10032, USA. Tel: +1-212-305-7976, E-mail: pcm6@cumc.columbia.edu

EPIDEMIOLOGY

Spinal cavernous malformations are relatively rare lesions. They represent 5–12% of all spinal vascular malformations, and just 3–5% of spinal cord lesions (McCormick et al., 1988; Canavero et al., 1994; Gross et al., 2010; El-Koussy et al., 2011; Mitha et al., 2011; Reitz et al., 2015). They are found in 1.86 per 100 000 population (El-Koussy et al., 2011). The mean age at presentation is 39.1 years, with a range of 2–80 years (Badhiwala et al., 2014). There is a slight male predominance (1.1:1.0), but gender and age are not effective ways to identify patients with this pathology.

The decision to search for a spinal lesion must be made based on neurologic examination. Further testing is indicated when clinical presentation localizes to a region of the spinal cord based on sensorimotor deficit, or other evidence of myelopathy such as hyperreflexia, gait changes, or dysfunction of the bowel or bladder. These lesions are distributed throughout the spinal column, found in a distribution roughly consistent with the volume of each segment of the spinal cord. The majority of lesions are thoracic (55%) or cervical (38%) (Badhiwala et al., 2014; Reitz et al., 2015). A smaller number are at the cervicothoracic (2.4%) or thoracolumbar (0.6%) junction. Finally, 2.1% are in the lumbar region, and 1.7% are found at the conus medullaris (Gross et al., 2010; El-Koussy et al., 2011; Badhiwala et al., 2014; Reitz et al., 2015).

A great deal of research has been done to identify genetic etiologies in patients with cavernous malformations (Bicknell et al., 1978; Lee and Spetzler, 1990; Del Curling et al., 1991; Maraire et al., 1999; Vishteh et al., 1999; Krings et al., 2005; Cohen-Gadol et al., 2006; Dashti et al., 2006). A genetic predisposition might explain multiple lesions in an individual, and 16.5% of patients with spinal cavernous malformations also had cerebral lesions (Krings et al., 2005; Cohen-Gadol et al., 2006; Badhiwala et al., 2014). Some might argue for imaging of the entire neuraxis in patients with cavernous malformations, but as there is little indication to intervene for asymptomatic lesions, there is not a justification for this practice.

The study of families in which multiple people have cavernous malformations has revealed some genetic forms of the disease. Of the patients reported in the literature to have spinal cavernous malformations, 10–15% had a family history (Bicknell et al., 1978; Zabramski et al., 1994; Detwiler et al., 1997; Labauge et al., 1998; Maraire et al., 1999; Vishteh et al., 1999; Aoyagi et al., 2003; Campeau and Lane, 2005; Cohen-Gadol et al., 2006; Choi et al., 2011; El-Koussy et al., 2011; Reitz et al., 2015). These data suggest that family history can be helpful in the workup of patients with neurologic deficits, but the vast majority of cavernous malformations still occur in patients without history among their family members.

PATHOPHYSIOLOGY

Cavernous malformations are low-flow, vascular lesions, which are made up of endothelium-lined chambers, and have sluggish blood flow. On gross appearance, because they are made up of sinusoidal chambers, they tend to resemble a mulberry. The mural elements of mature vasculature are absent, and there is little or no intervening neural tissue. Because of the slow blood flow through cavernous malformations, they tend to have intralesional thrombosis, calcifications, and recanalization (Batra et al., 2009). These can manifest as hemosiderin-laden macrophages within the lesions, and in the surrounding glia (Clatterbuck et al., 2001). Hematoxylin and eosin slides are included in Figure 22.1, and demonstrate the sinusoidal, thin-walled channels.

The exact etiology of cavernous malformations is probably variable, and not known with certainty. Although the majority of cavernous malformations are thought to be congenital, some have been discovered following radiation, or in association with other lesions, such as developmental venous anomalies (Bicknell et al., 1978;

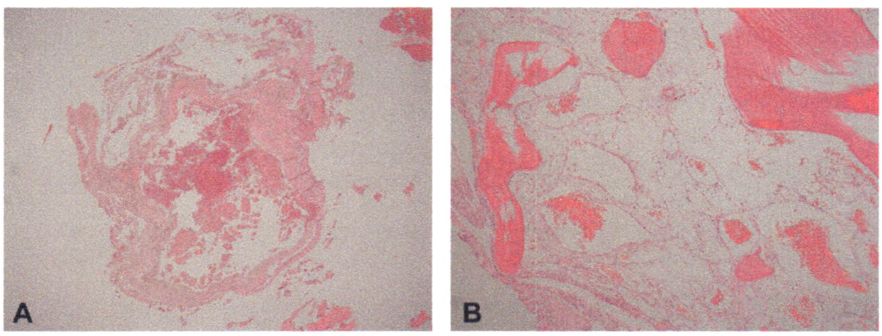

Fig. 22.1. Low- (**A**) and high- (**B**) power hematoxylin and eosin images of spinal cavernous malformations. Note the sinusoidal cavities without the muscular or adventitial layers of normal blood vessels.

Maraire et al., 1999; Vishteh et al., 1999; Campeau and Lane, 2005; Cohen-Gadol et al., 2006; Dashti et al., 2006; Abla et al., 2008; Hong et al., 2010; Mujagic and Becirovic-Ibrisevic, 2012). Most of the genetics has been studied with cranial cavernous malformations, and is beyond the scope of this chapter.

IMAGING

The slow flow and low pressure of cavernous malformations give them their characteristic imaging appearance. Hemorrhage, or widening of the spinal cord, can be appreciated on computed tomography (CT) scans, but they rarely have calcifications, and CT is not sensitive or specific for diagnosing cavernous malformations. Because of the relatively low blood flow to these vascular lesions, cavernous malformations are typically occult on digital subtraction angiogram. The development and wide availability of MRI increased the sensitivity of noninvasive diagnosis of these lesions (Hegde et al., 2012a).

Because cavernous malformations tend to have low-pressure hemorrhages, and have characteristic channels, they have a relatively predictable appearance on MRI. Zabramski et al. (1994) described different types of lesions based on imaging and pathologic characteristics, although there are some common characteristics on MRI. The T1-weighted, T2-weighted, and gradient-recalled echo (GRE) sequences are the most useful ways to make a noninvasive diagnosis. The heterogeneity is one of the characteristics of these lesions, and this gives them the classic "popcorn" or "mulberry" appearance. On T1-weighted images, these lobulated masses tend to be isointense and hyperintense with an edge of hypointensity, which represents the reactive, gliotic capsule (Turjman et al., 1995; Aoyagi et al., 2003; Hegde et al., 2012a; Kivelev et al., 2012a). On T2-weighted images, the lesions are also heterogeneous, with a predominant hyperintensity surrounded by a hypointense rim after hemorrhage (Hegde et al., 2012a). On the GRE, cavernous malformations are generally hypointense, but larger lesions can have a hyperintense core (Hegde et al., 2012a). GRE images are more commonly used with cerebral imaging, and so play less of a role in the diagnosis of spinal lesions. There is generally minimal or no enhancement on post-contrast studies, also a result of the relatively low blood flow to these lesions. The average size is 9.2 mm in maximum transverse diameter (Badhiwala et al., 2014; Reitz et al., 2015). Figure 22.2 demonstrates an example of some of the classic imaging findings.

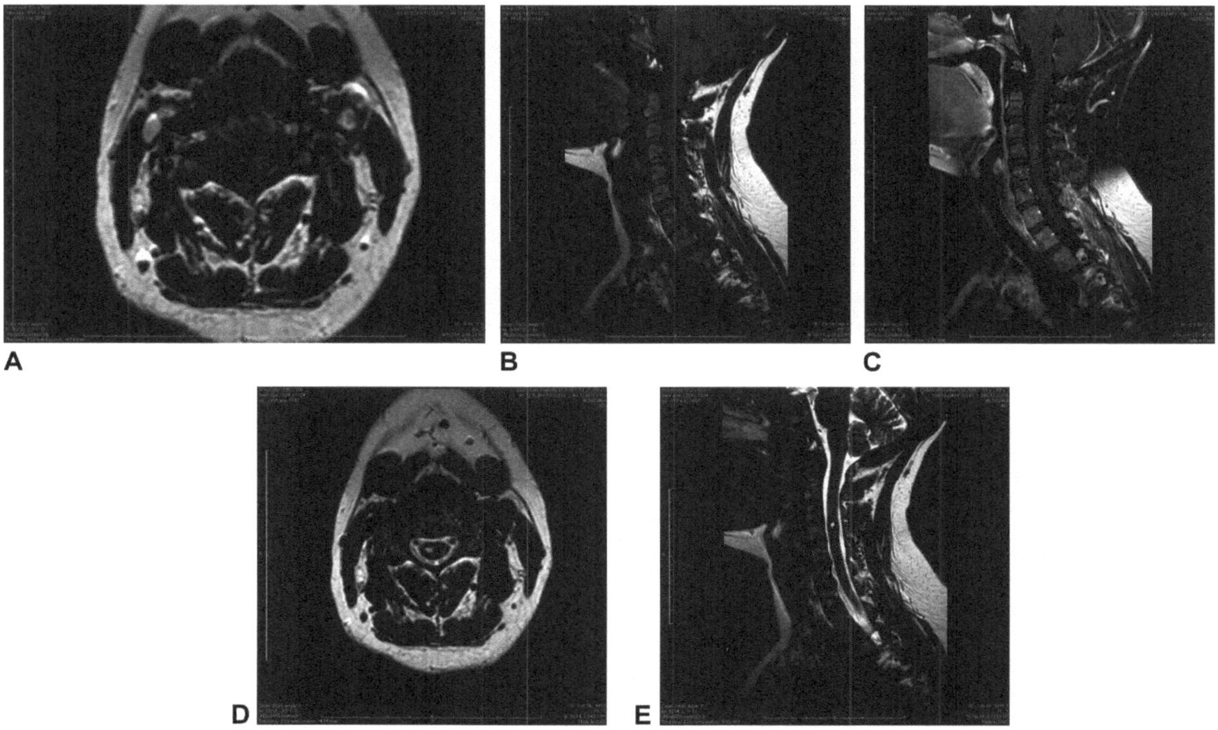

Fig. 22.2. Magnetic resonance imaging shows T1-weighted axial (**A**) and sagittal (**B**) views, along with postgadolinium views (**C**). The heterogeneous isointense and hyperintense signal is characteristic, with minimal enhancement. The T2-weighted axial (**D**) and sagittal (**E**) views of this cervical spine cavernous malformation demonstrate the typical heterogeneous, predominantly hyperintense signal with a hypointense rim.

CLINICAL PRESENTATION

Some believed that the widespread access to MRI since the 1990s would lead to many incidentally discovered cavernous malformations. In most published series, however, the majority of patients do present with symptoms (Kivelev et al., 2010, 2012a, b; Lu and Lawton, 2010; Steiger et al., 2010; Aoyama et al., 2011; El-Koussy et al., 2011; Liang et al., 2011; Mitha et al., 2011; Babu et al., 2013; Kim et al., 2013; Badhiwala et al., 2014; Reitz et al., 2015). This pattern of diagnosis may reflect a bias in which patients included in surgical series tend to have symptoms. Although more than 99% of reported patients are symptomatic, it is worth noting that cavernous malformations may be incidental, asymptomatic findings (Hegde et al., 2012b; Badhiwala et al., 2014).

Symptomatic patients with cavernous malformations typically present with deficits related to the anatomic location of the malformation. Approximately 60.5% present with motor deficits, 57.8% present with sensory deficits, 33.8% present with pain, 23.6% present with bowel and/or bladder dysfunction, and 0.5% present with respiratory distress (Badhiwala et al., 2014). The percentages add up to more than 100%, because many people present with more than one type of symptom.

Several groups have attempted to categorize the patterns in which people present with spinal cavernous malformations. One system separates patients into one of five categories (Ogilvy et al., 1992; Mitha et al., 2011): (1) discrete episodes of neurologic deterioration with varying degrees of recovery between episodes; (2) slow, progressive neurologic decline; (3) acute onset of symptoms with rapid decline; (4) acute onset of mild symptoms with subsequent gradual decline lasting weeks to months; and (5) sudden onset of back pain without neurologic symptoms. Although these categories can be useful when speaking with patients about what to expect, no characteristics of a lesion have been identified that predict into which category a patient will fall. Some have, therefore, distinguished between acute/stepwise deficits and progressive neurologic deterioration. These broader categories can be easier to explain, and include the vast majority of patients.

In a 2014 meta-analysis by Badhiwala et al., which combined 40 case series and 632 patients, there was documentation about clinical presentation in 522 patients. Of these patients, 45.4% presented with acute/stepwise symptoms, and 54.6% presented with progressive deficits. In the two largest published series, the pattern of presentation was quite different. Liang et al. (2011) presented 96 patients, 73 (76%) of whom presented with progressive decline, and 23 (24%) of whom presented with an acute decline. Mitha et al. (2011) found, in their series of 80 patients, 50 (63%) presented with acute symptoms, whereas 30 (37%) presented with a more gradual, progressive decline. In the most recent study of 48 cases, Reitz et al. (2015) found that just 18.8% suffered a slow, progressive neurologic decline, whereas the remainder had some sudden, acute episode, either followed by repetitive acute deteriorations (33.3%) over months to years, or suffering a more gradual decline (41.7%). It is not clear why patients present in different ways, or why various studies showed statistically significant patterns of presentation that conflict. We will discuss further the natural history of spinal cavernous malformations, but the many ways that they can present force the clinician to exercise caution when discussing prognosis with patients.

NATURAL HISTORY

More than 40 studies have been published on the outcomes and natural history of intramedullary, spinal cord cavernous malformations. The vast majority of the literature relates to patients who have had surgery, or to those referred to neurologic surgeons. Nearly 90% of reported outcomes are, therefore, for patients who have had surgery (Badhiwala et al., 2014). So, there is some selection bias for patients who have had surgery, or who had symptoms. This bias might overestimate the risks, selecting for patients with more serious forms of the disease. The hemorrhage rates may be underestimated, however, as only symptomatic bleeds would be clinically detected.

Much of what is known about hemorrhage rates assumes that cavernous malformations are congenital lesions. The published annual rate of hemorrhage for spinal cavernous malformations ranges from 1.4% to 6.8%, and a meta-analysis from Badhiwala et al. (2014) puts that rate at 2.1% (95% confidence interval (CI) 1.3–3.3%) (Zevgaridis et al., 1999; Sandalcioglu et al., 2003; Kharkar et al., 2007; Bian et al., 2009; Park et al., 2009; Kivelev et al., 2010; Lu and Lawton, 2010; Choi et al., 2011). Although it is useful to discuss hemorrhage rates with patients, we have already addressed the unpredictable neurologic progression of individual patients. Furthermore, there is evidence of spontaneous, *de novo* formation of cavernous malformations, as well as the development of these malformations in patients after radiation, or in patients with the familial form of the disease (Bicknell et al., 1978; Pozzati et al., 1996; Detwiler et al., 1997; Houtteville, 1997; Labauge et al., 1998; Larson et al., 1998; Rosahl et al., 1998; Brunken et al., 1999; Maraire et al., 1999; Labauge et al., 2001; Narayan and Barrow, 2003; Campeau and Lane, 2005; Cohen-Gadol et al., 2006; Dashti et al., 2006; Perrini and Lanzino, 2006; Safavi-Abbasi et al., 2006; El-Koussy et al., 2011; Mujagic and Becirovic-Ibrisevic, 2012; Meng et al., 2014). One would underestimate the rate of hemorrhage if one assumed that all lesions were congenital.

As it is difficult to estimate hemorrhage rates, most studies have focused on neurologic outcomes in patients with spinal cavernous malformations. Reitz et al. (2010) published 48 cases that underwent surgery, with follow-up of 79.3 ± 35.2 months. They found that 70.8% had stable neurologic function, 22.9% improved, and 6.3% worsened after surgery. They defined an unfavorable outcome as an American Spinal Injury Association (ASIA) grade of C or worse, which means that less than half of muscle groups below the level of injury were antigravity. They found a statistically significant association between poor outcome and: thoracolumbar location, low preoperative Epstein and Cooper lower-extremity grade, and preoperative function of ASIA grade A–C (Cooper and Epstein, 1985).

Prior to the series from Reitz et al. (2010), Badhiwala et al. (2014) presented their own 24-patient series, along with a meta-analysis of 40 studies (632 patients) of spinal cavernous malformations. The primary outcome was change in neurologic function between initial presentation and last follow-up. They found statistically significant correlation between improved neurologic function and resection within 3 months of symptom (odds ratio (OR) of 2.11, 95% CI of 1.31–3.41, $p=0.0002$). They also found a correlation between neurologic improvement and: surgical hemilaminectomy (OR 3.20, 95% CI 1.16–8.86, $p=0.02$), gross total resection (OR 3.61, 95% CI 1.24–10.52, $p=0.02$), presentation with motor symptoms (OR 1.76, 95% CI 1.08–2.86, $p=0.04$), and acute course (OR 1.72, 95% CI 1.10–2.68, $p=0.02$). They found a correlation between worse outcome in patients who presented with sensory symptoms (OR 0.58, 95% CI 0.35–0.98, $p=0.04$). They did not find any correlation with age, gender, spinal level, lesion size, presence of cerebral cavernous malformations, or family history of cavernomatosis (Badhiwala et al., 2014).

The advantage of a meta-analysis, despite the inconsistent methodologies between different studies, is that it provides a cohort of sufficient size to produce some statistically significant results. Most of the data about outcomes relates to patients who have had surgery, because 90% of patients in published series had surgery (Badhiwala et al., 2014). In this recent meta-analysis from Badhiwala et al. (2014), 10.1% of patients did not have surgery. They found that neurologic function improved in 30.2% of nonoperative patients vs. 51.5% in the surgical population. Neurologic function was stable in 58.5% vs. 37.8%, and it was worse in 11.3% vs. 10.7%. Deep lesions have historically been more likely to undergo observation, but for those that were surgically resected, the location and depth did not have a significant impact on neurologic outcome.

These data could be interpreted in very different ways. One might argue that surgery increased the rate of neurologic improvement. Another might point out that 88.7% were stable or improved without surgery, and 89.3% were stable or improved with surgery. Yet another way to interpret these results is that, with or without surgery, roughly the same percentages of patients deteriorated. However we consider the data, there are some limitations to them. Because the data are retrospective, and studies were without randomization, it is possible that the conservatively managed population had other characteristics that could contribute to poor neurologic outcomes, or at least a failure to improve. Also, the meta-analysis found that a clinical history less than 5 years was associated with good outcome (OR 10.0, 95% CI 1.07–93.44, $p=0.04$) (Badhiwala et al., 2014). This suggests that longer follow-up could uncover different long-term outcomes in the nonoperative group.

CONCLUSION

When a patient presents with a spinal cavernous malformation, a clinician must understand the likely natural history of this pathology to discuss treatment options. The rate of hemorrhage is thought to be approximately 2%, although the data behind this number have various biases. In an asymptomatic patient, very little is known about the risk of neurologic deterioration, and surgery should likely be delayed in favor of observation.

For patients who are symptomatic from spinal cavernous malformations, there is some evidence, as presented above, that intervention improves neurologic recovery. There are biases that limit the certainty of these data. Nevertheless, the existing evidence does argue for surgical intervention within 3 months, particularly in patients with motor symptoms, an acute course, and in whom a gross total resection is deemed possible. Before proceeding with surgery in symptomatic patients, the severity of symptoms, a patient's general medical health, and lesion location must be considered to determine whether safe resection is possible.

ACKNOWLEDGMENTS

Thank you for assistance with acquisition of pathology slides to Andrew Teich, M.D., Ph.D., Assistant Professor of Pathology and Cell Biology, Columbia University College of Physicians and Surgeons.

References

Abla A, Wait SD, Uschold T et al. (2008). Developmental venous anomaly, cavernous malformation, and capillary telangiectasia: spectrum of a single disease. Acta Neurochir (Wien) 150: 487–489; discussion 489.

Aoyagi N, Kojima K, Kasai H (2003). Review of spinal epidural cavernous hemangioma. Neurol Med Chir (Tokyo) 43: 471–475; discussion 476.

Aoyama T, Hida K, Houkin K (2011). Intramedullary cavernous angiomas of the spinal cord: clinical characteristics of 13 lesions. Neurol Med Chir (Tokyo) 51: 561–566.

Babu R, Owens TR, Karikari IO et al. (2013). Spinal cavernous and capillary hemangiomas in adults. Spine (Phila Pa 1976) 38: E423–E430.

Badhiwala JH, Farrokhyar F, Alhazzani W et al. (2014). Surgical outcomes and natural history of intramedullary spinal cord cavernous malformations: a single-center series and meta-analysis of individual patient data. J Neurosurg Spine 21: 662–676.

Batra S, Lin D, Recinos PF et al. (2009). Cavernous malformations: natural history, diagnosis and treatment. Nat Rev Neurol 5: 659–670.

Bian LG, Bertalanffy H, Sun QF et al. (2009). Intramedullary cavernous malformations: clinical features and surgical technique via hemilaminectomy. Clin Neurol Neurosurg 111: 511–517.

Bicknell JM, Carlow TJ, Kornfeld M et al. (1978). Familial cavernous angiomas. Arch Neurol 35: 746–749.

Brunken M, Sagehorn S, Leppien A et al. (1999). De novo formation of a cavernoma in association with a preformed venous malformation during immunosuppressive treatment. Zentralbl Neurochir 60: 81–85.

Campeau NG, Lane JI (2005). De novo development of a lesion with the appearance of a cavernous malformation adjacent to an existing developmental venous anomaly. AJNR Am J Neuroradiol 26: 156–159.

Canavero S, Pagni CA, Duca S et al. (1994). Spinal intramedullary cavernous angiomas: a literature meta-analysis. Surg Neurol 41: 381–388.

Choi GH, Kim KN, Lee S et al. (2011). The clinical features and surgical outcomes of patients with intramedullary spinal cord cavernous malformations. Acta Neurochir (Wien) 153: 1677–1684; discussion 1685.

Clatterbuck RE, Eberhart CG, Crain BJ et al. (2001). Ultrastructural and immunocytochemical evidence that an incompetent blood–brain barrier is related to the pathophysiology of cavernous malformations. J Neurol Neurosurg Psychiatry 71: 188–192.

Cohen-Gadol AA, Jacob JT, Edwards DA et al. (2006). Coexistence of intracranial and spinal cavernous malformations: a study of prevalence and natural history. J Neurosurg 104: 376–381.

Cooper PR, Epstein F (1985). Radical resection of intramedullary spinal cord tumors in adults. Recent experience in 29 patients. J Neurosurg 63: 492–499.

Dashti SR, Hoffer A, Hu YC et al. (2006). Molecular genetics of familial cerebral cavernous malformations. Neurosurg Focus 21: e2.

Del Curling Jr O, Kelly Jr DL, Elster AD et al. (1991). An analysis of the natural history of cavernous angiomas. J Neurosurg 702–708.

Detwiler PW, Porter RW, Zabramski JM et al. (1997). De novo formation of a central nervous system cavernous malformation: implications for predicting risk of hemorrhage. Case report and review of the literature. J Neurosurg 87: 629–632.

El-Koussy M, Stepper F, Spreng A et al. (2011). Incidence, clinical presentation and imaging findings of cavernous malformations of the CNS. A twenty-year experience. Swiss Med Wkly 141: w13172.

Gross BA, Du R, Popp AJ et al. (2010). Intramedullary spinal cord cavernous malformations. Neurosurg Focus 29:E14.

Hegde A, Mohan S, Tan KK et al. (2012a). Spinal cavernous malformations: magnetic resonance imaging and associated findings. Singapore Med J 53: 582–586.

Hegde AN, Mohan S, Lim CC (2012b). CNS cavernous haemangioma: "popcorn" in the brain and spinal cord. Clin Radiol 67: 380–388.

Hong YJ, Chung TS, Suh SH et al. (2010). The angioarchitectural factors of the cerebral developmental venous anomaly; can they be the causes of concurrent sporadic cavernous malformation? Neuroradiology 52: 883–891.

Houtteville JP (1997). Brain cavernoma: a dynamic lesion. Surg Neurol 48: 610–614.

Kharkar S, Shuck J, Conway J et al. (2007). The natural history of conservatively managed symptomatic intramedullary spinal cord cavernomas. Neurosurgery 60: 865–872; discussion 872.

Kim KM, Chung CK, Huh W et al. (2013). Clinical outcomes of conservative management of spinal cord cavernous angiomas. Acta Neurochir (Wien) 155: 1209–1214.

Kivelev J, Niemela M, Hernesniemi J (2010). Outcome after microsurgery in 14 patients with spinal cavernomas and review of the literature. J Neurosurg Spine 13: 524–534.

Kivelev J, Niemela M, Hernesniemi J (2012a). Characteristics of cavernomas of the brain and spine. J Clin Neurosci 19: 643–648.

Kivelev J, Niemela M, Hernesniemi J (2012b). Treatment strategies in cavernomas of the brain and spine. J Clin Neurosci 19: 491–497.

Krings T, Mull M, Gilsbach JM et al. (2005). Spinal vascular malformations. Eur Radiol 15: 267–278.

Labauge P, Laberge S, Brunereau L et al. (1998). Hereditary cerebral cavernous angiomas: clinical and genetic features in 57 French families. Societe Francaise de Neurochirurgie. Lancet 352: 1892–1897.

Labauge P, Brunereau L, Laberge S et al. (2001). Prospective follow-up of 33 asymptomatic patients with familial cerebral cavernous malformations. Neurology 57: 1825–1828.

Larson JJ, Ball WS, Bove KE et al. (1998). Formation of intracerebral cavernous malformations after radiation treatment for central nervous system neoplasia in children. J Neurosurg 88: 51–56.

Lee KS, Spetzler RF (1990). Spinal cord cavernous malformation in a patient with familial intracranial cavernous malformations. Neurosurgery 26: 877–880.

Liang JT, Bao YH, Zhang HQ et al. (2011). Management and prognosis of symptomatic patients with intramedullary spinal cord cavernoma: clinical article. J Neurosurg Spine 15: 447–456.

Lu DC, Lawton MT (2010). Clinical presentation and surgical management of intramedullary spinal cord cavernous malformations. Neurosurg Focus 29: E12.

Maraire JN, Abdulrauf SI, Berger S et al. (1999). De novo development of a cavernous malformation of the spinal cord following spinal axis radiation. Case report. J Neurosurg 90: 234–238.

McCormick PC, Michelsen WJ, Post KD et al. (1988). Cavernous malformations of the spinal cord. Neurosurgery 23: 459–463.

Meng G, Bai C, Yu T et al. (2014). The association between cerebral developmental venous anomaly and concomitant cavernous malformation: an observational study using magnetic resonance imaging. BMC Neurol 14: 50.

Mitha AP, Turner JD, Abla AA et al. (2011). Outcomes following resection of intramedullary spinal cord cavernous malformations: a 25-year experience. J Neurosurg Spine 14: 605–611.

Mujagic S, Becirovic-Ibrisevic J (2012). The developmental venous anomaly associated with the cavernous malformation. Acta Med Acad 41: 219–220.

Narayan P, Barrow DL (2003). Intramedullary spinal cavernous malformation following spinal irradiation. Case report and review of the literature. J Neurosurg 98: 68–72.

Ogilvy CS, Louis DN, Ojemann RG (1992). Intramedullary cavernous angiomas of the spinal cord: clinical presentation, pathological features, and surgical management. Neurosurgery 31: 219–229; discussion 29–30.

Park SB, Jahng TA, Chung CK (2009). The clinical outcomes after complete surgical resection of intramedullary cavernous angiomas: changes in motor and sensory symptoms. Spinal Cord 47: 128–133.

Perrini P, Lanzino G (2006). The association of venous developmental anomalies and cavernous malformations: pathophysiological, diagnostic, and surgical considerations. Neurosurg Focus 21: e5.

Pozzati E, Giangaspero F, Marliani F et al. (1996). Occult cerebrovascular malformations after irradiation. Neurosurgery 39: 677–682; discussion 82–84.

Reitz M, Burkhardt T, Vettorazzi E et al. (2015). Intramedullary spinal cavernoma: clinical presentation, microsurgical approach, and long-term outcome in a cohort of 48 patients. Neurosurg Focus 39: E19.

Rosahl SK, Vorkapic P, Eghbal R et al. (1998). Ossified and de novo cavernous malformations in the same patient. Clin Neurol Neurosurg 100: 138–143.

Safavi-Abbasi S, Feiz-Erfan I, Spetzler RF et al. (2006). Hemorrhage of cavernous malformations during pregnancy and in the peripartum period: causal or coincidence? Case report and review of the literature. Neurosurg Focus 21: e12.

Sandalcioglu IE, Wiedemayer H, Gasser T et al. (2003). Intramedullary spinal cord cavernous malformations: clinical features and risk of hemorrhage. Neurosurg Rev 26: 253–256.

Steiger HJ, Turowski B, Hanggi D (2010). Prognostic factors for the outcome of surgical and conservative treatment of symptomatic spinal cord cavernous malformations: a review of a series of 20 patients. Neurosurg Focus 29:E13.

Turjman F, Joly D, Monnet O et al. (1995). MRI of intramedullary cavernous haemangiomas. Neuroradiology 37: 297–302.

Vishteh AG, Zabramski JM, Spetzler RF (1999). Patients with spinal cord cavernous malformations are at an increased risk for multiple neuraxis cavernous malformations. Neurosurgery 45: 30–32; discussion 33.

Zabramski JM, Wascher TM, Spetzler RF et al. (1994). The natural history of familial cavernous malformations: results of an ongoing study. J Neurosurg 80: 422–432.

Zevgaridis D, Medele RJ, Hamburger C et al. (1999). Cavernous haemangiomas of the spinal cord. A review of 117 cases. Acta Neurochir (Wien) 141: 237–245.

Handbook of Clinical Neurology, Vol. 143 (3rd series)
Arteriovenous and Cavernous Malformations
R.F. Spetzler, K. Moon, and R.O. Almefty, Editors
http://dx.doi.org/10.1016/B978-0-444-63640-9.00023-0

Chapter 23

Epidemiology of cavernous malformations

HANNAH E. GOLDSTEIN AND ROBERT A. SOLOMON*
Department of Neurosurgery, The Neurological Institute, Columbia University Medical Center, New York, NY, USA

Abstract

Cavernous malformations, accounting for approximately 5–15% of all vascular abnormalities in the central nervous system, are angiographically occult lesions which most often present with seizures, rather than acute hemorrhage. Widely variable across populations, the incidence of cavernous malformations has been reported to be 0.15–0.56 per 100 000 persons per year, with an annual hemorrhage rate of 0.6–11% per patient-year. Seen in 0.17–0.9% of the population, up to one-half are familial, and at least three gene loci have been associated with a familial form, more common among Hispanic Americans. Most cavernous malformations are supratentorial, with 10–23% in the posterior fossa, and approximately 5% found in the spine.

Cavernous malformations, also known as cavernous angiomas, are relatively common, angiographically occult, benign vascular malformations. Made up of irregular vascular channels without intervening neural parenchyma, they are generally well-circumscribed, multilobulated lesions, with endothelin-lined sinusoidal vascular spaces lacking tight junctions (Wong et al., 2000). Cavernous malformations are thought to "leak" blood, as evidenced by a surrounding hemosiderin ring. However, most of these bleeds are microhemorrhages, and may remain clinically silent for years, or even decades. When they do present clinically, cavernous malformations most commonly present with seizures, rather than symptoms related to acute hemorrhage.

OVERALL INCIDENCE AND LOCATIONS

Cavernous malformations account for approximately 5–15% of all vascular abnormalities in the central nervous system (CNS), making up the second most common cause of hemorrhage (Giombini and Morello, 1978; Martin and Vinters, 1990; Robinson et al., 1991; Lonjon et al., 1993; Di Rocco et al., 1997; Moriarty et al., 1999). However, cavernous malformations most commonly present with seizures, with an annual hemorrhage rate of 0.6–11% per patient-year (Robinson et al., 1991; Zabramski et al., 1994; Aiba et al., 1995a; Kondziolka et al., 1995; Kim et al., 1997; Moriarty et al., 1999; Porter et al., 1999; Labauge et al., 2000, 2001; Kupersmith et al., 2001; Mathiesen et al., 2003).

Widely variable across populations, the incidence of cavernous malformations has been reported to be 0.15–0.56 per 100 000 persons per year, with a prevalence of 0.17–0.9% (Table 23.1) (McCormick, 1984; Simard et al., 1986; Otten et al., 1989; Del Curling et al., 1991; Robinson et al., 1991; Brown et al., 1996a, b; Kim et al., 1997; Moriarty et al., 1999; Morris et al., 2009; Al-Shahi Salman et al., 2012). These estimates are largely based on autopsy and magnetic resonance imaging studies, as 70–95% of lesions remain asymptomatic (McCormick, 1984; Jellinger, 1986; Rigamonti et al., 1988; Otten et al., 1989; Del Curling et al., 1991; Requena et al., 1991; Robinson et al., 1991; Sage et al., 1993; Zabramski et al., 1994). Prior to the invention of magnetic resonance imaging in the mid-1980s, cavernous malformations were thought to be extremely rare lesions (Voigt and Yasargil, 1976). Overall, there is no significant difference in prevalence between males and females (Yamasaki et al., 1986; Del Curling et al., 1991; Robinson et al., 1991); however,

*Correspondence to: Robert A. Solomon, MD, Department of Neurosurgery, The Neurological Institute, Columbia University Medical Center, 710 West 168th Street, 4th floor, New York NY 10032, USA. Tel: +1-212-305-4118, Fax: +1-212-305-2026, E-mail: ras5@columbia.edu

Table 23.1

Cavernous malformation incidence rate

Study	Incidence rate
McCormick, 1984	0.34%
Otten et al., 1989	0.53%
Del Curling et al., 1991	0.39%
Robinson et al., 1991	0.47%
Brown et al., 1996b	0.17%
Al-Shahi et al., 2003	0.56%
Morris et al., 2009	0.16%

males tend to present earlier, during the second and third decades, whereas females often present later, between the fourth and sixth decades of life (Yamasaki et al., 1986; Robinson et al., 1991). Additionally, there is conflicting and scarce evidence on whether or not pregnancy is associated with an increased risk of hemorrhage (Yamasaki et al., 1986; Lopate et al., 1990; Robinson et al., 1991; Canavero et al., 1994; Katayama et al., 1994; Pozzati et al., 1996; Awada et al., 1997; Porter et al., 1999; Safavi-Abbasi et al., 2006).

Most cavernous malformations are supratentorial, with 10–23% found in the posterior fossa, most commonly in the pons and cerebellum (Kashiwagi et al., 1990; Del Curling et al., 1991; Robinson et al., 1991; Zabramski et al., 1994; Moriarty et al., 1999). Other less common locations include the pineal gland, the middle cerebral fossa, the cerebellopontine angle, the cavernous sinus, and the optic nerve or chiasm. Spinal cord cavernous malformations, which are histologically identical to cranial cavernous malformations, are found at a rate proportional to the volume of nervous tissue in the spinal cord – making up about 5% of cavernous malformations overall (Voigt and Yasargil, 1976). Most often occurring as intramedullary lesions, spinal cord cavernous malformations may also occur epidurally, and have been associated with a history of prior craniospinal radiation therapy (Novelli et al., 1997; Detwiler et al., 1998; Larson et al., 1998; Maraire et al., 1999; Massa-Micon et al., 2000).

Though most commonly found as single entities, cavernous malformations may be associated with venous angiomas and capillary telangiectasias. Venous malformations have been reported in 2.1–36% of patients with cavernous malformations (Rigamonti and Spetzler, 1988; Awad et al., 1993; Wilms et al., 1994; Abdulrauf et al., 1999). They are often initially mistaken for vascular tumors on radiographic workup (Diamond et al., 1976; Hirsh, 1981; Wilkins, 1985; Lobato et al., 1988; Rigamonti and Spetzler, 1988; Weber et al., 1989; Rapacki et al., 1990; Requena et al., 1991; Rigamonti et al., 1991; Robinson et al., 1991Hsu et al., 1993).

Prior cranial irradiation is a known risk factor for the development of cavernous malformations, known as radiation-induced cavernous hemangiomas (RICH) (Singla et al., 2013; Cha et al., 2015; Cutsforth-Gregory et al., 2015). Developing approximately a decade after radiation treatment, RICHs may have slightly different histologic and radiologic findings, compared to *de novo* cavernous malformations, developing from blood-filled spaces after tissue destruction by radiation therapy (Cha et al., 2015). RICHs are more likely to occur as multiple lesions and present at a younger age, yet their symptomatic presentation – with either seizures or overt hemorrhage – is similar to that of *de novo* cavernous malformations (Cutsforth-Gregory et al., 2015).

REGIONAL AND ETHNIC DIFFERENCES

In all, 30–50% of cavernous malformations are familial (Rigamonti et al., 1988), and at least three gene loci have been associated with a familial form of cavernous malformations: *CCM1* on chromosome 7q, *CCM2* on chromosome 7p, and *CCM3* on chromosome 3q (Dubovsky et al., 1995; Gunel et al., 1995; Johnson et al., 1995; Gil-Nagel et al., 1996; Notelet et al., 1997; Craig et al., 1998; Mindea et al., 2006). A mutation of the *CCM1* gene is by far the most common of the three, accounting for more than half of familial cavernous malformations. However, a *CCM1* mutation alone is not sufficient for cavernoma formation, resulting in an autosomal-dominant pattern of inheritance with incomplete penetrance and variable expression (Mindea et al., 2006; Revencu and Vikkula, 2006). Familial forms of cavernous malformations are more common in the Hispanic population, accounting for up to 50% of cavernous malformations in Hispanic Americans, as compared to only 10–20% of those found in non-Hispanic Caucasian Americans (Hayman et al., 1982; Gunel et al., 1995; Dashti et al., 2006).

Familial cavernous malformations behave very similarly to sporadic cavernous malformations, with the exception that they have a much greater propensity to present as multiple lesions (Zabramski et al., 1994). Recent series have found multiple lesions in up to 80–90% of familial cases (Gangemi et al., 1990; Zabramski et al., 1994; Denier et al., 2004), as compared to 10–30% of sporadic cases (Gangemi et al., 1990; Zabramski et al., 1994). However, even in cases of multiple lesions, symptoms can usually be attributed to a single dominant lesion. Furthermore, though there are usually multiple lesions in patients with familial cavernous malformations, approximately 40% of these patients never present clinically (Zabramski et al., 1994; Labauge et al., 2001).

Table 23.2

Cavernous malformation seizure and hemorrhage risk

Study	No. of lesions	Mean follow-up	Seizure risk	Hemorrhage risk
Robinson et al., 1991	76	2.2 years		0.7% per lesion-year
Zabramski et al., 1994	128	2.2 years		1.1% per lesion-year
Aiba et al., 1995b	110	3.1 years		22.9% per lesion-year in patients initially presenting with hemorrhage; 0.39% in patients initially presenting with seizures
Kondziolka et al., 1995	122	2.8 years	4.3% of patients who did not initially present with seizures developed seizures in the follow-up period	2.63% per person-year overall; 4.5% per person-year in patients with a prior hemorrhage; 0.6% per person-year in patients without a prior bleed
Kim et al., 1997	108	1.9 years		3.8% per person-year
Porter et al., 1997	173	3.8 years		4.2% per person-year
Moriarty et al., 1999	68	5.2 years	2.4% per person-year in patients not initially presenting with seizures	3.1% per person-year
Josephson et al., 2011	139	7.3 years	4–6% 5-year risk per person	
Flemming et al., 2012	292	7 years		6.19% per person-year in patients initially presenting with hemorrhage; 2.18% per person-year in patients initially presenting with symptoms not related to hemorrhage; 0.33% in patients whose diagnosis was an incidental finding
Li et al., 2014	331	6.5 years		12.4–15.9% per person-year in patients initially presenting with hemorrhage; 8.7% per person-year in patients not initially presenting with hemorrhage
Moore et al., 2014	107	12.3 years		0.08% per person-year

PRESENTATION

Most patients with cavernous malformations present during the third to fifth decades, with seizures as the most common presenting symptom. Earlier presentations are often attributable to acute hemorrhage, with symptoms dependent on the exact location of the lesion (Voigt and Yasargil, 1976; Yamasaki et al., 1986; Dobyns et al., 1987; Mazza et al., 1989; Robinson et al., 1991; Bergeson et al., 1992).

Though the exact epileptogenic mechanism is unknown, it is not surprising that seizures are the most common presenting symptom, seen in 40–80% of patients with cavernous malformations, or 1.4–2.4% of patients per person-years (Del Curling et al., 1991; Kondziolka et al., 1995; Moriarty et al., 1999; Labauge et al., 2001). The vast majority of cavernous malformations are supratentorial lesions associated with recurrent microhemorrhages causing hemosiderin deposition, calcification, and a local gliomatous reaction, creating a clear epileptognic focus (Voigt and Yasargil, 1976; Giombini and Morello, 1978; Vaquero et al., 1983; Simard et al., 1986; Tagle et al., 1986; Rigamonti et al., 1988; Del Curling et al., 1991; Robinson et al., 1991; Zabramski et al., 1994; Kondziolka et al., 1995; Kim et al., 1997). While seizures are by far the most common presenting symptom across all subpopulations, males have a higher propensity to present with seizures as compared to females (Voigt and Yasargil, 1976; Simard et al., 1986; Rigamonti et al., 1988; Del Curling et al., 1991; Robinson et al., 1991; Zabramski et al., 1994; Kondziolka et al., 1995; Kim et al., 1997).

A total of 8–37% of patients with cavernous malformations present with symptomatic hemorrhage as their initial presentation (Maraire and Awad, 1995), with an annual prospective hemorrhage risk of 0.6–11% per patient-year (Table 23.2) (Robinson et al., 1991; Zabramski et al., 1994; Aiba et al., 1995b; Kondziolka et al., 1995; Kim et al., 1997; Moriarty et al., 1999; Porter et al., 1999; Labauge et al., 2000, 2001; Kupersmith et al., 2001; Mathiesen et al., 2003). Acute or subacute focal deficits are reported in 15–46% of patients, usually due to intralesional or perilesional hemorrhage of brainstem lesions (Giombini and Morello, 1978; Vaquero et al., 1983; Yamasaki et al., 1986; Rigamonti et al., 1988; Abe et al., 1989; Robinson et al., 1991, 1993; Barrow and Krisht, 1993; Lonjon et al., 1993). Often, these deficits improve even without surgical intervention, as the hemorrhage is organized and absorbed. A number of clinical series have found a higher rate of hemorrhage in women and children as compared with men (Hubert et al., 1989; Mazza et al., 1989; Robinson et al., 1991; Scott et al., 1992).

Spinal cord lesions may also present with either acute or subacute symptoms, depending on the exact location of the lesion and extent of hemorrhage (Ogilvy et al., 1992; Sandalcioglu et al., 2003). Symptoms related to spinal cord lesions vary from pain alone, seen in about one-quarter of patients with spinal cord lesions (Kim et al., 2006), to frank signs of myelopathy and radiculopathy (Cosgrove et al., 1988; Ogilvy et al., 1992; Canavero et al., 1994; Cristante and Hermann, 1998; Zevgaridis et al., 1999; Sandalcioglu et al., 2003; Kim et al., 2006; Kharkar et al., 2007). Of note, 40–47% of patients with spinal cord cavernous malformations have been found to have intracranial lesions as well, when undergoing screening imaging (Vishteh et al., 1999; Cohen-Gadol et al., 2006).

CONCLUSION

With the advent of magnetic resonance imaging, cavernous malformations have been recognized as a much more common entity than originally believed; they are accountable for the second leading cause of CNS hemorrhage. Making up 5–15% of all CNS vascular abnormalities, and seen in approximately 0.17–0.9% of the population, the vast majority of cavernous malformations remain clinically silent. When they do present, patients are much more likely to present with seizures, the pathophysiology of which is thought to be due to repetitive microhemorrhages, resulting in a surrounding gliomatous reaction, than with symptoms directly attributable to a large bleed.

REFERENCES

Abdulrauf SI, Kaynar MY, Awad IA (1999). A comparison of the clinical profile of cavernous malformations with and without associated venous malformations. Neurosurgery 44: 41–46; discussion 46–47.

Abe M, Kjellberg RN, Adams RD (1989). Clinical presentations of vascular malformations of the brain stem: comparison of angiographically positive and negative types. J Neurol Neurosurg Psychiatry 52: 167–175.

Aiba T, Koike T, Takeda N et al. (1995a). Intracranial cavernous malformations and skin angiomas associated with middle fossa arachnoid cyst: a report of three cases. Surg Neurol 43: 31–33; discussion 34.

Aiba T, Tanaka R, Koike T et al. (1995b). Natural history of intracranial cavernous malformations. J Neurosurg 83: 56–59.

Al-Shahi R, Bhattacharya JJ, Currie DG et al. (2003). Prospective, population-based detection of intracranial vascular malformations in adults: the Scottish Intracranial Vascular Malformation Study (SIVMS). Stroke 34: 1163–1169.

Al-Shahi Salman R, Hall JM, Horne MA et al. (2012). Untreated clinical course of cerebral cavernous malformations: a prospective, population-based cohort study. Lancet Neurol 11: 217–224.

Awad IA, Robinson Jr JR, Mohanty S et al. (1993). Mixed vascular malformations of the brain: clinical and pathogenetic considerations. Neurosurgery 33: 179–188; discussion 188.

Awada A, Watson T, Obeid T (1997). Cavernous angioma presenting as pregnancy-related seizures. Epilepsia 38: 844–846.

Barrow D, Krisht A (1993). Cavernous malformations and hemorrhage. In: I Awad, D Barrow (Eds.), Cavernous Malformations. American Association of Neurological Surgeons, Park Ridge, IL.

Bergeson PS, Rekate HL, Tack ED (1992). Cerebral cavernous angiomas in the newborn. Clin Pediatr (Phila) 31: 435–437.

Brown Jr RD, Wiebers DO, Torner JC et al. (1996a). Frequency of intracranial hemorrhage as a presenting symptom and subtype analysis: a population-based study of intracranial vascular malformations in Olmsted Country, Minnesota. J Neurosurg 85: 29–32.

Brown Jr RD, Wiebers DO, Torner JC et al. (1996b). Incidence and prevalence of intracranial vascular malformations in Olmsted County, Minnesota, 1965 to 1992. Neurology 46: 949–952.

Canavero S, Pagni CA, Duca S et al. (1994). Spinal intramedullary cavernous angiomas: a literature meta-analysis. Surg Neurol 41: 381–388.

Cha YJ, Nahm JH, Ko JE et al. (2015). Pathological evaluation of radiation-induced vascular lesions of the brain: distinct from de novo cavernous hemangioma. Yonsei Med J 56: 1714–1720.

Cohen-Gadol AA, Jacob JT, Edwards DA et al. (2006). Coexistence of intracranial and spinal cavernous malformations: a study of prevalence and natural history. J Neurosurg 104: 376–381.

Cosgrove GR, Bertrand G, Fontaine S et al. (1988). Cavernous angiomas of the spinal cord. J Neurosurg 68: 31–36.

Craig HD, Gunel M, Cepeda O et al. (1998). Multilocus linkage identifies two new loci for a mendelian form of stroke, cerebral cavernous malformation, at 7p15-13 and 3q25.2-27. Hum Mol Genet 7: 1851–1858.

Cristante L, Hermann HD (1998). Radical excision of intramedullary cavernous angiomas. Neurosurgery 43: 424–430; discussion 430–431.

Cutsforth-Gregory JK, Lanzino G, Link MJ et al. (2015). Characterization of radiation-induced cavernous malformations and comparison with a nonradiation cavernous malformation cohort. J Neurosurg 122: 1214–1222.

Dashti SR, Hoffer A, Hu YC et al. (2006). Molecular genetics of familial cerebral cavernous malformations. Neurosurg Focus 21. e2.

Del Curling Jr O, Kelly Jr DL, Elster AD et al. (1991). An analysis of the natural history of cavernous angiomas. J Neurosurg 75: 702–708.

Denier C, Labauge P, Brunereau L et al. (2004). Clinical features of cerebral cavernous malformations patients with KRIT1 mutations. Ann Neurol 55: 213–220.

Detwiler PW, Porter RW, Zabramski JM et al. (1998). Radiation-induced cavernous malformation. J Neurosurg 89: 167–169.

Di Rocco C, Iannelli A, Tamburrini G (1997). Cavernous angiomas of the brain stem in children. Pediatr Neurosurg 27: 92–99.

Diamond C, Torvik A, Amundsen P (1976). Angiographic diagnosis of teleangiectases with cavernous angioma of the posterior fossa. Report of two cases. Acta Radiol Diagn (Stockh) 17: 281–288.

Dobyns WB, Michels VV, Groover RV et al. (1987). Familial cavernous malformations of the central nervous system and retina. Ann Neurol 21: 578–583.

Dubovsky J, Zabramski JM, Kurth J et al. (1995). A gene responsible for cavernous malformations of the brain maps to chromosome 7q. Hum Mol Genet 4: 453–458.

Flemming KD, Link MJ, Christianson TJ et al. (2012). Prospective hemorrhage risk of intracerebral cavernous malformations. Neurology 78: 632–636.

Gangemi M, Maiuri F, Donati P et al. (1990). Familial cerebral cavernous angiomas. Neurol Res 12: 131–136.

Gil-Nagel A, Dubovsky J, Wilcox KJ et al. (1996). Familial cerebral cavernous angioma: a gene localized to a 15-cM interval on chromosome 7q. Ann Neurol 39: 807–810.

Giombini S, Morello G (1978). Cavernous angiomas of the brain. Account of fourteen personal cases and review of the literature. Acta Neurochir (Wien) 40: 61–82.

Gunel M, Awad IA, Anson J et al. (1995). Mapping a gene causing cerebral cavernous malformation to 7q11.2-q21. Proc Natl Acad Sci U S A 92: 6620–6624.

Hayman LA, Evans RA, Ferrell RE et al. (1982). Familial cavernous angiomas: natural history and genetic study over a 5-year period. Am J Med Genet 11: 147–160.

Hirsh LF (1981). Combined cavernous-arteriovenous malformation. Surg Neurol 16: 135–139.

Hsu F, Rigamonti D, Huhn S (1993). Epidemiology of cavernous malformations. In: I Awad, D Barrow (Eds.), Cavernous Malformations. American Association of Neurological Surgeons, Park Ridge, IL.

Hubert P, Choux M, Houtteville JP (1989). Cerebral cavernomas in infants and children. Neurochirurgie 35: 104–105.

Jellinger K (1986). Vascular malformations of the central nervous system: a morphological overview. Neurosurg Rev 9: 177–216.

Johnson EW, Iyer LM, Rich SS et al. (1995). Refined localization of the cerebral cavernous malformation gene (CCM1) to a 4-cM interval of chromosome 7q contained in a well-defined YAC contig. Genome Res 5: 368–380.

Josephson CB, Leach JP, Duncan R et al. (2011). Seizure risk from cavernous or arteriovenous malformations: prospective population-based study. Neurology 76: 1548–1554.

Kashiwagi S, van Loveren HR, Tew Jr JM et al. (1990). Diagnosis and treatment of vascular brain-stem malformations. J Neurosurg 72: 27–34.

Katayama Y, Tsubokawa T, Maeda T et al. (1994). Surgical management of cavernous malformations of the third ventricle. J Neurosurg 80: 64–72.

Kharkar S, Shuck J, Conway J et al. (2007). The natural history of conservatively managed symptomatic intramedullary spinal cord cavernomas. Neurosurgery 60: 865–872; discussion 865–872.

Kim DS, Park YG, Choi JU et al. (1997). An analysis of the natural history of cavernous malformations. Surg Neurol 48: 9–17; discussion 17–18.

Kim LJ, Klopfenstein JD, Zabramski JM et al. (2006). Analysis of pain resolution after surgical resection of intramedullary spinal cord cavernous malformations. Neurosurgery 58: 106–111; discussion 106–111.

Kondziolka D, Lunsford LD, Kestle JR (1995). The natural history of cerebral cavernous malformations. J Neurosurg 83: 820–824.

Kupersmith MJ, Kalish H, Epstein F et al. (2001). Natural history of brainstem cavernous malformations. Neurosurgery 48: 47–53; discussion 53–44.

Labauge P, Brunereau L, Levy C et al. (2000). The natural history of familial cerebral cavernomas: a retrospective MRI study of 40 patients. Neuroradiology 42: 327–332.

Labauge P, Brunereau L, Laberge S et al. (2001). Prospective follow-up of 33 asymptomatic patients with familial cerebral cavernous malformations. Neurology 57: 1825–1828.

Larson JJ, Ball WS, Bove KE et al. (1998). Formation of intracerebral cavernous malformations after radiation treatment for central nervous system neoplasia in children. J Neurosurg 88: 51–56.

Li D, Hao SY, Jia GJ et al. (2014). Hemorrhage risks and functional outcomes of untreated brainstem cavernous malformations. J Neurosurg 121: 32–41.

Lobato RD, Perez C, Rivas JJ et al. (1988). Clinical, radiological, and pathological spectrum of angiographically occult intracranial vascular malformations. Analysis of 21 cases and review of the literature. J Neurosurg 68: 518–531.

Lonjon M, Roche JL, George B et al. (1993). Intracranial cavernoma. 30 cases. Presse Med 22: 990–994.

Lopate G, Black JT, Grubb Jr RL (1990). Cavernous hemangioma of the spinal cord: report of 2 unusual cases. Neurology 40: 1791–1793.

Maraire JN, Awad IA (1995). Intracranial cavernous malformations: lesion behavior and management strategies. Neurosurgery 37: 591–605.

Maraire JN, Abdulrauf SI, Berger S et al. (1999). De novo development of a cavernous malformation of the spinal cord following spinal axis radiation. Case report. J Neurosurg 90: 234–238.

Martin NA, Vinters H (1990). Pathology and grading of intracranial vascular malformations. In: I Awad, D Barrow (Eds.), Intracranial Vascular Malformations. American Association of Neurological Surgeons, Park Ridge, IL.

Massa-Micon B, Luparello V, Bergui M et al. (2000). De novo cavernoma case report and review of literature. Surg Neurol 53: 484–487.

Mathiesen T, Edner G, Kihlstrom L (2003). Deep and brainstem cavernomas: a consecutive 8-year series. J Neurosurg 99: 31–37.

Mazza C, Scienza R, Dalla Bernardin B et al. (1989). Cerebral cavernous malformations (cavernomas) in children. Neurochirurgie 35: 106–108.

McCormick WF (1984). Pathology of vascular malformations of the brain. In: CB Wilson, BM Stein (Eds.), Intracranial Arteriovenous Malformations. Williams & Wilkins, Baltimore.

Mindea SA, Yang BP, Shenkar R et al. (2006). Cerebral cavernous malformations: clinical insights from genetic studies. Neurosurg Focus 21. e1.

Moore SA, Brown Jr RD, Christianson TJ et al. (2014). Long-term natural history of incidentally discovered cavernous malformations in a single-center cohort. J Neurosurg 120: 1188–1192.

Moriarty JL, Wetzel M, Clatterbuck RE et al. (1999). The natural history of cavernous malformations: a prospective study of 68 patients. Neurosurgery 44: 1166–1171; discussion 1172–1163.

Morris Z, Whiteley WN, Longstreth Jr WT et al. (2009). Incidental findings on brain magnetic resonance imaging: systematic review and meta-analysis. BMJ 339: b3016.

Notelet L, Chapon F, Khoury S et al. (1997). Familial cavernous malformations in a large French kindred: mapping of the gene to the CCM1 locus on chromosome 7q. J Neurol Neurosurg Psychiatry 63: 40–45.

Novelli PM, Reigel DH, Langham Gleason P et al. (1997). Multiple cavernous angiomas after high-dose whole-brain radiation therapy. Pediatr Neurosurg 26: 322–325.

Ogilvy CS, Louis DN, Ojemann RG (1992). Intramedullary cavernous angiomas of the spinal cord: clinical presentation, pathological features, and surgical management. Neurosurgery 31: 219–229; discussion 229–230.

Otten P, Pizzolato GP, Rilliet B et al. (1989). 131 cases of cavernous angioma (cavernomas) of the CNS, discovered by retrospective analysis of 24,535 autopsies. Neurochirurgie 35: 82–83, 128–131.

Porter PJ, Willinsky RA, Harper W et al. (1997). Cerebral cavernous malformations: natural history and prognosis after clinical deterioration with or without hemorrhage. J Neurosurg 87: 190–197.

Porter RW, Detwiler PW, Spetzler RF et al. (1999). Cavernous malformations of the brainstem: experience with 100 patients. J Neurosurg 90: 50–58.

Pozzati E, Acciarri N, Tognetti F et al. (1996). Growth, subsequent bleeding, and de novo appearance of cerebral cavernous angiomas. Neurosurgery 38: 662–669; discussion 669–670.

Rapacki TF, Brantley MJ, Furlow Jr TW et al. (1990). Heterogeneity of cerebral cavernous hemangiomas diagnosed by MR imaging. J Comput Assist Tomogr 14: 18–25.

Requena I, Arias M, Lopez-Ibor L et al. (1991). Cavernomas of the central nervous system: clinical and neuroimaging manifestations in 47 patients. J Neurol Neurosurg Psychiatry 54: 590–594.

Revencu N, Vikkula M (2006). Cerebral cavernous malformation: new molecular and clinical insights. J Med Genet 43: 716–721.

Rigamonti D, Spetzler RF (1988). The association of venous and cavernous malformations. Report of four cases and discussion of the pathophysiological, diagnostic, and therapeutic implications. Acta Neurochir (Wien) 92: 100–105.

Rigamonti D, Hadley MN, Drayer BP et al. (1988). Cerebral cavernous malformations. Incidence and familial occurrence. N Engl J Med 319: 343–347.

Rigamonti D, Johnson PC, Spetzler RF et al. (1991). Cavernous malformations and capillary telangiectasia: a spectrum within a single pathological entity. Neurosurgery 28: 60–64.

Robinson JR, Awad IA, Little JR (1991). Natural history of the cavernous angioma. J Neurosurg 75: 709–714.

Robinson Jr JR, Awad IA, Magdinec M et al. (1993). Factors predisposing to clinical disability in patients with cavernous malformations of the brain. Neurosurgery 32: 730–735; discussion 735–736.

Safavi-Abbasi S, Feiz-Erfan I, Spetzler RF et al. (2006). Hemorrhage of cavernous malformations during pregnancy and in the peripartum period: causal or coincidence? Case report and review of the literature. Neurosurg Focus 21. e12.

Sage MR, Brophy BP, Sweeney C et al. (1993). Cavernous haemangiomas (angiomas) of the brain: clinically significant lesions. Australas Radiol 37: 147–155.

Sandalcioglu IE, Wiedemayer H, Gasser T et al. (2003). Intramedullary spinal cord cavernous malformations: clinical features and risk of hemorrhage. Neurosurg Rev 26: 253–256.

Scott RM, Barnes P, Kupsky W et al. (1992). Cavernous angiomas of the central nervous system in children. J Neurosurg 76: 38–46.

Simard JM, Garcia-Bengochea F, Ballinger Jr WE et al. (1986). Cavernous angioma: a review of 126 collected and 12 new clinical cases. Neurosurgery 18: 162–172.

Singla A, Brace O'Neill JE, Smith E et al. (2013). Cavernous malformations of the brain after treatment for acute lymphocytic leukemia: presentation and long-term follow-up. J Neurosurg Pediatr 11: 127–132.

Tagle P, Huete I, Mendez J et al. (1986). Intracranial cavernous angioma: presentation and management. J Neurosurg 64: 720–723.

Vaquero J, Leunda G, Martinez R et al. (1983). Cavernomas of the brain. Neurosurgery 12: 208–210.

Vishteh AG, Zabramski JM, Spetzler RF (1999). Patients with spinal cord cavernous malformations are at an increased risk for multiple neuraxis cavernous malformations. Neurosurgery 45: 30–32; discussion 33.

Voigt K, Yasargil MG (1976). Cerebral cavernous haemangiomas or cavernomas. Incidence, pathology, localization, diagnosis, clinical features and treatment. Review of the literature and report of an unusual case. Neurochirurgia (Stuttg) 19: 59–68.

Weber M, Vespignani H, Bracard S et al. (1989). Intracerebral cavernous angioma. Rev Neurol (Paris) 145: 429–436.

Wilkins RH (1985). Natural history of intracranial vascular malformations: a review. Neurosurgery 16: 421–430.

Wilms G, Bleus E, Demaerel P et al. (1994). Simultaneous occurrence of developmental venous anomalies and cavernous angiomas. AJNR Am J Neuroradiol 15: 1247–1254; discussion 1255–1247.

Wong JH, Awad IA, Kim JH (2000). Ultrastructural pathological features of cerebrovascular malformations: a preliminary report. Neurosurgery 46: 1454–1459.

Yamasaki T, Handa H, Yamashita J et al. (1986). Intracranial and orbital cavernous angiomas. A review of 30 cases. J Neurosurg 64: 197–208.

Zabramski JM, Wascher TM, Spetzler RF et al. (1994). The natural history of familial cavernous malformations: results of an ongoing study. J Neurosurg 80: 422–432.

Zevgaridis D, Medele RJ, Hamburger C et al. (1999). Cavernous haemangiomas of the spinal cord. A review of 117 cases. Acta Neurochir (Wien) 141: 237–245.

Handbook of Clinical Neurology, Vol. 143 (3rd series)
Arteriovenous and Cavernous Malformations
R.F. Spetzler, K. Moon, and R.O. Almefty, Editors
http://dx.doi.org/10.1016/B978-0-444-63640-9.00024-2

Chapter 24

Radiology and imaging for cavernous malformations

KEVIN Y. WANG[1,2], OLUWATOYIN R. IDOWU[2], AND DORIS D.M. LIN[2]*
[1]*Department of Radiology, Baylor College of Medicine, Houston, TX, USA*
[2]*Division of Neuroradiology, Russell H. Morgan Department of Radiology and Radiological Science, Johns Hopkins University School of Medicine, Baltimore, MD, USA*

Abstract

Cavernous malformations are low-flow vascular malformations that are histologically characterized by the lack of mural elements of mature vascular structures and intervening parenchymal neural tissue. They are often clinically quiescent, and may grow, bleed, and regress, but can also manifest clinically as neurologic deficits or seizures in the setting of an acute hemorrhage. The low-flow nature of cavernous malformations renders them inherently occult on cerebral angiography. Magnetic resonance imaging has become the mainstay imaging modality in evaluating cavernous malformations, producing characteristic imaging features that usually provide a straightforward diagnosis. Features on magnetic resonance imaging include a reticulated pattern of mixed hyper- and hypointensity on T1- and T2-weighted imaging, with a characteristic hypointense rim best appreciated on T2-weighted imaging or gradient-echo sequences. Contrast enhancement is useful for revealing coexisting developmental venous anomalies that are frequently associated with sporadic cavernous malformations, and may further support the diagnosis. Susceptibility-weighted imaging is highly sensitive for cavernous malformations and accompanying developmental venous anomalies, and is superior to gradient-echo sequences in screening for multifocal, familial cavernous malformations.

INTRODUCTION

Cavernous malformations (CMs) are low-flow vascular malformations found in both the supratentorial and infratentorial regions of brain, and less commonly in the spinal cord. They are also often referred to as cavernous angiomas, cavernous hemangiomas, cryptic vascular malformations, or cavernomas. Accounting for approximately 5–15% of all central nervous system (CNS) vascular malformations, CMs are the second most common vascular malformations in the CNS after developmental venous anomalies (DVAs), and the two are often found concurrently (Maraire and Awad, 1995; Moriarity et al., 1999a, b). These lesions on histology characteristically lack mural elements of mature vascular structures and intervening parenchymal neural tissue (Tomlinson et al., 1994). Multiple lesions are more often associated with familial forms of CMs (Zabramski et al., 1994). CMs are often clinically quiescent, but can be dynamic lesions that grow, bleed, and regress, and may manifest clinically as neurologic deficits or seizures. Hemorrhage is not an uncommon complication with a varying reported annual rate of bleed (Batra et al., 2009). Neuroimaging plays a crucial role in diagnosis and monitoring. The increasing number of incidental CMs diagnosed preoperatively with the advent of magnetic resonance imaging (MRI) has prompted the need to better elucidate the natural history in order to provide optimal timing for treatment of such lesions. In this chapter, we start by reviewing past and current imaging modalities used in the diagnosis of CMs, provide illustrative examples of typical and atypical cases, review alternative entities with resembling imaging features, and conclude with key points from a neuroimaging standpoint.

*Correspondence to: Doris D.M. Lin, M.D., Ph.D., 600 North Wolfe Street, Phipps B-100, Baltimore MD 21287, USA. Tel: +1-443-287-3079, E-mail: ddmlin@jhmi.edu

ROLE OF NEUROIMAGING: PAST AND CURRENT MODALITIES

Prior to the advent of computed tomography (CT) and MRI, detection of CMs was often challenging and rarely were CMs diagnosed preoperatively. Skull radiographs and radionuclide scanning were neither sensitive nor specific. The low-flow nature of CMs rendered them inherently occult on cerebral angiography (McCormick, 1966). The significantly increased sensitivity on CT provided a first glimpse at a more accurate incidence rate of CMs, given up to 40% of cases were clinically quiescent (Batra et al., 2009). MRI is now the most important and sensitive diagnostic technique for CMs, producing characteristic imaging features that usually provide a straightforward diagnosis. In this section, we highlight each of the above-mentioned modalities as well as advances in techniques in MRI that have further enhanced detection and evaluation of CMs.

Skull radiographs

The first report of the use of conventional skull radiograph for CM was described in 1969 (Runnels et al., 1969). Skull radiographs may reveal granular or coarse macroscopic calcifications in approximately 7–40% of cases (Voigt and Yasargil, 1976; Giombini and Morello, 1978; Savoiardo et al., 1983b; Yamasaki et al., 1986; Rigamonti et al., 1987), and are thought to reflect microscopic calcifications not infrequently seen on histology. Associated sella turcica osseous changes suggestive of increased intracranial pressure may also be appreciated in patients with CMs (Savoiardo et al., 1983b). Overall, conventional skull radiographs are neither sensitive nor specific for CMs.

Radionuclide scanning

Radionuclide brain scans may demonstrate increased isotope uptake of Technetium 99m in CMs, but are generally normal, and fail to characterize the vascular nature of the abnormality. Its utility diminished with the introduction of CT. Sensitivity for CMs has been variable, with abnormally increased uptake reported in 42–100% of cases (Kramer and Wing, 1977; Savoiardo and Passerini, 1978; Becker et al., 1979; Savoiardo et al., 1983b; Yamasaki et al., 1986). It has been suggested that lesions must be larger than 20 mm in diameter to be sufficiently detectable by radionuclide scans (Kawai et al., 1978). The size of the lesion estimated on radionuclide scans is often smaller than those on CT (Savoiardo and Passerini, 1978). Radionuclide scans may have been useful as a complementary study during the time when CT was the mainstay for diagnosis, as reported by Yamasaki et al. (1986); in this study additional diagnostic information was obtained in 40% of lesions.

Cerebral angiography

Angiographic detection of CMs remains challenging, and is thought to be secondary to small feeding vessels, low-flow circulation, and thrombi within vascular spaces in CMs (Savoiardo and Passerini, 1978). Advances in techniques such as rapid serial angiography, magnification angiography, and film subtraction technique have made detection increasingly possible, albeit nevertheless difficult (Jensen et al., 1963).

Early reports described an avascular mass with displacement of adjacent vessels, but without evidence of pathologic vasculature (Schneider and Liss, 1958; Kamrin and Buchsbaum, 1965; Ishijima et al., 1966; Jain and Robertson, 1970). The first reported case with definitive angiographic abnormality was noted by Jonutis et al. (1971), demonstrating early venous filling and a persistent capillary blush into the late venous phase. Increased sensitivity for a vascular blush has been reported with prolonged contrast injection (Numaguchi et al., 1979; Yamasaki et al., 1986). Bogren et al. (1970) also reported the occasional presence of dilated draining veins. However, the latter two appear to be rare findings (Vaquero et al., 1983). Rather, the most common angiographic abnormality is ultimately avascular regions with displacement of vessels (Savoiardo et al., 1983b; Tagle et al., 1986).

Despite advances in angiographic techniques, studies report approximately 20–85% of cases failed to exhibit abnormal findings, suggesting the limited efficacy of angiography in detection and diagnosis (Hayman et al., 1982; Savoiardo et al., 1983b; Vaquero et al., 1983; Tagle et al., 1986; Yamasaki et al., 1986; Rigamonti et al., 1988). Variability in sensitivity may be attributed to differences in technique, such as the amount of contrast used or the rate of injection (Savoiardo et al., 1983a). Furthermore, the angiographic findings are also frequently nonspecific, with blushing and early draining associated with other vascular malformations, neoplasms, ischemia, and inflammatory processes (Savoiardo et al., 1983b). Not surprisingly, the avascular finding commonly seen on angiography leads to the characteristic designation of CMs as angiographically occult lesions.

It is not uncommon for multiple types of vascular malformations to coexist within the CNS. MRI may be sensitive, but less specific and straightforward compared to angiography in the diagnosis. Such cases may benefit from the use of angiography in excluding other lesions that are characteristic and specific on angiography, such as arteriovenous malformations (AVMs) and DVAs.

Computed tomography

With the advent of CT, the sensitivity in the preoperative diagnosis of CMs increased significantly. CT is superior to angiography in delineation of the morphology, boundaries, and location of lesions (Savoiardo et al., 1983b). Early series reported detection rates upwards of 100% utilizing CT (Kramer and Wing, 1977; Savoiardo et al., 1983b; Vaquero et al., 1983; Tagle et al., 1986). However, CT scanners at the time were limited in resolution to detect small lesions (Terao et al., 1979; Rigamonti et al., 1987) and sometimes even relatively large ones (Rigamonti et al., 1987). False negatives have also been reported in approximately 10% of cases (Rigamonti et al., 1988).

Images without contrast typically demonstrate hyperdense lesions, and less commonly, mixed hyper- and isodense lesions. Calcification within the lesion, minimal mass effect, and poor contrast enhancement are additional findings associated with CMs (Vaquero et al., 1983). Reported descriptions of occasional faint contrast enhancement have been shown to improve the delineation of and increase the sensitivity for isodense lesions (Savoiardo et al., 1983b; Vaquero et al., 1983). Pozzati et al. (1980) proposed the following features as indicative of CMs: (1) roundish shape, well-defined margin, and a slightly uneven high density in the calcification range; (2) minimal or no enhancement; (3) the absence of surrounding edema and mass effect (unless an intracerebral hemorrhage occurs).

Findings on CT are, nevertheless, nonspecific. For example, there is significant challenge in differentiating CMs from partially calcified, avascular low-grade gliomas (Kramer and Wing, 1977). Tagle et al. and Vaquero et al. reported that CT findings provided a confident, preoperative diagnosis in only 1 of 13 cases (Tagle et al., 1986) and 7 of 16 cases (Vaquero et al., 1983), respectively. Rigamonti et al. reported that CT correctly identified CMs in 14 of 27 lesions (Rigamonti et al., 1987) and 20 of 56 lesions (Rigamonti et al., 1988), respectively. Angiography in combination with CT may help with differentiation and increase confidence in the preoperative diagnosis (Savoiardo et al., 1983a; Servo et al., 1984), with Servo et al. (1984) reporting a correct preoperative diagnosis in all five of such cases based on histologic verification. Ultimately, it is noteworthy to mention that the incidence of nonsurgical and incidental cases of CMs increased significantly with the introduction of CT technology, and consequently raised a fundamental question regarding the decision for surgical and/or radiosurgical intervention.

Magnetic resonance imaging

Spin-echo sequences

In 1987, Rigamonti et al. demonstrated that MRI at 1.5 T is both sensitive and specific for the detection of CMs. T2-weighted imaging (T2WI) was 100% sensitive, whereas T1-weighted imaging (T1WI) was less so, detecting 23 of 27 lesions. Horowitz and Kondziolka (1995) has proposed that the characteristic and consistent MR appearance of histologically confirmed CMs may warrant the diagnosis on the basis of imaging characteristics alone with acceptable confidence. MRI has since become and continues to be the imaging modality of choice for identification and characterization of CMs.

Hemosiderin deposition in and around CMs, thought to be secondary to either repeated subclinical hemorrhages or lysis of erythrocytes (Pozzati et al., 1981; Bellotti et al., 1985), results in magnetic susceptibility, particularly at high field strength, and consequently manifests as a rim of prominently decreased signal intensity on T2WI. A rim of hypointensity on T1WI also may be observed, but to a lesser extent (Gomori et al., 1986). The magnetic field inhomogeneity in the presence of hemosiderin is also useful for distinguishing blood flow in AVMs from the hemosiderin effects in CMs (Rigamonti et al., 1987). Overall, a combination of a rim of decreased signal with a reticulated-appearing core of mixed hyper- and hypointensity on T2WI is of high diagnostic likelihood of CMs. For smaller lesions, a punctate area of T2 hypointensity may be appreciated instead (Rigamonti et al., 1987). Accompanying vasogenic edema, seen as elevated signal on T2WI, and mass effect are usually not present unless recent bleeding has occurred (Batra et al., 2009). CMs may appear as a region of isointensity or hypointensity on T1WI (Rigamonti et al., 1987), although the presence of methemoglobin in CMs can occasionally contribute to the appearance of mixed internal hyperintensity on both T1WI and T2WI (Gomori et al., 1985; Batra et al., 2009). More recently, a report of T1 hyperintense perilesional signal has been described and associated with CMs (Yun et al., 2008), and is thought to be due to extravascular leakage of red blood cells and plasma during edema formation. The perilesional signal appears to be more commonly found in CMs associated with recent overt hemorrhaging, and in such cases was reported to be highly specific and predictive for the diagnosis of CM.

Gadolinium contrast enhancement is variable, ranging from none to moderate (Kattapong et al., 1995; Batra et al., 2009). However, contrast-enhanced MRI may be useful for delineating other entities such as neoplasms, AVMs, or coexisting DVAs. Rigamonti and Spetzler (1988) initially determined an association between DVAs and CMs, which have been subsequently reported in nearly one-third of CM cases (Dillon, 1997; Batra et al., 2009), but have only been associated in sporadic but not familial forms of CMs (Abdulrauf et al., 1999; Petersen et al., 2010). The coexistence of CMs and DVAs was initially thought to be incidental. However, based on the observation that most CMs appear at the distal radical vein of DVAs, it has been postulated that DVAs may contribute to the formation of CMs (Dillon, 1997). The contradiction between reports of DVAs causing neurologic symptoms and pathologic studies reporting isolated DVAs as truly benign lesions may be explained by coexisting yet undetected CMs, particularly in the pre-MRI era (Rigamonti and Spetzler, 1988). Therefore, patients with "symptomatic" DVAs may benefit from high-field MRI to rule out the presence of coexisting CMs that may be a confounding cause of neurologic symptoms.

Gradient-echo sequence

In 1998, Labauge et al. reported findings that suggested higher sensitivity of gradient-echo (GRE) sequence for the detection of CMs. The CMs demonstrated greater signal loss, particularly on T2*-weighted GRE, when compared to conventional spin-echo sequences. This finding was confirmed in subsequent studies (Porter et al., 1997; Brunereau et al., 2000, 2001; Lehnhardt et al., 2005). In the study of 16 patients by Labauge et al. (1998), 5 demonstrated multiple lesions on GRE when only a single lesion was detected on conventional spin-echo sequence. Three patients also demonstrated lesions exclusively on GRE. The false-negative rate of spin-echo sequence for detecting CM lesions was approximately 5%. Mean number of lesions per person on spin-echo sequence was 5, compared to 16 for GRE. In the study by Brunereau et al. (2000), mean number of lesions detected by spin-echo was 7.2, when compared to 20.2 on GRE. Lehnhardt et al. (2005) reported in a study of five family members with familial CMs an underestimation of lesion number in 2 patients and failure to detect multiple lesions in 1 patient on spin-echo sequence.

With increased utilization of MRI, it has become apparent that familial CMs are more common than previously thought (Zabramski et al., 1994), accounting for 50% of patients with symptomatic CMs in one study (Rigamonti et al., 1988). Moreover, the increased sensitivity of GRE is particularly useful for detection of multiple intracerebral CMs, as there is a decline in sensitivity on spin-echo sequence from the detection of single to multiple lesions (97% vs. 93%) (Brunereau et al., 2000). Brunereau et al. (2000) proposed that any patient who demonstrates a single lesion on spin-echo but multiple lesions on GRE images should be suspected of having the familial form of CMs. Because familial cases often present with multiple small intracerebral lesions that may be inconspicuous on conventional spin-echo sequences, the use of GRE as the definitive modality for diagnosing familial cases of CMs has been proposed by multiple authors (Kattapong et al., 1995; Brunereau et al., 2000; Lehnhardt et al., 2005). However, susceptibility-induced blooming artifacts on GRE sequences may limit the characterization of the lesion (Sigal et al., 1990). Conventional spin-echo sequences are useful in the analysis of the size and signal of lesions, whereas GRE is advantageous in assessing extent of lesion burden by the means of its higher sensitivity (Brunereau et al., 2001). Higher field strengths would further enhance lesion detection, as the susceptibility effect would be correspondingly accentuated (Batra et al., 2009).

Zabramski et al. (1994) proposed a classification system with four different categories based on spin-echo and GRE features and histopathologic correlation. Type I lesions are characterized by a hyperintense core on T1WI and a hypo- or hyperintense core on T2WI depending on the intracellular or extracellular stage of methemoglobin, and are intended to reflect CMs complicated by subacute hemorrhages. Type II lesions are characterized by features that are now commonly ascribed in the literature as being pathognomonic of the MRI features of CMs, and entails a reticulated core of mixed high and low T2 signal intensity with a surrounding hypointense ring, thought to correlate pathologically to regions of ongoing thomboses and hemorrhages of different ages. Type III lesions are characterized by marked hypointensity on T2WI and magnification of size of the hypointensity on GRE, with iso- or hypointensity seen on T1WI, and are intended to reflect chronic hemorrhage with residual hemosiderin residing within and surrounding the lesion. Type IV lesions are not well understood, and are poorly visualized on conventional spin-echo sequences. They appear as small punctate hypointense lesions on GRE, and are thought to reflect small deposits of hemosiderin in either minute CMs or perhaps capillary telangiectasias.

The significance of the MR classification of these lesions is unclear, although Zabramski et al. (1994) report that the clinical severity of CMs may be related to the appearance on MRI. Patients with type I or II lesions were nearly uniformly symptomatic, whereas only one-third of patients with type III or IV lesions were symptomatic. Exacerbation of symptoms was more frequently associated with type I lesions. Type I or II lesions were at significantly higher risk of recurrent hemorrhage, both clinically and subclinically. Brunereau et al. (2000) also reported that type I and II lesions were more numerous in the symptomatic group, whereas type III and IV were equally distributed between symptomatic and asymptomatic groups. On the other hand, Willinsky et al. (1996) found that the classification was not useful for predicting future bleeds. Interestingly, none of the type I or II lesions in the study by Zabramski et al. (1994) exhibited a conversion to type III lesions as one may expect with the natural evolution and resorption of blood to hemosiderin, supporting the theory that there is perhaps a perpetual process of ongoing bleeding occurring.

Susceptibility-weighted imaging

In 1999, Essig et al. and Lee et al. reported a MRI technique using three-dimensional GRE, known as susceptibility-weighted imaging (SWI) or high-resolution venography. SWI detects CMs by taking advantage of the blood oxygen level dependent (BOLD) induced phase effects between venous blood and the surrounding brain parenchyma, which allows for characterization of small, low-flow venous vessels at the submillimeter resolution, and thus, is particularly useful for screening for suspected CMs, capillary telangiectasias, and DVAs without the need for contrast enhancement (Essig et al., 1999; Lee et al., 1999). In the study by Lee et al. (1999), two of 10 cases demonstrated additional lesions on SWI that were not apparent on T2*GRE, and overall, margins of lesions were better delineated on SWI. Subsequent studies have also demonstrated higher sensitivity of SWI in detecting multifocal, familial CMs when compared to T2*GRE (Cooper et al., 2008; de Souza et al., 2008; de Champfleur et al., 2011; Bulut et al., 2014). The superiority of SWI to T2*GRE for sporadic CMs is less clear, and thus far, de Champfleur et al. (2011) failed to find significantly higher sensitivity on SWI. Bulut et al. (2014) recently proposed incepting a type V lesion to the Zabramski classification for lesions detected by SWI but not T2*GRE.

SWI has the advantage of detecting CM and telangiectasias that have not yet demonstrated overt hemorrhage (Campbell et al., 2010). However, due to prominent susceptibility artifact in the presence of paramagnetic hemosiderin from chronic vascular stasis or prior bleeds, SWI typically overestimates the size of lesions. It may also misrepresent cortically based CMs as blood vessels (Essig et al., 1999). Small lesions can be difficult to differentiate from small hemorrhages (Campbell et al., 2010). Correlation with conventional spin-echo sequences is thus useful for further delineating anatomic detail. Pinker et al. (2007) demonstrated the use of high-resolution SWI at 3 T, and despite the increased susceptibility effects of the higher field strength, it allowed for significant conspicuity of intralesional tubular structures of CMs, which corresponded pathologically to hyaline, collagenous vascular channels. Serial SWIs may be useful in establishing whether CMs that increase in number or size with time arise *de novo* or merely represent subsequent bleeding from previously unrecognized small malformations (Lee et al., 1999).

ILLUSTRATIVE CASES

In this section, case-based illustrations of neuroimaging findings with modalities most commonly used in current clinical practice in the evaluation of CMs are presented. CT and MRI findings of typical features of intraparenchymal CMs, extra-axial CMs, spinal cord CMs, familial CMs, and CMs in the setting of an acute or recent bleed will be discussed.

Typical features of cavernous malformations

On noncontrast CT, CMs can be occult isodense lesions or appear as high-density lesions with associated calcification. They may mildly enhance with administration of intravenous iodinated contrast. Perilesional edema and mass effect are not typical of uncomplicated CMs. Figure 24.1 shows the appearance of CMs on noncontrast CT in three different cases, all demonstrating mildly hyperattenuated lesions without surrounding edema or mass effect. At times, associated calcification may be seen, such as in Figure 24.1C. Both small lesions in Figure 24.1A and B were identified incidentally. The appearance of CM on CT remains nonspecific and may mimic a previous granulomatous infection, other vascular lesions such as an aneurysm or AVMs, a subacute hematoma, or a hyperattenuated neoplasm such as a meningioma.

Features on MRI are highly characteristic and diagnostic for CMs. A characteristic hypointense rim surrounding the lesion is exhibited on all sequences, but most prominently on T2WI, GRE, and SWI. Figure 24.2 illustrates a rim of peripheral hypointensity that is easily noted on T2WI and T2 fluid-attenuated inversion recovery (FLAIR), but not T1WI. The hypointensity may magnify, extend into, and "stain" the adjacent brain parenchyma when imaged with GRE and SWI due to the "blooming effect" of hemosiderin, effectively increasing the apparent size of the lesion, and can also be depicted by the diffusion-weighted echoplanar sequence on b_0 (Fig. 24.2A). When very small, CM can be completely inconspicuous on most of these conventional sequences and be evident only on high-resolution SWI (Fig. 24.2B).

The center of the lesion may be hypointense on T2WI if the lesion is small and punctate, and there has been no recent hemorrhage. The center of larger CMs will exhibit

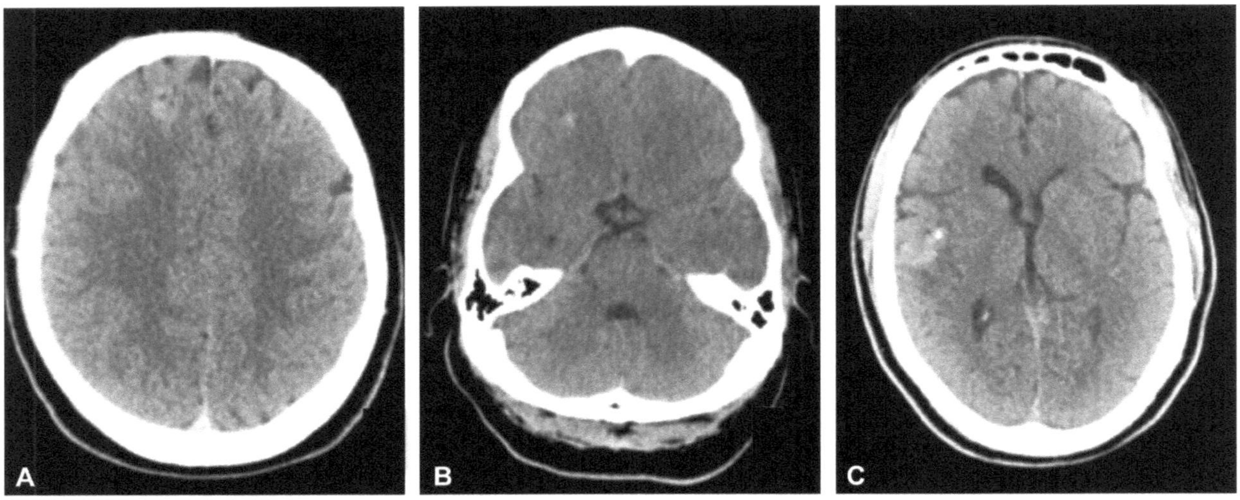

Fig. 24.1. Computed tomography (CT) appearance of cavernous malformation (CM). (**A**) An 18-year-old man with a known seizure disorder: head CT shows a mildly hyperattenuated lesion in the right frontal lobe without adjacent edema or mass effect. (**B**) A 27-year-old woman with polymyositis on chronic steroids and after a recent fall. Small hyperattenuated lesion was incidentally noted in the right frontal lobe. (**C**) A 40-year-old man complaining of headache. CT shows a moderate-sized (3 cm) hyperattenuation in the right temporal region, including a small and more pronounced density medially, suggestive of focal calcification or acute hemorrhage. Minimal edema is present. The lesion was resected and pathology revealed a CM.

the characteristic reticulated pattern of hypo- and hyperintense signal on T1WI and T2WI. Figure 24.3 shows a right frontal CM with the classic "mulberry-like" or "popcorn" pattern comprising small reticular spaces of mixed hypo- and hyperintense signals on T2WI and FLAIR, with similar reticulated pattern of intermediate and hyperintense signal seen on T1WI. This mixed, heterogeneous signal pattern is thought to reflect blood products of different ages from ongoing repeated subclinical intralesional hemorrhages, and in this case, is a reflection of methemoglobin at the subacute stage.

Gadolinium contrast enhancement is variable, ranging from none to moderate. It is more useful for revealing coexisting DVAs, seen in nearly a third of CMs, and if present, may increase the diagnostic confidence of the CM in question. Figure 24.4 shows the MRI of the same case as in Figure 24.1B, with rim enhancement of the CM on T1WI, and with a coexisting DVA exhibiting the classic caput medusae enhancement pattern, making the diagnosis of CM of high likelihood.

SWI is particularly useful for screening for suspected CMs and DVAs due to its high sensitivity. When compared to GRE, SWI allows for better delineation of margins of lesions as well as sensitivity for lesions in familial CMs. Figure 24.5A shows a lesion in the left middle cerebellar peduncle that is poorly delineated on GRE, but more conspicuously seen and well defined on SWI. SWI also shows an additional punctate lesion in the midpons that is not detected on GRE. Figure 24.5B demonstrates mixed intermediate but predominantly hypointense signal near the right lentiform nucleus on GRE. On SWI, a coexisting DVA is depicted that is not seen on GRE, and the CM itself is demonstrating the "blooming effect" with a magnification of the area of dark signal.

Extra-axial cavernous malformations

The nomenclature for CMs is confusing, as they are often alternatively referred to as cavernous hemangiomas or cavernous angiomas. Whereas cavernous hemangiomas in soft tissues are defined as encapsulated, hypervascular tumors, the "cavernous hemangiomas" of the CNS often refer to true vascular malformations, as described in this chapter, and are not thought to be neoplastic despite the misleading terminology (Gonzalez et al., 2006).

The interchangeability of the terms "CM" and "cavernous hemangioma" is especially confusing in the context of discussing extra-axial lesions. Extra-axial "cavernous hemangiomas" have been reported in the cerebellopontine angle (Brunori and Chiappetta, 1996; Kim et al., 1997; Beskonakli et al., 2002), cavernous sinus (Maruishi et al., 1994; Bristot et al., 1997), other dural-based locations (Meyer et al., 1990; Vogler and Castillo, 1995), cranial nerves (Adachi et al., 2008; Manjila et al., 2011), intraosseously (Dorner et al., 2005), and intraventricularly (Reyns et al., 1999; Suess et al., 2002; Nieto et al., 2003). However, extra-axial lesions that are likely originating from or associated with the dura mater (cavernous sinus, spinal epidural

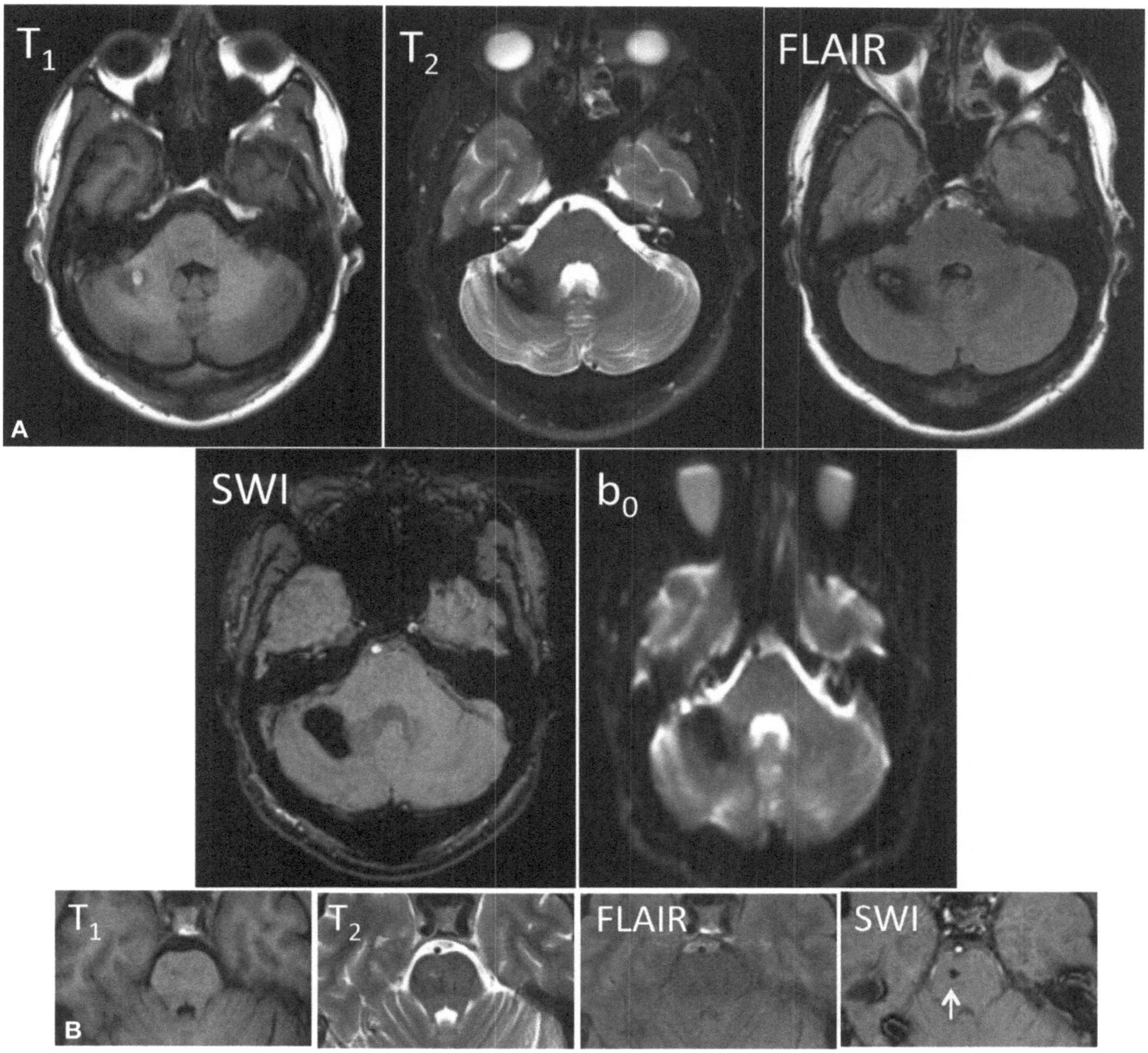

Fig. 24.2. Magnetic resonance imaging appearance of cavernous malformation on various pulse sequences in a 24-year-old man. (**A**) The larger lesion in the right middle cerebellar peduncle shows a small T1 hyperintensity with dark peripheral rim, and is much more conspicuous on T_2 and fluid-attenuated inversion recovery (FLAIR) sequences with prominent dark peripheral signal reflecting hemosiderin staining. The hemosiderin stain is further exaggerated on susceptibility-weighted imaging (SWI), as well as the b_0-echo planar imaging diffusion-weighted sequence, due to increased susceptibility or dephasing artifact – so-called "blooming" artifact. (**B**) A second punctate lesion is inconspicuous on conventional images, but shows up as a punctate dark signal on SWI in the right paramedian pons (arrow).

space, and dural convexities) are clinically, radiologically, and histopathologically distinct from intraparenchymal CMs (Gonzalez et al., 2006), raising uncertainty over the exact pathology of such lesions, and as to whether they are true vascular malformations similar to intraparenchymal CMs.

Clinically, dural-based extra-axial lesions exhibit behavior similar to those of neoplastic lesions, as they are often noted to enlarge over time, compress adjacent structures, cause mass effect, and encase neurovascular structures (Sohn et al., 2003; Gonzalez et al., 2006; Jinhu et al., 2008). In contrast to intra-axial CMs, the rates of hemorrhage are low (Kanaan et al., 2001; Deshmukh et al., 2003; Sohn et al., 2003), and symptomatic manifestation is thought not to be secondary to acute bleeds. Rather, in the case of cavernous sinus

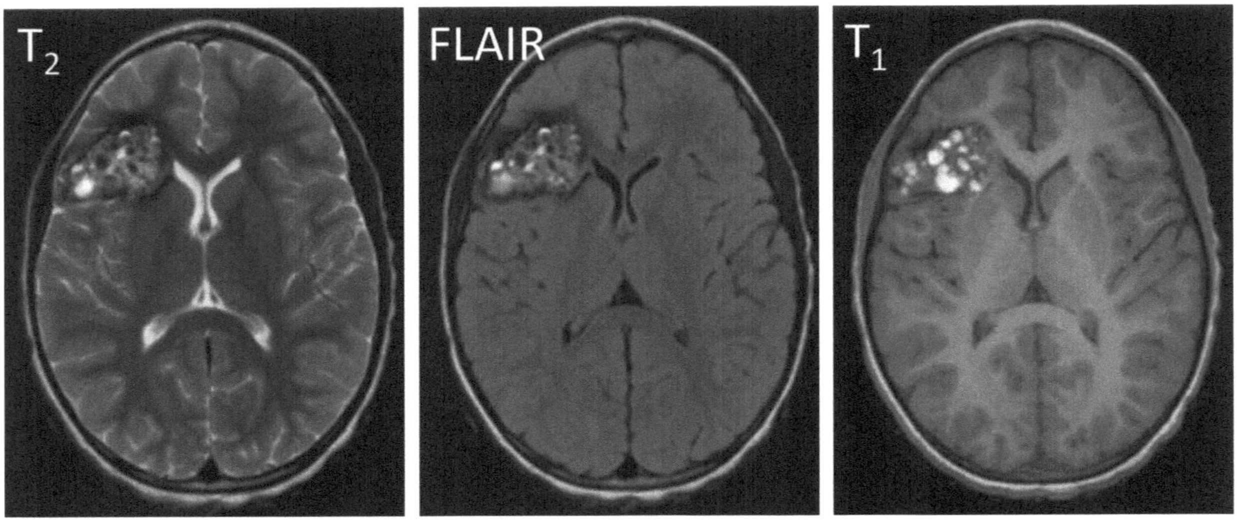

Fig. 24.3. Classic magnetic resonance appearance of cavernous malformation (Zabramski type II). Right frontal reticulated lesion (mulberry-like) containing blood of different ages (in this case, mostly extracellular methemoglobin, of subacute stage) and dark hemosiderin rim, without surrounding edema or substantial mass effect. FLAIR, fluid-attenuated inversion recovery.

lesions, symptoms manifest as cranial nerve palsies, hypopituitarism, other findings secondary to compression, or exophthalmos (Gonzalez et al., 2006).

Neuroimaging features of dural-based extra-axial lesions allow differentiation from the intra-axial counterpart. Lesions are often hypo- to isointense on T1WI, and markedly hyperintense on T2WI, and do not exhibit the classic dark hemosiderin rim seen in intra-axial CMs (Kanaan et al., 2001; Sohn et al., 2003; Salanitri et al., 2004; Jinhu et al., 2008). While intra-axial CMs show variable degrees of contrast enhancement, dural-based extra-axial lesions tend to consistently exhibit avid homogeneous enhancement (Vogler and Castillo, 1995; Kanaan et al., 2001; Sohn et al., 2003; Salanitri et al., 2004; Dorner et al., 2005; Jinhu et al., 2008), reflecting the high-flow, hypervascular state of these lesions. A pathologically documented cavernous sinus hemangioma with findings that typify such dural-based extra-axial lesions is depicted in Figure 24.6, which shows a large, well-defined T2 hyperintense parasellar cavernous sinus mass with mass effect on the surrounding structures. Lastly, on cerebral angiography, such extra-axial lesions appear highly vascular, contrasting them to intra-axial CMs, which are low-flow and angiographically occult (Kanaan et al., 2001; Sohn et al., 2003).

Histopathologically, dural-based extra-axial lesions may exhibit very similar features to those of intra-axial CMs, further obfuscating the means to distinguish extra-axial lesions as either neoplasms or vascular malformations. Both extra- and intra-axial lesions exhibit dilated vessels lined with a single layer of endothelium without smooth muscle or elastic lamina (Gonzalez et al., 2006). There is also no evidence of hemorrhage or hemosiderin-laden macrophages on histology in dural-based lesions (Gonzalez et al., 2006). Evidence of ongoing, repeated microhemorrhages and resulting deposition of hemosiderin is a consistent feature of CMs (Zabramski et al., 1994). The absence of hemosiderin-laden macrophages and corresponding absence of dark hemosiderin rims on MRI further distinguish dural-based lesions from intraparenchymal CMs.

In contrast to the more distinct findings of dural-based extra-axial lesions, radiologic findings of extra-axial lesions likely originating from neural tissue (lesions at the cerebellopontine angle, intraventricularly, or associated with cranial nerves) are less straightforward, more variable, and less extensively described in the literature. Authors have reported cases with MRI findings that closely resemble those of intraparenchymal CMs, including mixed T2 hypo- and hyperintensity pattern, hemosiderin rim staining, variable levels of contrast enhancement, and signal voids reflecting intracellular methemoglobin (Brunori and Chiappetta, 1996; Kim et al., 1997; Ferrante et al., 1998; Beskonakli et al., 2002; Deshmukh et al., 2003; Kivelev et al., 2010). Intraventricular lesions were consistently described as either Zabramski type I or II lesions in one study (Kivelev et al., 2010). More strikingly, 2 cases within the lateral ventricle in that study were found to have coexisting DVAs on MRI. Yet, extra-axial lesions that are associated with cranial nerves have also exhibited imaging findings similar to dural-based lesions (T2 hyperintensity with homogeneous contrast enhancement) (Chow et al., 2002; Manjila

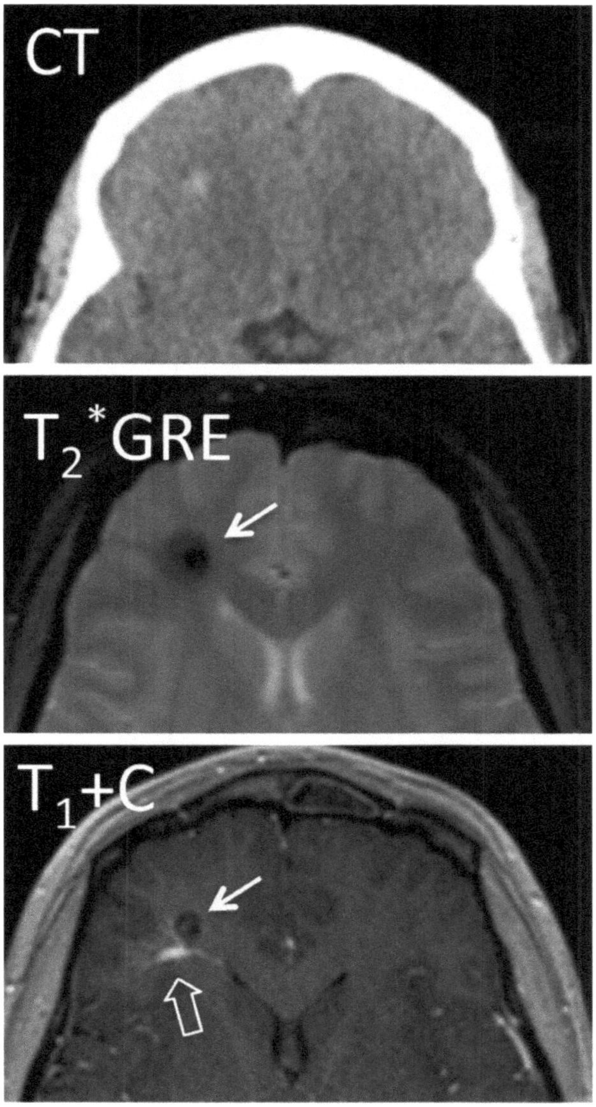

Fig. 24.4. Same case as in Figure 24.1B, demonstrating a small, mildly hyperattenuated lesion on computed tomography (CT) that was incidentally noted. It is confirmed to represent a cavernous malformation on magnetic resonance imaging, best seen on gradient-echo (GRE) with blooming artifact (arrow), and postgadolinium T1-weighted imaging showing the associated developmental venous anomaly (block arrow) makes the diagnosis definitive.

et al., 2011). Authors have also reported cases of low signal on both T1WI and T2WI (Beskonakli et al., 2002; Albanese et al., 2009), high signal on both T1WI and T2WI (Barrera et al., 2004), and T1 hypointensity and T2 isointensity (Aquilina et al., 2004).

Histopatholgically, organized thrombi within vascular spaces and hemosiderin deposition, which are thought to be observed exclusively in intra-axial CMs, have also been reported in cranial nerve-associated extra-axial lesions (Kim et al., 1997; Manjila et al., 2011). Clinically, cranial nerve lesions may resemble intra-axial CMs as they may present as acute intralesional bleeds, manifested as chiasmal apoplexy in the case of optic nerves (Chow et al., 2002; Deshmukh et al., 2003; Manjila et al., 2011), or facial palsy and sudden hearing loss in the case of cranial nerves VII and VIII, respectively (Manjila et al., 2011). In contrast, other cranial nerve lesions may resemble neoplasms in that they enlarge with time and lead to cranial nerve palsies secondary to compression and mass effect (Manjila et al., 2011). With intraventricular lesions, intraventricular hemorrhage was seen in all 12 patients in one study, with 8 experiencing rehemorrhages, whereas the literature prior to this demonstrates a significantly lower rate of bleed (11%) and more frequent presentations due to mass effect (Kivelev et al., 2010). Furthermore, intraventricular lesions have a reported preponderance of rapid growth (Kivelev et al., 2010).

Overall, while the findings of dural-based extra-axial lesions seem to suggest an entity distinct from intra-axial CMs and more akin to vascular neoplasms, neural-based extra-axial lesions exhibit highly variable findings that share characteristics seen in both intra-axial CMs and extra-axial dural-based lesions. It is therefore recommended by many authors that dural-based extra-axial lesions be referred to as "cavernous hemangiomas" rather than extra-axial CMs to more accurately reflect the neoplastic nature of such lesions. This reiterates the proposal from Kanaan et al. (2001) that extra-axial lesions likely represent an entity distinct from intra-axial CMs. On the other hand, the nature of neural-based extra-axial lesions remains more uncertain until further characterization from a clinical and radiologic perspective is possible.

Spinal cord cavernous malformations

CMs of the spinal cord may be extradural, intradural and extramedullary, or intramedullary. The incidence of multiple intracranial CMs appears higher in those with associated spinal cord CMs (Vishteh et al., 1999). Similar to cerebral CMs, a high incidence of intramedullary cases is associated with DVAs (Gross et al., 2010). Intramedullary CMs have similar MRI characteristics as those seen in cerebral CMs, with reticulated appearance of mixed intensities in both T1WI and T2WI, surrounded by a dark hemosiderin rim, and blooming on GRE (Turjman et al., 1995; Jallo et al., 2006). Calcification and hemosiderin rim are seen at lower frequencies (Gross et al., 2010). Infrequently, lesions may take on a more homogeneously hypo-, iso-, or hyperintense appearance on T1WI or

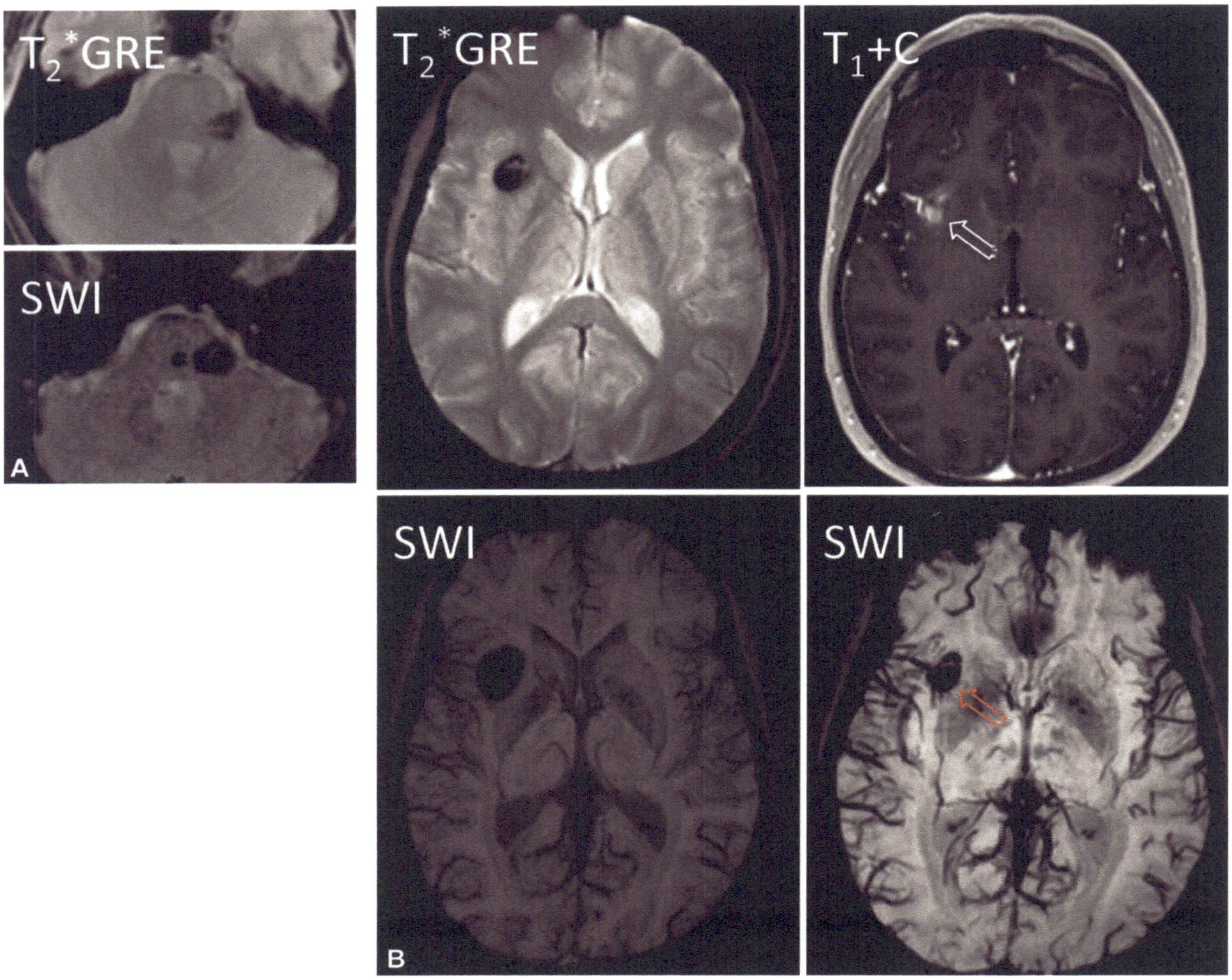

Fig. 24.5. Susceptibility-weighted imaging (SWI) is superior to T_2* gradient-echo (GRE) for delineation of cavernous malformation (CM). (**A**) SWI, being of higher resolution and increased contrast compared to GRE, depicts the left pontine CM with better definition and conspicuity, as well as an additional small lesion in the midpons. (**B**) Right lentiform nucleus CM is better depicted on SWI than on T_2*GRE scan, acquired on the same scan at 3 T. On the adjacent axial image section, SWI additionally shows the coexisting developmental venous anomaly (block arrow) that demonstrates contrast enhancement (T_1+C).

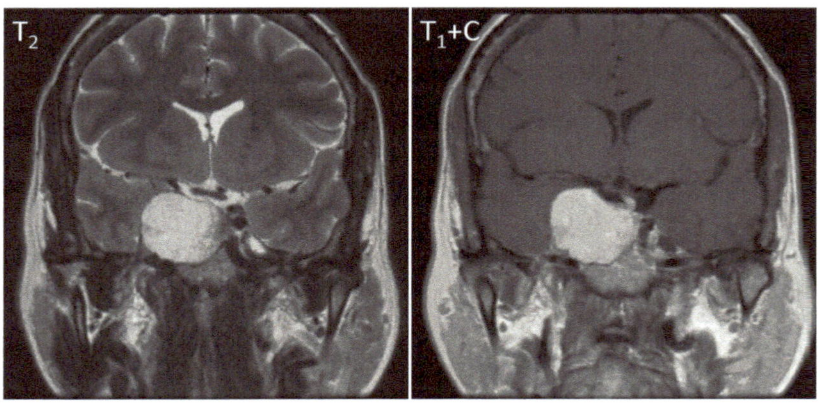

Fig. 24.6. Cavernous sinus cavernous hemangioma, considered an encapsulated, hypervascular tumor rather than a vascular malformation. Magnetic resonance imaging shows a rounded T_2 hyperintense mass centered in the right cavernous sinus with avid and homogeneous contrast enhancement. Pathology revealed bland vascular and spindle cell proliferation consistent with a cavernous sinus cavernous hemangioma. The cavernous sinus location and avidly enhancing pattern may also suggest a meningioma or schwannoma.

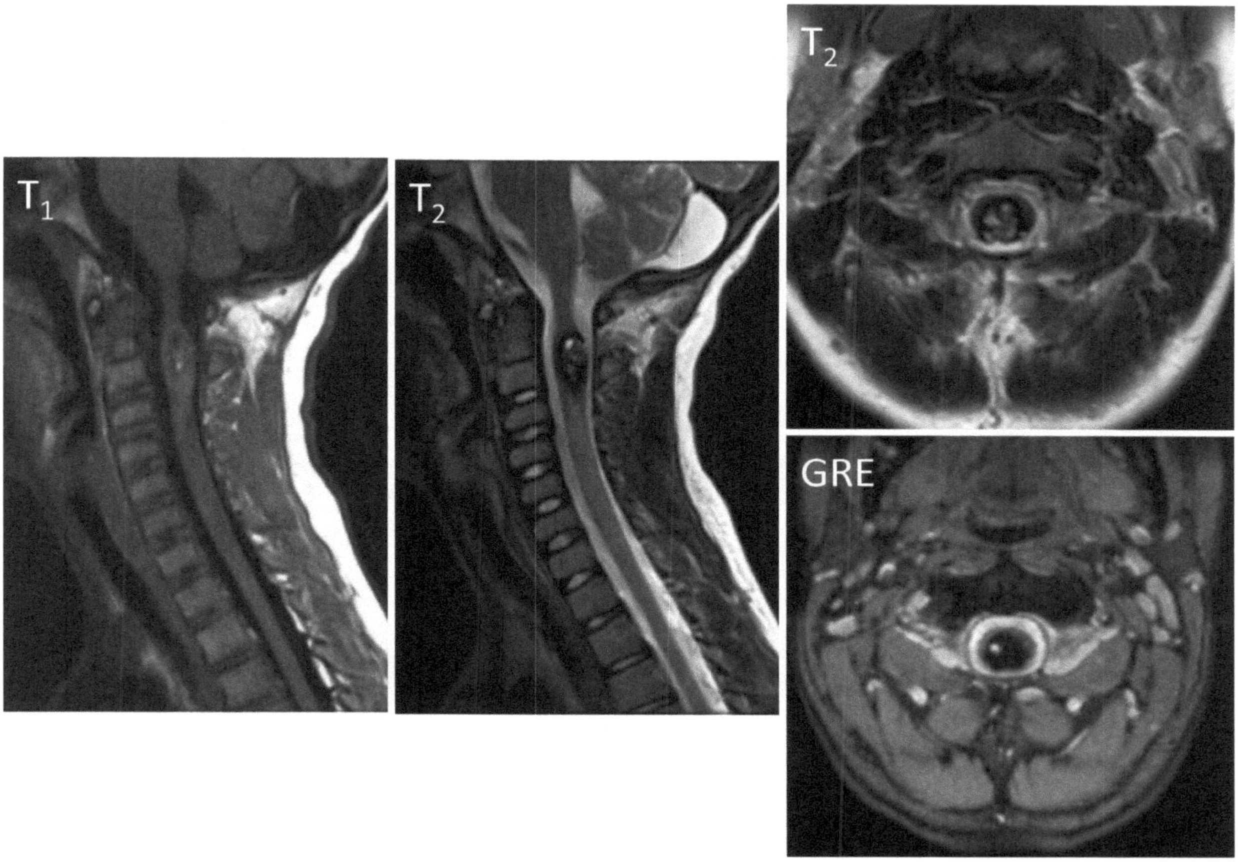

Fig. 24.7. Spinal cord cavernous malformation (CM). A 10-year-old boy presented for magnetic resonance imaging (MRI) after being hit by a basketball pole with lingering concussion symptoms. MRI shows a mildly expansile intramedullary lesion with reticulated core of T_2 hyperintensity and T_1 iso- to hyperintensity, with peripheral T_2 dark hemosiderin rim, in an appearance characteristic of a CM. The lesion was excised and confirmed on pathology.

T2WI (Turjman et al., 1995; Gross et al., 2010). Figure 24.7 shows a 10-year-old boy with an intramedullary cervical cord CM at C2–C3 with predominant isointensity to the spinal cord on T1WI and mixed hyperintensities with a hemosiderin ring on T2WI.

Primary epidural lesions without adjacent vertebral involvement are rare (Shin et al., 2001), and exhibit imaging findings resembling those seen in extra-axial dural-based cavernous hemangiomas, with isointensity to the spinal cord seen on T1WI, hyperintensity on T2WI, and avid, homogeneous contrast enhancement (Shin et al., 2001; Hatiboglu et al., 2006). Presence of a dark peripheral rim is rare. Similar to intracranial dural-based cavernous hemangiomas, such findings may support the hypothesized neoplastic nature of such lesions, or may be related to different subtypes of vascular malformations that tend to occur in the extra-axial location. Lesions occur most often in the thoracic spine, and less frequently in the cervical and lumbar spine. Within the spinal canal, lesions typically demonstrate a large posterolateral component, thought to be due to the larger, accommodating posterior spinal canal space (Shin et al., 2001; Hatiboglu et al., 2006).

Familial cavernous malformations

Multiple lesions are common in familial forms of CMs, with over 50% of familial patients harboring more than one malformation. By contrast, only 12–20% patients with sporadic CMs have multiple lesions (Batra et al., 2009). The hallmarks of familial CMs include MRI findings of multiple CMs, appearance of new lesions over time, and a positive family history (Zabramski et al., 1994). GRE and SWI are the mainstay methods to identify cases of familial CMs, as they are most sensitive to small lesions undetected on conventional sequences. The failure to note DVAs in familial cases should not necessarily raise doubt in the diagnostic confidence of CMs, as an association of DVAs with CMs in familial cases has not been found (Abdulrauf et al., 1999; Petersen et al., 2010). Figure 24.8 shows a case of familial CM on noncontrast CT with multiple lesions of mild hyperdensities

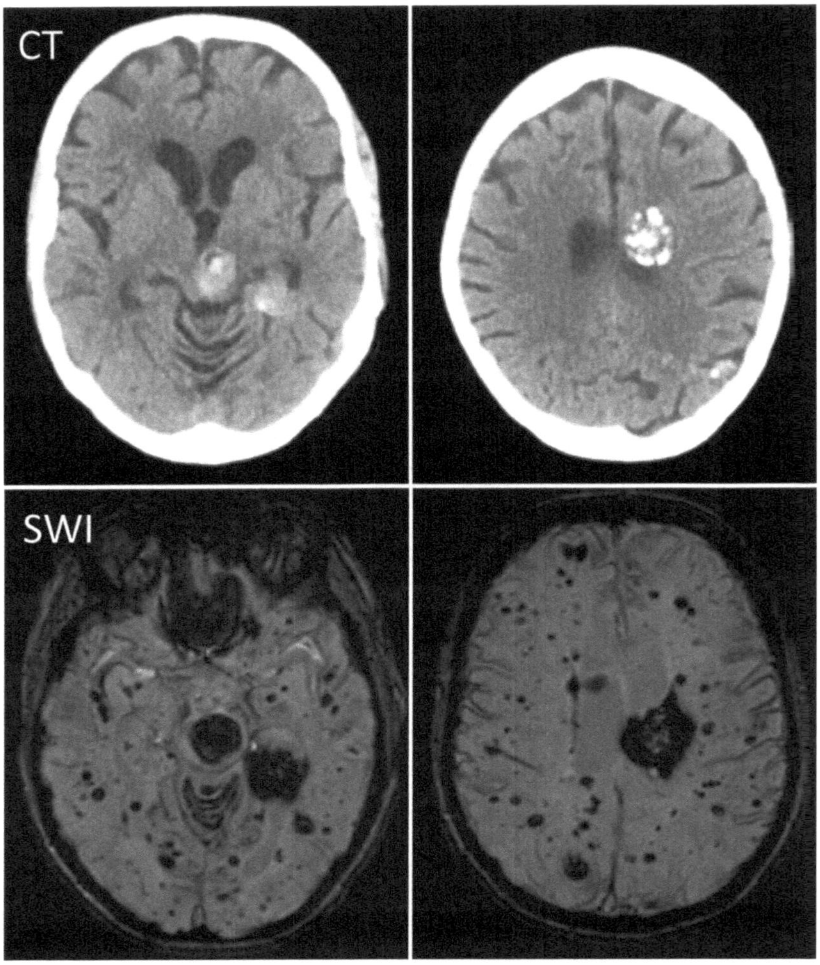

Fig. 24.8. Familial case of multiple intracranial cavernous malformations (CMs). Computed tomography (CT) shows multiple lesions with a mixture of mild hyperattenuation and calcific density, none associated with substantial edema. Susceptibility-weighted imaging (SWI) shows innumerable CMs, most of which are not evident on CT. Note also that there is hemosiderosis along the folia of superior vermis suggesting previous, likely repeated hemorrhage.

and calcific densities, none associated with substantial edema. SWI depicts innumerable additional punctate CMs throughout the supra- and infratentorial brain. It is not uncommon to have a few of these lesions undergoing different stages of evolution from recent or acute hemorrhage.

Cavernous malformations with acute to subacute hemorrhage

In the setting of an acute or recent hemorrhage, the imaging appearance of CMs may become less recognizable as hemorrhage may obscure the underlying malformation. In addition to the typical features of CMs, such lesions frequently demonstrate an overlying large, sharply marginated, spherical lesion with MRI features consistent with acute to subacute hematoma. Only in instances of bleeding are there typically mass effect and perilesional edema in CMs. Presence of significant T1 hyperintensity often suggests subacute hemorrhage, whereas surrounding edema is more suggestive of acute hemorrhage. Figure 24.9 shows a case of acute hemorrhage due to an underlying CM on noncontrast CT and FLAIR MRI. There is a mild rim of vasogenic edema. This lesion was resected with pathology demonstrating CM. The pons is a frequent location for a hypertensive bleed, but is also a common site for CMs, and such etiologies need to be considered, particularly in a young individual.

Alternative causes of hemorrhages, such as vasculopathy, infection, neoplasm, or AVMs, must be considered and worked up in the setting of a recent isolated bleed. Multiple episodes of repeated bleeding, a hemosiderin ring, and encapsulation are evidence for a CM, while the presence of mass effect, single-aged blood products, and expansile hemorrhage is more indicative of an AVM (Batra et al., 2009). Figure 24.10 illustrates another case

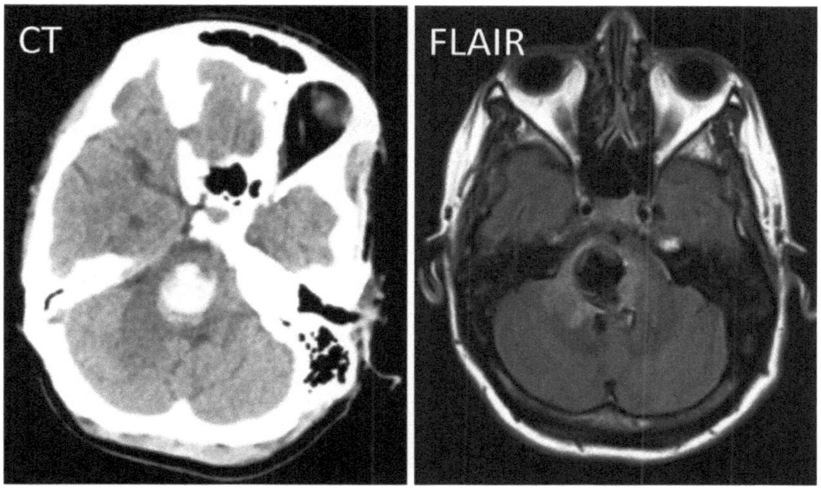

Fig. 24.9. Acute hemorrhage due to underlying cavernous malformation (CM). A 33-year-old previously healthy woman presented with 2 days of headaches, slurred speech, and left-sided weakness. The pontine hematoma is associated with mild surrounding edema and mass effect, better depicted on fluid-attenuated inversion recovery (FLAIR) sequence than computed tomography (CT). This lesion was excised and pathology showed vascular abnormality consistent with a CM.

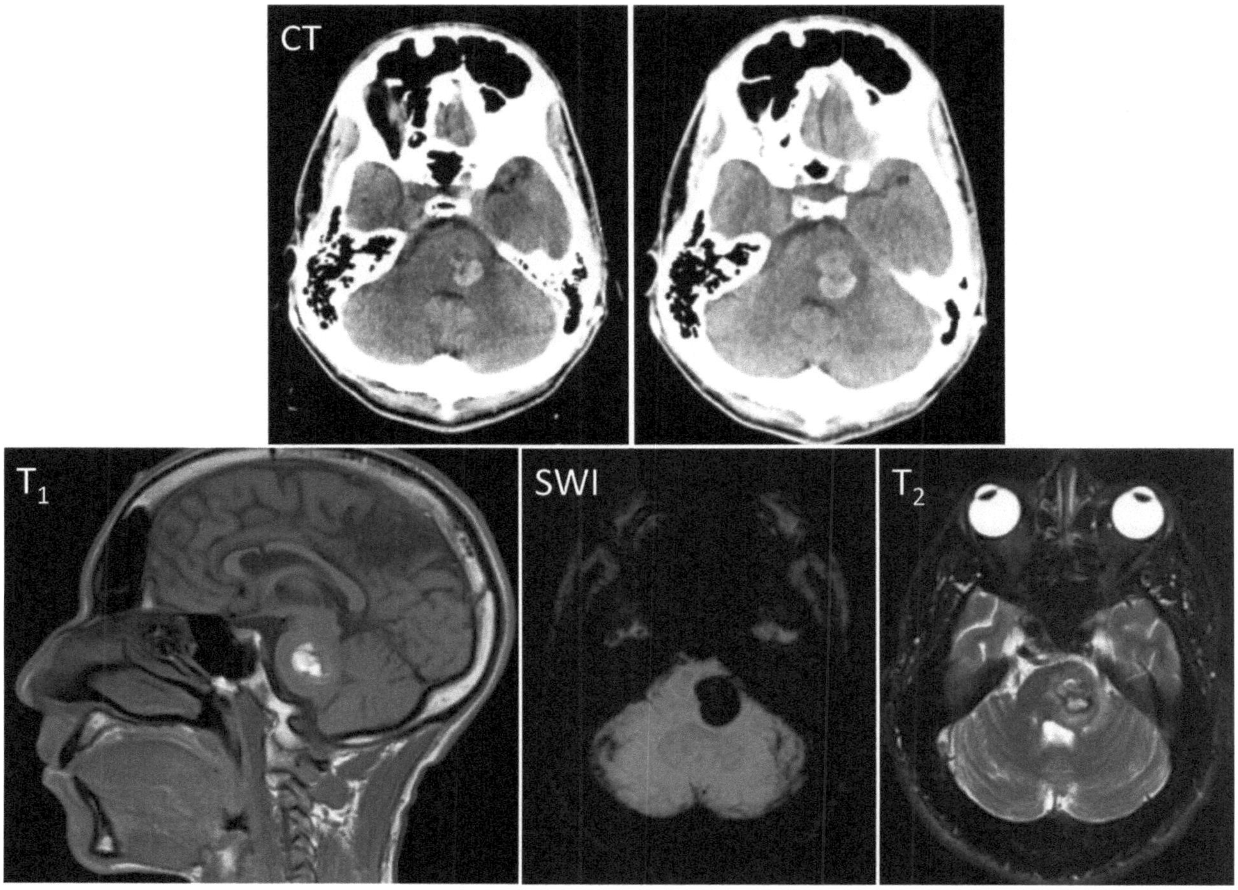

Fig. 24.10. Pontine cavernous malformation with recent hemorrhage. A 39-year-old man with headaches. Computed tomography (CT) shows a lobular lesion in the left pons containing blood, with only minimal rim of surrounding edema. Magnetic resonance imaging of the same lesion shows small multilocular areas of intrinsic T_1 hyperintensity and T_2 hyperintense core suggestive of subacute (extracellular methemoglobin) blood products. SWI, susceptibility-weighted imaging.

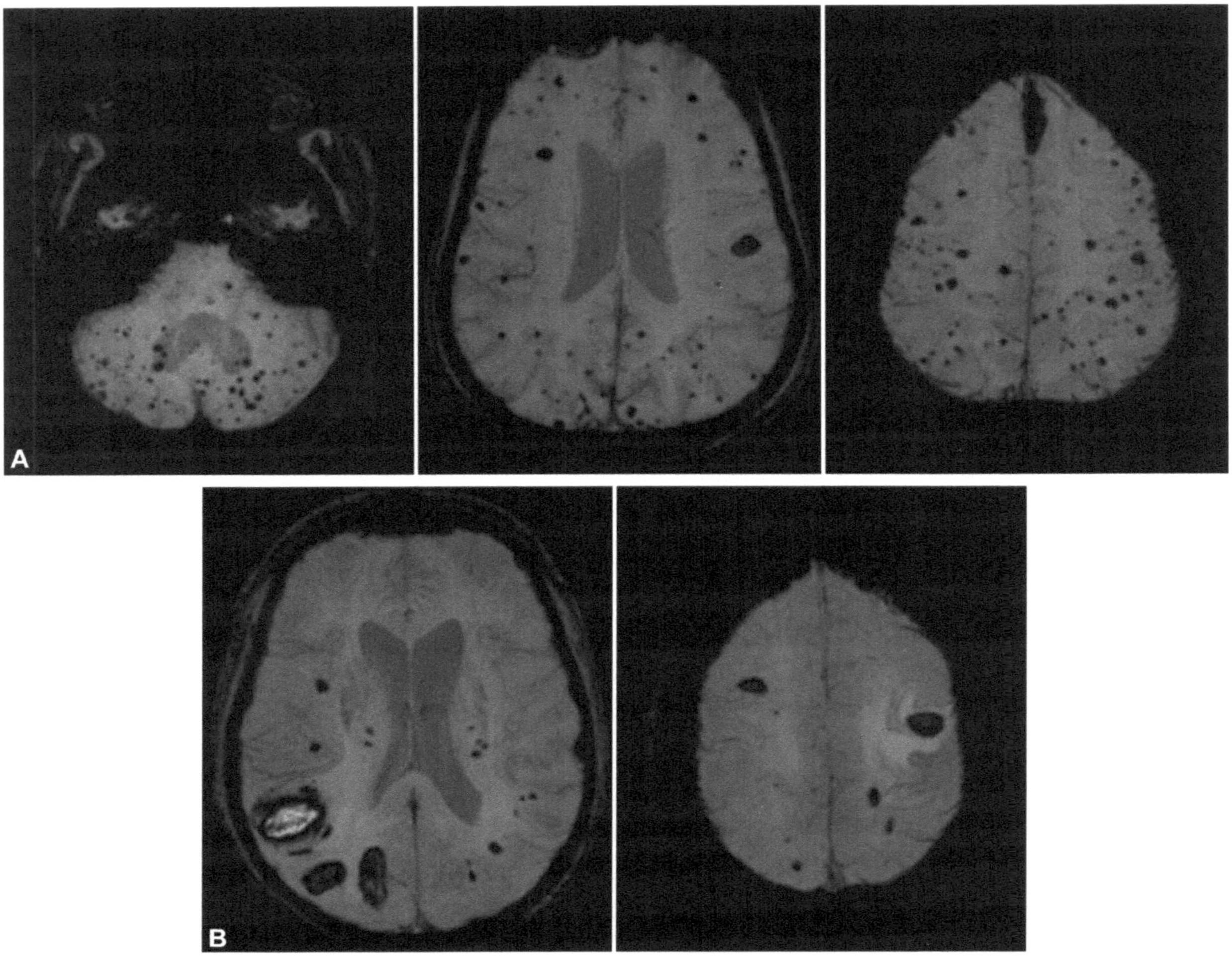

Fig. 24.11. Multiple microhemorrhages can be mimicked by various etiologies other than multiple (typically familial) cavernous malformations. (**A**) A 60-year-old woman with lung adenocarcinoma and multiple brain metastases showing associated microhemorrhages best depicted on susceptibility-weighted imaging. (**B**) A 56-year-old woman status post liver and kidney transplant presents acutely with altered metal status and seizure-like activities. Magnetic resonance imaging shows multifocal hemorrhages of different sizes and variable amount of surrounding edema, clinically correlated with cardiogenic emboli and underlying thrombocytopenia. Pathology showed macrophage-rich lesions without microorganisms or neoplastic cells.

of hyperattenuated pontine hematoma on noncontrast CT. On MRI, the lesion is hyperintense on T1WI with mixed intensities and a surrounding hemosiderin ring on T2WI, suggestive of a characteristic subacute hemorrhage by a CM.

DIFFERENTIAL DIAGNOSIS AND MIMICKERS

Although the MR findings of typical intra-axial CMs are sufficiently characteristic to confidently diagnose such lesions, extra-axial and multifocal lesions may mimic imaging findings suggestive of other disease entities. Diffuse, bilateral punctate lesions on SWI may alternatively represent hemorrhagic metastatic disease (Fig. 24.11A), hemorrhagic emboli (Fig. 24.11B), hypertensive or amyloid angiopathy, radiation-induced telangiectasia (Batra et al., 2009), neurovasculitis, or traumatic diffuse axonal injury. The classic features of CM are seen in an extra-axial intraventricular lesion (Fig. 24.12) at the foramen of Monro, but this was ultimately determined to be a colloid cyst on pathology, likely owing its appearance to hemorrhage. Cavernous sinus hemangiomas are most often misdiagnosed as meningomas and schwannomas on the basis of similar high T2 signal and strong enhancement characteristics (Sohn et al., 2003; Salanitri et al., 2004). Chondromas, chordomas, and chondrosarcomas may demonstrate hypo- or isointensity on T1WI with heterogeneous enhancement and markedly homogeneous signal intensity on T2WI that may resemble a cavernous sinus cavernous hemangioma.

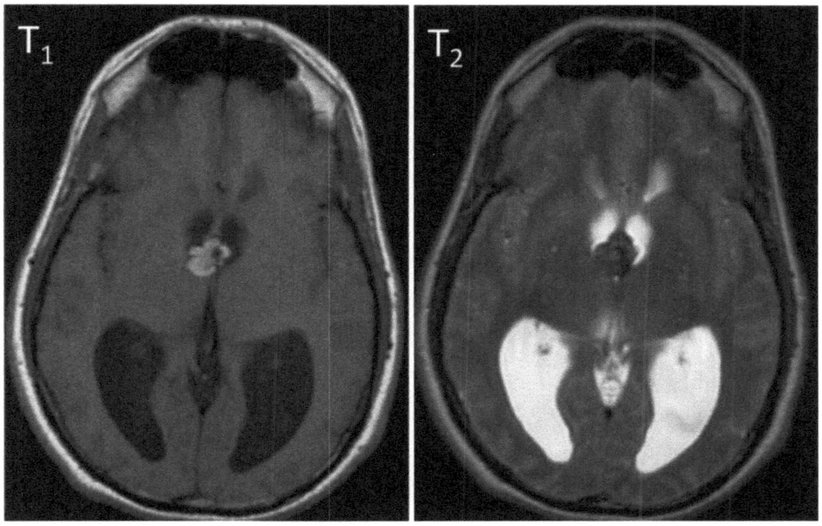

Fig. 24.12. Intraventricular lesion mimicking cavernous malformation (CM). A reticulated T_1 hyperintense and T_2 intermediate signal lesion near the foramen of Monro, at a classic location for colloid cyst, but the multilocular feature is reminiscent of a CM. There is obstructive hydrocephalus. The lesion was excised and pathology revealed a colloid cyst.

CONCLUSIONS

MRI has become the mainstay imaging modality in evaluating and diagnosing CMs. Features on MRI are highly characteristic, including a reticulated pattern of mixed hyper- and hypointensity on T1WI and T2WI, and a surrounding hypointense rim best appreciated on T2WI or GRE. Contrast enhancement is useful for revealing coexisting DVAs in aiding the diagnosis, as DVAs are frequently associated with sporadic CMs. SWI is highly sensitive for screening for suspected CMs and accompanying DVAs, and is superior to GRE in detecting multifocal, familial CMs.

References

Abdulrauf SI, Kaynar MY, Awad IA (1999). A comparison of the clinical profile of cavernous malformations with and without associated venous malformations. Neurosurgery 44: 41–46; discussion 46–47.

Adachi K, Yoshida K, Akiyama T et al. (2008). Cavernous angioma of the vestibular nerve: case report and literature review. Surg Neurol 70: 82–86; discussion 86.

Albanese A, Sturiale CL, D'Alessandris QG et al. (2009). Calcified extra-axial cavernoma involving lower cranial nerves: technical case report. Neurosurgery 64: onsE135–onsE136; discussion onsE136.

Aquilina K, Nanra JS, Brett F et al. (2004). Cavernous angioma of the internal auditory canal. J Laryngol Otol 118: 368–371.

Barrera JE, Jenkins H, Said S (2004). Cavernous hemangioma of the internal auditory canal: a case report and review of the literature. Am J Otolaryngol 25: 199–203.

Batra S, Lin D, Recinos PF et al. (2009). Cavernous malformations: natural history, diagnosis and treatment. Nat Rev Neurol 5: 659–670.

Becker DH, Townsend JJ, Kramer RA et al. (1979). Occult cerebrovascular malformations. A series of 18 histologically verified cases with negative angiography. Brain 102: 249–287.

Bellotti C, Medina M, Oliveri G et al. (1985). Cystic cavernous angiomas of the posterior fossa. Report of three cases. J Neurosurg 63: 797–799.

Beskonakli E, Kaptanoglu E, Okutan O et al. (2002). Extra-axial cavernomas of the cerebellopontine angle involving the seventh-eighth nerve complex. Neurosurg Rev 25: 222–224.

Bogren H, Svalander C, Wickbom I (1970). Angiography in intracranial cavernous hemangiomas. Acta Radiol Diagn (Stockh) 10: 81–89.

Bristot R, Santoro A, Fantozzi L et al. (1997). Cavernoma of the cavernous sinus: case report. Surg Neurol 48: 160–163.

Brunereau L, Labauge P, Tournier-Lasserve E et al. (2000). Familial form of intracranial cavernous angioma: MR imaging findings in 51 families. French Society of Neurosurgery. Radiology 214: 209–216.

Brunereau L, Leveque C, Bertrand P et al. (2001). Familial form of cerebral cavernous malformations: evaluation of gradient-spin-echo (GRASE) imaging in lesion detection and characterization at 1.5 T. Neuroradiology 43: 973–979.

Brunori A, Chiappetta F (1996). Cystic extra-axial cavernoma of the cerebellopontine angle. Surg Neurol 46: 475–476.

Bulut HT, Sarica MA, Baykan AH (2014). The value of susceptibility weighted magnetic resonance imaging in evaluation of patients with familial cerebral cavernous angioma. Int J Clin Exp Med 7: 5296–5302.

Campbell PG, Jabbour P, Yadla S et al. (2010). Emerging clinical imaging techniques for cerebral cavernous malformations: a systematic review. Neurosurg Focus 29. E6.

Chow M, Addas B, Sangalang V et al. (2002). Cavernous malformation of the hypoglossal nerve: case report and review of the literature. Can J Neurol Sci 29: 191–194.

Cooper AD, Campeau NG, Meissner I (2008). Susceptibility-weighted imaging in familial cerebral cavernous malformations. Neurology 71: 382.

de Champfleur NM, Langlois C, Ankenbrandt WJ et al. (2011). Magnetic resonance imaging evaluation of cerebral cavernous malformations with susceptibility-weighted imaging. Neurosurgery 68: 641–647; discussion 647–648.

Deshmukh VR, Albuquerque FC, Zabramski JM et al. (2003). Surgical management of cavernous malformations involving the cranial nerves. Neurosurgery 53: 352–357; discussion 357.

de Souza JM, Domingues RC, Cruz Jr LC et al. (2008). Susceptibility-weighted imaging for the evaluation of patients with familial cerebral cavernous malformations: a comparison with t2-weighted fast spin-echo and gradient-echo sequences. AJNR Am J Neuroradiol 29: 154–158.

Dillon WP (1997). Cryptic vascular malformations: controversies in terminology, diagnosis, pathophysiology, and treatment. AJNR Am J Neuroradiol 18: 1839–1846.

Dorner L, Buhl R, Hugo HH et al. (2005). Unusual locations for cavernous hemangiomas: report of two cases and review of the literature. Acta Neurochir (Wien) 147: 1091–1096; discussion 1096.

Essig M, Reichenbach JR, Schad LR et al. (1999). High-resolution MR venography of cerebral arteriovenous malformations. Magn Reson Imaging 17: 1417–1425.

Ferrante L, Acqui M, Trillo G et al. (1998). Cavernous angioma of the VIIIth cranial nerve. A case report. Neurosurg Rev 21: 270–276.

Giombini S, Morello G (1978). Cavernous angiomas of the brain. Account of fourteen personal cases and review of the literature. Acta Neurochir (Wien) 40: 61–82.

Gomori JM, Grossman RI, Goldberg HI et al. (1985). Intracranial hematomas: imaging by high-field MR. Radiology 157: 87–93.

Gomori JM, Grossman RI, Goldberg HI et al. (1986). Occult cerebral vascular malformations: high-field MR imaging. Radiology 158: 707–713.

Gonzalez LF, Lekovic GP, Eschbacher J et al. (2006). Are cavernous sinus hemangiomas and cavernous malformations different entities? Neurosurg Focus 21. e6.

Gross BA, Du R, Popp AJ et al. (2010). Intramedullary spinal cord cavernous malformations. Neurosurg Focus 29E14.

Hatiboglu MA, Iplikcioglu AC, Ozcan D (2006). Epidural spinal cavernous hemangioma. Neurol Med Chir (Tokyo) 46: 455–458.

Hayman LA, Evans RA, Ferrell RE et al. (1982). Familial cavernous angiomas: natural history and genetic study over a 5-year period. Am J Med Genet 11: 147–160.

Horowitz M, Kondziolka D (1995). Multiple familial cavernous malformations evaluated over three generations with MR. AJNR Am J Neuroradiol 16: 1353–1355.

Ishijima Y, Matsumura H, Kageyama N (1966). Intracranial cavernous hemangioma: report of two cases. Nihon Geka Hokan 35: 748–754.

Jain KK, Robertson E (1970). Recurrence of an excised cavernous hemangioma in the opposite cerebral hemisphere. Case report. J Neurosurg 33: 453–456.

Jallo GI, Freed D, Zareck M et al. (2006). Clinical presentation and optimal management for intramedullary cavernous malformations. Neurosurg Focus 21. e10.

Jensen HP, Brumlik J, Boshes B (1963). The application of serial angiography to diagnosis of the smallest cerebral angiomatous malformations. J Nerv Ment Dis 136: 1–14.

Jinhu Y, Jianping D, Xin L et al. (2008). Dynamic enhancement features of cavernous sinus cavernous hemangiomas on conventional contrast-enhanced MR imaging. AJNR Am J Neuroradiol 29: 577–581.

Jonutis AJ, Sondheimer FK, Klein HZ et al. (1971). Intracerebral cavernous hemangioma with angiographically demonstrated pathologic vasculature. Neuroradiology 3: 57–63.

Kamrin RB, Buchsbaum HW (1965). Large vascular malformations of the brain not visualized by serial angiography. Arch Neurol 13: 413–420.

Kanaan I, Jallu A, Alwatban J et al. (2001). Extra-axial cavernous hemangioma: two case reports. Skull Base 11: 287–295.

Kattapong VJ, Hart BL, Davis LE (1995). Familial cerebral cavernous angiomas: clinical and radiologic studies. Neurology 45: 492–497.

Kawai K, Fukui M, Tanaka A et al. (1978). Extracerebral cavernous hemangioma of the middle fossa. Surg Neurol 9: 19–25.

Kim M, Rowed DW, Cheung G et al. (1997). Cavernous malformation presenting as an extra-axial cerebellopontine angle mass: case report. Neurosurgery 40: 187–190.

Kivelev J, Niemela M, Kivisaari R et al. (2010). Intraventricular cerebral cavernomas: a series of 12 patients and review of the literature. J Neurosurg 112: 140–149.

Kramer RA, Wing SD (1977). Computed tomography of angiographically occult cerebral vascular malformations. Radiology 123: 649–652.

Labauge P, Laberge S, Brunereau L et al. (1998). Hereditary cerebral cavernous angiomas: clinical and genetic features in 57 French families. Societe Francaise de Neurochirurgie. Lancet 352: 1892–1897.

Lee BC, Vo KD, Kido DK et al. (1999) MR high-resolution blood oxygenation level-dependent venography of occult (low-flow) vascular lesions. AJNR Am J Neuroradiol 20: 1239–1242.

Lehnhardt FG, von Smekal U, Ruckriem B et al. (2005). Value of gradient-echo magnetic resonance imaging in the diagnosis of familial cerebral cavernous malformation. Arch Neurol 62: 653–658.

Manjila S, Moon K, Weiner MA et al. (2011). Cavernous malformation of the trochlear nerve: case report and review of the literature on cranial nerve cavernomas. Neurosurgery 69: E230–E238; discussion E238.

Maraire JN, Awad IA (1995). Intracranial cavernous malformations: lesion behavior and management strategies. Neurosurgery 37: 591–605.

Maruishi M, Shima T, Okada Y et al. (1994). Cavernous sinus cavernoma treated with radiation therapy – case report. Neurol Med Chir (Tokyo) 34: 773–777.

McCormick WF (1966). The pathology of vascular ("arteriovenous") malformations. J Neurosurg 24: 807–816.

Meyer FB, Lombardi D, Scheithauer B et al. (1990). Extra-axial cavernous hemangiomas involving the dural sinuses. J Neurosurg 73: 187–192.

Moriarity JL, Clatterbuck RE, Rigamonti D (1999a). The natural history of cavernous malformations. Neurosurg Clin N Am 10: 411–417.

Moriarity JL, Wetzel M, Clatterbuck RE et al. (1999b). The natural history of cavernous malformations: a prospective study of 68 patients. Neurosurgery 44: 1166–1171; discussion 1172–1163.

Nieto J, Hinojosa J, Munoz MJ et al. (2003). Intraventricular cavernoma in pediatric age. Childs Nerv Syst 19: 60–62.

Numaguchi Y, Kishikawa T, Fukui M et al. (1979). Prolonged injection angiography for diagnosing intracranial cavernous hemangiomas. Radiology 131: 137–138.

Petersen TA, Morrison LA, Schrader RM et al. (2010). Familial versus sporadic cavernous malformations: differences in developmental venous anomaly association and lesion phenotype. AJNR Am J Neuroradiol 31: 377–382.

Pinker K, Stavrou I, Szomolanyi P et al. (2007). Improved preoperative evaluation of cerebral cavernomas by high-field, high-resolution susceptibility-weighted magnetic resonance imaging at 3 Tesla: comparison with standard (1.5 T) magnetic resonance imaging and correlation with histopathological findings – preliminary results. Invest Radiol 42: 346–351.

Porter PJ, Willinsky RA, Harper W et al. (1997). Cerebral cavernous malformations: natural history and prognosis after clinical deterioration with or without hemorrhage. J Neurosurg 87: 190–197.

Pozzati E, Padovani R, Morrone B et al. (1980). Cerebral cavernous angiomas in children. J Neurosurg 53: 826–832.

Pozzati E, Gaist G, Poppi M et al. (1981). Microsurgical removal of paraventricular cavernous angiomas. Report of two cases. J Neurosurg 55: 308–311.

Reyns N, Assaker R, Louis E et al. (1999). Intraventricular cavernomas: three cases and review of the literature. Neurosurgery 44: 648–654; discussion 654–645.

Rigamonti D, Spetzler RF (1988). The association of venous and cavernous malformations. Report of four cases and discussion of the pathophysiological, diagnostic, and therapeutic implications. Acta Neurochir (Wien) 92: 100–105.

Rigamonti D, Drayer BP, Johnson PC et al. (1987). The MRI appearance of cavernous malformations (angiomas). J Neurosurg 67: 518–524.

Rigamonti D, Hadley MN, Drayer BP et al. (1988). Cerebral cavernous malformations. Incidence and familial occurrence. N Engl J Med 319: 343–347.

Runnels JB, Gifford DB, Forsberg PL et al. (1969). Dense calcification in a large cavernous angioma. Case report. J Neurosurg 30: 293–298.

Salanitri GC, Stuckey SL, Murphy M (2004). Extracerebral cavernous hemangioma of the cavernous sinus: diagnosis with MR imaging and labeled red cell blood pool scintigraphy. AJNR Am J Neuroradiol 25: 280–284.

Savoiardo M, Passerini A (1978). CT, angiography, and RN scans in intracranial cavernous hemangiomas. Neuroradiology 16: 256–260.

Savoiardo M, Strada L, Passerini A (1983a). Cavernous hemangiomas of the orbit: value of CT, angiography, and phlebography. AJNR Am J Neuroradiol 4: 741–744.

Savoiardo M, Strada L, Passerini A (1983b). Intracranial cavernous hemangiomas: neuroradiologic review of 36 operated cases. AJNR Am J Neuroradiol 4: 945–950.

Schneider RC, Liss L (1958). Cavernous hemangiomas of the cerebral hemispheres. J Neurosurg 15: 392–399.

Servo A, Porras M, Raininko R (1984). Diagnosis of cavernous haemangiomas by computed tomography and angiography. Acta Neurochir (Wien) 71: 273–282.

Shin JH, Lee HK, Rhim SC et al. (2001). Spinal epidural cavernous hemangioma: MR findings. J Comput Assist Tomogr 25: 257–261.

Sigal R, Krief O, Houtteville JP et al. (1990). Occult cerebrovascular malformations: follow-up with MR imaging. Radiology 176: 815–819.

Sohn CH, Kim SP, Kim IM et al. (2003). Characteristic MR imaging findings of cavernous hemangiomas in the cavernous sinus. AJNR Am J Neuroradiol 24: 1148–1151.

Suess O, Hammersen S, Brock M (2002). Intraventricular cavernoma: unusual occurrence in the region of the foramen of Monro. Br J Neurosurg 16: 78–79.

Tagle P, Huete I, Mendez J et al. (1986). Intracranial cavernous angioma: presentation and management. J Neurosurg 64: 720–723.

Terao H, Hori T, Matsutani M et al. (1979). Detection of cryptic vascular malformations by computerized tomography. Report of two cases. J Neurosurg 51: 546–551.

Tomlinson FH, Houser OW, Scheithauer BW et al. (1994). Angiographically occult vascular malformations: a correlative study of features on magnetic resonance imaging and histological examination. Neurosurgery 34: 792–799; discussion 799–800.

Turjman F, Joly D, Monnet O et al. (1995). MRI of intramedullary cavernous haemangiomas. Neuroradiology 37: 297–302.

Vaquero J, Leunda G, Martinez R et al. (1983). Cavernomas of the brain. Neurosurgery 12: 208–210.

Vishteh AG, Zabramski JM, Spetzler RF (1999). Patients with spinal cord cavernous malformations are at an increased risk for multiple neuraxis cavernous malformations. Neurosurgery 45: 30–32; discussion 33.

Vogler R, Castillo M (1995). Dural cavernous angioma: MR features. AJNR Am J Neuroradiol 16: 773–775.

Voigt K, Yasargil MG (1976). Cerebral cavernous haemangiomas or cavernomas. Incidence, pathology, localization, diagnosis, clinical features and treatment. Review of the literature and report of an unusual case. Neurochirurgia (Stuttg) 19: 59–68.

Willinsky R, Harper W, Wallace MC et al. (1996). Follow-up MR of intracranial cavernomas: the relationship between haemorrhagic events and morphology. Interv Neuroradiol 2: 127–135.

Yamasaki T, Handa H, Yamashita J et al. (1986). Intracranial and orbital cavernous angiomas. A review of 30 cases. J Neurosurg 64: 197–208.

Yun TJ, Na DG, Kwon BJ et al. (2008). A T1 hyperintense perilesional signal aids in the differentiation of a cavernous angioma from other hemorrhagic masses. AJNR Am J Neuroradiol 29: 494–500.

Zabramski JM, Wascher TM, Spetzler RF et al. (1994). The natural history of familial cavernous malformations: results of an ongoing study. J Neurosurg 80: 422–432.

Handbook of Clinical Neurology, Vol. 143 (3rd series)
Arteriovenous and Cavernous Malformations
R.F. Spetzler, K. Moon, and R.O. Almefty, Editors
http://dx.doi.org/10.1016/B978-0-444-63640-9.00025-4

Chapter 25

Pathology of cavernous malformations

EFREM M. COX[1]*, NICHOLAS C. BAMBAKIDIS[1], AND MARK L. COHEN[2]

[1]*Department of Neurosurgery, University Hospitals, Case Western Reserve University, Cleveland, OH, USA*

[2]*Department of Pathology, University Hospitals, Case Western Reserve University, Cleveland, OH, USA*

Abstract

Cavernous malformations (CMs) are low-pressure angiographically occult lesions, composed of blood-filled sinusoidal locules known as "caverns." Although these lesions were once believed to be congenital in nature, there is compelling evidence to support *de novo* formation of CMs as well. They can occur as sporadic lesions or be inherited in an autosomal-dominant phenotype in familial forms of the disease. The pathophysiology of CMs is commonly believed to be due to abnormal vascular pathology. Three genes, *CCM1*, *CCM2*, and *CCM3*, have been extensively studied for their role in vascular pathology, resulting in abnormal angiogenesis and compromising the structural integrity of vessel endothelial cell. The expression of growth factors has been researched to gain insight into the dynamic behavior of CM lesions. Gross and microscopic images are utilized in this chapter to illustrate the pathologic findings of these lesions. Ultrastructural analysis demonstrates the aberrations in CM endothelial cells and structural integrity that may provide better understanding into how and why these lesions have a propensity to hemorrhage.

PATHOLOGY OF CAVERNOUS MALFORMATIONS

Cerebrovascular vascular malformations (CVMs) are a diverse group of disorders that differ from one another in both physiologic and pathologic character. There are CVMs that are characterized by arteriovenous shunting (e.g., arteriovenous malformations, dural arteriovenous fistulas, vein of Galen malformations, carotid-cavernous fistulas, and pial arteriovenous fistulas). In contrast, there are CVMs that do not exhibit arteriovenous shunting (e.g., cavernous malformations (CMs), developmental venous anomalies (DVAs), sinus pericranii, and capillary telangiectasias). There is a broad spectrum of clinical behavior amongst these vascular disorders. Some may remain clinically asymptomatic (e.g., DVAs or capillary telangectasias) for an individual's lifetime. Other CVMs, such as arteriovenous malformations and CMs, may hemorrhage and cause varying severity of neurologic symptoms depending upon location (e.g., seizure, focal neurologic deficit, or nonspecific headache). Lesions within deep nuclei or the brainstem are more likely to result in profound neurologic deficits with hemorrhage in comparison to superficial noneloquent cortex. For the scope of this chapter discussion will focus on the pathology of CMs.

CMs are among the most common vascular malformations of the brain and are detected in the population at a rate of approximately 0.6 per 100 000 (Al-Shahi et al., 2003; Leblanc et al., 2009). They are composed of thin-walled, blood-filled sinusoidal locules known as "caverns," as described by their name. These lesions are commonly referred to as angiographically occult CVMs due to the fact that they are low-pressure systems that do not shunt blood and are not generally apparent on diagnostic catheter angiography. They are a distinct form of vascular malformation in comparison to arteriovenous malformations, for they do not have any intervening brain parenchyma within the lesion. The terminology has evolved as we have gained a greater understanding

*Correspondence to: Efrem M. Cox, BS, MD, Neurological Surgery, University Hospitals Case Medical Center, 11100 Euclid Avenue, HAN 5042, Cleveland, OH 44106, USA. Tel: +1-216-844-3472, E-mail: Efrem.Cox@uhhospitals.org

of their etiology. These lesions were once known as "cavernomas," which was predicated by the misconception of them representing vascular tumors. The lesions were noted to grow over time; however, this increase in size is related to recurrent hemorrhages. It is believed that these hemorrhages occur as a result of the immature angiogenic nature of these lesions.

In most natural history studies, the overall rate of hemorrhage has been estimated to range from 0.7% to as high as 4.1% for cerebral CMs (Fritschi et al., 1994; Aiba et al., 1995; Dorsch and McMahon, 1998; Kupersmith et al., 2001; Ziyal and Ozgen, 2001; Gross et al., 2009). The annual rates of hemorrhage for incidental and unruptured CMs are lower – 0.08–0.2% and 0.3–0.6%, respectively (Fritschi et al., 1994; Aiba et al., 1995; Dorsch and McMahon, 1998; Kupersmith et al., 2001; Ziyal and Ozgen, 2001; Gross et al., 2009). Rehemorrhage rates have been reported to range from 3.8% to 22.9% (Fritschi et al., 1994; Aiba et al., 1995; Dorsch and McMahon, 1998; Kupersmith et al., 2001; Ziyal and Ozgen, 2001; Gross et al., 2009). The wide range in reported rehemorrhage rates is in part due to differing definitions of rehemorrhage in the literature. Some studies have based hemorrhage solely upon clinical manifestations alone, in comparison to confirmation with magnetic resonance imaging (MRI). Reporting clinical changes alone may overinflate the rates of rehemorrhage. Utilizing MRI to confirm rehemorrhage in patients with clinical alterations can differentiate true rehemorrhages from the effects of associated CM thrombosis (Gangemi et al., 1990). There is also discrepancy amongst studies whether or not to define microhemorrhages or intralesional hemorrhages as rehemorrhage, which can further add to the variance in the reported rates. The location of CMs and its effect on the rate of hemorrhage is also controversial. Some studies have suggested that deep CMs (e.g., thalamic, basal ganglia, brainstem) may have higher rates of hemorrhage (Robinson et al., 1991; Aiba et al., 1995; Kondziolka et al., 1997; Porter et al., 1997; Pandey et al., 2012), whereas some prospective studies do not support this claim (Al-Shahi Salman et al., 2008; Flemming et al., 2012). Further studies may better elucidate the role of location on the clinical behavior and pathophysiology of CMs. The natural history of CMs is discussed in further detail in Chapter 22.

ETIOLOGY AND PATHOPHYSIOLOGY

CMs were once generally believed to be congenital in nature; however, there is compelling evidence that they can occur through *de novo* formation as well (Pozzati et al., 1996; Amirjamshidi and Abbassioun, 2000; Duhem et al., 2005; Furuse et al., 2005). They can occur sporadically as isolated lesions or in clusters, as noted in the familial forms of the disease (Hayman et al., 1982; Robinson et al., 1991; Kondziolka et al., 1997). Sporadic CMs have been reported to have a more benign course and have been commonly associated with DVA (Perrini and Lanzino, 2006). The association of sporadic CMs and DVAs has led some to suggest that sporadic and familial CMs may have a different pathogenesis. One theory is that sporadic CMs may develop from a pre-existing DVA as a result of diapedesis of red blood cells (Dillon, 1997). There are reported cases and small case series that support this hypothesis (Pozzati et al., 1989; Maeder et al., 1998; Cakirer, 2003; Campeau and Lane, 2005; Wurm et al., 2005); however, this does not account for sporadic CMs that develop without DVAs. Familial CMs have been reported to have a more aggressive course (Maiuri et al., 2006; Petersen et al., 2010). Familial CMs account for 25–54% of CMs (Gangemi et al., 1990). They are commonly found in individuals of Hispanic descent (Bicknell et al., 1978; Hayman et al., 1982; Mason et al., 1988; Rigamonti et al., 1988; Zabramski et al., 1994). Individuals with familial CMs have a higher rate of multiple CMs than individuals with a sporadic case (Houtteville, 1997; Labauge et al., 2001).

There has been compelling evidence in the literature that some CMs may not be congenital. *De novo* formation of CMs has been well documented secondary to effects of radiation exposure (Pozzati et al., 1996; Amirjamshidi and Abbassioun, 2000; Heckl et al., 2002; Duhem et al., 2005; Furuse et al., 2005). In a review of 76 patients with *de novo* CMs following radiation therapy, Nimjee and colleagues (2006) found that there was a male and pediatric patient population predominance of cases in the literature. The average dose of radiation was 60.45 Gy with a latency period of 8.9 years before detection of the CM (Nimjee et al., 2006). There are several hypotheses as to why the *de novo* CMs develop. In regard to the predominance of these lesions arising in the pediatric population, some proposed that the developmentally immature brain is more sensitive to radiation than the adult brain (Gaensler et al., 1994). One theory is that radiation may initiate a reactive angiogenesis or cause local mutations of the *CCM* genes in the exposed brain tissue, thereby resulting in *CCM* gene loss of function and subsequent development of the *de novo* CM (Massa-Micon et al., 2000; Baumgartner et al., 2003; Pagenstecher et al., 2009). Some have suggested that the radiation therapy results in vessel wall necrosis and vessel wall changes (e.g., cellular swelling, vessel lumen dilatation, hyalinization, fibrosis, and mineralization) that lead to the development of *de novo* CMs (Gaensler et al., 1994; Humpl et al., 1997). One study proposed that the ionizing effects of radiation energy cause endothelial cell destruction within the vessel wall and also the surrounding connective tissue stroma, which predisposes to CM formation (Nyáry et al., 2005). A final theory that argues against *de novo* formation and rather supports a congenital hypothesis for the these lesions was proposed by Larson and colleagues (1998), who suggest the CM may have been present before radiation therapy but radiographically occult at the time of treatment and become detectable on MRI following radiation exposure.

According to the Zabramski classification of MRI signaling characteristics, type IV lesions may be poorly seen or not well visualized on T1 and T2 and appear as punctate hypointense lesions on gradient echo (Zabramski et al., 1994). This would support the possibility of CMs being present prior to radiation exposure and not being detected due to the limitation of the MRI available.

A recent retrospective study followed 18 children with very small type IV lesions known as "dot-like" or "black spot" CMs lesions and noted the annual hemorrhage rate was 0.7%, suggesting that such lesions should not be considered benign. The authors also acknowledge that a major limitation in establishing a definitive diagnosis of CM for these "dot-like" or "black spot" lesions is the lack of histopathologic examination, which is rarely available for such small lesions. The compelling evidence for *de novo* formation is not hindered by the theory of type IV lesions but is more of a reflection of the limitations of imaging modalities at the time of the classification.

Although not currently utilized or readily available in clinical practice, MRI at 9.4 or 14.1 T was able to able to demonstrate histologic validation of the angioarchitecture of excised cerebral CM lesions and also detect areas within the CM that were suggestive of angiogenic proliferation (Shenkar et al., 2008). With continued advancements in MRI and sequences there will be increasing sensitivity and specificity for detection of CMs, whether congenital or *de novo*. This will further aid in improved surveillance and management of CMs.

Irrespective of whether a CM is congenital, *de novo*, sporadic, or familial in origin, the underlying pathophysiology of CM development, proliferation, and their propensity to hemorrhage has been a keen area of research. There are three genes that have been associated with familial CMs: *CCM1* (K-Rev interaction trapped 1 (KRIT1)), *CCM2* (MGC4607/osmosensing scaffold for MEKK3/malcavernin), and *CCM3* (programmed cell death 10 (PDCD10)) genes are expressed as autosomal-dominant phenotypes with penetrance of 60–80%, 100%, and 63% respectively (Labauge et al., 2007). These three CCM genes/proteins are expressed in a variety of cell types, including neurons and astrocytes and particularly in vascular endothelium (Petit et al., 2006; Seker et al., 2006). The expression of CCM proteins can be observed on histologic specimens of CMs. (Van Gompel et al., 2009). Mutations in the *CCM* genes are hypothesized to result in a loss of function of cell signaling compromising endothelium integrity and abnormal angiogenesis, producing the vascular pathology observed in CM lesions (Boulday et al., 2009; Whitehead et al., 2009). The complex functions and roles of the CCM genes and protein products are beyond the scope of this chapter and are discussed in greater detail in Chapter 24.

There is growing evidence that the underlying pathophysiology observed in CMs may be related to faulty structural integrity of the vasculature of CM lesions. One theory is that the vessels are immature with poorly differentiated vascular smooth-muscle cells that give rise to the development of dysmorphic vessel walls that are prone to hemorrhage, thus further activating dysangiogenesis and lesion proliferation (Uranishi et al., 2001a). Ultrastructural analysis of the vasculature in CMs has demonstrated that the endothelial cells within the vessel wall have defective interendothelial tight junctions, rendering the vasculature permeable and blood–brain barrier incompetent (Wong et al., 2000; Clatterbuck et al., 2001; Tu et al., 2005; Tanriover et al., 2013).

Some studies have evaluated the expression of angiogenic growth factors and proliferative indices (vascular endothelial growth factor (VEGF), tenascin, transforming growth factor-beta (TGF-β), platelet-derived growth factor (PDGF), Ki-67, and bcl-2)) in order to gain a greater understanding into clinical progression of CMs (Rothbart et al., 1996; Kilic et al., 2000; Sure et al., 2001; Uranishi et al., 2001b; Maiuri et al., 2006). Tenascin is an extracellular matrix glycoprotein that is expressed during embryonic development and found in healing wounds and tumors (e.g., gliomas). It is considered a marker for vascular proliferation and has been found in CMs (Viale et al., 2002). TGF-β is a polypeptide member of a superfamily of growth factors (Massague, 1998). TGF-β is a signaling molecule with a diverse range of cellular functions (e.g., cell differentiation, proliferation, and embryologic development) and is involved in angiogenesis (Dickson et al., 1995; Moses and Serra, 1996). PDGF is known to stimulate connective tissue and neuroglial cell proliferation. It has also been shown to have significant expression in CMs and may play an important role in the angiogenesis of CMs (Maiuri et al., 2006; Yildirim et al., 2010).

VEGF is a glycoprotein that mediates various phases of blood vessel development in vasculogenesis. There has been a suggestion that VEGF may play a significant role in regulating the angiogenesis and clinical progression of the dynamic behavior of CMs (Jung et al., 2003). Expression of VEGF has been reported to be expressed at 40–90% within CMs (Kilic et al., 2000; Sure et al., 2001; Maiuri et al., 2006), but its correlation for determining clinical behavior is controversial. Somewhat surprisingly, Maiuri and colleagues (2006) found that VEGF, tenascin, and PDGF were not expressed significantly higher in aggressive lesions compared to stable lesions; however, they did note that TGF-β expression was significantly higher in more aggressive CM specimens (Maiuri et al., 2006). Interestingly, Ki-67 and bcl-2 expression was always absent in stable CM lesions and positively expressed in 72% of aggressive CM specimens (Maiuri et al., 2006). The observation of higher expression of proliferative indices (Ki-67 and bcl-2) within aggressive CMs is consistent with that of other proliferative disorders such as tumors. Further elucidation of the roles of the respective growth factors on angiogenesis will hopefully yield targets for potential therapies and increase our understanding of the dynamic behavior of CMs.

PATHOLOGY OF CAVERNOUS MALFORMATIONS CAUSING EPILEPSY

Although most CMs are diagnosed incidentally (48%), epileptic seizures are the second most common initial clinical presentation for patients with CMs, accounting for approximately 25% of cases (Del Curling et al., 1991; Maraire and Awad, 1995; Casazza et al., 1997; Moran et al., 1999; Moriarity et al., 1999; Bertalanffy et al., 2005). Seizure prevalence in patients with CMs has been reported in the literature to range from 30% to as high as 70% of patients with CMs (Chang et al., 2009; Flemming et al., 2012). The 5-year risks of epilepsy for incidentally found CMs is approximately 4% and does not warrant prophylactic treatment with antiepileptic drugs (AED) (Josephson et al., 2011). Patients who present with hemorrhage or focal neurologic deficit only increase the 5-year risk of epilepsy to 6% and do not warrant AED prophylaxis either (Josephson et al., 2011). However, if an individual with a CM experiences a single seizure, the 5-year risk for developing epilepsy rises significantly to 94% and AED use is recommended (Josephson et al., 2011). Some individuals with CM may have an isolated seizure, whereas some individuals may ultimately develop epilepsy that becomes medically refractory and surgical intervention may be indicated for potential seizure control.

The pathologic effects CMs have on the surrounding brain parenchyma are particularly important for the treatment of epilepsy secondary to CMs, especially with regard to focal cortical dysplasia (FCD). In 2011 the International League Against Epilepsy (ILAE) developed a new three-tiered classification system for FCDs (Blumcke et al., 2011). Type I (abnormal radial cortical lamination: FCD Ia; abnormal tangential cortical lamination: FCD Ib, both abnormal radial and tangential cortical lamination), and type II (dysmorphic neurons: FCD IIa, and dysmorphic neurons with balloon cells: FCD IIb) maintained the systematics of the prior classification (Palmini et al., 2004) but a new category was defined, type III, which comprises combinations of cortical lamination abnormalities and hippocampal sclerosis (FCD IIIa), epilepsy-associated tumors (FCD IIIb), vascular malformations (FCD IIIc), or other epileptogenic lesions obtained early in life (i.e., traumatic injury, ischemic injury, or encephalitis) (FCD IIId) (Blumcke et al., 2011). Cortical lamination abnormalities in combination with vascular malformations constitute FCD IIIc.

Preoperative planning for CM-related epilepsy surgery commonly might only entail MRI and noninvasive video surface electroencephalographic monitoring. The extent of resection required for seizure freedom is controversial and there are various treatment strategies, including lesionectomy, lesionectomy plus perilesional hemosiderin-stained parenchyma, or lobectomy for cases noted to have hippocampal sclerosis. Some studies have proposed that resection of the lesion and the surrounding hemosiderin rim of parenchyma may be beneficial for postoperative seizure control; however, it remains controversial (Baumann et al., 2006; Hammen et al., 2007; Stavrou et al., 2008; Englot et al., 2011). The overall results of CM-related epilepsy surgery are varied significantly among reports in the literature due to the heterogeneity of the patients, with approximately 25% of patients continuing to have seizures postoperatively (Englot et al., 2011).

According to the ILAE task force, factors that have positive predictive value for postoperative seizure freedom in patients include lesions of small size (<1.5 cm in diameter), fewer lesions, well-controlled partial seizures preoperatively, and shorter duration of epilepsy (McIntosh et al., 2004; Janszky et al., 2005; Jeong et al., 2005; Baumann et al., 2006; Yeon et al., 2009; Rosenow et al., 2013). The exact mechanism for how epilepsy develops in patients with CMs is complex. While mass effect from larger lesions is a plausible mechanism, the hemosiderin rim and gliosis of the surrounding parenchyma may also contribute to epileptic seizures. Lamination abnormalities may be noted within the cortex representing FCD (Blumcke et al., 2011). The cortex surrounding CMs is believed to be hyperexcitable, especially in the mesiotemporal

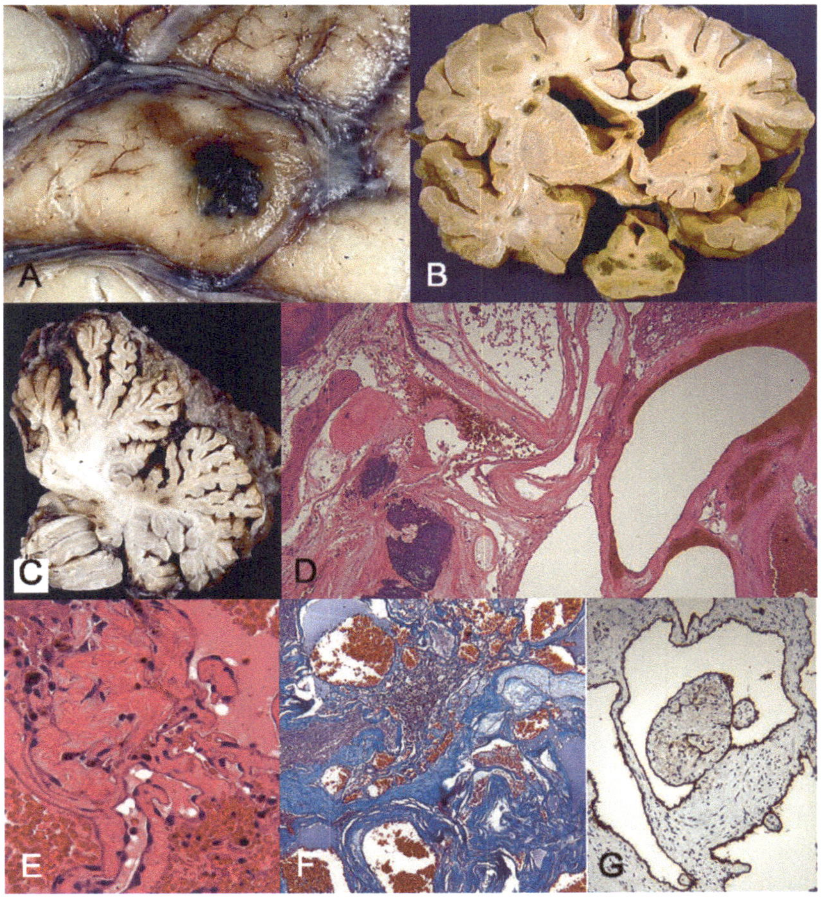

Fig. 25.1. Gross and histologic features of cavernous malformations. (**A**) Close-up view of the surface of the brain demonstrating a cortically based cavernous malformation. (**B**) Coronal section through the cerebrum at the level of the anterior commissures and transverse section of the pons demonstrating numerous cavernous malformations without evidence of hemorrhage. (**C**) Sagittal section through the cerebellum demonstrating diffuse atrophy secondary to Purkinje cell loss as a result of superficial siderosis. (**D**) Low-power hematoxylin and eosin-stained section showing cavernous blood-filled vessels accompanied by focal dystrophic calcification (lower left). (**E**) High-power hematoxylin and eosin-stained section showing hyalinized blood vessels lined by a single layer of attenuated endothelium. A few hemosiderin-laden macrophages, indicative of prior bleeding, are seen in the upper left-hand corner of the photograph. (**F**) Trichrome staining demonstrates that the vascular walls contain collagen only, without vascular smooth muscle. (**G**) CD31 immunohistochemical staining highlights endothelial cells lining the dilated vascular spaces.

archiocortex (Menzler et al., 2010). The epileptogenic zone can extend beyond the lesion itself, which has been demonstrated with electrocorticography and stereotactic encephalography (Baumann et al., 2006; Aubert et al., 2009; Amandine et al., 2014). Through a process known as secondary epileptogenesis, repeat seizure exposure can lead to synergistic synaptic alterations in cortex distant from the primary epileptogenic lesion or foci (Luders, 2001). This can be observed in patients with extrahippocampal epilepsy that spreads to the hippocampus (Surge et al., 2008). For these reasons, it is often useful to understand the seizure semiology and determine the extent of the epileptogenic zone for preoperative planning.

GROSS FEATURES

Gross specimens of CMs have the appearance of a reddish-purple raspberry due to the collection of blood-filled caverns. They are typically well-circumscribed and lobulated lesions. The surrounding cortex is usually hemosiderin stained as a result of prior hemorrhage. Recurrent hemorrhages lead to superficial siderosis of the leptominges that can be grossly apparent (Fig. 25.1). Hemosiderosis is also readily apparent on MRI. Thus, the hemosiderin staining of the surrounding cortex can be helpful for localization of CMs during resection. Within the lesion there is no intervening brain parenchyma. Blood in the caverns may be of various ages depending upon the

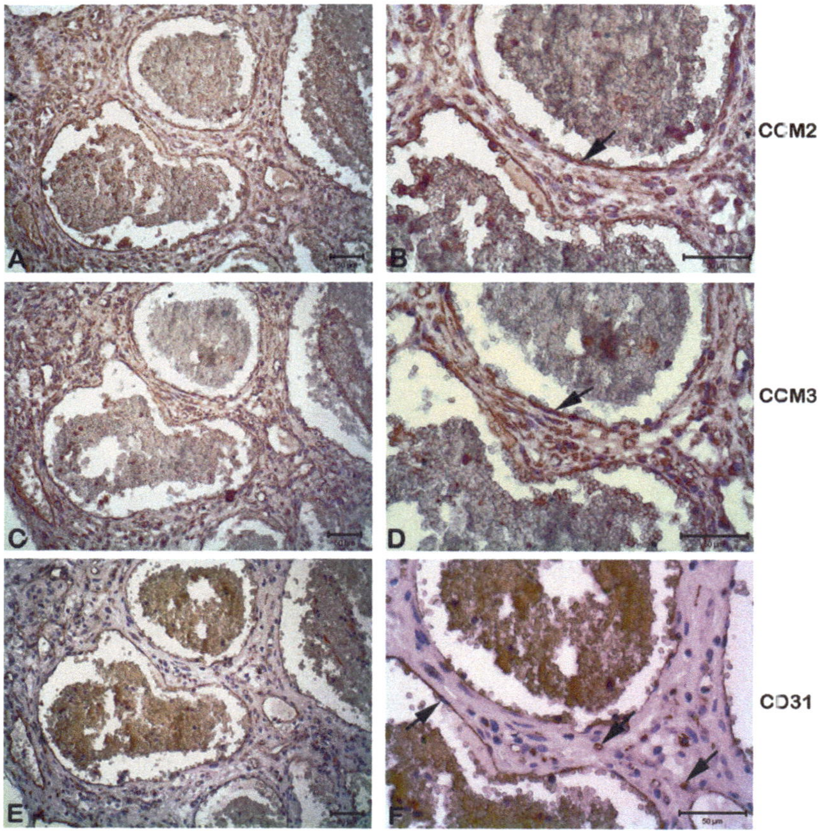

Fig. 25.2. Immunohistochemistry for CCM protein expression. Representative pictures of (**A**) CCM2, (**C**) CCM3, and (**E** and **F**) CD31/platelet endothelial cell adhesion molecule (PECAM-1) protein expression in cavernoma. (**B**) Higher-magnification view of (**A**); (**D**) higher-magnification view of (**C**); and (**F**) higher-magnification view of (**E**). Intense immunostaining is seen in vascular endothelium (arrows) and capillaries (**F**, arrows) in cavernoma tissues. Scale bars represent 50 µm. CCM2 and CCM3 proteins are detected strongly in the vascular endothelium with very weak staining pattern in stroma (Fig. 25.3**A–D**). To clarify the cell types, CD31/PECAM-1 staining was used to confirm the endothelial staining of CCM2 and CCM3 on serial sections of the CCM samples. CCM2 and CCM3 expression is detected in the vascular endothelium, also labeled with CD31/PECAM-1 in CCMs. In addition, CD31/PECAM immunoreactivity is detected in some of the endothelium of small capillaries in intercavernous areas (Fig. 25.3**E** and **F**). (Reproduced from Tanriover et al. (2013).)

temporal history of hemorrhage for the lesion. This can be elucidated radiographically (see Chapter 26).

HISTOLOGIC FEATURES AND ULTRASTRUCTURAL ANALYSIS

Microscopic examination of CMs demonstrates the typical features of vascular sinusoids or "caverns" lined by endothelial cells. Red blood cells are commonly visualized inside of the caverns (Fig. 25.1). In between the caverns there is usually a dense connective-tissue matrix that is composed of fibroblasts. Within the stroma and surrounding periphery of the CM lesion hemosiderin staining can be found as evidence of prior hemorrhage. Commonly, reactive gliosis can be observed in the brain parenchyma abutting the lesion. Occasionally, calcium deposition can also be noted. Immunohistochemistry is particularly useful in identifying the vascular endothelium and subsequently illustrates the presence of CCM proteins within the vascular endothelium (Fig. 25.2).

Ultrastructural characterization of CM lesions may be achieved with scanning electron micrographs. By doing so, the complex angioarchitecture of CMs can be visualized in a three-dimensional manner, illustrating a labyrinth-like structure with varying luminal morphology (Fig. 25.3) (Tanriover et al., 2013). Within the CM some cavernous areas demonstrate normal endothelium and some areas can demonstrate partial detachments of the endothelium in ruptured lumens, which may be indicative of compromised vascular integrity (Fig. 25.4).

Some studies have suggested that the blood–brain barrier is incompetent within CM lesions (Wong et al., 2000; Clatterbuck et al., 2001; Tu et al., 2005; Tanriover et al., 2013). The blood–brain barrier is

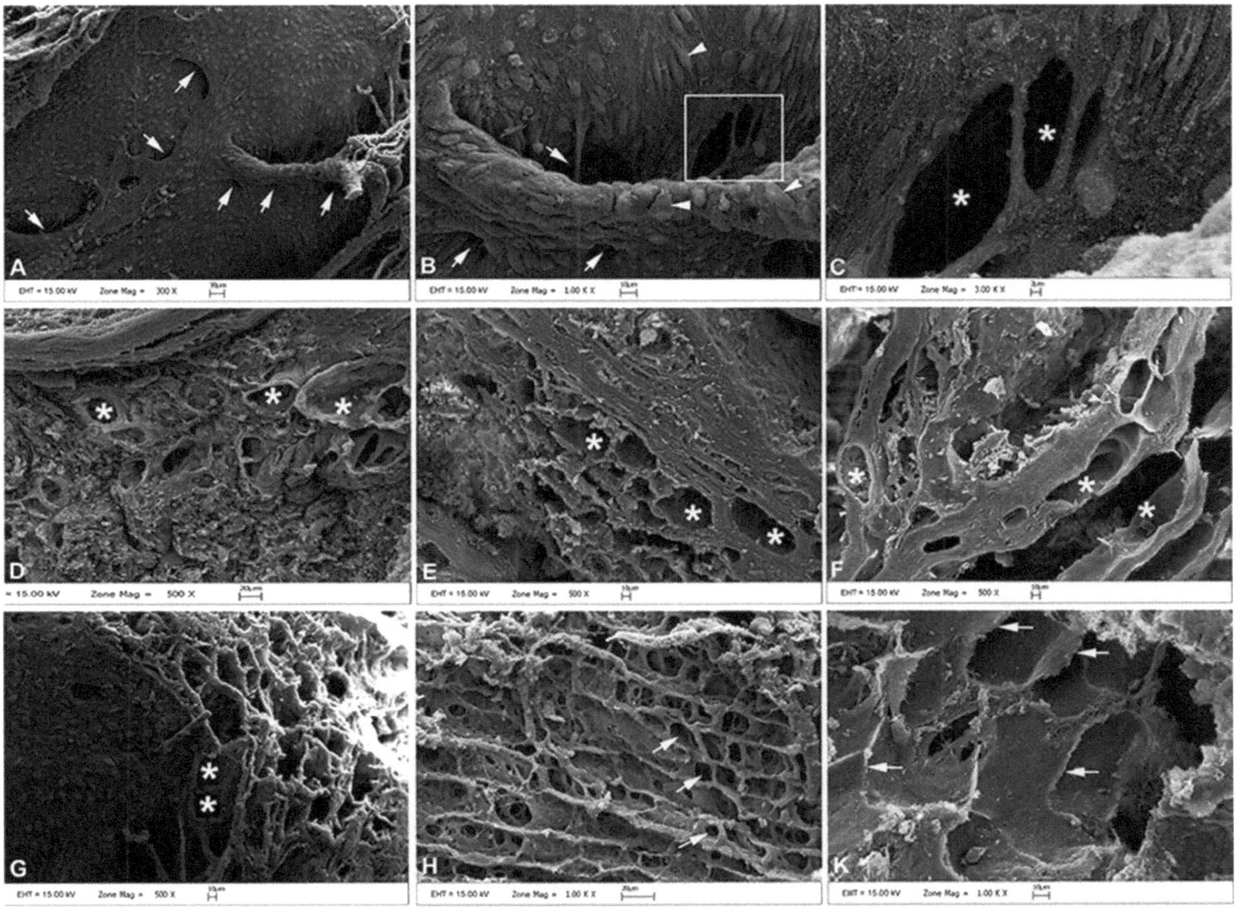

Fig. 25.3. Scanning electron microscopy of cavernous malformations. (**A** and **B**) The caverns are connected with each other constituting a three-dimensional labyrinth-like structure (arrows). The higher magnification of the area from **B** is presented in **C**. This image was shown that the cavern (asterisks). The caverns with different size and luminal morphology (**D–G**, asterisks). Some caverns are in regular endothelial structures (**B**, arrowheads); however, some are grid-like luminal surface (**H**, arrows). Ruptured luminal lining and partial detachments of the endothelium are demonstrated in different size and diameters (**K**, arrows). (Reproduced from Tanriover et al. (2013).)

composed of a cohesive association of endothelium, endothelial cell tight junctions, and astrocyte foot processes. The tight cell junctions between endothelial cells constitute the main permeability barrier. Within the tight junction are claudins and ocludins (ZO-1, ZO-2, ZO-3), in addition to several other proteins that maintain and restrict permeability between endothelial cells.

The physiology of the blood–brain barrier is comprised of the interactions between endothelial cell pericytes/smooth-muscle cells and the astrocyte foot processes covering most of the microvasculature (Pardridge, 1999). Ultrastructural analysis of CMs has demonstrated malformed endothelial cell tight junctions illustrated with appreciable gaps in between endothelial cells on electron microscopy and a lack of astrocytic endfoot processes (Wong et al., 2000; Clatterbuck et al., 2001; Tu et al., 2005). Stamatovic and colleagues (2015) have demonstrated significantly lower expression of tight junction proteins in familial CM lesions (PDCD10/CCM3) in comparison to the tight-junction proteins of sporadic CMs and normal brain vascular endothelial tight-junction complexes. This may provide insight into why CMs related to mutations in CCM3/PDCD10 are noted to exhibit significantly more aggressive clinical behavior (Shenkar et al., 2015). Subendothelial smooth-muscle cells and pericytes are also commonly noted to be absent or malformed in CMs (Uranishi et al., 2001a; Tu et al., 2005). These findings underscore the belief that the incompetency of the blood–brain barrier is an integral feature of CM pathophysiology and may be largely responsible for a lesion's propensity to hemorrhage (Clatterbuck et al., 2001; Tu et al., 2005; Tanriover et al., 2013). Furthermore, electron microscopy has demonstrated that the basal lamina is often thin, partial, or absent depending upon whether prior hemorrhage has occurred (Fig. 25.4) (Clatterbuck

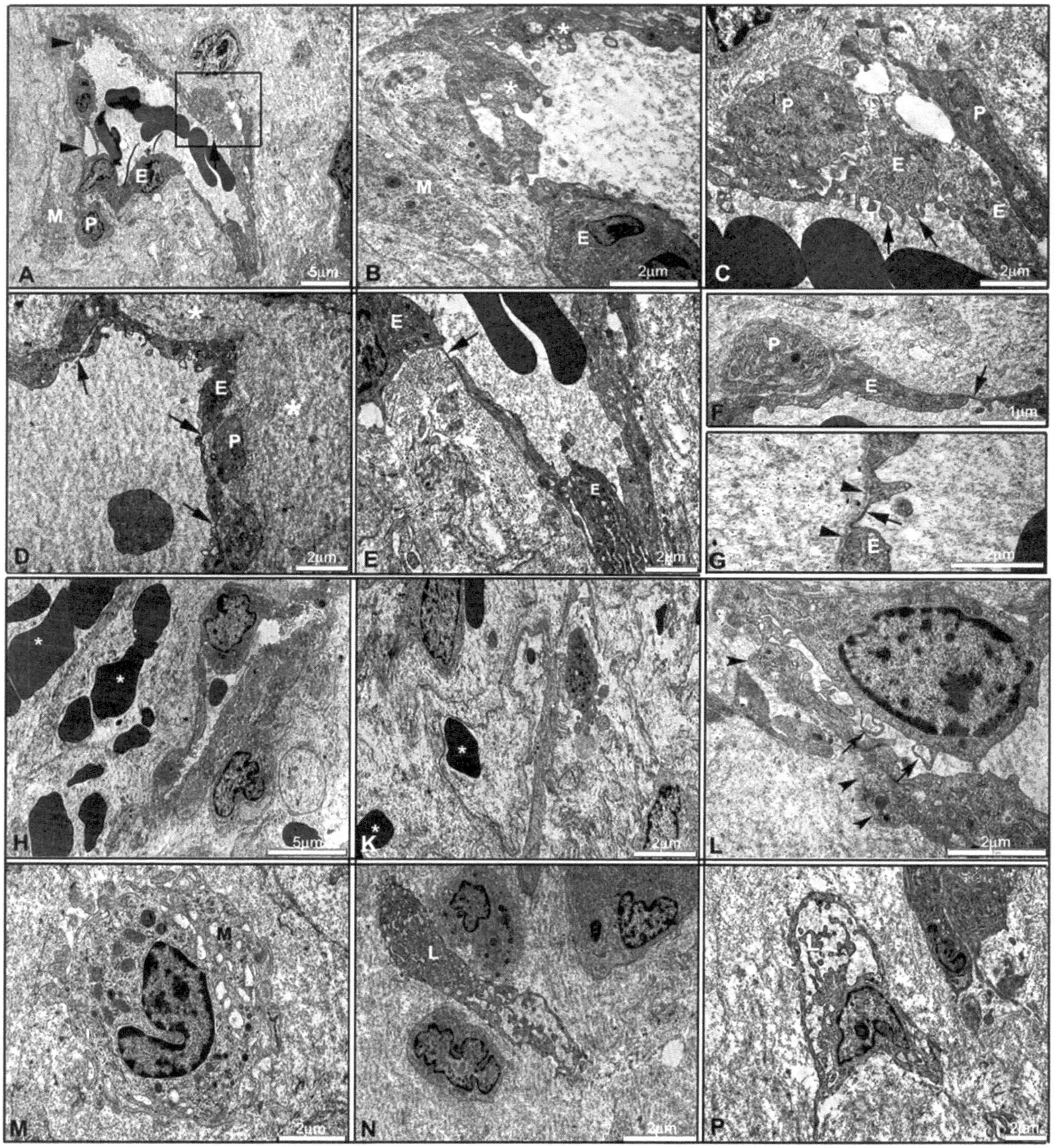

Fig. 25.4. Transmission electron microscopy of cavernous malformations. (**A**) The endothelium consists of enlarged pinocytotic vesicles and intracytoplasmic canaliculi (arrowheads). (**B**) The endothelial cells have dense chaotic protoplasmic protrusions (asterisks and C, L: arrows). (**C**) Representative electron micrograph is a higher magnification image of the selected area in **A**. (**D–G**) The endothelium was exposed directly to the amorphous matrix, which has only membranous wall (arrows). The considerable accumulations of amorphous substance with different sizes and irregular shapes were situated in the CMs (**D**, asterisk). (**G**) In some areas, the CM basal lamina is notably very thin and partially exposed to the lumen (arrowheads). (**H** and **K**) The CMs are characterized by caverns that are filled with erythrocytes (asterisk). (**M**, **N** and **P**) The intercavernous area was composed of an amorphous material and labyrinth-like cells (**L**) were detected as irregularly shaped with heterochromatic nucleus. E, endothelium; P, pericyte; M, macrophage. (Reproduced from Tanriover et al. (2013).)

et al., 2001; Tu et al., 2005; Tanriover et al., 2013). Aberrations in endothelial cell shape (e.g., increased number of filopodia, Weibel–Palade bodies, micropinocytic vesicles, filament, and other organelles) have also been noted in CMs that have had prior hemorrhages (Fig. 25.4) (Tu et al., 2005; Tanriover et al., 2013).

SUMMARY

Although these lesions were once believed to be solely congenital in nature, there is compelling evidence to support *de novo* formation as well. They can occur as sporadic lesions or be inherited in an autosomal-dominant phenotype in familial forms of the disease. Studies over the last few decades have yielded great insight into the pathophysiology of CMs. Most have centered on abnormal angiogenesis and effects of CMs on their surrounding tissue. Poor structural integrity of the vasculature of CMs predisposes lesions to red blood cell extravasation or hemorrhage, which subsequently leads to further dysangiogenesis and lesion proliferation. Ultrastructural analysis has demonstrated aberrations in the blood–brain barrier. The expression of growth factors and proliferative indices has been studied with regard to the clinical behavior of CMs. Future research on a molecular and genetic level will undoubtedly provide more insight into the pathophysiology of CMs.

References

Aiba T, Tanaka R, Koike T (1995). Natural history of intracranial cavernous malformations. J Neurosurg 83 (1): 56–59.

Al-Shahi R, Bhattacharya JJ, Currie DG et al. (2003). Prospective, population-based detection of intracranial vascular malformations in adults: the Scottish Intracranial Vascular Malformation Study (SVIMS). Stroke 34 (5): 1163–1169.

Al-Shahi Salman R, Berg MJ, Morrison L et al. (2008). Hemorrhage from cavernous malformations of the brain: definition and reporting standards. Angioma Alliance Scientific Advisory Board. Stroke 39 (12): 3222–3230.

Amandine S, Martine G, Agnes T et al. (2014). Beyond the lesion: the epileptogenic networks around cavernous angiomas. Epilepsy Res 108: 701–708.

Amirjamshidi A, Abbassioun K (2000). Radiation-induced tumors of the central nervous system occurring in childhood and adolescence. Four unusual lesions in three patients and a review of the literature. Childs Nerv Syst 16: 390–397.

Aubert S, Wendling F, Regis J et al. (2009). Local and remote epileptogenecity in focal cortical dysplasias and neurodevelopmental tumours. Brain 132: 3072–3086.

Baumann CR, Schuknecht B, Lo Russo G et al. (2006). Seizure outcome after resection of cavernous malformation is better when surrounding hemosiderin-stained brain is also is removed. Epilepsia 47: 563–566.

Baumgartner JE, Ater JL, Ha CS et al. (2003). Pathologically proven cavernous angiomas of the brain following radiation therapy for pediatric brain tumors. Pediatr Neurosurg 39 (4): 201–207.

Bertalanffy H, Benes L, Miyazawa T et al. (2005). Cerebral cavernomas in the adult. Review of the literature and analysis of 72 surgically treated. Neurosurg Rev 25: 1–53.

Bicknell JM, Carlow TJ, Kornfeld M et al. (1978). Familial cavernous angiomas. Arch Neurol 746–749.

Blumcke I, Thom M, Aronica E et al. (2011). The clinicopathologic spectrum of focal cortical dysplasias: a consensus classification proposed by an ad hoc task force of the ILAE Diagnostic Methods Commission. Epilepsia 52: 158–174.

Boulday G, Blécon A, Petit N et al. (2009). Tissue-specific conditional CCM2 knockout mice establish the essential role of endothelial CCM2 in angiogenesis: implications for human cerebral cavernous malformations. Dis Model Mech 2: 168–177.

Cakirer S (2003). De novo formation of a cavernous malformation of the brain in the presence of a developmental venous anomaly. Clin Radiol 58 (3): 251–256.

Campeau NG, Lane JI (2005). De novo development of a lesion with the appearance of a cavernous malformation adjacent to an existing developmental venous anomaly. AJNR Am J Neuroradiol 26 (1): 156–159.

Casazza M, Avanzini G, Ciceri E et al. (1997). Lesionectomy in epileptogenic temporal lobe lesions: preoperative seizure course and postoperative outcome. Acta Neurochir Suppl 68: 64–69.

Chang EF, Gabriel RA, Potts MB et al. (2009). Seizure characteristics and control after microsurgical resection of supratentorial cerebral cavernous malformations. Neurosurgery 65: 31–37.

Clatterbuck RE, Eberhart CG, Crain BJ et al. (2001). Ultrastructural and immunocytochemical evidence that an incompetent blood–brain barrier is related to the pathophysiology of cavernous malformations. J Neurol Neurosurg Psychiatry 71: 188–192.

Del Curling O, Kelly DL, Elster AD et al. (1991). An analysis of the natural history of cavernous angiomas. J Neurosurg 83: 237–242.

Dickson MC, Martin JS, Cousins FM et al. (1995). Defective haematopoiesis and vasculogenesis in transforming growth factor-β1 knockout mice. Development 121: 1845–1854.

Dillon WP (1997). Cryptic vascular malformations: controversies in terminology, diagnosis, pathophysiology, and treatment. AJNR Am J Neuroradiol 18 (10): 1839–1846.

Dorsch NWC, McMahon JHA (1998). Intracranial cavernous malformations – natural history and management. Crit Rev Neurosurg 8 (3): 154–168.

Duhem R, Vinchon M, Leblond P et al. (2005). Cavernous malformations after cerebral irradiation during childhood: report of nine cases. Childs Nerv Syst 21: 922–925.

Englot DJ, Han SJ, Lawton MT et al. (2011). Predictors of seizure freedom in the surgical treatment of supratentorial cavernous malformations. J Neurosurg 115: 1169–1174.

Flemming KD, Link MJ, Christianson TJ et al. (2012). Prospective hemorrhage risk of intracerebral cavernous malformations. Neurology 78 (9): 632–636.

Fritschi JA, Reulen HJ, Spetzler RF et al. (1994). Cavernous malformations of the brain stem. A review of 139 cases. Acta Neurochir (Wien) 130 (1-4): 35–46.

Furuse M, Miyatake SI, Kuroiwa T (2005). Cavernous malformation after radiation therapy for astrocytoma in adult patients: report of 2 cases. Acta Neurochir 147: 1097–1101.

Gaensler EH, Dillon WP, Edwards MS et al. (1994). Radiation-induced telangiectasia in the brain simulates cryptic vascular malformations at MR imaging. Radiology 193: 629–636.

Gangemi M, Longatti P, Maiuri F et al. (1990). Familial cerebral cavernous angiomas. Neurol Res 12: 131–136.

Gross BA, Batjer HH, Awad IA et al. (2009). Cavernous malformations of the basal ganglia and thalamus. Neurosurgery 65 (1): 7–18.

Hammen T, Romstock J, Dorfler A et al. (2007). Prediction of postoperative outcome with special respect to removal of hemosiderin fringe: a study in patients with cavernous haemangiomas associated with symptomatic epilepsy. Seizure 16: 248–253.

Hayman LA, Evans RA, Ferrell RE et al. (1982). Familial cavernous angiomas. Natural history and genetic study over a 5-year period. Am J Med Genet 11: 147–160.

Heckl S, Aschoff A, Kunze S (2002). Radiation-induced cavernous hemangiomas of the brain. A late effect predominantly in children. Cancer 94: 3285–3291.

Houtteville JP (1997). Brain cavernoma: a dynamic lesion. Surg Neurol 48: 610–614.

Humpl T, Bruhl K, Bohl J et al. (1997). Cerebral haemorrhage in long-term survivors of childhood acute lymphoblastic leukaemia. Eur J Pediatr 156: 367–370.

Janszky J, Janszky I, Schulz R et al. (2005). Temporal lobe epilepsy with hippocampal sclerosis: predictors for long-term surgical outcome. Brain 128: 395–404.

Jeong SW, Lee SK, Hong KS et al. (2005). Prognostic factors for the surgery for mesial temporal lobe epilepsy: longitudinal analysis. Epilepsia 46: 1273–1279.

Josephson CB, Leach JP, Duncan R et al. (2011). Seizure risk from cavernous or arteriovenous malformations: prospective population-based study. Neurology 76: 1548–1554.

Jung KH, Chu K, Jeong SW et al. (2003). Cerebral cavernous malformations with dynamic and progressive course. Correlation study with vascular endothelial growth factor. Arch Neurol 60: 1613–1618.

Kilic T, Pamir MN, Kullu S et al. (2000). Expression of structural proteins and angiogenic factors in cerebrovascular anomalies. Neurosurgery 46: 1179–1192.

Kondziolka D, Lunsford LD, Kestle JR (1997). The natural history of cerebral cavernous malformations. J Neurosurg 83 (5): 820–824.

Kupersmith MJ, Kalish H, Epstein F (2001). Natural history of brainstem cavernous malformations. Neurosurgery 48 (1): 47–53.

Labauge P, Brunereau L, Laberge S et al. (2001). Prospective follow-up of 33 asymptomatic patients with familial cerebral cavernous malformations. Neurology 57: 1825–1828.

Labauge P, Denier C, Bergametti F et al. (2007). Genetics of cavernous angiomas. Lancet Neurol 6: 237–244.

Larson JJ, Ball WS, Bove KE et al. (1998). Formation of intracerebral cavernous malformations after radiation treatment for central nervous system neoplasia in children. J Neurosurg 88: 51–56.

Leblanc GG, Golanov E, Awad IA et al. (2009). Biology of vascular malformations of the brain. Stroke 40: e694–e702.

Luders HO (2001). Clinical evidence for secondary epileptogenesis. Int Rev Neurobiol 45: 469–480.

Maeder P, Gudinchet F, Meuli R et al. (1998). Development of a cavernous malformation of the brain. AJNR Am J Neuroradiol 19: 1141–1145.

Maiuri F, Cappabianca P, Gangemi M et al. (2006). Clinical progression and familial occurrence of cerebral cavernous angiomas: the role of angiogenic and growth factors. Neurosurg Focus 21 (1): e3.

Maraire JN, Awad IA (1995). Intracranial cavernous malformations: lesion behavior and management strategies. Neurosurgery 37: 591–605.

Mason I, Aase JM, Orrison WW et al. (1988). Familial cavernous angiomas of the brain in an Hispanic family. Neurology 38: 324–326.

Massague J (1998). TGF-β signal transduction. Annu Rev Biochem 67: 753–791.

Massa-Micon B, Luparello V, Bergui M et al. (2000). De novo cavernoma case report and review of literature. Surg Neurol 53: 484–487.

McIntosh AM, Kalnins RM, Mitchell LA et al. (2004). Temporal lobectomy: long-term seizure outcome, late recurrence and risks for seizure recurrence. Brain 127: 2018–2030.

Menzler K, Chen X, Thiel P et al. (2010). Epileptogenicity of cavernomas depends on (archi-) cortical localization. Neurosurg 67: 918–924.

Moran NF, Fish DR, Kitchen N et al. (1999). Supratentorial cavernous haemangiomas and epilepsy: a review of the literature and case series. J Neurol Neurosurg Psychiatry 66: 561–568.

Moriarity JL, Wetzel M, Clatterbuck RE et al. (1999). The natural history of cavernous malformations: a prospective study of 68 patients. Neurosurgery 44: 1166–1171.

Moses HL, Serra R (1996). Regulation of differentiation by TGF-β. Curr Opin Genet Dev 6: 581–586.

Nimjee SM, Powers CJ, Bulsara KR (2006). Review of the literature on de novo formation of cavernous malformations of the central nervous system after radiation therapy. Neurosurg Focus 21 (1): E4.

Nyáry I, Major O, Hanzély Z et al. (2005). Histopathological findings in a surgically resected thalamic cavernous hemangioma 1 year after 40-Gy irradiation. J Neurosurg 102 (Suppl): 56–58.

Pagenstecher A, Stahl S, Sure U et al. (2009). A two- hit mechanism causes cerebral cavernous malformations: complete inactivation of CCM1, CCM2 or CCM3 in affected endothelial cells. Hum Mol Genet 18: 911–918.

Palmini A, Najm I, Avanzini G et al. (2004). Terminology and classification of the cortical dysplasias. Neurology 62: S2–S8.

Pandey P, Westbroek EM, Gooderham PA et al. (2012). Cavernous malformation of brainstem, thalamus, and basal ganglia: a series of 176 patients. Neurosurgery 72 (4): 573–589.

Pardridge WM (1999). Blood–brain barrier biology and methodology. J Neurovirol 5: 556–569.

Perrini P, Lanzino G (2006). The association of venous developmental anomalies and cavernous malformations: pathophysiological, diagnostic, and surgical consideration. Neurosurg Focus 21: 1e5.

Petersen TA, Morrison LA, Schrader RM et al. (2010). Familial versus sporadic cavernous malformations: differences in venous anomaly association and lesion phenotype. AJNR Am J Neuroradiol 31: 377–382.

Petit N, Blécon A, Denier C et al. (2006). Patterns of expression of the three cerebral cavernous malformation (CCM) genes during embryonic and postnatal brain development. Gene Expr Patterns 6: 495–503.

Porter PJ, Willinsky RA, Harper W et al. (1997). Cerebral cavernous malformations: natural history and prognosis after clinical deterioration with or without hemorrhage. J Neurosurg 87 (2): 190–197.

Pozzati E, Giuliani G, Nuzzo G et al. (1989). The growth of cerebral cavernous angiomas. Neurosurgery 25 (1): 92–97.

Pozzati E, Giangaspero F, Marliani F et al. (1996). Occult cerebrovascular malformations after irradiation. Neurosurgery 39: 677–684.

Rigamonti D, Hadley MN, Drayer BP et al. (1988). Cerebral cavernous malformations: incidence and familial occurrence. N Engl J Med 319: 343–347.

Robinson JR, Awad IA, Little JR (1991). Natural history of the cavernous angioma. J Neurosurg 75 (5): 709–714.

Rosenow F, Alonso-Vanegas MA, Baumgasrtner C et al. (2013). Cavernoma-related epilepsy:review and recommendations for management-report of the Surgical Task Force, Commission on Therapeutic Strategies. Epilepsia 54: 2025–2035.

Rothbart D, Awad IA, Lee J et al. (1996). Expression of angiogenic factors and structural proteins in central nervous system vascular malformations. Neurosurgery 38: 915–925.

Seker A, Pricola KL, Guclu B et al. (2006). CCM2 expression parallels that of CCM1. Stroke 37: 518–523.

Shenkar R, Venkatasubramanian PN, Zhao JC et al. (2008). Advanced magnetic resonance imaging of cerebral cavernous malformations: part I. High-field imaging of excised human lesions. Neurosurgery 63 (4): 782–789.

Shenkar R, Shi C, Rebeiz T et al. (2015). Exceptional aggressiveness of cerebral cavernous malformation disease associated with PDCD10 mutations. Genet Med 17: 188–196.

Stamatovic SM, Sladojevic N, Keep RF et al. (2015). PDCD10 (CCM3) regulates brain endothelial barrier integrity in cerebral cavernous malformations type 3: role of CCM3-ERK1/2-cortactin cross-talk. Acta Neuropathol 130: 731–750.

Stavrou I, Baumgartner C, Frischer JM et al. (2008). Long-term seizure control after resection of supratentorial cavernomas: a retrospective single-center study in 53 patients. Neurosurgery 63: 888–896.

Sure U, Butz N, Schlegel J et al. (2001). Endothelial proliferation, neoangiogenesis, and the potential de novo generation of cerebrovascular malformations. J Neurosurg 94: 972–977.

Surge R, Schulze-Bonhage A, Altenmuller DM (2008). Hippocampal Involvement in secondarily generalized seizures: surgical outcomes in 60 consecutive patients. Seizure 18: 14–20.

Tanriover G, Sozen B, Seker A et al. (2013). Ultrastructural analysis of vascular features in cerebral cavernous malformations. Clin Neurol Neurosurg 115 (4): 438–444.

Tu J, Stoodley MA, Morgan MK et al. (2005). Ultrastructural characteristics of hemorrhagic, nonhemorrhagic, and recurrent cavernous malformations. J Neurosurg 103: 903–909.

Uranishi R, Baev NI, Kim JH et al. (2001a). Vascular smooth muscle cell differentiation in human cerebral vascular malformations. Neurosurgery 49 (3): 671–679.

Uranishi R, Baev NI, Ng PY et al. (2001b). Expression of endothelial cell angiogenesis receptors in human cerebrovascular malformations. Neurosurgery 48: 359–368.

Van Gompel J, Rubio J, Cascino G et al. (2009). Electrocorticography-guided resection of temporal cavernoma: is electrocorticography warranted and does it alter the surgical approach? J Neurosurg 110: 1179–1185.

Viale GL, Castellani P, Dorcaratto A et al. (2002). Occurrence of a glioblastoma-associated tenascin-C isoform in cerebral cavernomas and neighboring vessels. Neurosurgery 50: 838–842.

Whitehead KJ, Chan AC, Navankasattusas S et al. (2009). The cerebral cavernous malformation signaling pathway promotes vascular integrity via Rho GTPases. Nat Med 15: 177–184.

Wong JH, Awad IA, Kim JH (2000). Ultrastructural pathological features of cerebrovascular malformations: a preliminary report. Neurosurgery 46: 1454–1459.

Wurm G, Schnizer M, Fellner FA (2005). Cerebral cavernous malformations associated with venous anomalies: surgical considerations. Neurosurgery 57 (1 Suppl): 42–58.

Yeon JY, Kim JS, Choi SJ et al. (2009). Supratentorial cavernous angiomas presenting with seizures: surgical outcomes in 60 consecutive patients. Seizure 18: 14–20.

Yildirim O, Bicer A, Ozkan A et al. (2010). Expression of platelet-derived growth factor ligand and receptor in cerebral arteriovenous and cavernous malformations. J Clin Neurosci 17 (12): 1557–1562.

Zabramski JM, Wascher TM, Spetzler RF (1994). The natural history of familial cavernous malformations: results of an ongoing study. J Neurosurg 80 (3): 422–432.

Ziyal IM, Ozgen T (2001). Natural history of brainstem cavernous malformations. Neurosurgery 49 (4): 1023–1024.

Handbook of Clinical Neurology, Vol. 143 (3rd series)
Arteriovenous and Cavernous Malformations
R.F. Spetzler, K. Moon, and R.O. Almefty, Editors
http://dx.doi.org/10.1016/B978-0-444-63640-9.00026-6

Chapter 26

Developmental venous anomalies

MICHAEL A. MOONEY AND JOSEPH M. ZABRAMSKI*

Department of Neurosurgery, Barrow Neurological Institute, St. Joseph's Hospital and Medical Center, Phoenix, AZ, USA

Abstract

Developmental venous anomalies (DVAs) are relatively common lesions, present in up to 3% of the population. The defining characteristic of these lesions is the confluence of radially oriented veins into a single dilated venous channel. DVAs are also known as cerebral venous angiomas, cerebral venous malformations, and cerebral venous medullary malformations. They are the most common type of cerebral vascular malformation found on autopsy studies, and they are often encountered as incidental findings on neuroimaging studies. DVAs are congenital lesions thought to arise from aberrations that occur during venous development, but continue to provide the normal venous drainage to the cerebral territory in which they reside. Although the natural history of DVAs is benign, they may be associated with cavernous malformations or other vascular abnormalities, which can lead to hemorrhage in the vicinity of the DVA. Surgical or endovascular obliteration of DVAs carries a significant risk of venous infarction; thus, conservative management is the treatment of choice for patients with these lesions.

INTRODUCTION

Developmental venous anomalies (DVAs) have been described in the medical literature under varying names, including cerebral venous angiomas, cerebral venous malformations, and cerebral venous medullary malformations. The defining feature of these lesions is the confluence of radially oriented veins into a single dilated venous channel. DVAs are the most common cerebral vascular malformation found on autopsy studies, and they are often encountered as incidental findings on neuroimaging studies. The natural history of DVAs is benign; however, they may be associated with other cerebrovascular abnormalities and can lead to hemorrhagic or ischemic pathology. The epidemiology, anatomy, pathophysiology, imaging characteristics, natural history, and treatment options for DVAs are reviewed here.

EPIDEMIOLOGY

The prevalence of DVAs is relatively high compared with that of other vascular malformations. DVAs are estimated to affect approximately 3% of the general population. One of the most comprehensive studies to date examining the prevalence of vascular malformations examined a series of 5754 consecutive autopsies (McCormick, 1984). In this series, a total of 270 vascular malformations were discovered, 173 of which were venous malformations (3.0%). In comparison, 30 arteriovenous malformations (0.5%) and 19 cavernous malformations (0.3%) were identified. These numbers have been supported by subsequent prevalence studies (Del Curling et al., 1991; Garner et al., 1991; Robinson et al., 1991; Al-Holou et al., 2012; Elhammady et al., 2015).

Although DVAs are most commonly found as isolated incidental lesions on neuroimaging studies, they can be associated with adjacent pathology. Their most common association is with cerebral cavernous malformations. An estimated 8–33% of DVAs occur in association with a cerebral cavernous malformation (Topper et al., 1999; San Millan Ruiz et al., 2007). Furthermore, it has been proposed that all cerebral cavernous malformations have an associated DVA and that the DVA may contribute to

*Correspondence to: Joseph M. Zabramski, MD, c/o Neuroscience Publications, Barrow Neurological Institute, St. Joseph's Hospital and Medical Center, 350 W. Thomas Rd, Phoenix AZ 85013, USA. Tel: +1-602-406-3593, Fax: +1-602-406-4104, E-mail: Neuropub@dignityhealth.org

the pathophysiology of these lesions (Rigamonti and Spetzler, 1988; Batra et al., 2009; Ruiz et al., 2009).

Multiple other associated conditions have been described in the literature, and the presence of a cerebral DVA should be considered when these conditions are encountered. For example, the condition known as sinus pericranii is an abnormal communication between extracranial veins and intracranial dural sinuses, and these lesions are often associated with an adjacent cerebral DVA. One series documented an adjacent DVA in 8 of 15 patients with sinus pericranii who underwent digital subtraction angiographic evaluation (Gandolfo et al., 2007). The presence of the DVA must be considered when planning for sinus pericranii repair. Cerebral DVAs also have a documented increased incidence in patients with head and neck DVAs (Boukobza et al., 1996), lymphatic malformations of the orbit and periorbital region (Bisdorff et al., 2007), and hemangiomas of the skin and gastrointestinal tract (blue rubber bleb nevus syndrome) (Gabikian et al., 2003). Of note, cerebral DVAs are usually solitary lesions, except in patients with blue rubber bleb nevus syndrome, who may have multiple cerebral DVAs.

ANATOMY AND PATHOPHYSIOLOGY

The normal venous drainage patterns of the cerebrum can be separated into superficial and deep venous systems. Typically, venous channels progressively join one another and enlarge until they enter the venous sinuses of the superficial or deep systems. In the case of a DVA, cortical, subcortical, or white-matter venous tributaries drain centrifugally toward an abnormal collecting vein of the DVA, which can drain into the superficial venous system, deep venous system, or both (Fig. 26.1). This drainage pattern gives a characteristic appearance on imaging studies that has been described as the caput medusae sign (Saba, 1998). It must be noted that DVAs represent a confluence of venous drainage from normal cerebral gray and white matter, and sacrifice or compromise of the DVA can lead to venous infarction in the draining territories. Care must be taken to preserve a DVA when operating on an adjacent cavernous malformation or associated lesion.

DVAs occur throughout the brain in rough proportion to cerebral volume. Published reports cite a frontoparietal location in 36–64% of DVAs, whereas 14–27% occur in the cerebellar hemispheres. Given the smaller volume of the cerebellum, the presence of this proportion of DVAs there may represent a predilection of DVAs for this region, but this finding has not been consistently reported to date.

The pathophysiology of DVA formation is controversial, and numerous theories to explain it have been proposed. The unifying theme of most current theories is that an alteration in venous drainage during fetal development leads to DVA formation. Lasjaunias et al. (1986) proposed that "hemodynamic need" during development leads to the formation of alternative drainage patterns and DVA formation; this theory highlights the importance of the DVA for normal venous drainage of the developed brain. An alternative theory proposed by Saito and Kobayashi (1981) suggests that thrombosis of normal venous drainage patterns during development leads to the shunting of drainage into alternative pathways, resulting in subsequent DVA formation. A single theory for DVA formation has not been confirmed, and thus it is possible that multiple pathways lead to the same end result of DVA formation during cerebral venous development.

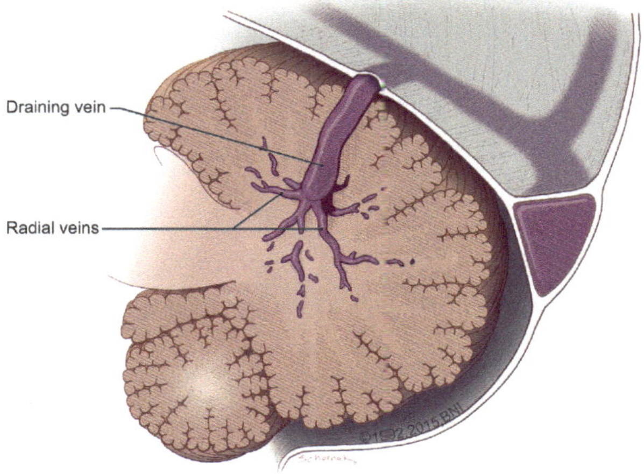

Fig. 26.1. Cerebellar developmental venous anomaly draining to the tentorium and straight sinus. (Courtesy of Barrow Neurological Institute, Phoenix, AZ.)

IMAGING CHARACTERISTICS

DVAs are often found incidentally on neuroimaging studies, such as computed tomography (CT) and magnetic resonance imaging (MRI), and their discovery can prompt further workup with vascular imaging studies. The extent of the DVA can range from a single draining vein to an anomalous vein draining an entire lobe or hemisphere.

The characteristic appearance of the caput medusae can be appreciated on a CT scan with contrast, as well as on an MRI scan of the brain with or without contrast (Fig. 26.2). Further evaluation of the DVA with vascular imaging (i.e., CT angiography, MR angiography, or formal angiography) is only warranted to rule out an associated arterial component that may require treatment. The typical appearance of the caput medusae on angiographic

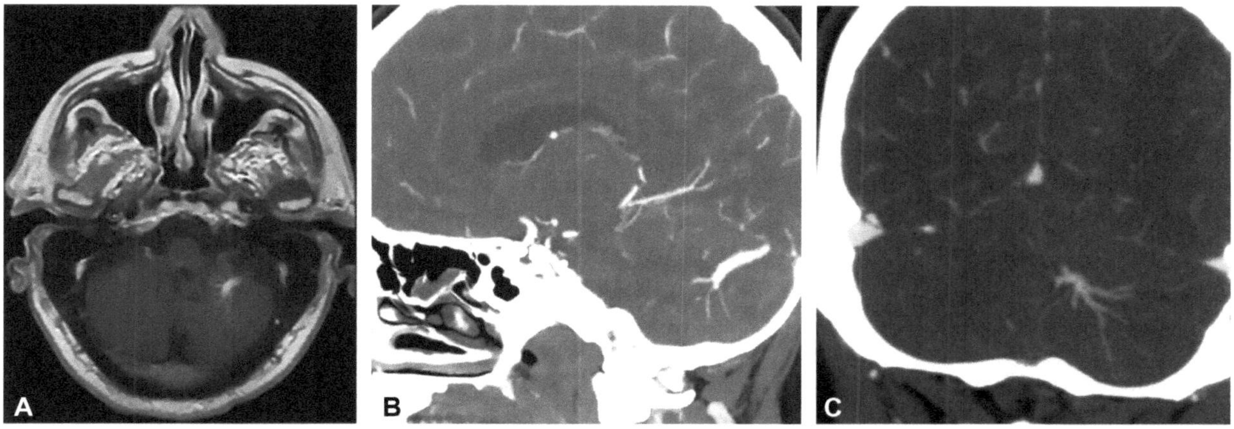

Fig. 26.2. (**A**) Axial T1-weighted magnetic resonance imaging scan with contrast demonstrating a left cerebellar developmental venous anomaly (DVA) with the typical caput medusae appearance. (**B**) Sagittal and (**C**) coronal computed tomography angiograms demonstrating a left cerebellar DVA in a different patient. (Courtesy of Barrow Neurological Institute, Phoenix, AZ.)

studies represents deep medullary veins draining into a single large draining vein that then connects to a superficial cortical vein, subependymal vein, or dural venous sinus (Truwit, 1992).

Given the current quality of MR studies and the characteristic appearance of most DVAs, formal angiography is rarely needed to evaluate these lesions. If the diagnosis of a simple DVA has been confirmed, no follow-up imaging is necessary. In cases where an associated vascular abnormality has been identified (e.g., a cavernous malformation), follow-up is recommended on the basis of the associated lesion.

NATURAL HISTORY AND TREATMENT

DVAs are now widely accepted as lesions with a benign natural history. In a longitudinal study of 63 patients with venous malformations who were followed over a mean of >4 years, hemorrhage was encountered in 2 patients, resulting in a calculated annual hemorrhage risk of 0.15% per lesion per year (Naff et al., 1998). In a separate prospective study, 80 patients were followed conservatively over a 10-year period, and an annual hemorrhage rate of 0.68% was found (McLaughlin et al., 1998). Given our current understanding of these lesions, it is thought that these hemorrhages represent the presence of an adjacent cavernous malformation that ruptured, rather than being related to the DVA itself.

An array of symptoms has been attributed to the presence of a cerebral DVA. Numerous reports of DVAs in relation to seizure foci or as the etiology of headaches exist (Striano et al., 2000; Morioka et al., 2006), but this correlation is not widely accepted. Venous thrombosis of a DVA resulting in venous infarction has been reported (Masson et al., 2000); however, this is a rare complication and the incidence has not been documented in the medical literature.

As mentioned above, DVAs should never be surgically obliterated because of their contribution to normal venous drainage and the risk for venous infarction (Senegor et al., 1983; Abla et al., 2011). Conservative management is recommended for all DVAs. When resection is necessary for a cavernous malformation associated with a DVA, the surgical approach must take into consideration how best to preserve the DVA.

CONCLUSIONS

DVAs are relatively common lesions present in up to 3% of the population. DVAs result from aberrations in venous development and provide normal venous drainage to the cerebral territory in which they arise. While they pose no significant risk themselves, they may be associated with cavernous malformations or other vascular abnormalities, which can lead to hemorrhage in the vicinity of the DVA. Surgical or endovascular obliteration of DVAs carries a significant risk of venous infarction, and conservative management is the treatment of choice for these lesions.

References

Abla AA, Lekovic GP, Turner JD et al. (2011). Advances in the treatment and outcome of brainstem cavernous malformation surgery: a single-center case series of 300 surgically treated patients. Neurosurgery 68: 403–414; discussion 414–415.

Al-Holou WN, O'Lynnger TM, Pandey AS et al. (2012). Natural history and imaging prevalence of cavernous malformations in children and young adults. J Neurosurg Pediatr 9: 198–205.

Batra S, Lin D, Recinos PF et al. (2009). Cavernous malformations: natural history, diagnosis and treatment. Nat Rev Neurol 5: 659–670.

Bisdorff A, Mulliken JB, Carrico J et al. (2007). Intracranial vascular anomalies in patients with periorbital lymphatic and lymphaticovenous malformations. AJNR Am J Neuroradiol 28: 335–341.

Boukobza M, Enjolras O, Guichard JP et al. (1996). Cerebral developmental venous anomalies associated with head and neck venous malformations. AJNR Am J Neuroradiol 17: 987–994.

Del Curling Jr O, Kelly Jr DL, Elster AD et al. (1991). An analysis of the natural history of cavernous angiomas. J Neurosurg 75: 702–708.

Elhammady MS, Hayes S, Heros RC (2015). Cerebral arteriovenous malformations. In: RF Spetzler, MYS Kalani, P Nakaji (Eds.), Neurovascular Surgery, 2nd edn. Thieme, New York.

Gabikian P, Clatterbuck RE, Gailloud P et al. (2003). Developmental venous anomalies and sinus pericranii in the blue rubber-bleb nevus syndrome. Case report. J Neurosurg 99: 409–411.

Gandolfo C, Krings T, Alvarez H et al. (2007). Sinus pericranii: diagnostic and therapeutic considerations in 15 patients. Neuroradiology 49: 505–514.

Garner TB, Del Curling Jr O, Kelly Jr DL et al. (1991). The natural history of intracranial venous angiomas. J Neurosurg 75: 715–722.

Lasjaunias P, Burrows P, Planet C (1986). Developmental venous anomalies (DVA): the so-called venous angioma. Neurosurg Rev 9: 233–242.

Masson C, Godefroy O, Leclerc X et al. (2000). Cerebral venous infarction following thrombosis of the draining vein of a venous angioma (developmental abnormality). Cerebrovasc Dis 10: 235–238.

McCormick WF (1984). Pathology of vascular malformations of the brain. In: CB Wilson, BM Stein (Eds.), Intracranial Arteriovenous Malformations. Williams & Wilkins, Baltimore, MD.

McLaughlin MR, Kondziolka D, Flickinger JC et al. (1998). The prospective natural history of cerebral venous malformations. Neurosurgery 43: 195–200; discussion 200–201.

Morioka T, Hashiguchi K, Nagata S et al. (2006). Epileptogenicity of supratentorial medullary venous malformation. Epilepsia 47: 365–370.

Naff NJ, Wemmer J, Hoenig-Rigamonti K et al. (1998). A longitudinal study of patients with venous malformations: documentation of a negligible hemorrhage risk and benign natural history. Neurology 50: 1709–1714.

Rigamonti D, Spetzler RF (1988). The association of venous and cavernous malformations. Report of four cases and discussion of the pathophysiological, diagnostic, and therapeutic implications. Acta Neurochir (Wien) 92: 100–105.

Robinson JR, Awad IA, Little JR (1991). Natural history of the cavernous angioma. J Neurosurg 75: 709–714.

Ruiz DS, Yilmaz H, Gailloud P (2009). Cerebral developmental venous anomalies: current concepts. Ann Neurol 66: 271–283.

Saba PR (1998). The caput medusae sign. Radiology 207: 599–600.

Saito Y, Kobayashi N (1981). Cerebral venous angiomas: clinical evaluation and possible etiology. Radiology 139: 87–94.

San Millan Ruiz D, Delavelle J, Yilmaz H et al. (2007). Parenchymal abnormalities associated with developmental venous anomalies. Neuroradiology 49: 987–995.

Senegor M, Dohrmann GJ, Wollmann RL (1983). Venous angiomas of the posterior fossa should be considered as anomalous venous drainage. Surg Neurol 19: 26–32.

Striano S, Nocerino C, Striano P et al. (2000). Venous angiomas and epilepsy. Neurol Sci 21: 151–155.

Topper R, Jurgens E, Reul J et al. (1999). Clinical significance of intracranial developmental venous anomalies. J Neurol Neurosurg Psychiatry 67: 234–238.

Truwit CL (1992). Venous angioma of the brain: history, significance, and imaging findings. AJR Am J Roentgenol 159: 1299–1307.

Handbook of Clinical Neurology, Vol. 143 (3rd series)
Arteriovenous and Cavernous Malformations
R.F. Spetzler, K. Moon, and R.O. Almefty, Editors
http://dx.doi.org/10.1016/B978-0-444-63640-9.00027-8

Chapter 27

Supratentorial cavernous malformations

JASON A. ELLIS AND DANIEL L. BARROW*
Department of Neurosurgery, Emory University School of Medicine, Atlanta, GA, USA

Abstract

Supratentorial cavernous malformations are uncommon cerebral vascular lesions that may present many unique challenges for treating physicians. The vast majority will be discovered during workup for seizures or after symptomatic intracerebral hemorrhage. Supratentorial cavernous malformations are increasingly being discovered incidentally in patients who obtain brain imaging for unrelated reasons. Management strategies including watchful waiting, antiepileptic drug therapy, microsurgery, or an expanding array of "minimally invasive" therapies. Thus, clinical decision making is not always straightforward and a nuanced approach tempered by experience and good judgment is critical to achieving excellent clinical results.

INTRODUCTION

Cerebral cavernous malformations were first described by Plenck in 1776 (Heros and Morcos, 2000). Later reports by Virchow in 1863 and modern classification by McCormick in 1966 form the basis for our current understanding of this unique vascular pathology (McCormick and Nofzinger, 1966). Adapted from McCormick's classic schema, cavernous malformations are recognized as one of the four classes of cerebral vascular malformations which also include arteriovenous malformation (AVM), venous malformation (developmental venous anomalies), and capillary telangiectasia. Histopathologic analysis shows that cavernous malformations are epithelial-lined sinusoidal channels that are devoid of smooth muscle and lack intervening brain parenchyma. Hemosiderin-laden gliotic brain may form a pseudocapsule and serve as a focus for epileptogenic activity. Cavernous malformations may be fed by small peripheral arteries but, unlike AVMs, have little blood flow and are characterized by low intralesional pressures.

The magnetic resonance imaging (MRI) and gross appearance of cavernous malformations has been described as "mulberry-like." Both spontaneous and familial varieties are known to exist, with much recent work being focused on delineating the molecular-genetic basis of this entity (Leblanc et al., 2009). They are found throughout the nervous system, with the highest proportion in the supratentorial compartment. It is the comprehensive description, clinical workup, and definitive management of cavernous malformations found within the supratentorial compartment that we shall be concerned with in this chapter. These so-called lobar cavernous malformations are in many ways distinct from those found elsewhere in the central nervous system. Surprisingly, supratentorial cavernous malformations can often be managed conservatively or with pharmacotherapy, in contradistinction to symptomatic infratentorial and spinal cavernous malformations. Additionally, their relationship to eloquent cortex as well as to subcortical fiber tracts heavily influences surgical decision making. Deep supratentorial malformations, including those of the diencephalon and basal ganglia, require separate specific considerations and are thus discussed elsewhere.

EPIDEMIOLOGY

Cavernous malformations are found in approximately 0.5% of the population and comprise 10–15% of all central nervous system vascular malformations (Del Curling et al., 1991; Robinson et al., 1991; Batra et al., 2009). The vast majority are found in the supratentorial

*Correspondence to: Daniel L. Barrow, M.D., Department of Neurosurgery, Emory University School of Medicine, Atlanta, GA, USA. Tel: +1-404-778-5770, E-mail: dbarr01@emory.edu

compartment in predictable proportion to the volume of neural tissue present. Thus, about 75% of all cavernous malformations are found above the tentorium. In terms of lobar distribution, approximately 20% of all cavernous malformations are frontal, 15% are parietal, 5% are occipital, 20% are temporal, and 5% are in the diencephalon. In addition, large multilobar cavernous malformations are recognized and may represent less than 5% of cases (Robinson et al., 1991; Moran et al., 1999; Rosenow et al., 2013).

CLINICAL PRESENTATION AND RADIOGRAPHIC ASSESSMENT

Although supratentorial cavernous malformations do not represent a distinct pathophysiologic entity from those located elsewhere, their clinical presentation and management are distinct. Patients with supratentorial cavernous malformations may present with seizures, headache, focal neurologic deficits secondary to mass effect, or after incidental discovery during radiographic evaluation for unrelated reasons. Presenting symptoms may occur either in an isolated fashion or as part of a more complex syndrome. Nonetheless, seizure is the most common presentation for patients with supratentorial cavernous malformations. In one systematic review of the literature by Moran et al. (1999), 79% of patients with supratentorial cavernous malformations were found to have seizures alone or in addition to other manifestations.

Cerebral cavernous malformations are hyperdense with a nonspecific appearance on computed tomography (CT) scan. They are angiographically occult save for the frequently associated developmental venous anomalies. MRI is the gold-standard modality for the radiographic assessment of cavernous malformations. T2-weighted MRI typically demonstrates a characteristic "salt-and-pepper" reticulated core with a hypointense rim (Fig. 27.1). Zabramski and coworkers (1994) proposed an MRI classification including four types of cavernous malformations that has been widely cited (Table 27.1). Although radiographic evidence of prior hemorrhage is universally seen in patients with cavernous malformations, clinically correlated symptomatic hemorrhage is less common. This is particularly so in the supratentorial compartment where small lobar hemorrhages often remain silent or manifest with delayed seizure onset.

With increasing widespread utilization of high-resolution brain MRI for a variety of clinical and research indications, cavernous malformations are more commonly being detected in asymptomatic patients. In various MRI studies, it is an incidental finding in 0.2–0.4% of the cohorts evaluated (Katzman et al., 1999; Vernooij et al., 2007; Morris et al., 2009). Thus it is of utmost importance that the natural history of this entity be thoroughly understood by clinicians, lest patients be subjected to unnecessary and potentially morbid interventions.

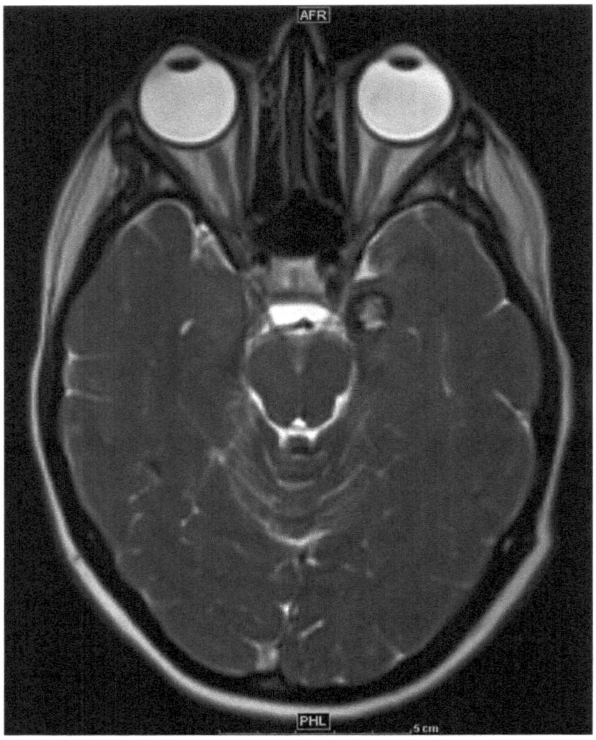

Fig. 27.1. Classic cavernous malformation magnetic resonance imaging (MRI) appearance. This cavernous malformation, situated in the left uncus, demonstrates classic T2 MRI findings, including a mixed-intensity core, "popcorn" appearance, and hypointense rim.

Table 27.1

Zabramski magnetic resonance imaging (MRI) classification

Type	MRI signal	Description
I	T1: hyperintense core T2: hypointense rim	Subacute hemorrhage
II	T1: mixed signal core T2: mixed signal core; hypointense rim	Loculated mixed-age hemorrhage and thrombosis
III	T1: isointense or hypointense T2: hypointense GRE: hypointense	Chronic hemorrhage
IV	T1: poorly seen T2: poorly seen GRE: punctate hypointense	Small cavernous malformation or other hemorrhagic lesion

GRE, gradient-recalled echo.

NATURAL HISTORY

Location appears to be the most important factor that determines the natural history of cavernous malformations. Supratentorial lobar cavernous malformations have a much more benign prognosis than deep lesions located in thalamus, basal ganglia, or posterior fossa. In one study addressing this issue the overall event rate was found to be 4.2% per year (Porter et al., 1997). However the event rate for superficial lesions was 0% per year while that for deep lesions was 10.6% per year ($p = 0.0001$).

We have already noted that the vast majority of supratentorial cavernous malformations will present with seizure activity. In those that do not initially present this way, seizures may nonetheless develop as irritative blood breakdown products accumulate in the surrounding brain. The estimated risk of developing new-onset seizures in patients with cavernous malformations, presumably all supratentorial in location, is 1.5-2.5% per year (Del Curling et al., 1991; Moriarity et al., 1999).

Pooled analyses have reported conflicting data regarding cavernous malformation location-specific hemorrhage rates. In one such analysis by Taslimi et al. (2016) nonbrainstem hemorrhage rates were reported to be 0.3% per year versus 2.8% per year for brainstem lesions. Another similar study by Gross and Du (2016) reported an overall hemorrhage rate of 2.5% per year, with location not being an independent risk factor. There is consensus however that prior hemorrhage is a significant risk factor for future hemorrhagic events. Furthermore, cavernous malformations display a phenomenon termed temporal clustering wherein rehemorrhage tends to occur, most commonly within the first 2–3 years after a prior hemorrhage. After this initial clustering of hemorrhage events, a relatively quiescent period where no overt hemorrhages occur may be seen (Barker et al., 2001).

INDICATIONS FOR CONSERVATIVE MANAGEMENT

Incidental lesion

Supratentorial cavernous malformations that are discovered incidentally and found to be clinically asymptomatic are typically candidates for conservative follow-up. Although there is a small risk of eventual seizure development and symptomatic hemorrhage, prophylactic treatment with either antiepileptic drugs or resection is not standard. This philosophy is in contrast to what is typically advocated for other intracranial vascular lesions, such as aneurysms and AVMs, where hemorrhage events are much more likely to result in serious morbidity or death. Both the supratentorial location and the properties of cavernous malformations in general allow patients to tolerate and more readily recover from associated events should they occur.

Imaging follow-up is not standardized and should be based on clinical judgment. For example, lesions that exhibit classic MRI features, such as a reticulated, mixed-signal, "popcorn" core with T2 hypointense rim with an associated developmental venous anomaly, may be imaged on an as-needed basis. On the other hand, lesions which are merely suspicious for cavernous malformation but are lacking in some classic hallmarks should be imaged sooner and more regularly to rule out nefarious pathologies such as neoplasm. Additionally, *de novo* development of cavernous malformations may occur in up to 30% of familial cases, making close follow-up especially important in this group (Washington et al., 2010).

The question often arises whether it is safe to initiate antiplatelet or anticoagulant medications in patients who harbor incidental cavernous malformations. What can be said of this is that there is no evidence that such therapy increases the risk of malformation hemorrhage (Schneble et al., 2012). However, in the event that a bleed were to spontaneously occur, it is possible that the resulting intracerebral hemorrhage may be larger as a consequence of such medications. Nonetheless, we do not recommending withholding antithrombotic therapy if there is a compelling medical indication.

Eloquent or inaccessible location

Symptomatic supratentorial cavernous malformations that either reside within eloquent cortex or are otherwise in a location where its access would incur morbidity are relatively indicated for conservative management. Certainly such cases require expert clinician judgment and calculated risk assessments.

Multiple lesions

Up to 20% of sporadic and over 50% of familial cases of cerebral cavernous malformation will present with multiple lesions (Batra et al., 2009). In some familial cases, the presence of innumerable, tiny cavernous malformations detectable by gradient-recalled echo MRI render it impossible to identify the symptomatic lesion with confidence (Brunereau et al., 2000; Nicholas-Bublick and Koffman, 2012). In such cases, where the patient does not present in extremis, conservative management may be the best management strategy. Alternatively, excellent surgical outcomes can be obtained if a single symptomatic lesion can be definitively identified after comprehensive presurgical assessment (Rocamora et al., 2009).

INDICATIONS FOR MICROSURGICAL RESECTION

General considerations

Englehart reported one of the first successful cases of brain cavernous malformation excision in 1904 (Huhn et al., 1993). In 1928 Walter Dandy published his surgical series of cerebral vascular lesions which included several cases of cavernous malformations that were also successfully removed (Dandy, 1928; Kretzer et al., 2010). Modern cavernous malformation resections have certainly become much safer and more complete with the routine utilization of microsurgical techniques. Selective use of adjuncts such as frameless stereotactic guidance, electrocorticography, and intraoperative ultrasound and/or MRI further facilitates effective lesion resection.

The resection technique employed must be tailored to the specific location and architecture of cavernous malformation. Piecemeal excision may be necessary for large lesions, whereas en bloc resection may be readily feasible for small ones. If the adjacent brain tissue is not eloquent, consideration should be given to resecting the hemosiderin-laden gliotic brain which surrounds the malformation (Fig. 27.2). This may be even more important if the treatment indication is epilepsy. Care should be taken in all cases of cavernous malformation resection to spare any associated developmental venous anomalies which typically drain the surrounding normal brain (Fig. 27.3).

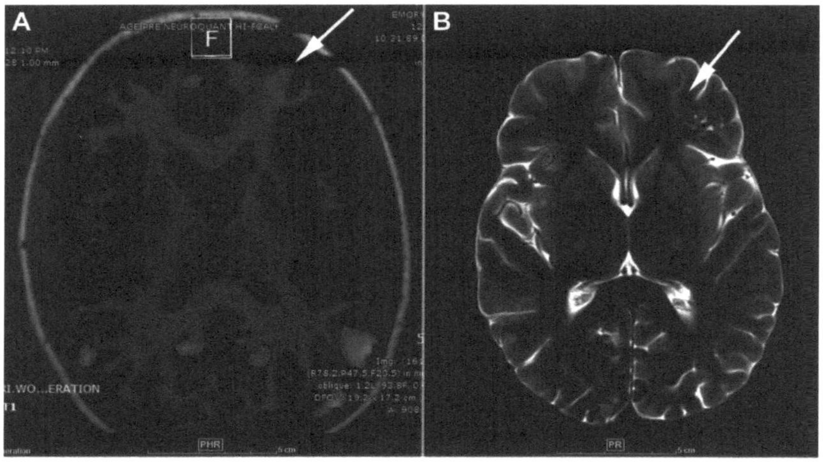

Fig. 27.2. Eloquent region cavernous malformation. This 26-year-old woman presented with intractable seizures. Prior to undergoing surgical resection of her left inferior frontal gyrus cavernous malformation, functional magnetic resonance imaging was utilized to localize active speech areas (arrow) in the region of the lesion (**A**). The cavernous malformation is seen to be located one gyrus lateral to an area of language activation, indicating that complete resection is feasible without trespass into this eloquent zone (**B**).

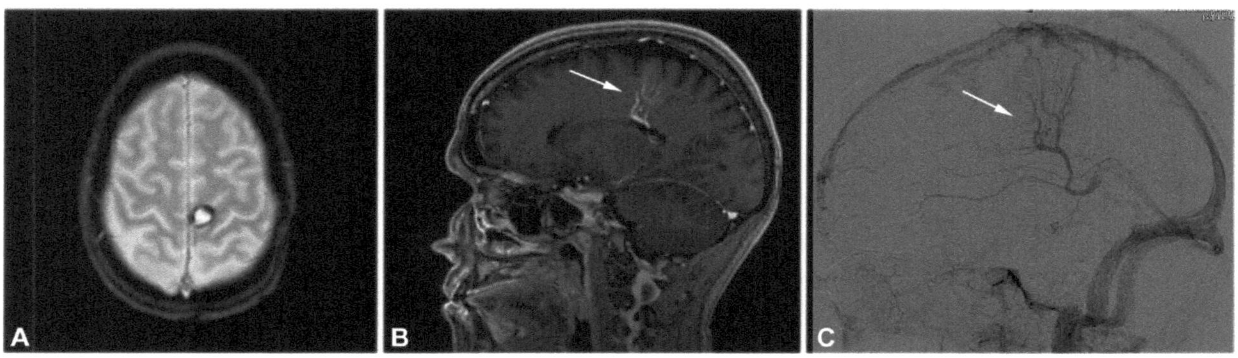

Fig. 27.3. Developmental venous anomaly. This patient presented after hemorrhage from a left posterior frontal parasagittal cavernous malformation (**A**). Sagittal magnetic resonance imaging shows that the malformation is associated with a prominent developmental venous anomaly displaying the characteristic caput medusae of medullary veins (arrow) converging on a dilated central trunk (**B**). This can be more readily appreciated in the venous phase of a catheter cerebral angiogram (**C**).

Lesional epilepsy

Medically intractable seizure, defined as failure of at least two antiepileptic drugs to control seizures, is the most common indication for microsurgical resection of supratentorial cavernous malformations. In some cases, initial surgical intervention is indicated if medication compliance is predicted to be problematic or there is a strong patient desire to not utilize antiepileptic drugs long term (Rosenow et al., 2013). A strategy of initial surgery after a single cavernous malformation-related seizure has been evaluated retrospectively and may be a reasonable option for some patients (Dammann et al., 2016). Comprehensive evaluation by an epileptologist as well as electroencephalogram monitoring to confirm that the cavernous malformation serves as the seizure focus are essential prior to surgery. The technical goal of surgery should be, at a minimum, complete lesionectomy (Fig. 27.4) (Yeon et al., 2009). As previously alluded to, seizure outcomes seem to be better if the surrounding hemosiderin-stained brain is also resected (Baumann et al., 2006). Such a strategy must be utilized with caution, however, especially when adjacent to eloquent brain regions.

Symptomatic hemorrhage

Patients with cavernous malformations who present in extremis with active or impending herniation should be treated in a fashion similar to patients with intracranial hemorrhage due to other etiologies. Fortunately, most supratentorial cavernous malformation hemorrhages are not overt and will be contained within the margins of the lesion capsule. Furthermore, the neurologic decline that is observed after such contained hemorrhages often rapidly improves due to the fact that the cavernous malformation compresses rather than destroys surrounding brain tissue. On the other hand, persistent or progressive neurologic decline from a symptomatic hemorrhage is a strong indication for microsurgical resection if a minimally disruptive operative corridor is available. If the patient's condition permits, waiting several days for the hematoma to partially liquefy may facilitate complete removal.

Paraventricular location

While it has been advocated by some that cavernous malformations in a ventricular or paraventricular location are strong candidates for microsurgical removal, we have adopted a more measured strategy in which each case is judged independently. The argument for aggressive upfront surgery holds that the surrounding tissue pressure is insufficient to tamponade a bleed should it occur, permitting the genesis of a potentially devastating intraventricular hemorrhage. This argument is certainly plausible and should be discussed with patients and their families prior to pursuing elective intervention (Fig. 27.5).

Lesion expansion

Cavernous malformations are dynamic lesions that are apt to exhibit radiographic growth and regression during various intervals of follow-up. Such minor expected perturbations in lesion size should not prompt a change in course unless there is verified clinical correlation. Significant lesion expansion, however, is a relative

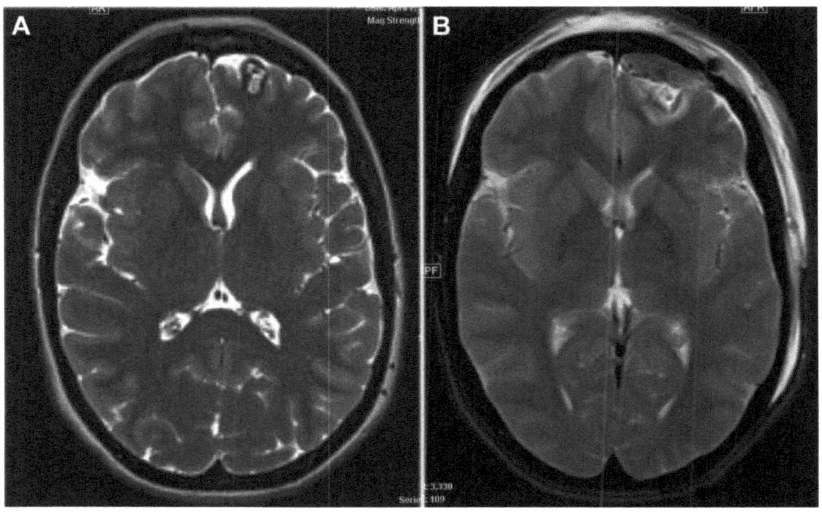

Fig. 27.4. Cavernous malformation excision. This superficial frontal pole cavernous malformation was discovered during the workup for epilepsy of several years' duration (**A**). Ictal waveforms were observed during electroencephalogram monitoring in region of the malformation, implicating it as causative. A left frontal craniotomy with lesionectomy as well as resection of perilesional hemosiderin-stained brain tissue was performed (**B**). This patient has remained seizure-free since surgery.

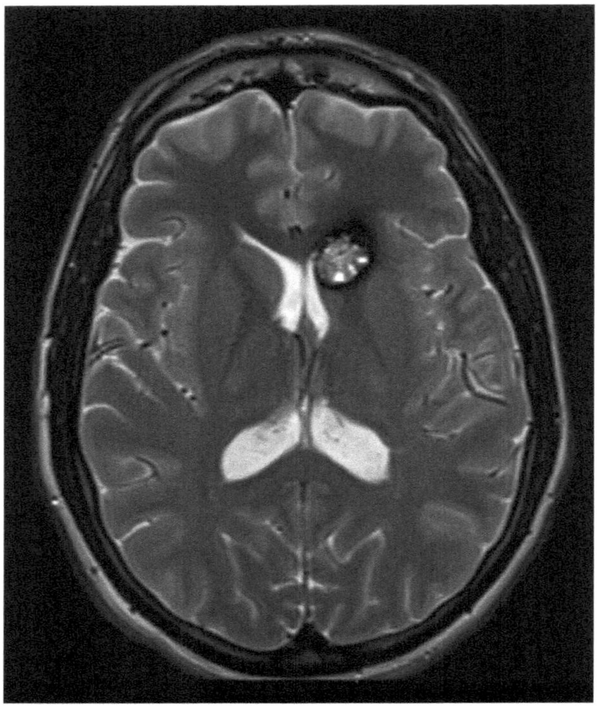

Fig. 27.5. Paraventricular cavernous malformation. This 18-year-old patient presented to the neurovascular clinic with an incidental left caudate head cavernous malformation. In areas, the malformation apposed and outwardly bowed the ependymal lining of the lateral ventricle. The patient and his parents were engaged in extensive discussions addressing the possibility of spontaneous intraventricular hemorrhage, subsequent hydrocephalus, and neurologic decline versus operative intervention with the potential for morbidity secondary to deep structure injury. A decision was ultimately made to follow this patient conservatively.

indication for supratentorial cavernous malformation resection. In most cases, a significant expansion will be associated with neurologic manifestations, making the decision to intervene relatively straightforward.

Intractable headaches

Headaches are a common reason for many patients to obtain a brain MRI. Statistically, discovery of a cavernous malformation during such a workup is improbable and such a finding would be considered incidental. In a minority of cases, headache onset can be clearly correlated to a recent hemorrhage event. Conservative management should be the initial strategy for treating isolated headaches in patients harboring supratentorial cavernous malformations. In rare circumstances, resection may be considered in an attempt to treat patients with medically intractable, debilitating headaches if there is reasonable suspicion that the cavernous malformation is causative.

ALTERNATIVE AND EMERGING INTERVENTIONS

Radiosurgery

There is no consensus regarding the utility of radiosurgery in the management of patients with supratentorial cavernous malformations (Sheehan and Schlesinger, 2010). Poor initial results have been attributed to the use of CT rather than MRI for planning, improper dosing, and poor patient selection (Barrow and Schuette, 2011). Modern series indicate results that may at least be on par with the natural history of cavernous malformations. That benefit is not conferred for at least 2–3 years after treatment, concurrent with the time period of temporal hemorrhage clustering, allows one to view radiosurgery results with an eye of suspicion (Bertalanffy and Gerganov, 2013). Data from randomized studies would certainly be helpful for clarifying the role of radiosurgery in the management of cavernous malformations.

Laser ablation

MRI-guided laser interstitial thermal therapy is an emerging technique that has been utilized for the treatment of a variety of focal brain lesions (Missios et al., 2015; Ellis et al., 2016; McGovern et al., 2016). Promising early results have been published from at least one small series of cavernous malformation-related epilepsy patients (McCracken et al., 2015). While this therapy must necessarily be considered investigational at this time, further utilization and expansion of such minimally invasive strategies seem assured in the near future.

CONCLUSION

Supratentorial cavernous malformations are uncommon cerebrovascular lesions. They most often present with seizures but are more and more frequently being discovered incidentally. Conservative management either with radiographic and clinical observation alone or in addition to antiepileptic drugs is the current first-line management strategy for most patients. Microsurgical resection is curative for patients who cannot be effectively managed conservatively. The role of minimally invasive strategies such as radiosurgery and laser ablation in the comprehensive management of patients with cavernous malformations awaits future studies.

References

Barker 2nd FG, Amin-Hanjani S, Butler WE et al. (2001). Temporal clustering of hemorrhages from untreated cavernous malformations of the central nervous system. Neurosurgery 49 (1): 15–24; discussion 25.

Barrow DL, Schuette AJ (2011). Cavernous malformations: a paradigm for progress. Clin Neurosurg 58: 27–41.

Batra S, Lin D, Recinos PF (2009). Cavernous malformations: natural history, diagnosis and treatment. Nat Rev Neurol 5 (12): 659–670.

Baumann CR, Schuknecht B, Lo Russo G et al. (2006). Seizure outcome after resection of cavernous malformations is better when surrounding hemosiderin-stained brain also is removed. Epilepsia 47 (3): 563–566.

Bertalanffy H, Gerganov VM (2013). Microsurgical or radiosurgical management of intracranial cavernomas. Acta Neurochir Suppl 116: 103–106.

Brunereau L, Labauge P, Tournier-Lasserve E et al. (2000). Familial form of intracranial cavernous angioma: MR imaging findings in 51 families. French Society of Neurosurgery Radiology 214 (1): 209–216.

Dammann P, Wrede K, Jabbarli R et al. (2016). Outcome after conservative management or surgical treatment for new-onset epilepsy in cerebral cavernous malformation. J Neurosurg 1–9.

Dandy WE (1928). Venous abnormalities and angiomas of the brain. Arch Surg 17 (5): 715–793.

Del Curling Jr O, Kelly Jr DL, Elster AD et al. (1991). An analysis of the natural history of cavernous angiomas. J Neurosurg 75 (5): 702–708.

Ellis JA, Mejia Munne JC, Wang SH et al. (2016). Staged laser interstitial thermal therapy and topectomy for complete obliteration of complex focal cortical dysplasias. J Clin Neurosci 31: 224–228.

Gross BA, Du R (2016). Hemorrhage from cerebral cavernous malformations: a systematic pooled analysis. J Neurosurg 1–9.

Heros RC, Morcos JJ (2000). Cerebrovascular surgery: past, present, and future. Neurosurgery 47 (5): 1007–1033.

Huhn SL, Rigamonti D, Hsu F (1993). Indications for surgical intervention. In: IA Awad, DL Barrow (Eds.), Cavernous malformations. American Association of Neurological Surgeons, Park Ridge, IL.

Katzman GL, Dagher AP, Patronas NJ (1999). Incidental findings on brain magnetic resonance imaging from 1000 asymptomatic volunteers. Jama 282 (1): 36–39.

Kretzer RM, Coon AL, Tamargo RJ (2010). Walter E. Dandy's contributions to vascular neurosurgery. J Neurosurg 6: 1182–1191.

Leblanc GG, Golanov E, Awad IA et al. (2009). Biology of vascular malformations of the brain. Stroke 40 (12): e694–e702.

McCormick WF (1966). The pathology of vascular ("arteriovenous") malformations. J Neurosurg 24 (4): 807–816.

McCormick WF, Nofzinger JD (1966). "Cryptic" vascular malformations of the central nervous system. J Neurosurg 24 (5): 865–875.

McCracken DJ, Willie JT, Fernald BA et al. (2015). Magnetic resonance thermometry-guided stereotactic laser ablation of cavernous malformations in drug-resistant epilepsy: imaging and clinical results. Neurosurgery. http://dx.doi.org/10.1227/NEU.0000000000001033.

McGovern RA, Banks GP, McKhann 2nd GM (2016). New techniques and progress in epilepsy surgery. Curr Neurol Neurosci Rep 16 (7): 65.

Missios S, Bekelis K, Barnett GH (2015). Renaissance of laser interstitial thermal ablation. Neurosurg Focus 38 (3): E13.

Moran NF, Fish DR, Kitchen N et al. (1999). Supratentorial cavernous haemangiomas and epilepsy: a review of the literature and case series. J Neurol Neurosurg Psychiatry 66 (5): 561–568.

Moriarity JL, Wetzel M, Clatterbuck RE et al. (1999). The natural history of cavernous malformations: a prospective study of 68 patients. Neurosurgery 44 (6): 1166–1171; discussion 1172–1173.

Morris Z, Whiteley WN, Longstreth Jr WT et al. (2009). Incidental findings on brain magnetic resonance imaging: systematic review and meta-analysis. BMJ 339: b3016.

Nicholas-Bublick S, Koffman BM (2012). Varying clinical presentations of familial cerebral cavernous malformations (CCMs) and spinal cord cavernous malformations (SCCMs). Radiol Case Rep 7 (3): 678.

Porter PJ, Willinsky RA, Harper W et al. (1997). Cerebral cavernous malformations: natural history and prognosis after clinical deterioration with or without hemorrhage. J Neurosurg 87 (2): 190–197.

Robinson JR, Awad IA, Little JR (1991). Natural history of the cavernous angioma. J Neurosurg 75 (5): 709–714.

Rocamora R, Mader I, Zentner J et al. (2009). Epilepsy surgery in patients with multiple cerebral cavernous malformations. Seizure 18 (4): 241–245.

Rosenow F, Alonso-Vanegas MA, Baumgartner C et al. (2013). Cavernoma-related epilepsy: review and recommendations for management – report of the Surgical Task Force of the ILAE Commission on Therapeutic Strategies. Epilepsia 54 (12): 2025–2035.

Schneble HM, Soumare A, Herve D et al. (2012). Antithrombotic therapy and bleeding risk in a prospective cohort study of patients with cerebral cavernous malformations. Stroke 43 (12): 3196–3199.

Sheehan J, Schlesinger D (2010). Editorial. Radiosurgery and cavernous malformations. J Neurosurg 113 (4): 689–690; discussion 690.

Taslimi S, Modabbernia A, Amin-Hanjani S et al. (2016). Natural history of cavernous malformation: systematic review and meta-analysis of 25 studies. Neurology 86 (21): 1984–1991.

Vernooij MW, Ikram MA, Tanghe HL et al. (2007). Incidental findings on brain MRI in the general population. N Engl J Med 357 (18): 1821–1828.

Virchow R (1863). Die krankhaften Geschwülste: dressig Vorlesungen gehalted während des Wintersemesters 1862–1863 an der Universität zu Berlin, August Hirschwald, Berlin.

Washington CW, McCoy KE, Zipfel GJ (2010). Update on the natural history of cavernous malformations and factors predicting aggressive clinical presentation. Neurosurg Focus 29 (3): E7.

Yeon JY, Kim JS, Choi SJ et al. (2009). Supratentorial cavernous angiomas presenting with seizures: surgical outcomes in 60 consecutive patients. Seizure 18 (1): 14–20.

Zabramski JM, Wascher TM, Spetzler RF et al. (1994). The natural history of familial cavernous malformations: results of an ongoing study. J Neurosurg 80 (3): 422–432.

Handbook of Clinical Neurology, Vol. 143 (3rd series)
Arteriovenous and Cavernous Malformations
R.F. Spetzler, K. Moon, and R.O. Almefty, Editors
http://dx.doi.org/10.1016/B978-0-444-63640-9.00028-X

Chapter 28

Brainstem and cerebellar cavernous malformations

GURSANT S. ATWAL, CHRISTINA E. SARRIS, AND ROBERT F. SPETZLER*
Department of Neurosurgery, Barrow Neurological Institute, St. Joseph's Hospital and Medical Center, Phoenix, AZ, USA

Abstract

Cavernous malformations are vascular lesions that occur throughout the central nervous system, most commonly in the supratentorial location, with brainstem and cerebellar cavernous malformations occurring more rarely. Cavernous malformations are associated with developmental venous anomalies that occur sporadically or in familial form. Patients with a cavernous malformation can present with headaches, seizures, sensorimotor disturbances, or focal neurologic deficits based on the anatomic location of the lesion. Patients with infratentorial lesions present more commonly with a focal neurologic deficit. Cavernous malformations are increasingly discovered incidentally due to the increasing use of magnetic resonance imaging. Understanding the natural history of these lesions is essential to their management. Observation and surgical resection are both reasonable options in the treatment of patients with these lesions. The clinical presentation of the patient, the location of the lesion, and the surgical risk assessment all play critical roles in management decision-making.

INTRODUCTION

Intracranial cavernous malformations (CMs) are vascular lesions composed of sinusoidal channels lined by a single layer of endothelium. They occur in various locations throughout the central nervous system (CNS) and have a variable presentation. Brainstem and cerebellar CMs are rare, and only a few reports in the medical literature outline their natural history and treatment outcomes. The location of these critical lesions poses significant challenges in their management. In this chapter, we focus on the natural history and management of brainstem and cerebellar CMs.

EPIDEMIOLOGY

CMs occur in about 0.5% of the population. There is no sex predilection. They can occur anywhere in the CNS. Most CMs are supratentorial in location (Washington et al., 2010). A meta-analysis on the natural history of CMs found that brainstem CMs comprised 18% of the infratentorial CMs, with 6% of them occurring in the cerebellum (Gross et al., 2009). They can also be associated with developmental venous anomalies. They occur sporadically or in familial form. Pediatric brainstem CMs are extraordinarily rare, comprising 13.3–14.5% of all brainstem CMs (Abla et al., 2010a, 2011; Li et al., 2013) and occurring even less frequently in the cerebellum (Knerlich-Lukoschus et al., 2015). In the cerebellum, the incidence ranges from 3% to 13.5% of all pediatric intracranial CMs (Knerlich-Lukoschus et al., 2015).

CLINICAL PRESENTATION

With the increasing availability of magnetic resonance imaging (MRI), a growing number of CMs are discovered incidentally (Washington et al., 2010). The clinical presentations of patients with symptomatic CMs vary by location. Patients with symptomatic CMs may present with headaches, seizure, sensorimotor disturbances, or focal neurologic deficits. In one meta-analysis, seizure was the most common presentation for patients with supratentorial lesions, whereas focal neurologic deficit

*Correspondence to: Robert F. Spetzler, MD, c/o Neuroscience Publications, Barrow Neurological Institute, St. Joseph's Hospital and Medical Center, 350 W. Thomas Rd., Phoenix AZ 85013, USA. Tel: +1-602-406-3593, Fax: +1-602-406-4104, E-mail: Neuropub@dignityhealth.org

or ataxia was the most common presentation of patients with infratentorial lesions (Washington et al., 2010).

Brainstem CMs

Because of the size and central location of the pons, pontine CMs account for most of the CMs found in the brainstem (Abla et al., 2014). Lesions abutting the ventricle can result in hydrocephalus as the presenting condition of affected patients. In the series by Abla et al. (2011) on brainstem CMs, the presentation correlated with lesion location. In that series, most of the patients presented with radiographic or clinical hemorrhage. Cranial neuropathy was the most common presenting sign, followed by sensory disturbances.

In a retrospective review of 52 pediatric patients with brainstem CMs, Li et al. (2014) documented 78 hemorrhages in these patients, who had a 12.3% preoperative annual hemorrhage rate. After the first hemorrhage, 26.9% of the children experienced a full recovery. However, after the second hemorrhage, only 13% achieved full recovery. At presentation, the pediatric patients had preoperative symptoms that corresponded to lesion location, with the most common symptoms being motor and sensory deficits; the next most common symptoms were headache, facial palsy, abnormal gait, and vertigo.

Cerebellar CMs

Cerebellar CMs are traditionally categorized with other intracranial CMs, having a relatively benign natural history. Data on hemorrhage risk and outcomes are limited. In a 2015 study of 58 patients with cerebellar CMs, Wu et al. found that most patients presented with hemorrhage (93.1%) and associated headache, nausea, vomiting, or vertigo. Just over one-third (36%) of these patients presented with focal neurologic deficits. In a study of pediatric patients with cerebellar CMs, Knerlich-Lukoschus et al. (2015) found that patients commonly presented with sudden-onset occipital headaches, vomiting, gait ataxia, and double vision.

NATURAL HISTORY

There is mounting literature on the natural history of CMs. However, the retrospective nature of most of these studies makes it challenging to predict the hemorrhage rate accurately. An understanding of the natural history of this disease is essential for making management decisions, and such studies serve as useful guidelines. A recent meta-analysis reported a hemorrhage rate of 2.4% per patient-year (Washington et al., 2010). Another study on the clinical course of untreated CMs found that the initial presentation of patients with intracranial hemorrhage (ICH) or focal neurologic deficit and brainstem location was independently associated with a hemorrhage over the 5-year period after the initial diagnosis. In this study, the 5-year hemorrhage rate in patients presenting with ICH, or focal neurologic deficit and brainstem location, was 30.8% (Horne et al., 2016). The 5-year hemorrhage rate for hemorrhages in a brainstem location without ICH or focal neurologic deficit at presentation was 8.0% compared to a rate of 3.8% for lesions located elsewhere. In the study of cerebellar CMs by Wu et al. (2015), the annual hemorrhage rate was 2.9%.

ANATOMY AND HISTOLOGY

CMs are well-circumscribed, multilobulate lesions. Histologically, they are lined by endothelium and lack smooth muscle. There is no intervening brain parenchyma. Reactive gliosis and hemosiderin can commonly be seen around CMs (Batra et al., 2011). Calcifications may also be associated with these lesions.

In a series of 104 patients with brainstem CMs, Garcia et al. (2015) found that the mean maximal diameter of the lesions was 19.5 mm; 54.8% also had an associated developmental venous anomaly. Understanding the surface and intrinsic anatomy of the brainstem is critical for devising a surgical plan. The deep-seated location and the associated vascular structures make the approaches to this region challenging. However, an understanding of the anatomy and advanced imaging techniques can facilitate surgical resection of these lesions. Giliberto et al. (2010) detailed the intrinsic anatomy of these brainstem lesions and outlined the safe entry zones to approach them. Abla et al. (2011) described a "2-point method" and several surgical approaches to brainstem CMs. In this method, a point is placed in the center of the lesion, and another point is placed where it comes closest to the pial surface to outline the ideal craniotomy for that lesion. However, this method should be used in conjunction with the safe entry zones for best outcomes. A detailed discussion of the intrinsic anatomy of brainstem lesions and the surgical approaches used to treat them is beyond the scope of this chapter.

TREATMENT

Observation

Possible surgical morbidity must be weighed against the natural history of the disease. The clinical presentation, location of the lesion, number of hemorrhages, and medical comorbidities should be considered when deciding on a treatment option. For incidentally discovered small lesions, observation is reasonable.

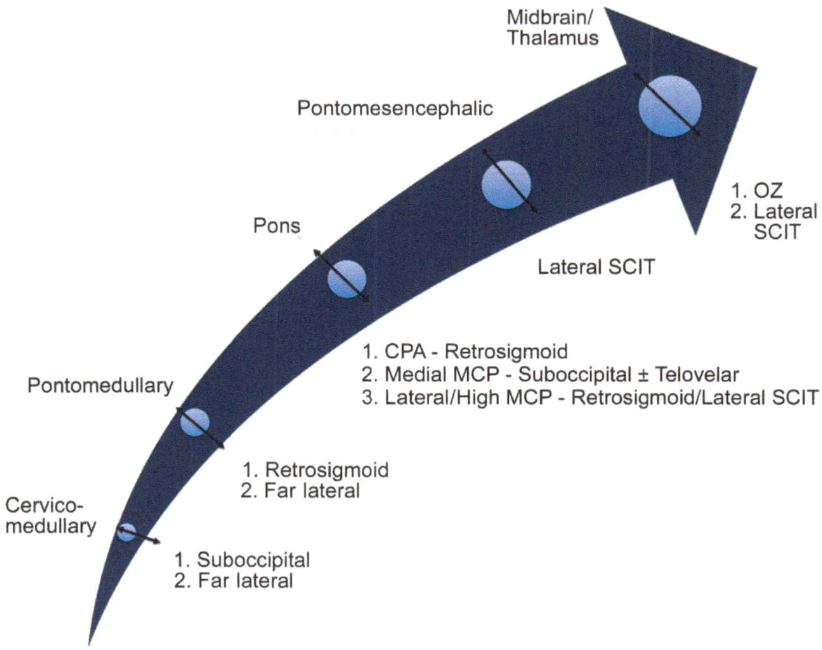

Fig. 28.1. Schematic of preferred surgical approaches based on location of brainstem cavernous malformations. CPA, cerebellopontine angle; MCP, middle cerebellar peduncle; OZ, orbitozygomatic; SCIT, supracerebellar infratentorial; ± with or without. (Reproduced from Abla et al. (2010b), with permission.)

Radiosurgery

Given the deep-seated location of brainstem CMs, some authors have proposed stereotactic radiosurgery (SRS) as an appropriate minimally invasive option or as a complementary modality to microsurgery for treatment of these lesions. However, several series on SRS for CMs seem to indicate that the greatest risk reduction in hemorrhage usually happens after a 2-year latency period. In a study of 54 patients with symptomatic brainstem CMs who had not undergone prior resection, the annual hemorrhage rate after Gamma Knife surgery was 7.06% within the first 2 years for those who received 13.1 Gy and 9.84% for those receiving 13.7 Gy (Lee et al., 2014). Adverse effects of radiation, including 1 permanent case of diplopia, developed in 8.2%. In a study of the clinical courses of untreated CMs, brainstem location was independently associated with a higher 5-year hemorrhage rate (Horne et al., 2016). Additionally, the annual risk of hemorrhage was highest during the first year and decreased subsequently. Because radiation-induced damage in the brainstem can be devastating, SRS is not recommended as a treatment option for brainstem CMs.

Surgery

Innovative techniques, including intraoperative MRI and frameless stereotactic navigation, are increasingly being utilized in the surgical management of brainstem CMs. Flores et al. (2015) described their use of diffusion tensor imaging and diffusion tensor tractography to help identify functionally important white-matter tracts for surgical planning. These techniques showed that 82% of patients with brainstem CMs had involvement of the corticospinal tract and other major fiber tracts, and the surgeons used these findings to develop more appropriate surgical selection and planning.

Abla et al. (2010b) described the approaches to the brainstem CMs based on location (Fig. 28.1) and applied their two-point method of lesion location. The proximity of the lesion to the pial surface, the presence or absence of hemorrhage, and the results of the neurologic examination must be taken into account when deciding on a particular approach. Neuromonitoring and navigation are valuable tools to maximize patient safety. The orbitozygomatic approach allows safe access to the cerebral peduncles (Fig. 28.2), while the supracerebellar infratentorial approach works well for tectum or midbrain-thalamic lesions (Fig. 28.3). For lesions primarily in the pons, the retrosigmoid approach is suitable (Fig. 28.4). The supracerebellar infratentorial approach provides visualization of midbrain or high pontomesencephalic lesions (Fig. 28.3). The far lateral approach (Fig. 28.5) allows access to lesions of the pontomedullary junction, medulla, and cervicomedullary junction (Fig. 28.6).

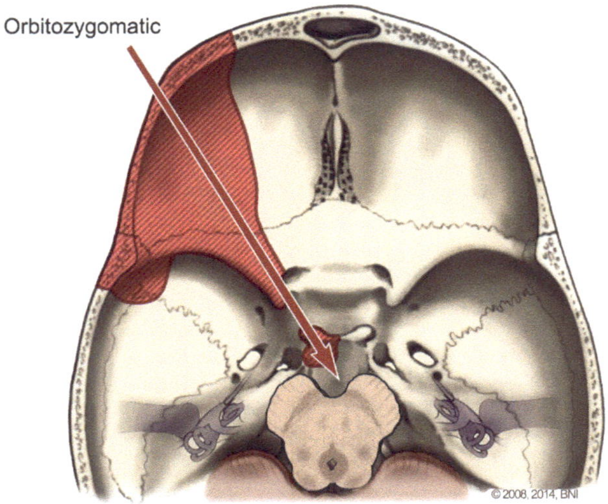

Fig. 28.2. Orbitozygomatic approach. This illustration shows how the orbitozygomatic approach provides access to the interpeduncular fossa. (Courtesy of Barrow Neurological Institute, Phoenix, AZ.)

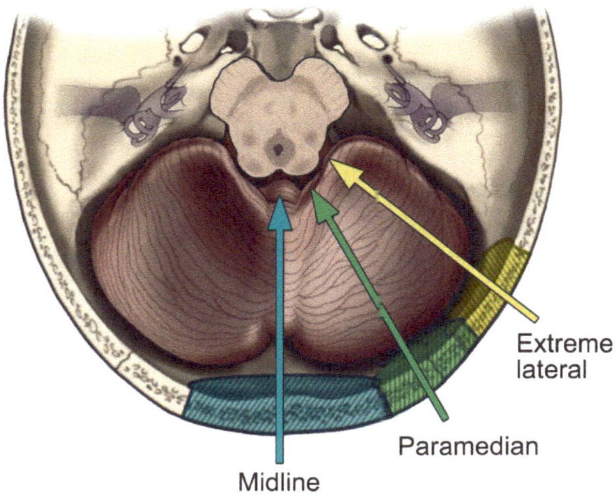

Fig. 28.3. Midline (blue), paramedian (green), and extreme lateral (yellow) approaches. The supracerebellar infratentorial approach allows for various modifications to access the upper pons and midbrain. (Courtesy of Barrow Neurological Institute, Phoenix, AZ.)

For pediatric brainstem CMs, Li et al. (2014) recommended surgery only if the patient had a single hemorrhage, poor neurologic function, or large lesions abutting the pial membrane. The most frequently used approach to remove midbrain lesions was a subtemporal transtentorial approach; suboccipital posterior midline approaches were used for lesions of the medulla and pons. Gross total resection was obtained in 94.2% of patients; failure to resect such lesions was attributed to multiple lobes or tight adhesions.

Wu et al. (2015) reported removing 70% of the cerebellar CMs through a median suboccipital approach. Retrosigmoid approaches were used for lesions of the petrosal surface of the cerebellum or the cerebellar peduncle, and far lateral approaches were used for lesions in the inferior cerebellar peduncle.

OUTCOMES

In the largest series of brainstem CMs ($n = 260$ patients), Abla et al. (2011) reported a mortality rate of less than 1%. Although 53% of patients in this series developed new neurologic deficits immediately following resection, 40% of them had improved to neurologic baseline at last follow-up. Development of new deficits did not correlate with the location of the lesion in the brainstem (medulla, pons, or midbrain). All patients had gross total resections of their brainstem CMs using diffusion tensor imaging,

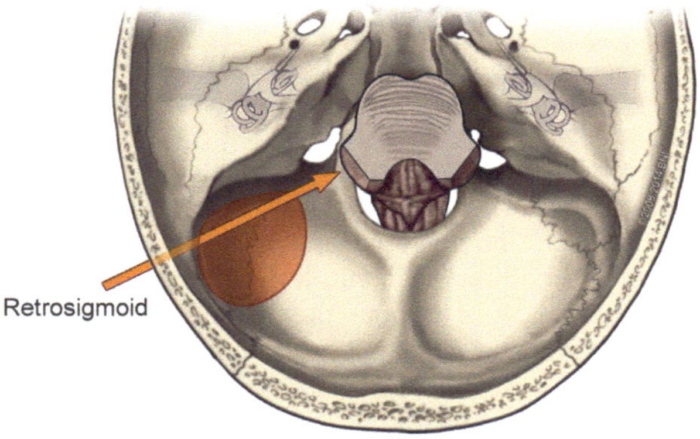

Fig. 28.4. Retrosigmoid approach. This approach provides trajectories to lesions primarily in the pons, with minimal retraction. (Courtesy of Barrow Neurological Institute, Phoenix, AZ.)

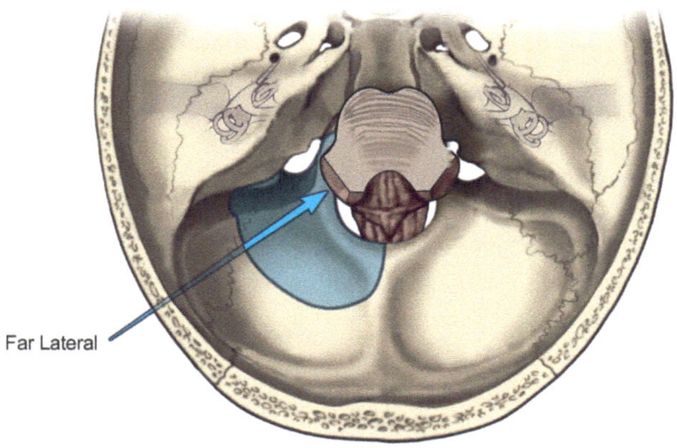

Fig. 28.5. Far lateral approach. Sagittal view demonstrating far lateral access to the pontomedullary junction and medulla. (Courtesy of Barrow Neurological Institute, Phoenix, AZ.)

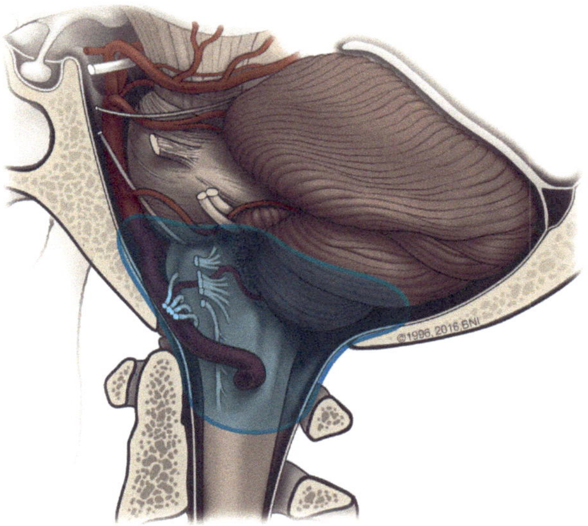

Fig. 28.6. Far lateral approach. Lateral or anterolateral lesions of the medulla are visualized through the far lateral approach. (Courtesy of Barrow Neurological Institute, Phoenix, AZ.)

and their mean overall hospital stay was 13.09 days. Notably, about one-half (45.5%) required admission to an inpatient rehabilitation facility after their hospital stay.

Although no surgery-related deaths occurred in the pediatric series, 48.1% of these patients developed surgically related morbidities, including facial nerve deficits (19.2%) and cranial nerve VI and IX–X palsies (9.6%) (Li et al., 2014). Postoperatively, less than one-fourth (9.6%) of the patients had improved functional status compared with their preoperative status, and in 32.7%, it declined immediately postoperatively. However, at the 6-month follow-up, 34.6% had improved functional status. In comparison to their adult counterparts, children have a longer lifespan; thus, the likelihood of near-full recovery must be considered, along with particular attention paid to their ability to attend school and participate in normal social activities.

In the cerebellar series, the most common postoperative complication was facial paralysis (Wu et al., 2015). Complete resection was achieved in all the patients. There was no postoperative mortality, and at 6-month follow-up, 77.6% of patients had improved neurologic status or remained the same.

CONCLUSION

Brainstem and cerebellar CMs can cause significant morbidity due to repeated hemorrhaging. With the utilization of image guidance, appropriate patient and approach selection, and detailed knowledge of the intrinsic brainstem anatomy, these lesions can often be safely resected with good outcomes.

References

Abla AA, Lekovic GP, Garrett M et al. (2010a). Cavernous malformations of the brainstem presenting in childhood: surgical experience in 40 patients. Neurosurgery 67: 1589–1598; discussion 1598–1599.

Abla AA, Turner JD, Mitha AP et al. (2010b). Surgical approaches to brainstem cavernous malformations. Neurosurg Focus 29 (3): E8.

Abla AA, Lekovic GP, Turner JD et al. (2011). Advances in the treatment and outcome of brainstem cavernous malformation surgery: a single-center case series of 300 surgically treated patients. Neurosurgery 68: 403–414; discussion 414–415.

Abla AA, Benet A, Lawton MT (2014). The far lateral transpontomedullary sulcus approach to pontine cavernous malformations: technical report and surgical results. Neurosurgery 10 (Suppl 3): 472–480.

Batra S, Crain B, Engelman R et al. (2011). Pathology of cavernous malformations. In: D Rigamonti (Ed.), Cavernous Malformations of the Nervous System. Cambridge University Press, New York.

Flores BC, Whittemore AR, Samson DS et al. (2015). The utility of preoperative diffusion tensor imaging in the surgical management of brainstem cavernous malformations. J Neurosurg 122: 653–662.

Garcia RM, Ivan ME, Lawton MT (2015). Brainstem cavernous malformations: surgical results in 104 patients and a proposed grading system to predict neurological outcomes. Neurosurgery 76: 265–277; discussion 277–278.

Giliberto G, Lanzino DJ, Diehn FE et al. (2010). Brainstem cavernous malformations: anatomical, clinical, and surgical considerations. Neurosurg Focus 29: E9.

Gross BA, Batjer HH, Awad IA et al. (2009). Brainstem cavernous malformations. Neurosurgery 64: E805–E818; discussion E818.

Horne MA, Flemming KD, Su IC et al. (2016). Clinical course of untreated cerebral cavernous malformations: a meta-analysis of individual patient data. Lancet Neurol 15 (2): 166–173.

Knerlich-Lukoschus F, Steinbok P, Dunham C et al. (2015). Cerebellar cavernous malformation in pediatric patients: defining clinical, neuroimaging, and therapeutic characteristics. J Neurosurg Pediatr 16: 256–266.

Lee SH, Choi HJ, Shin HS et al. (2014). Gamma Knife radiosurgery for brainstem cavernous malformations: should a patient wait for the rebleed? Acta Neurochir (Wien) 156: 1937–1946.

Li D, Yang Y, Hao SY et al. (2013). Hemorrhage risk, surgical management, and functional outcome of brainstem cavernous malformations. J Neurosurg 119: 996–1008.

Li D, Hao SY, Tang J et al. (2014). Surgical management of pediatric brainstem cavernous malformations. J Neurosurg Pediatr 13: 484–502.

Washington CW, McCoy KE, Zipfel GJ (2010). Update on the natural history of cavernous malformations and factors predicting aggressive clinical presentation. Neurosurg Focus 29: E7.

Wu H, Yu T, Wang S et al. (2015). Surgical treatment of cerebellar cavernous malformations: a single-center experience with 58 cases. World Neurosurg 84: 1103–1111.

Handbook of Clinical Neurology, Vol. 143 (3rd series)
Arteriovenous and Cavernous Malformations
R.F. Spetzler, K. Moon, and R.O. Almefty, Editors
http://dx.doi.org/10.1016/B978-0-444-63640-9.00029-1

Chapter 29

Thalamic cavernous malformations

CHRISTINA E. SARRIS, GURSANT S. ATWAL, AND PETER NAKAJI*
Department of Neurosurgery, Barrow Neurological Institute, St. Joseph's Hospital and Medical Center, Phoenix, AZ, USA

Abstract

Cavernous malformations of the thalamus represent a particularly complex subset of cavernous malformations because of the highly eloquent nature of the involved tissue and their deep location. The decision about whether to operate on any individual lesion depends on the specific location of the lesion within the thalamus, the nature of the patient's symptoms, and the patient's history. When surgery is recommended, the approach must be chosen carefully. Each part of the thalamus is reached by a different surgical approach. These approaches include the orbitozygomatic approach to the anteroinferior thalamus, the anterior interhemispheric transcallosal approach to the medial thalamus, the anterior contralateral interhemispheric transcallosal approach to the lateral thalamus, the posterior interhemispheric transcallosal approach to the posterosuperior thalamus, the parieto-occipital transventricular approach to the lateral posteroinferior thalamus, and the suboccipital supracerebellar infratentorial/transtentorial approach to the medial posteroinferior thalamus. Careful attention to safe entry zones and image guidance can allow safe removal of these lesions when necessary.

INTRODUCTION

Thalamic cavernous malformations (CMs) are relatively rare, deep-seated lesions that can cause significant neurologic morbidity and mortality. Their infrequent occurrence has led to few reports in the literature of their natural history, treatment, and outcomes (Roda et al., 1990; Cappabianca et al., 1991; Steinberg et al., 2000; Mathiesen et al., 2003; Mhoyan et al., 2005; Gross et al., 2009; Abla and Spetzler, 2013; Li et al., 2013; Pandey et al., 2013; Rangel-Castilla and Spetzler, 2015). These lesions are particularly challenging for the neurosurgeon because they occur in critical locations that are often difficult to access surgically. Many surgeons still believe that most of these lesions are inoperable due to the risks involved. Accordingly, there remains much debate over the optimal treatment of thalamic CMs, with recommended options varying from conservative management to radiosurgery to open resection. A comprehensive understanding of the natural history of these lesions is critical to decision-making. This chapter focuses on the natural history of thalamic CMs, options for treating them, and surgical approaches to these lesions.

In 1979, Becker et al. described one of the first reported cases of thalamic CM in a 34-year-old woman who presented with hemiparesis and hemihypesthesia. The lesion was resected surgically, but the patient failed to improve after surgery, and she also developed speech difficulties. A subsequent case of thalamic CM was reported by Roda et al. in Spain in 1990. In this case, a 50-year-old man presented with hemiparesis, and his neurologic deficit worsened after a transcallosal interhemispheric approach for resection. These early reports of poor surgical outcomes following thalamic CM resection set the stage for their notoriety as "inoperable" lesions, a perception that persists today.

EPIDEMIOLOGY

CMs affect about 0.5% of the population and occur in males and females equally. They usually present in the

*Correspondence to: Peter Nakaji, MD, c/o Neuroscience Publications, Barrow Neurological Institute, St. Joseph's Hospital and Medical Center, 350 W. Thomas Rd., Phoenix, AZ 85013, USA. Tel: +1-602-406-3593, Fax: +1-602-406-4104, E-mail: Neuropub@dignityhealth.org

fourth or fifth decade of life. They occur throughout the central nervous system, with the cerebral hemispheres being the most common location, followed by the brainstem, basal ganglia and thalamus, and cerebellum, in decreasing order of frequency. About 8% of CMs are located in the basal ganglia and thalamus (Gross et al., 2011). In a large study from Stanford of 176 patients (Pandey et al., 2013), 9.1% of deep CMs (i.e., those of the brainstem, basal ganglia, and thalamus) occurred in the thalamus.

Much of the reported epidemiologic data for thalamic CMs comes from analyses that have grouped these lesions with brainstem CMs, and, more broadly, basal ganglia CMs, making it difficult to draw conclusions that are solely applicable to thalamic lesions. Thalamic CMs share many similarities with their brainstem counterparts, but thalamic lesions are further complicated by their intimate relationship with the brain's deep venous drainage. The incidence of incidentally found deep lesions has been estimated at 0.8 per million persons per year, whereas the incidence of symptomatic deep lesions is 3.2 per million persons per year (Mathiesen et al., 2003). Of patients who are symptomatic from thalamic CMs, 60% present with mass lesions, and 15% present with symptomatic hydrocephalus.

CLINICAL PRESENTATION

The clinical presentation of CMs is highly variable and depends on their location. Patients with CMs usually present with headaches, seizures, or sensorimotor disturbances. In one recent analysis, patients with supratentorial lesions most commonly presented with seizures, whereas those with infratentorial lesions usually presented with focal neurologic deficits (Washington et al., 2010). The occurrence of seizures associated with CMs is thought to be related to perilesional gliosis secondary to repeated microhemorrhages. Although CMs may present with symptoms that result from hemorrhage, about 40% are discovered incidentally.

Symptomatic patients with CMs of the thalamus most commonly present with contralateral sensorimotor deficits (Bertalanffy et al., 1991; Houtteville, 1995; Lawton et al., 1996; Heffez, 1997; Pozzati, 2000; Mathiesen et al., 2003; Otani et al., 2008; Gross et al., 2009). In the series of 27 patients by Li et al. (2013), the most common presenting symptoms were weakness (63%), paresthesia (56%), and headache (48%). In the largest series (n=46) of thalamic CMs reported to date, 59% presented with hemorrhage and 11% presented with hydrocephalus (Rangel-Castilla and Spetzler, 2015). Dorsal or medial thalamic CMs, because of their close proximity to the cerebral ventricles, have a higher likelihood of causing obstructive hydrocephalus after bleeding (Pozzati, 2000; Lanzino et al., 2002; Mathiesen et al., 2003). Less frequently, patients with thalamic CMs have presented with oculomotor paresis, hemianopsia, and thalamic pain syndrome (Cohen et al., 1982; Pozzati, 2000; Otani et al., 2008).

In the Stanford study (Pandey et al., 2013), 50% of the 16 patients with thalamic CMs presented with preoperative modified Rankin Scale scores of 3–5, indicating poor neurologic status. This rate was worse than that of CMs of the basal ganglia (3.8% [1/26] poor) and midbrain (28.6% [8/28] poor) but on a par with the presentation of CMs in the pons and medulla.

NATURAL HISTORY

It is essential for clinicians to have an understanding of the natural history of these lesions in order to manage patients' care effectively. However, as Abla and Spetzler (2013) remark, the natural history of these lesions is difficult to assess accurately given that all data assessing annual bleeding rates before surgery are from retrospective studies. A recent meta-analysis on the natural history of CMs reported a hemorrhage rate of 2.4% per patient-year, considering all locations of CMs. Additionally, this study determined that female sex and prior hemorrhage were risk factors for hemorrhage (Gross et al., 2011). However, size and multiplicity of lesions did not affect the hemorrhage rate. Furthermore, some studies suggest that the higher risk of rehemorrhage may be time-limited, with the risk decreasing within the first 2 years after the initial hemorrhage (Wang et al., 2003).

The natural history of thalamic CMs has been more difficult to elucidate than that of other lesions, with much of our current knowledge originating from case reports and larger retrospective reviews that combine thalamic lesions with basal ganglia CMs. In the study by Li et al. (2013), the preoperative bleeding rate was 5.2% annually, and two or more clinical or radiographic hemorrhages occurred in 63% of patients. In the Stanford study (Pandey et al., 2013), 47.6% (n=20) of the 42 patients with thalamic and ganglionic CMs presented with more than one clinical hemorrhage.

In the past, CMs were considered to be congenital lesions. However, recent studies have demonstrated the dynamic nature of these lesions, which exhibit growth and regression over time (Washington et al., 2010). In addition, the *de novo* development of CMs occurs at a rate of 0.1–0.6% per patient-year (Zabramski et al., 1994; Kattapong et al., 1995; Labauge et al., 2000). *De novo* development can occur in both sporadic and familial forms.

ANATOMY AND HISTOLOGY

The thalamus is in close proximity to the internal capsule, basal ganglia, midbrain, foramen of Monro, and internal cerebral veins. Its critical location relative to such important neurovascular structures makes it more difficult to approach thalamic lesions surgically compared to most other locations within the central nervous system. However, the subregions of the thalamus each require their own distinct approaches, and CMs can arise within any of them.

The six different subregions into which Rangel-Castilla and Spetzler (2015) classified the thalamus were defined based on the region's anatomy and the most suitable surgical approach for accessing lesions in that region. They applied this classification to the largest series of thalamic CMs to date (n=46) and presented their operative results when using these trajectories. They described six anatomic thalamic regions: (1) anteroinferior, (2) medial, (3) lateral, (4) posterosuperior, (5) lateral posteroinferior, and (6) medial posteroinferior. The most common location for thalamic CMs was found to be the lateral thalamus, and the least common was the posterosuperior thalamus. In their series, developmental venous anomalies were found intraoperatively in all patients (Rangel-Castilla and Spetzler, 2015). Because of their deep location, image guidance is highly recommended, particularly for those lesions not presenting to a pial surface.

Aside from gross thalamic anatomy, it is important to understand the histology of the lesions of the thalamus. CMs are well-circumscribed vascular lesions often described as having a "mulberry-like" appearance. Histologically, CMs consist of packed endothelium-lined blood vessels with no intervening brain parenchyma. These vessels lack a smooth-muscle layer and can be surrounded by loose connective tissue. Thrombotic material of varying ages is a common finding in these lesions. Calcifications can be seen within the CMs or in the surrounding tissue. Hemosiderin and fibrosis can often be found surrounding CMs (Batra et al., 2011). Additionally, reactive gliosis can be seen adjacent to these lesions. At the cellular level, the endothelial cells of CMs lack tight junctions, indicating the lack of a blood-brain barrier (Batra et al., 2011). Several angiogenetic markers, including vascular endothelial growth factor, basic fibroblast growth factor, and hypoxia-inducible factor-1alpha, have been studied in CM (Batra et al., 2011).

TREATMENT

Observation

The risks of surgical morbidity should be weighed against the natural history of the disease. In addition, the number of hemorrhages and the patient's clinical status are critical in decision-making. For incidentally found small lesions, observation and follow-up with serial imaging are reasonable options. Most patients with a single hemorrhage are managed conservatively unless they are acutely symptomatic.

Radiosurgery

It is not surprising that radiosurgery, a "minimally invasive" treatment, has served as an attractive option for the treatment of thalamic CMs, in particular those in deep locations such as the thalamus that render surgical resection more challenging. However, whether radiosurgery lessens the risk of future hemorrhage is debatable, and for comparably sized arteriovenous malformations the risk of radiation complications appears higher. Gross et al. (2009) found that permanent radiation-induced complication rates have ranged from 0% to 75% in the treatment of CMs of the basal ganglia and thalamus. However, they suggest that lesions in these areas exhibit less radiosurgical morbidity compared with brainstem CMs. Furthermore, there is a tendency to reduce the radiation dose given the eloquence of the deep-seated CMs, thereby possibly reducing the effectiveness of the thrombo-obliteration that the procedure induces (Amin-Hanjani et al., 1998).

In a series that contained both basal ganglia and thalamic CMs, Amin-Hanjani et al. (1998) noted a decrease from 16.4% to 2.2% in the annual hemorrhage rate after radiosurgery in 16 CMs. Interestingly, the improvement in patients' conditions after surgery on CMs in the above locations was greater than that in patients with brainstem CMs. Other studies have not found such regionally specific responses to CMs from radiosurgery (Karlsson et al., 1998). In a more recent series specific to radiosurgery for thalamic and brainstem CMs, the bleeding rate was 3.72% per year during the first 2 years and 3.59% per year in subsequent years (Jay et al., 2012).

Surgery

Surgical resection remains an option for thalamic CMs. The neurosurgeon must consider a wide variety of factors in deciding whether surgery is the best option, in particular the number of times the lesion has bled and the timing of surgery (Pandey et al., 2013). Also to be considered is the location of the lesion in relation to the ependymal or pial surface when deciding on a surgical approach. In the series by Rangel-Castilla and Spetzler (2015), the thalamus was divided into six regions (Fig. 29.1), each with its corresponding preferred surgical approach (Table 29.1). Of these, the anterior contralateral interhemispheric transcallosal approach was the most commonly used approach (Fig. 29.2). The lateral supracerebellar infratentorial

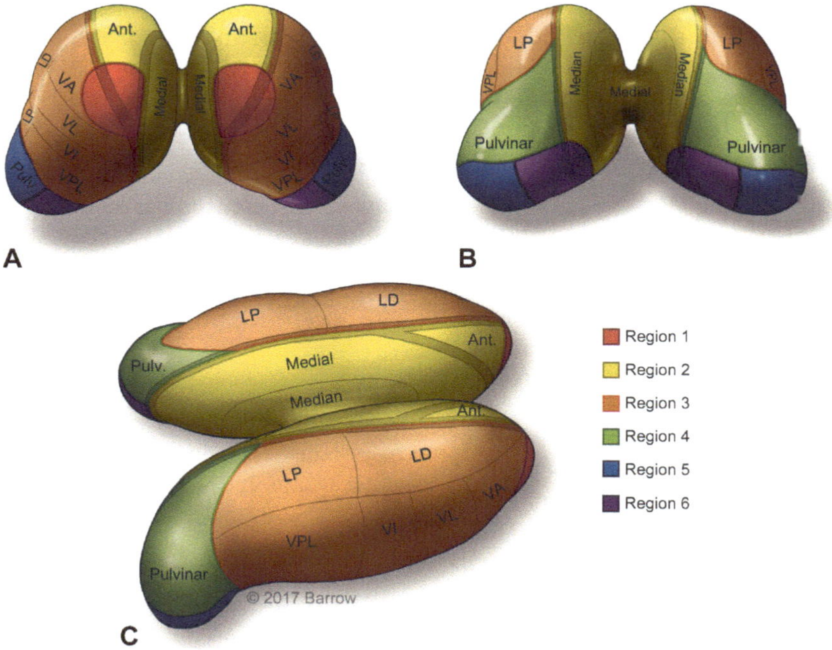

Fig. 29.1. (**A**) Anterior, (**B**) posterior, and (**C**) superior views showing the thalamic nuclei in each of the six regions of the thalamus. Ant, anterior; LD, lateral dorsal; LP, lateral posterior; Pulv, pulvinar; VA, ventral anterior; VI, ventral intermediate; VL, ventral lateral; VPL, ventral posterior lateral. (Courtesy of Barrow Neurological Institute, Phoenix, AZ.)

Table 29.1

Preferred surgical approach for each thalamic region

Region	Preferred surgical approach
Anteroinferior	Orbitozygomatic
Medial	Anterior ipsilateral interhemispheric transcallosal
Lateral	Anterior contralateral interhemispheric transcallosal
Posterosuperior	Posterior interhemispheric transcallosal
Lateral posteroinferior	Parieto-occipital transventricular
Medial posteroinferior	Supracerebellar infratentorial

Modified with permission from Rangel-Castilla L and Spetzler RF (2015).

was also commonly used. Li et al. (2013) describe the posterior interhemispheric transsplenial, supracerebellar infratentorial, and anterior transchoroidal approaches for access to thalamic CMs. At Barrow Neurological Institute, Dr. Spetzler advocates all the aforementioned approaches, except the posterior interhemispheric approach. Instead, he prefers the parieto-occipital approach through the atrium to the posterior thalamus, in order to avoid manipulation of the vein of Galen and retraction of the occipital lobe (personal communication).

While most surgeons who tackle deep CMs surgically report positive long-term outcomes, a high incidence of immediate postoperative morbidity does exist. In the Stanford study (Pandey et al., 2013), thalamic CMs were not operated on if the lesions were incidental and asymptomatic. In addition, surgery was preferentially performed within 4–8 weeks of the last hemorrhagic episode. The optimal timing for surgery after the last instance of hemorrhage has not been clearly defined. We have operated on some patients immediately and some within the first 3 weeks, both with good results.

POSTOPERATIVE OUTCOMES

In the series of 46 patients reported by Rangel-Castilla and Spetzler (2015), 87% of patients had good outcomes, and 22% noted temporary worsening of preoperative symptoms. Ten patients with preoperative hemiparesis or hemiparesthesia had worsening of symptoms, and 6 of them recovered over the 6–8 weeks following surgery. In Li et al.'s (2013) series of 27 patients with surgically resected thalamic CMs, the only significant predictor of favorable outcome was patient age younger than 40 years. Good outcomes were seen in 78% of their patients.

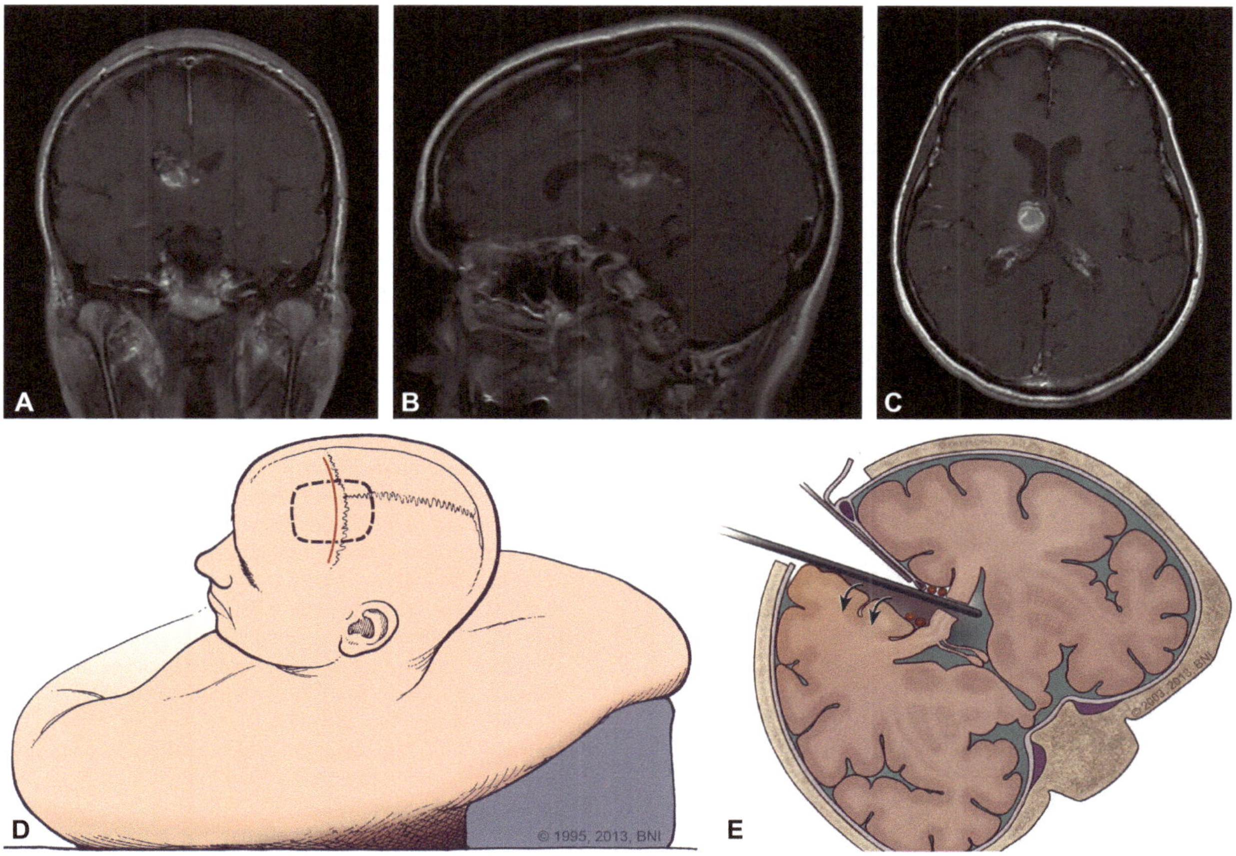

Fig. 29.2. (**A**) Coronal, (**B**) sagittal, and (**C**) axial magnetic resonance images showing a right lateral thalamic cavernous malformation on the superior surface of the thalamus. The location of this lesion makes it appropriate to approach via a contralateral interhemispheric approach for resection. (**D**) Patient positioning for an anterior interhemispheric transcallosal approach using a left frontal craniotomy. (**E**) A coronal cutaway view shows a left anterior interhemispheric transcallosal contralateral approach to the right lateral ventricle en route to the right thalamus. (Courtesy of Barrow Neurological Institute, Phoenix, AZ.)

Sixty percent of the thalamic CMs surgically resected by Pandey et al. (2013) had improved long-term outcomes. The rate of surgical morbidity found by Gross et al. (2009) was 10%, with a mortality rate of 1.9%.

With regard to specific preoperative deficits, Li et al. (2013) found that 53% of patients (9/17) had postoperative improvement of their weakness, 60% (9/15) had improvement in paresthesias, and 100% had improvement of headache (13/13), vomiting (9/9), nausea (7/7), and dizziness (6/6). Rangel-Castilla and Spetzler (2015) found that 91% (42/46) of patients were unchanged or improved compared to preoperatively, and 9% were worse.

CONCLUSIONS

Thalamic CMs, although rare, can be a source of significant neurologic morbidity. It is important to acknowledge them as distinct entities from their brainstem counterparts, but with an added complexity related to their deep venous drainage. Observation and surgical resection are reasonable options for the management of these lesions when patients are symptomatic. Clinical presentation, location within the thalamus, natural history, and surgical morbidity must be taken into account when deciding on a management strategy. Given the infrequency with which surgeons encounter thalamic CMs, it is critically important that the neurosurgical community continues to share its experiences and innovate in order to help guide and advance treatment for this select group of patients.

ACKNOWLEDGMENTS

The authors thank the Neuroscience Publications staff at Barrow Neurological Institute for assistance with manuscript preparation.

References

Abla AA, Spetzler RF (2013). Cavernous malformations of the thalamus: a relatively rare but controversial entity. World Neurosurg 79: 641–644.

Amin-Hanjani S, Ogilvy CS, Candia GJ et al. (1998). Stereotactic radiosurgery for cavernous malformations: Kjellberg's experience with proton beam therapy in 98 cases at the Harvard Cyclotron. Neurosurgery 42: 1229–1236; discussion 1236–1238.

Batra S, Crain B, Engelmann R et al. (2011). Pathology of cavernous malformations. In: D Rigamonti (Ed.), Cavernous Malformations of the Nervous System. Cambridge University Press, New York.

Becker DH, Townsend JJ, Kramer RA et al. (1979). Occult cerebrovascular malformations. A series of 18 histologically verified cases with negative angiography. Brain 102: 249–287.

Bertalanffy H, Gilsbach JM, Eggert HR et al. (1991). Microsurgery of deep-seated cavernous angiomas: report of 26 cases. Acta Neurochir (Wien) 108: 91–99.

Cappabianca P, Spaziante R, De Divitiis E et al. (1991). Thalamic cavernous malformations. J Neurosurg 75: 169–171.

Cohen HC, Tucker WS, Humphreys RP et al. (1982). Angiographically cryptic histologically verified cerebrovascular malformations. Neurosurgery 10: 704–714.

Gross BA, Batjer HH, Awad IA et al. (2009). Cavernous malformations of the basal ganglia and thalamus. Neurosurgery 65: 7–18; discussion 18–19.

Gross BA, Lin N, Du R et al. (2011). The natural history of intracranial cavernous malformations. Neurosurg Focus 30: E24.

Heffez DS (1997). Stereotactic transsylvian, transinsular approach for deep-seated lesions. Surg Neurol 48: 113–124.

Houtteville JP (1995). The surgery of cavernomas both supra-tentorial and infra-tentorial. Adv Tech Stand Neurosurg 22: 185–259.

Jay SM, Chandran H, Blackburn TP (2012). Gamma Knife stereotactic radiosurgery for thalamic & brainstem cavernous angiomas. Br J Neurosurg 26: 367–370.

Karlsson B, Kihlstrom L, Lindquist C et al. (1998). Radiosurgery for cavernous malformations. J Neurosurg 88: 293–297.

Kattapong VJ, Hart BL, Davis LE (1995). Familial cerebral cavernous angiomas: clinical and radiologic studies. Neurology 45: 492–497.

Labauge P, Brunereau L, Levy C et al. (2000). The natural history of familial cerebral cavernomas: a retrospective MRI study of 40 patients. Neuroradiology 42: 327–332.

Lanzino G, Wanebo JE, Spetzler RF (2002). Contralateral interhemispheric resection of thalamic cavernous malformations with frameless stereotaxy. Operative Techniques in Neurosurg 5: 191–197.

Lawton MT, Golfinos JG, Spetzler RF (1996). The contralateral transcallosal approach: experience with 32 patients. Neurosurgery 39: 729–734; discussion 734–735.

Li D, Zhang J, Hao S et al. (2013). Surgical treatment and long-term outcomes of thalamic cavernous malformations. World Neurosurg 79: 704–713.

Mathiesen T, Edner G, Kihlstrom L (2003). Deep and brainstem cavernomas: a consecutive 8-year series. J Neurosurg 99: 31–37.

Mhoyan A, Fulbright R, Bannykh SI (2005). Cavernous hemangioma of the thalamus. J Neurooncol 71: 257.

Otani N, Fujioka M, Oracioglu B et al. (2008). Thalamic cavernous angioma: paraculminar supracerebellar infratentorial transtentorial approach for the safe and complete surgical removal. Acta Neurochir Suppl 103: 29–36.

Pandey P, Westbroek EM, Gooderham PA et al. (2013). Cavernous malformation of brainstem, thalamus, and basal ganglia: a series of 176 patients. Neurosurgery 72: 573–589; discussion 588–589.

Pozzati E (2000). Thalamic cavernous malformations. Surg Neurol 53: 30–39; discussion 39–40.

Rangel-Castilla L, Spetzler RF (2015). The 6 thalamic regions: surgical approaches to thalamic cavernous malformations, operative results, and clinical outcomes. J Neurosurg 123: 676–685.

Roda JM, Alvarez F, Isla A et al. (1990). Thalamic cavernous malformation. Case report. J Neurosurg 72: 647–649.

Steinberg GK, Chang SD, Gewirtz RJ et al. (2000). Microsurgical resection of brainstem, thalamic, and basal ganglia angiographically occult vascular malformations. Neurosurgery 46: 260–270; discussion 270–271.

Wang CC, Liu A, Zhang JT et al. (2003). Surgical management of brain-stem cavernous malformations: report of 137 cases. Surg Neurol 59: 444–454; discussion 454.

Washington CW, McCoy KE, Zipfel GJ (2010). Update on the natural history of cavernous malformations and factors predicting aggressive clinical presentation. Neurosurg Focus 29. E7.

Zabramski JM, Wascher TM, Spetzler RF et al. (1994). The natural history of familial cavernous malformations: results of an ongoing study. J Neurosurg 80: 422–432.

Handbook of Clinical Neurology, Vol. 143 (3rd series)
Arteriovenous and Cavernous Malformations
R.F. Spetzler, K. Moon, and R.O. Almefty, Editors
http://dx.doi.org/10.1016/B978-0-444-63640-9.00030-8

Chapter 30

Spinal cavernous malformations

AARON J. CLARK, DORIS D. WANG, AND MICHAEL T. LAWTON*
Department of Neurological Surgery, University of California, San Francisco, CA, USA

Abstract

Spinal cavernous malformations are rare intramedullary vascular lesions of the central nervous system. Most are located in the thoracic spine. Patients present with either acute neurologic deficit or gradual deterioration. Weakness is the most common presenting symptom. The annual hemorrhage risk is 2.1%. Diagnosis is made by magnetic resonance imaging as these lesions are occult on angiography. Surgical removal is indicated in patients with hemorrhage and neurologic deficit. All lesions are approached posteriorly by laminectomy. Dorsal cavernous malformations are exposed by focused laminectomy of the level or levels overlying the lesion with minimally facet violation. Ventral and lateral lesions are approached by laminectomy including a level above and level below as well as unilateral radical facetectomy. After midline dural opening, the dentate ligament is divided and retracted to allow up to 90° of rotation of the spinal cord. Microsurgical treatment is associated with 42% symptom improvement and 50% symptom stabilization. Postoperative worsening is associated with longer preoperative duration of symptoms. Therefore we recommend consideration of early surgery for cavernous malformation removal in patients with symptoms attributable to the lesion.

INTRODUCTION AND DEMOGRAPHICS

Spinal cavernous malformations are histologically identical to their intracranial counterpart. They are characterized by lobulated dilated sinusoidal channels of dense capillary-like vessels. There is no intervening neural tissue. Grossly, they are well-circumscribed lesions with a mulberry-like appearance.

Cavernous malformations in the central nervous system are rare entities with a reported prevalence of 0.4–0.6%. Spinal cavernous malformations account for 5% of all cavernous malformations and 5–12% of intraspinal vascular malformations (Lu and Lawton, 2010; Badhiwala et al., 2014). The average age at presentation is 39–42 years with an equal male-to-female distribution (Gross et al., 2010; Badhiwala et al., 2014). The most common location for spinal cavernous malformation is thoracic (55–57%), followed by cervical (38%). Lumbar (2–4%) and conus medullaris (1.7%) lesions are significantly less common. A total of 2.4% span the cervicothoracic junction and 0.6% span the thoracolumbar junction. Twelve percent of patients have familial cavernous malformation and 17–27% have concomitant intracranial cavernous malformation.

PRESENTATION AND DIAGNOSIS

Lesions are rarely asymptomatic (0.9%), likely due to the high density of eloquent tissue in the spinal cord. About half of the patients present with acute, stepwise neurologic decline and the other half present with slow progressive neurologic deterioration. Progressive deterioration is thought to occur due to microhemorrhage, gliosis, microcirculatory changes, and partial thrombosis (Gross et al., 2010). Acute or recurrent presentations are thought be attributable to larger hemorrhage within the cavernous malformation or rupturing beyond its capsule (Jallo et al., 2006). The reported rates of hemorrhage range from 1.4% to 6.8% per year, with an average estimated annual hemorrhage risk of 2.1% (Badhiwala et al., 2014). The most common presenting symptoms include motor weakness (61%), sensory disturbances (58%),

*Correspondence to: Michael T. Lawton, MD, University of California, San Francisco, Department of Neurological Surgery, 505 Parnassus Ave. Rm. M779, San Francisco CA 94143-0112, USA. Tel: +1-415-353-3998, E-mail: Michael.Lawton@ucsf.edu

pain (34%), urinary and bowel dysfunction (24%), and respiratory distress (0.5%) with involvement of upper cervical spine. Additionally, a small prospective study suggests a very high annual re-hemorrhage rate (Sandalcioglu et al., 2003). While there may exist a subset of patients who remain neurologically and clinically stable after initial presentation, the predictors of this group have not been elucidated (Kharkar et al., 2007). The literature suggests that progressive neurologic decline is characteristic of the disease process (Ogilvy et al., 1992).

Spinal cavernous malformations are not visualized by conventional angiography. Diagnosis of spinal cavernous malformation can be made by magnetic resonance imaging (MRI). Spinal cavernous malformations are 1 cm on average at presentation (Lu and Lawton, 2010). Like their intracranial counterparts, spinal cavernous malformations have the typical appearance of circumscribed, multilobulated lesions with a reticulated core of heterogeneous signal intensity and associated hypointense rim on T2-weighted imaging (Hegde et al., 2012). Associated venous malformations may be found in up to 34% of patients and calcification is seen less frequently compared to intracranial cavernous malformations (Gross et al., 2010). Based on previous histologic studies, methemoglobin and thrombosis are responsible for high-intensity signal while calcifications, fibrosis, and acute or subacute blood products are responsible for the low-signal areas within the lesion (Deutsch et al., 2000). Susceptibility imaging has extremely high sensitivity for even small amounts of blood products and can be helpful for diagnosis.

SURGICAL TREATMENT

Complete surgical removal can be accomplished safely and is the treatment of choice for spinal cavernous malformations (McCormick et al., 1988; Lu and Lawton, 2010; Mitha et al., 2011). In our practice, spinal cavernous malformations are surgically treated through a posterior approach (Lu and Lawton, 2010). This eliminates the need for a thoracotomy and minimizes risk to the lung, heart, great vessels, and other critical mediastinal structures. Furthermore, watertight dural closure is significantly more likely through a posterior exposure. For cervical lesions, the head is fixed in the Mayfield head holder. The posterior spine is exposed through a midline incision. The paraspinal musculature is reflected off the spine using a subperiosteal dissection to expose the spinous processes and laminae bilaterally. For dorsal lesions, care is taken to preserve the facet capsule. Laminectomies are performed in the standard fashion at the levels overlying the dorsal lesion. Minimal medial facetectomy is required for dorsally located malformations. Prior to opening the dura, Surgicel is packed under the laminectomy edges to maintain hemostasis during the intradural component of the operation. The dura is opened in the midline and reflected laterally. The remainder of the resection is performed under microscopic visualization (Fig. 30.1).

For lesions that come to the surface, the myelotomy is made directly over the lesion. For deeper lesions, the myelotomy is made in the midline posterior median sulcus. Microdissectors are used to separate the cavernous malformation from surrounding tissue. In cases with a hematoma present, the hematoma is first drained and the lesion is then delivered into the resulting cavity. Smaller lesions are removed en bloc while larger lesions must be removed in a piecemeal fashion. After resection, the cavity is meticulously explored for residual malformation. The gliotic wall of the resection cavity is not removed. The dura is then closed in a watertight fashion using a running stitch under microscopic visualization.

Modifications of this procedure are required for ventral and lateral lesions (Lu and Lawton, 2010). For these lesions, the facet is exposed unilaterally on the side to which the lesion is eccentric. Furthermore, laminectomy is performed to include one level above and one level below when the lesion is lateral or ventral. A unilateral radical facetectomy is performed ipsilateral to the lesion by removing the superior and inferior articular process until the bony removal is flush with the pedicle. This provides the more lateral to medial trajectory required to visualize and remove the lesion (Fig. 30.2). Some studies advocate hemilaminectomy or laminoplasty approaches; however, these reports did not include significant numbers of ventrally located lesions (Bian et al., 2009; Choi et al., 2011). The dura is opened in the midline as before. Again, ipsilateral to the lesion, the dentate ligament is divided. Release of the dentate ligament safely allows up to 90° of rotation of the spinal cord. To facilitate this maneuver, a stitch is placed in the free end of the dentate ligament and lateral traction is applied. Myelotomy is then made over the cavernous malformation. In our series, no patients experienced new or worsening motor deficit following this treatment protocol, indicating that sectioning of the dentate ligament, rotation of the spinal cord, and myelotomy over a lateral or ventral lesion are safe (Lu and Lawton, 2010). This additional exposure is associated with longer operative time and higher estimated blood loss. Some authors argue for watchful waiting for patients with ventral lesions (Liang et al., 2011); however our data suggest that these lesions can be safely and efficiently removed using our exposure technique.

Our protocol includes performing intraoperative neurophysiologic monitoring during all spinal cavernous

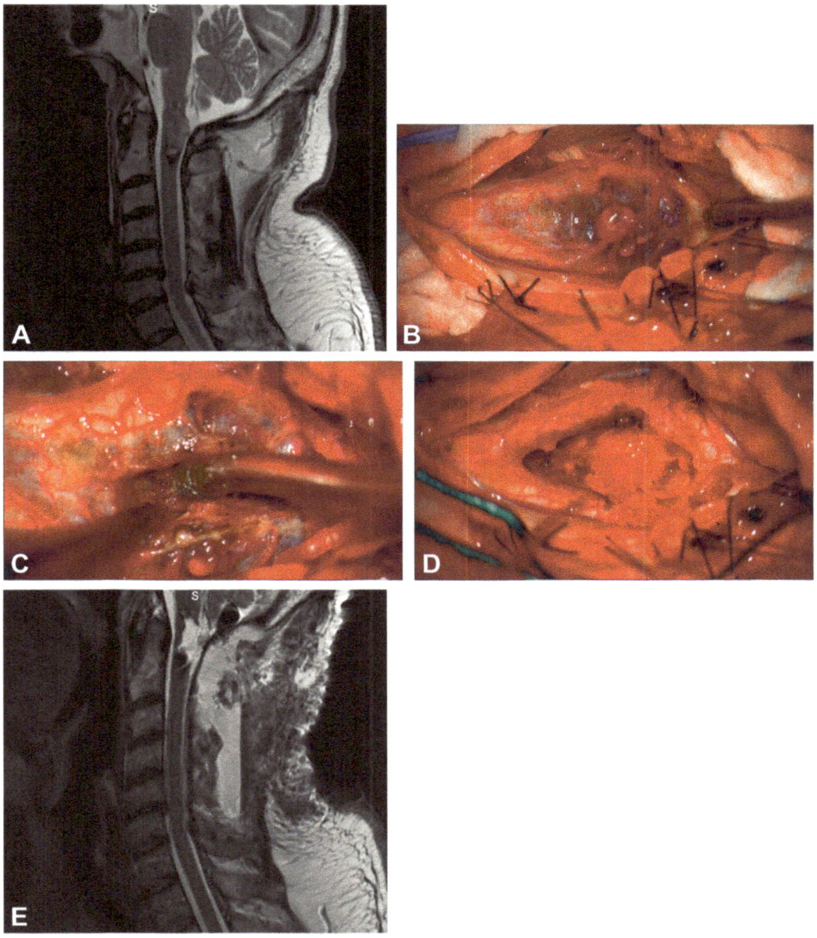

Fig. 30.1. Pre-operative magnetic resonance imaging (MRI) (**A**) of a 50-year-old male patient with several months of weakness and difficulty with ambulation who had undergone an aborted C1–2 laminectomy at a referring institution for a dorsally located spinal cavernous malformation. The previous operative exposure was extended with a C3 laminectomy. Upon opening the dura, the cavernous malformation was visualized at the dorsal surface of the spinal cord with surrounding hemosiderin-stained gliotic tissue (**B**). Myelotomy was performed directly over the lesion and the lesion was removed (**C**). Careful inspection of the cavity demonstrates complete removal without residual cavernous malformation (**D**). Postoperative MRI demonstrates complete removal (**E**) and the patient remained neurologically stable.

malformation operations. Should a change in motor (MEP) or somatosensory (SSEP) evoked potentials occur, the resection is paused, the operative field is gently irrigated with saline solution, and rotation of the spinal cord is relaxed. Our department has published a checklist of response to changes in neuromonitoring during spine surgery which describes an algorithmic team-based approach (Ziewacz et al., 2012). However, in our cavernous malformation series, neuromonitoring accurately predicted postoperative deficit in only 1 out of 3 patients who awoke with a new deficit (Lu and Lawton, 2010). This is consistent with a recent report from our department indicating lower sensitivity and specificity of MEPs in predicting new deficits in patients with myelopathy of nondegenerative causes such as tumor compared to those with degenerative spondylotic myelopathy (Clark et al., 2013, 2016). This may be due to the severity of preexisting disruption of sensory and/or motor pathways which lead to a lack of sensitivity of neuromonitoring. However, in the Clark et al. (2013) series, spinal cavernous malformations and other intramedullary lesions were not included. Others routinely monitor MEP and SSEP for all surgeries (Liang et al., 2011; Maslehaty et al., 2011). As such, the current literature is limited to small case series. Until larger-scale higher-level data are available with respect to the use of intraoperative neuromonitoring for spinal cavernous malformation resection, we encourage using neuromonitoring changes to guide surgery but not as a reason to abort the resection.

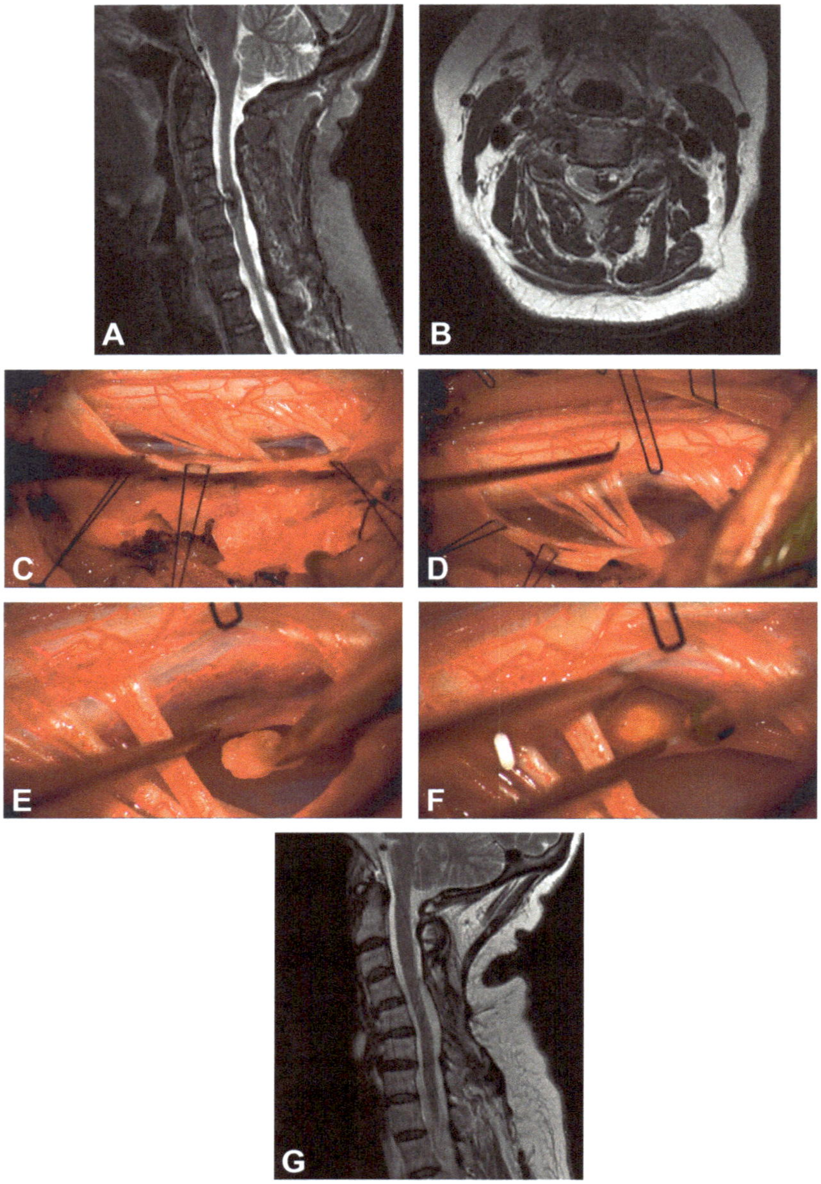

Fig. 30.2. Preoperative magnetic resonance imaging (**A**, **B**) of a 61-year-old female with acute-onset left-arm weakness and right-arm paresthesias demonstrates a ventrolaterally located C4–5 cavernous malformation. The patient underwent C3–5 laminectomy with left C4 radical facetectomy allowing left lateral exposure of the spinal cord (**C**). The dentate ligament was divided and a suture was used to retract the spinal cord to allow visualization of the cavernous malformation (**D**). The lesion was removed using microsurgical techniques (**E**). The cavity was explored and found to be free of residual lesion (**F**), which was confirmed on routine postoperative imaging (**G**). The patient made a full neurologic recovery.

There is no evidence regarding the need for spinal fusion for stabilization after laminectomy for removal of cavernous malformation. Spine tumor literature reports up to 18% of patients develop postlaminectomy deformity after removal of spinal cord tumors (McGirt et al., 2010). A meta-analysis showed that 8% of patients required fusion and indications included deformity present prior to surgery, laminectomies involving three or more levels, removing 50% or more of the facet joints, surgery encompassing the cervicothoracic or thoracolumbar junctions, or laminectomy of the C2 vertebra (Avila et al., 2015). In our experience, our patients have not developed postlaminectomy deformity or instability following spinal cavernous malformation resection (Lu and Lawton, 2010). This may be due to our limited laminectomies (usually one to three levels) and preservation of facet joints (minimal facet violation for dorsally located lesions and unilateral facetectomy for ventrally located lesions). However, one larger study with long-term follow-up reported a 6% rate of postlaminectomy kyphosis requiring anterior stabilization and fusion (Mitha et al., 2011). Future studies with long-term follow-up can elucidate the rate of postsurgical deformity after spinal cavernous malformation surgery.

SURGICAL OUTCOMES

Although transient postoperative worsening may occur, at final follow-up patients experience excellent outcomes after surgical removal of spinal cavernous malformations (Anson and Spetzler, 1993; Lu and Lawton, 2010; Mitha et al., 2011; Tong et al., 2012). In our series of patients treated by posterior approach, 41% improved after surgery and 50% remained functionally and neurologically stable (Lu and Lawton, 2010). The proportion worsening postoperatively was 9%, all of which were characterized by worsening dysesthetic pain. Clinical worsening is associated with longer duration of symptoms prior to surgery. Furthermore, data suggest that pain may be more difficult to treat when related to spinal cavernous malformations (Kim et al., 2006). Postoperative MRI confirmed complete removal of the lesion in 100% of our patients, even those with ventral cavernous malformations (Lu and Lawton, 2010). In the immediate postoperative period, there was a 5% complication rate manifested as wound infection requiring reoperation.

A large meta-analysis pooling 40 individual case series included 567 patients who underwent operative resection and 64 who underwent conservative therapy (Badhiwala et al., 2014). Of the patients who received operations, gross total resection was achieved in 94% of the patients. Early transient postoperative complications were estimated to be 27%. At last follow-up, neurologic function was improved from preoperative baseline in 52% of patients, remained the same for 38%, and was worse for 11% of patients. For those who underwent conservative management with observation, only 30% of patients had improved neurologic function and 59% remained the same, while 11% showed decline. Comparing surgical and conservative management, resection was significantly associated with favorable neurologic outcome (odds ratio 2.79, 95% confidence interval 1.46–5.33, $p=0.002$) (Badhiwala et al., 2014). In addition, concordant with our experience, resection of the lesion within 3 months of symptom onset is correlated with favorable neurologic outcome, while a larger proportion of patients who underwent later resection (>3 months) had worse neurologic function (Badhiwala et al., 2014). Other factors associated with favorable neurologic function included acute/stepwise presentation, presence of motor symptoms, hemilaminectomy approach, and gross total resection of the lesion. Interestingly, patient age, level of lesion, and dorsal versus ventral location of the lesion were not significant predictors of outcome. Others have reported that long-term improvement and worsening are correlated with size of the lesion, in particular the anteroposterior dimension (Mitha et al., 2011). This evidence strongly supports early surgical intervention and complete resection for management of spinal cavernous malformations. A large series and others advocated that surgery should be delayed 4–6 weeks after neurologic decline to allow for a gliotic plane to develop around the lesion, facilitating safe removal (Liang et al., 2011; Tong et al., 2012). Our experience suggests that safe removal can be attained even within 4–6 weeks of hemorrhage and may be associated with improved outcomes. Urgent surgery in the setting of acute neurologic decline is our standard protocol.

CONCLUSION

Early surgical removal of symptomatic spinal cavernous malformations is indicated in patients with evidence of hemorrhage and neurologic deficit. Removal by posterior laminectomy with or without unilateral facetectomy is associated with excellent clinical and radiographic outcomes.

References

Anson JA, Spetzler RF (1993). Surgical resection of intramedullary spinal cord cavernous malformations. J Neurosurg 78: 446–451.

Avila MJ, Walter CM, Skoch J et al. (2015). Fusion after intradural spine tumor resection in adults: a review of evidence and practices. Clin Neurol Neurosurg 138: 169–173.

Badhiwala JH, Farrokhyar F, Alhazzani W et al. (2014). Surgical outcomes and natural history of intramedullary spinal cord cavernous malformations: a single-center series and meta-analysis of individual patient data: clinic article. J Neurosurg Spine 21: 662–676.

Bian LG, Bertalanffy H, Sun QF et al. (2009). Intramedullary cavernous malformations: clinical features and surgical technique via hemilaminectomy. Clin Neurol Neurosurg 111: 511–517.

Choi GH, Kim KN, Lee S et al. (2011). The clinical features and surgical outcomes of patients with intramedullary spinal cord cavernous malformations. Acta Neurochir (Wien) 153: 1677–1684; discussion 1685.

Clark AJ, Ziewacz JE, Safaee M et al. (2013). Intraoperative neuromonitoring with MEPs and prediction of postoperative neurological deficits in patients undergoing surgery for cervical and cervicothoracic myelopathy. Neurosurg Focus 35: E7.

Clark AJ, Safaee M, Chou D et al. (2016). Comparative sensitivity of intraoperative neuromonitoring with MEP in predicting postoperative neurological deficits: non-degenerative vs. degenerative myelopathy. Global Spine Journal 6: 452–458.

Deutsch H, Jallo GI, Faktorovich A et al. (2000). Spinal intramedullary cavernoma: clinical presentation and surgical outcome. J Neurosurg 93: 65–70.

Gross BA, Du R, Popp AJ et al. (2010). Intramedullary spinal cord cavernous malformations. Neurosurg Focus 29: E14.

Hegde A, Mohan S, Tan KK et al. (2012). Spinal cavernous malformations: magnetic resonance imaging and associated findings. Singapore Med J 53: 582–586.

Jallo GI, Freed D, Zareck M et al. (2006). Clinical presentation and optimal management for intramedullary cavernous malformations. Neurosurg Focus 21: e10.

Kharkar S, Shuck J, Conway J et al. (2007). The natural history of conservatively managed symptomatic intramedullary spinal cord cavernomas. Neurosurgery 60: 865–872.

Kim LJ, Klopfenstein JD, Zabramski JM et al. (2006). Analysis of pain resolution after surgical resection of intramedullary spinal cord cavernous malformations. Neurosurgery 58: 106–111.

Liang JT, Bao YH, Zhang HQ et al. (2011). Management and prognosis of symptomatic patients with intramedullary spinal cord cavernoma: clinical article. J Neurosurg Spine 15: 447–456.

Lu DC, Lawton MT (2010). Clinical presentation and surgical management of intramedullary spinal cord cavernous malformations. Neurosurg Focus 29: E12.

Maslehaty H, Barth H, Petridis AK et al. (2011). Symptomatic spinal cavernous malformations: indication for microsurgical treatment and outcome. Eur Spine J 20: 1765–1770.

McCormick PC, Michelsen WJ, Post KD et al. (1988). Cavernous malformations of the spinal cord. Neurosurgery 23: 459–463.

Mcgirt MJ, Garces-Ambrossi GL, Parker SL et al. (2010). Short-term progressive spinal deformity following laminoplasty versus laminectomy for resection of intradural spinal tumors: analysis of 238 patients. Neurosurgery 66: 1005–1012.

Mitha AP, Turner JD, Abla AA et al. (2011). Outcomes following resection of intramedullary spinal cord cavernous malformations: a 25-year experience. J Neurosurg Spine 14: 605–611.

Ogilvy CS, Louis DN, Ojemann RG (1992). Intramedullary cavernous angiomas of the spinal cord: clinical presentation, pathological features, and surgical management. Neurosurgery 31: 219–229; discussion 229–230.

Sandalcioglu IE, Wiedemayer H, Gasser T et al. (2003). Intramedullary spinal cord cavernous malformations: clinical features and risk of hemorrhage. Neurosurg Rev 26: 253–256.

Tong X, Deng X, Li H et al. (2012). Clinical presentation and surgical outcome of intramedullary spinal cord cavernous malformations. J Neurosurg Spine 16: 308–314.

Ziewacz JE, Berven SH, Mummaneni VP et al. (2012). The design, development, and implementation of a checklist for intraoperative neuromonitoring changes. Neurosurg Focus 33: E11.

Handbook of Clinical Neurology, Vol. 143 (3rd series)
Arteriovenous and Cavernous Malformations
R.F. Spetzler, K. Moon, and R.O. Almefty, Editors
http://dx.doi.org/10.1016/B978-0-444-63640-9.00031-X

Chapter 31

Pharmacotherapy for cavernous malformations

ROBERT F. RUDY AND ROSE DU*

Department of Neurosurgery, Brigham and Women's Hospital and Harvard Medical School, Boston, MA, USA

Abstract

Cerebral cavernous malformations, vascular abnormalities comprised of endothelial cells in the absence of connective tissue or muscle, are often epileptogenic and often treated initially with antiepileptic drugs. This chapter discusses the role of pharmacotherapy in managing focal epilepsy secondary to cavernous malformations in adults, children, and pregnant women. Several drugs are available and potentially efficacious in suppressing seizures stemming from cavernous malformations. In addition, antiepileptic pharmacotherapy following surgical resection is indicated to decrease the risk of postoperative seizures. Finally, potential novel therapeutics targeting cavernous malformations are discussed.

INTRODUCTION

Neuraxial cavernous malformations are dilated, endothelial cell-lined vascular channels in the central nervous system that lack a smooth-muscle and elastin lining (Rigamonti et al., 1988; Awad et al., 1993). While many follow a benign course, cavernous malformations can hemorrhage, result in focal neurologic deficits, and/or act as a nidus for seizures (Gross et al., 2011). Treatment strategies depend on the symptomatology as well as location of the lesion. Medical management and routine follow-up without surgical intervention are indicated as the first-line therapy for asymptomatic lesions (Awad and Jabbour, 2006). Lesions that are symptomatic, specifically those with symptomatic hemorrhages or associated with seizures, are considered for microsurgical resection.

Currently, pharmacotherapy is used to manage symptoms associated with cavernous malformations, namely focal epilepsy secondary to cavernous malformations. Despite advances in the understanding of the basic biology of cavernous malformations, no approved pharmaceutical agents for the treatment of cavernous malformations are currently available. However, there are considerable efforts under way to develop drugs targeting the pathogenesis of these lesions. This chapter discusses the role of pharmacotherapy in the modern management of seizures originating from parenchymal tissue surrounding cavernomas and reviews ongoing investigations into potential therapeutics for these potentially dangerous vascular malformations.

CURRENT MEDICAL MANAGEMENT OF SEIZURES ASSOCIATED WITH CEREBRAL CAVERNOUS MALFORMATIONS

Cavernous malformations and seizures

Intracerebral cavernous malformations can be highly epileptogenic and often present with seizures. In fact, seizures are the most common presentation in adults with intracranial cavernous malformations (Gross et al., 2011) and the second most common in children (Gross et al., 2015). Cavernous malformations cause focal epilepsy, meaning the seizures begin within an area confined to a single hemisphere (Berg et al., 2010). Focal seizures are also known as partial seizures and can be simple, complex, or can generalize (secondary generalization). A 1999 literature review found that most patients with cavernous malformations presenting with seizures had secondary generalization (43%), whereas complex (37%) and simple (21%) seizures were slightly less common (Moran et al., 1999). Lesions located in the cortex

*Correspondence to: Rose Du, M.D., Ph.D., Department of Neurosurgery, Brigham and Women's Hospital and Harvard Medical School, 75 Francis Street, Boston MA 02115, USA. Tel: +1-617-525-8132, E-mail: rdu@bwh.harvard.edu

and, in particular, those located in the temporal lobe pose an increased risk for developing epilepsy (Chang et al., 2009).

The pathogenesis of focal epilepsy secondary to cavernous malformations primarily stems from damage to parenchymal tissue bordering the lesion. Intracerebral hemorrhage is a known risk factor for seizures (Claassen et al., 2007). Cavernomas can hemorrhage and lead to large intracranial hematomas, but they may also slowly ooze blood into nearby parenchyma, which is subsequently reabsorbed. However, repetitive microbleeds may cause inflammatory processes, accumulation of hemosiderin, and nearby alterations in blood flow that contribute to parenchymal damage (Awad and Jabbour, 2006). More recently, the role of cavernous malformations in blood–brain barrier breakdown leading to storage of albumin in proximate astrocyctes has emerged as a mechanism for epilepsy (Raabe et al., 2012).

In this section, the use of antiepileptic drugs to manage epilepsy associated with cavernous malformations is discussed and current recommendations for selection of antiepileptic drugs are reviewed.

Pharmacotherapy for epilepsy

Antiepileptic pharmacotherapy dates back to the mid 19th century, when the use of potassium bromide to treat epilepsy was reported (Clouston, 1868). However, it was the use of phenobarbital, beginning in 1912, that initiated the era of antiepileptic drugs in the USA (Porter et al., 1984). Over the course of the 20th century, several more antiepileptic agents were identified, including phenytoin, ethosuximide, and carbamazepine (Brodie, 2010). In 1978, valproic acid entered the market and marked the beginning of a period of rapid antiepileptic drug development (Meldrum and Rogawski, 2007). Modern drugs, while not always more effective than older iterations, generally have fewer side-effects and are thus commonly used. Currently, there are over 20 antiepileptic drugs approved for use in the USA for patients with epilepsy (Sirven et al., 2012).

Antiepileptic drugs act on one or more targets, often inhibiting sodium, calcium, and/or potassium conductance across neuronal membranes by antagonizing voltage-gated ion channels or glutamatergic receptors, or by potentiating gamma-aminobutyric acid (GABA) channels. Additional mechanisms of action of antiepileptic drugs involve targeting neurotransmitter trafficking and enzymatic processing of transmitters. However, despite advances in our understanding of antiepileptic drugs, the precise mechanism of action of some is still undetermined (Meldrum and Rogawski, 2007).

Treatment of epilepsy associated with cavernous malformations

Management of focal epilepsy secondary to cavernous malformations is shaped in large part by the presenting symptoms and the characteristics of the lesion itself. However, barring additional symptoms that require surgical removal, medical management with antiepileptic drugs is indicated for patients with newly diagnosed epilepsy from a cavernous malformation (Awad and Jabbour, 2006). Pharmacologic management of epilepsy resulting from a cavernous malformation utilizes those antiepileptic drugs that are most efficacious for focal seizures, of which there are many.

Several guidelines have been published regarding treatment of focal epilepsy in adults. The National Institute for Health and Care Excellence (NICE), a UK-based organization, recommends monotherapy with carbamazepine or lamotrigine as first-line therapy for focal epilepsy, with levetiracetam, oxcarbazepine, or sodium valproate as possibilities if needed (Bhattacharjee et al., 2012). In contrast, the International League Against Epilepsy (ILAE) recommends carbamazepine, phenytoin, levetiracetam, and zonisamide for first-line treatment of adult partial-onset epilepsy (Glauser et al., 2013). The American Academy of Neurology (AAN) published guidelines in 2004 recommending gabapentin, lamotrigine, topiramate, and oxcarbazepine for new-onset seizures in adults (French et al., 2004a). The AAN reports separate guidelines for antiepileptic drug treatment of resistant partial epilepsy in which they recommend monotherapy with lamotrigine, oxcarbazepine, or topiramate. They also indicate that those three, as well as gabapentin, tiagabine, levetiracetam, and zonisamide, can be used for add-on therapy but that these latter drugs are not recommended for monotherapy in refractory cases (French et al., 2004b). Guideline recommendations are summarized in Table 31.1. Although the first-line recommendations differ between these guidelines, it is reasonable to consider any of these drugs for a patient with focal epilepsy secondary to cavernous malformations.

Approved by the Food and Drug Administration (FDA) in 1999 for adjunctive therapy in partial epilepsy, levetiracetam is a modern antiepileptic drug with relatively milder side-effects and comparable efficacy to first-generation agents. While it is indicated as an adjunct therapy in the USA, levetiracetam is approved as a first-line therapy for partial-onset epilepsy in Europe. When compared to phenytoin, levetiracetam has fewer adverse effects in patients with nontraumatic subarachnoid hemorrhage (Shah and Husain, 2009) and in patients who underwent a craniotomy (Milligan et al., 2008). Furthermore, levetiracetam has an easier dosing

Table 31.1

Recommendations for focal epilepsy treatment

Guideline	First-line adult	Alternative adult	Refractory adult	Pediatric
NICE (Bhattacharjee et al., 2012)	Carbamazepine* Lamotrigine	Levetiracetam Oxcarbazepine Sodium valproate*		Carbamazepine* Lamotrigine
ILAE (Glauser et al., 2013)	Carbamazepine* Levetiracetam Phenytoin* Zonisamide			Oxcarbazepine
AAN (French et al., 2004a, b)	Gabapentin Lamotrigine Oxcarbazepine Topiramate*		Gabapentin† Lamotrigine‡ Topiramate*‡ Tiagabine† Oxcarbazepine‡ Levetiracetam† Zonisamide†	Refractory pediatric: Gabapentin Lamotrigine Oxcarbazepine Topiramate*

*Food and Drug Administration category D: known risk to fetus when used during pregnancy.
†Recommended as adjunctive therapy for refractory epilepsy.
‡Recommended as monotherapy or as adjunctive therapy for refractory epilepsy.
NICE, National Institute for Health and Care Excellence; ILAE, International League Against Epilepsy; AAN, American Academy of Neurology.

schedule than other antiepileptic drugs, which, along with the milder side-effect profile, may increase compliance. Levetiracetam was not deemed first-line therapy in the NICE guidelines due to its high cost (NICE, 2012) and, at the time of the AAN report, insufficient data were available to support its use (French et al., 2004a).

Another option for first-line antiepileptic therapy for focal epilepsy secondary to cavernous malformations is lamotrigine. Lamotrigine, first approved for use in the USA as an adjunctive therapy for partial epilepsy in 1994, is thought to prevent seizures via antagonism of voltage-gated sodium and N- and P-type voltage-sensitive calcium channels (White et al., 2007). A large, randomized controlled trial found lamotrigine to be more effective than carbamazepine, gabapentin, and topiramate in controlling partial epilepsy (Marson et al., 2007). Lamotrigine is generally well tolerated in patients, but does carry a warning for severe skin reactions that can progress to Stevens–Johnson syndrome and toxic epidermal necrolysis. Nonetheless, lamotrigine is an effective option for antiepileptic drug monotherapy in the context of focal epilepsy secondary to cavernous malformations.

Carbamazepine and phenytoin are also effective agents for focal epilepsy via their blockade of sodium channels (White et al., 2007). Both of these drugs are enzyme inducers, meaning they increase the activity of several liver enzymes, including cytochrome P450 monoxygenases that are involved in the metabolism of many other drugs. As a result, by increasing the rate at which many drugs are degraded, carbamazepine and phenytoin pose a significant risk for drug–drug interaction (Brodie et al., 2013). Phenytoin's use as a prophylactic agent is recommended in the American Heart Association guidelines, along with benzodiazepines for in-hospital seizures following intracranial hemorrhage (Broderick et al., 2007). Oxcarbazepine, a more recent addition to the antiepileptic repertoire modeled off the carbamazepine structure, does not induce hepatic enzymes to the degree of phenytoin or carbamazepine, thereby reducing the risk of drug–drug interactions (Shorvon, 2000). Oxcarbazepine does, however, pose a significantly increased risk of hyponatremia (Dong et al., 2005).

Zonisamide is a voltage-gated sodium and T-type calcium channel antagonist (Oommen and Mathews, 1999), introduced first in Japan and approved for use as an adjunctive agent in the USA in 2000, that is effective for the treatment of multiple types of epilepsy. In adults, zonisamide is an effective add-on therapeutic for drug-resistant epilepsy (Carmichael et al., 2013) and was found to be noninferior to carbamazepine when used alone in new-onset partial epilepsy (Baulac et al., 2012). Zonisamide has relatively fewer major adverse effects as compared to other antiepileptic drugs, but is associated with several central nervous system symptoms (Leppik, 2006).

Topiramate is another antiepileptic drug with efficacy in the treatment of focal epilepsy. It has several actions, including but not limited to voltage-gated sodium channel blockade, potentiation of GABA-mediated chloride conductance, and inhibition of AMPA and kainite channels (Shank et al., 2000). Treatment with topiramate alone has been shown to be effective for focal epilepsy (Ben-Menachem et al., 2008). Furthermore, topiramate is efficacious as an adjunctive therapy for drug-resistant focal epilepsy (Pulman et al., 2014b).

Tiagabine, a GABA uptake inhibitor, is another relatively recent drug approved by the FDA in 1997 that has been shown to be efficacious as an adjunctive therapy for focal epilepsy (Pulman et al., 2014a). The final drug recommended in the AAN guidelines, gabapentin, is considered to be an effective add-on therapy (Al-Bachari et al., 2013), but was less effective in controlling focal epilepsy when prescribed alone than lamotrigine, topiramate, and carbamazepine (Marson et al., 2007).

Individuals who present with seizures secondary to a cavernous malformation who do not undergo surgical resection of the lesion may require antiepileptic prophylaxis for life, as the lesion will continue to act as a nidus for potential seizure development. Therefore, the long-term effects of antiepileptic therapy should be taken into account when considering treatment options.

Antiepileptic therapy for cavernous malformations in pediatric patients

Focal epilepsy is also a common manifestation of cavernous malformations in children. Pharmacotherapy for focal epilepsy differs in the pediatric population as compared to adults. As in adults, treatment guidelines differ on recommended first-line agents. NICE recommends carbamazepine and lamotrigine for first-line therapy in children with focal epilepsy whereas the ILAE recommends oxcarbazepine as the only established initial treatment in children (Glauser et al., 2013). The AAN guidelines indicate lamotrigine is effective in the treatment of childhood absence seizures but they do not report recommendations for new-onset focal epilepsy in childhood (French et al., 2004a). However, AAN guidelines do recommend the use of gabapentin, lamotrigine, topiramate, and oxcarbazepine in the treatment of refractory childhood partial epilepsy (French et al., 2004b). As in adults, there is not a clear choice for first-line therapy of focal epilepsy secondary to cavernous malformations in children. However, while several antiepileptic drugs are recommended in these guidelines and have evidence of efficacy in the pediatric population, a 2013 systematic review found oxcarbazepine to be the only agent with class I evidence for use in children with focal epilepsy (Arya and Glauser, 2013).

In addition to efficacy, toxicity and ease of use are important considerations. Many antiepileptic agents have toxicity specific to children. Valproic acid hinders growth in children, possibly through direct inhibition of the growth plate, reduces bone formation, and decreases bone mass (Guo et al., 2001; Oner et al., 2004; Lee et al., 2013). Carbamazepine treatment in children increases total cholesterol by increasing both low- and high-density lipoprotein (Eiris et al., 2000).

Pregnancy, epilepsy, and cavernous malformation treatment

Pregnancy poses unique challenges for women with cavernous malformations. The natural history of cerebral cavernous malformations during pregnancy is controversial. Despite early suspicion that pregnancy predisposed women to cerebral cavernous malformation hemorrhage (Robinson et al., 1991; Flemming et al., 2003) , recent natural-history studies have not found an increased risk of hemorrhage in pregnant women (Witiw et al., 2012; Kalani and Zabramski, 2013). However, focal epilepsy secondary to cavernous management during pregnancy, either *de novo* or existing, does carry risks for the mother and fetus. Women with epilepsy are at an elevated risk of obstetric complications (Kaplan et al., 2007) and pharmacologic management of epilepsy in pregnant women is complicated by potential harm to the fetus. Epilepsy treatment in the pregnant woman requires particular care to the balance between effective management of the mother's epilepsy and the minimization of fetal exposure to teratogenic antiepileptic drugs.

Under the former US FDA letter-based pregnancy risk classification, four of the recommended antiepileptic drugs (valproic acid, carbamazepine, phenytoin, and topiramate) were category D, meaning they pose a significant risk of harm to the fetus. The FDA no longer uses this letter-based classification system, but its rating highlights the risk of these drugs. Valproic acid during the first trimester is strongly associated with congenital malformations in the fetus (Vajda et al., 2004; Wyszynski et al., 2005; Morrow et al., 2006; Meador et al., 2008). Furthermore, valproate use during pregnancy leads to decreased cognitive function early in life as compared to exposure to other antiepileptic drugs (Meador et al., 2009), and an association between exposure of the fetus to valproic acid and neurodevelopmental disorders, the majority of which were autism spectrum disorders, has been reported (Bromley et al., 2013).

In contrast, lamotrigine and carbamazepine exposure during pregnancy was not found to be associated with neurodevelopmental disorders during childhood (Bromley et al., 2013). Treatment with levetiracetam, lamotrigine, and oxcarbazepine carries a much lower risk of congenital malformations, with a similar risk profile to treatment of women who do not have epilepsy (Meischenguiser et al., 2004; Morrow et al., 2006; Molgaard-Nielsen and Hviid, 2011; Mawhinney et al., 2013). A large study on 4899 women who took antiepileptic drugs during the first trimester found that the risk of major malformations was 9.3% for valproate, 5.5% for phenobarbital, 4.2% for topiramate, 3.0% for carbamazepine, 2.9% for phenytoin, 2.4% for levetiracetam, 2.2% for oxcarbazepine, and 2.0% for lamotrigine, compared to 1.1% for those who were unexposed (Hernandez-Diaz et al., 2012).

Higher doses of antiepileptic drugs, particularly valproic acid, are associated with an increased risk of congenital malformations (Tomson et al., 2011). Furthermore, multidrug treatment during pregnancy is associated with significantly more risk of congenital malformations than monotherapy (Morrow et al., 2006). Attention to antiepileptic drug dosing is of critical importance as well since the rate of drug metabolism changes in pregnancy, resulting in faster clearance and lower drug levels (Pennell, 2008).

Medical management versus procedural intervention for patients with cavernous malformations presenting with seizures

Surgery for focal epilepsy is an effective and increasingly safe treatment. Surgical versus pharmacological management of focal epilepsy secondary to cavernous malformations is a complicated decision that is not without debate. A recent study reported that patients who underwent surgical resection for a cavernous malformation had an increased risk of poor outcome using the Oxford Handicap Scale (adjusted hazard ratio (HR) 2.2) and an increased risk of new intracranial hemorrhage or focal neurologic deficit (HR 3.6) compared to those treated with conservative management over 5 years (Moultrie et al., 2014). The increased risk of poor outcomes associated with surgery was a reflection of postoperative complications, as any and all new-onset intracranial hemorrhage or focal neurologic deficits occurred soon after surgery (Moultrie et al., 2014). However, this study does not sufficiently account for the natural history of symptomatic cavernous malformations, given that the majority (51%) of patients who were observed had incidental cavernous malformations while most patients who underwent surgery were symptomatic (88%).

Moreover, epileptogenic cavernous malformations can cause seizures that are intractable to pharmacotherapy but may be amenable to surgical resection. Surgical resection of cavernous malformations has been demonstrated to decrease the likelihood of future seizures, from 80% with medical management to 19% with surgery in a series of 31 patients with a mean follow-up time of 5.3 years (Noto et al., 2005). Therefore, surgical resection is generally recommended for patients with intractable epilepsy if it can be performed safely. Finally, one should also consider the long-term effects of antiepileptic medications, which may be avoided with surgical resection. First-line therapy for epilepsy secondary to cavernous malformations is therefore controversial and will differ based on lesion location, symptomatology, and patient characteristics.

Antiepileptic therapy following surgical resection

Although there are no guidelines on pharmacotherapy following epileptogenic intracranial cavernous malformation resection, patients with intractable epilepsy are typically maintained on antiepileptic monotherapy following surgery. The likelihood of a seizure-free outcome following surgery is multifactorial and depends on the frequency of seizures preoperation, extent of cavernous malformation resection, extent of hemosiderin resection, and on the timing of surgery relative to the initial presentation (Stavrou et al., 2008; Englot et al., 2011). Given the encouraging outcomes following surgical resection, there is a paucity of data on the requirement of antiepileptic drugs following excision of cavernous malformations. Discontinuation of antiepileptic drug therapy following epilepsy surgery has been reported to significantly increase the risk of seizure recurrence (Schiller et al., 2000; Schmidt et al., 2004; Yardi et al., 2014). However, the majority of patients whose antiepileptic drugs are withdrawn do not develop seizures within the first 5 years following surgery (Schmidt et al., 2004) and a prospective trial found that planned antiepileptic drug withdrawal, with the appropriate dosing taper, was safe following surgery for mesial temporal-lobe epilepsy (Rathore et al., 2011). Future research on the safety of planned antiepileptic discontinuation following surgery for cavernous malformations is warranted to define appropriate guidelines.

PHARMACOTHERAPY FOR THE TREATMENT OF CAVERNOUS MALFORMATIONS

The final section of this chapter will focus on developments in pharmacotherapy for the treatment of cavernous malformations. While there are no currently approved pharmacotherapeutics for the treatment of cavernomas, efforts to identify targets and potential compounds are under way. The RhoA/Rho kinase pathway is an example of a potential target. Cerebral cavernous malformation protein 2 (OSM, CCM2) normally acts in a complex with KRIT1 (CCM1) to suppress RhoA (Stockton et al., 2010). Loss of one or two copies of CCM2, as seen in cerebral cavernous malformations, is associated with a dose-dependent increase in activity of the RhoA GTPase *in vivo* (Whitehead et al., 2009). The RhoA effector, Rho kinase (ROCK), is constitutively active in the setting of CCM1 and 2 deficiency, leading to destabilization of endothelial cell junctions and vascular permeability (Stockton et al., 2010). Suppression of ROCK with either shRNA knockdown or small-molecule inhibition rescues CCM1-, 2-, and 3-deficient cells in culture, enabling them to form tube-like structures necessary for vasculogenesis (Borikova et al., 2010). Furthermore, the ROCK inhibitor fasudil has been shown to decrease the number of cavernomas in CCM1-deficient mice (McDonald et al., 2012).

Another avenue of research into cavernous malformation therapy involves Notch, transforming growth factor-beta (TGF-β), and bone morphogenic protein (BMP). CCM1, 2, or 3 deficiencies are associated with endothelial-to-mesenchymal transition (EndMT) secondary to activation of BMP and TGF-β and inhibition of either TGF-β or BMP was found to decrease the number and size of lesions in CCM1-deficient mice (Maddaluno et al., 2013). The group then investigated the role of Wnt/β-catenin in EndMT, and found that suppression of β-catenin also reduced the number and size of cavernous malformations in CCM3-deficient mice (Bravi et al., 2015). While still ongoing, these findings illustrate the importance of a robust understanding of the basic biology underpinning cavernous malformations and provide a promising avenue for future research.

Another strategy for targeting cavernous malformations has been to screen existing drugs for efficacy in CCM2-deficient cell lines and mice (Gibson et al., 2015). These efforts are complementary to a targeted approach and may bring pharmacotherapy for cavernomas to bear in a shorter period of time by avoiding the lengthy drug development pipeline.

CONCLUSION

Pharmacotherapy has an important role in the treatment of focal epilepsy secondary to intracranial cavernous malformations as either the sole therapy or as a complement to surgical resection. Several antiepileptic drugs are available and effective for focal epilepsy as monotherapy or as add-on agents for refractory disease. Treatment plans may differ in children and in pregnant women, to minimize adverse effects unique to these populations. While antiepileptic drugs may adequately control focal seizures secondary to cavernous malformations, in the event of intractable epilepsy requiring surgical resection, antiepileptic drugs are also indicated to minimize the risk of postoperative seizures. Ongoing research into the pathogenesis of cavernous malformations is yielding potential targets for drug development, which may be the future of cavernous malformation treatment.

References

Al-Bachari S, Pulman J, Hutton JL et al. (2013). Gabapentin add-on for drug-resistant partial epilepsy. Cochrane Database Syst Rev 7: CD001415.

Arya R, Glauser TA (2013). Pharmacotherapy of focal epilepsy in children: a systematic review of approved agents. CNS Drugs 27: 273–286.

Awad I, Jabbour P (2006). Cerebral cavernous malformations and epilepsy. Neurosurg Focus 21: e7.

Awad IA, Robinson Jr JR, Mohanty S et al. (1993). Mixed vascular malformations of the brain: clinical and pathogenetic considerations. Neurosurgery 33: 179–188; discussion 188.

Baulac M, Brodie MJ, Patten A et al. (2012). Efficacy and tolerability of zonisamide versus controlled-release carbamazepine for newly diagnosed partial epilepsy: a phase 3, randomised, double-blind, non-inferiority trial. Lancet Neurol 11: 579–588.

Ben-Menachem E, Sander JW, Stefan H et al. (2008). Topiramate monotherapy in the treatment of newly or recently diagnosed epilepsy. Clin Ther 30: 1180–1195.

Berg AT, Berkovic SF, Brodie MJ et al. (2010). Revised terminology and concepts for organization of seizures and epilepsies: report of the ILAE Commission on Classification and Terminology, 2005-2009. Epilepsia 51: 676–685.

Bhattacharjee S, Chatterjee N, Han S et al. (2012). CGEN: an R package for analysis of case-control studies in genetic epidemiology. R package version 3.10.0.

Borikova AL, Dibble CF, Sciaky N et al. (2010). Rho kinase inhibition rescues the endothelial cell cerebral cavernous malformation phenotype. J Biol Chem 285: 11760–11764.

Bravi L, Rudini N, Cuttano R et al. (2015). Sulindac metabolites decrease cerebrovascular malformations in CCM3-knockout mice. Proc Natl Acad Sci U S A 112: 8421–8426.

Broderick J, Connolly S, Feldmann E et al. (2007). Guidelines for the management of spontaneous intracerebral hemorrhage in adults: 2007 update: a guideline from the American Heart Association/American Stroke Association Stroke Council, High Blood Pressure Research Council, and the Quality of Care and Outcomes in Research Interdisciplinary Working Group. Circulation 116: e391–e413.

Brodie MJ (2010). Antiepileptic drug therapy the story so far. Seizure 19: 650–655.

Brodie MJ, Mintzer S, Pack AM et al. (2013). Enzyme induction with antiepileptic drugs: cause for concern? Epilepsia 54: 11–27.

Bromley RL, Mawer GE, Briggs M et al. (2013). The prevalence of neurodevelopmental disorders in children prenatally exposed to antiepileptic drugs. J Neurol Neurosurg Psychiatry 84: 637–643.

Carmichael K, Pulman J, Lakhan SE et al. (2013). Zonisamide add-on for drug-resistant partial epilepsy. Cochrane Database Syst Rev 12: CD001416.

Chang EF, Gabriel RA, Potts MB et al. (2009). Seizure characteristics and control after microsurgical resection of supratentorial cerebral cavernous malformations. Neurosurgery 65: 31–37; discussion 37–38.

Claassen J, Jette N, Chum F et al. (2007). Electrographic seizures and periodic discharges after intracerebral hemorrhage. Neurology 69: 1356–1365.

Clouston TS (1868). Experiments to determine the precise effect of bromide of potassium in epilepsy. J Ment Sci: 305–321.

Dong X, Leppik IE, White J et al. (2005). Hyponatremia from oxcarbazepine and carbamazepine. Neurology 65: 1976–1978.

Eiris J, Novo-Rodriguez MI, Del Rio M et al. (2000). The effects on lipid and apolipoprotein serum levels of long-term carbamazepine, valproic acid and phenobarbital therapy in children with epilepsy. Epilepsy Res 41: 1–7.

Englot DJ, Han SJ, Lawton MT et al. (2011). Predictors of seizure freedom in the surgical treatment of supratentorial cavernous malformations. J Neurosurg 115: 1169–1174.

Flemming KD, Goodman BP, Meyer FB (2003). Successful brainstem cavernous malformation resection after repeated hemorrhages during pregnancy. Surg Neurol 60: 545–547; discussion 547–548.

French JA, Kanner AM, Bautista J et al. (2004a). Efficacy and tolerability of the new antiepileptic drugs, I: Treatment of new-onset epilepsy: report of the TTA and QSS Subcommittees of the American Academy of Neurology and the American Epilepsy Society. Epilepsia 45: 401–409.

French JA, Kanner AM, Bautista J et al. (2004b). Efficacy and tolerability of the new antiepileptic drugs, II: Treatment of refractory epilepsy: report of the TTA and QSS Subcommittees of the American Academy of Neurology and the American Epilepsy Society. Epilepsia 45: 410–423.

Gibson CC, Zhu W, Davis CT et al. (2015). Strategy for identifying repurposed drugs for the treatment of cerebral cavernous malformation. Circulation 131: 289–299.

Glauser T, Ben-Menachem E, Bourgeois B et al. (2013). Updated ILAE evidence review of antiepileptic drug efficacy and effectiveness as initial monotherapy for epileptic seizures and syndromes. Epilepsia 54: 551–563.

Gross BA, Lin N, Du R et al. (2011). The natural history of intracranial cavernous malformations. Neurosurg Focus 30: E24.

Gross BA, Du R, Orbach DB et al. (2015). The natural history of cerebral cavernous malformations in children. J Neurosurg Pediatr: 1–6.

Guo CY, Ronen GM, Atkinson SA (2001). Long-term valproate and lamotrigine treatment may be a marker for reduced growth and bone mass in children with epilepsy. Epilepsia 42: 1141–1147.

Hernandez-Diaz S, Smith CR, Shen A et al. (2012). Comparative safety of antiepileptic drugs during pregnancy. Neurology 78: 1692–1699.

Kalani MY, Zabramski JM (2013). Risk for symptomatic hemorrhage of cerebral cavernous malformations during pregnancy. J Neurosurg 118: 50–55.

Kaplan PW, Norwitz ER, Ben-Menachem E et al. (2007). Obstetric risks for women with epilepsy during pregnancy. Epilepsy Behav 11: 283–291.

Lee HS, Wang SY, Salter DM et al. (2013). The impact of the use of antiepileptic drugs on the growth of children. BMC Pediatr 13: 211.

Leppik IE (2006). Practical prescribing and long-term efficacy and safety of zonisamide. Epilepsy Res 68 (Suppl 2): S17–S24.

Maddaluno L, Rudini N, Cuttano R et al. (2013). EndMT contributes to the onset and progression of cerebral cavernous malformations. Nature 498: 492–496.

Marson AG, Al-Kharusi AM, Alwaidh M et al. (2007). The SANAD study of effectiveness of carbamazepine, gabapentin, lamotrigine, oxcarbazepine, or topiramate for treatment of partial epilepsy: an unblinded randomised controlled trial. Lancet 369: 1000–1015.

Mawhinney E, Craig J, Morrow J et al. (2013). Levetiracetam in pregnancy: results from the UK and Ireland epilepsy and pregnancy registers. Neurology 80: 400–405.

McDonald DA, Shi C, Shenkar R et al. (2012). Fasudil decreases lesion burden in a murine model of cerebral cavernous malformation disease. Stroke 43: 571–574.

Meador K, Reynolds MW, Crean S et al. (2008). Pregnancy outcomes in women with epilepsy: a systematic review and meta-analysis of published pregnancy registries and cohorts. Epilepsy Res 81: 1–13.

Meador KJ, Baker GA, Browning N et al. (2009). Cognitive function at 3 years of age after fetal exposure to antiepileptic drugs. N Engl J Med 360: 1597–1605.

Meischenguiser R, D'Giano CH, Ferraro SM (2004). Oxcarbazepine in pregnancy: clinical experience in Argentina. Epilepsy Behav 5: 163–167.

Meldrum BS, Rogawski MA (2007). Molecular targets for antiepileptic drug development. Neurotherapeutics 4: 18–61.

Milligan TA, Hurwitz S, Bromfield EB (2008). Efficacy and tolerability of levetiracetam versus phenytoin after supratentorial neurosurgery. Neurology 71: 665–669.

Molgaard-Nielsen D, Hviid A (2011). Newer-generation antiepileptic drugs and the risk of major birth defects. JAMA 305: 1996–2002.

Moran NF, Fish DR, Kitchen N et al. (1999). Supratentorial cavernous haemangiomas and epilepsy: a review of the literature and case series. J Neurol Neurosurg Psychiatry 66: 561–568.

Morrow J, Russell A, Guthrie E et al. (2006). Malformation risks of antiepileptic drugs in pregnancy: a prospective study from the UK Epilepsy and Pregnancy Register. J Neurol Neurosurg Psychiatry 77: 193–198.

Moultrie F, Horne MA, Josephson CB et al. (2014). Outcome after surgical or conservative management of cerebral cavernous malformations. Neurology 83: 582–589.

NICE (2012). The epilepsies: the diagnosis and management of the epilepsies in adults and children in primary and secondary care: pharmacological update of clinical guideline 20, National Clinical Guideline Centre, London.

Noto S, Fujii M, Akimura T et al. (2005). Management of patients with cavernous angiomas presenting epileptic seizures. Surg Neurol 64: 495–498; discussion 498–499.

Oner N, Kaya M, Karasalihoglu S et al. (2004). Bone mineral metabolism changes in epileptic children receiving valproic acid. J Paediatr Child Health 40: 470–473.

Oommen KJ, Mathews S (1999). Zonisamide: a new antiepileptic drug. Clin Neuropharmacol 22: 192–200.

Pennell PB (2008). Antiepileptic drugs during pregnancy: what is known and which AEDs seem to be safest? Epilepsia 49 (Suppl 9): 43–55.

Porter RJ, Cereghino JJ, Gladding GD et al. (1984). Antiepileptic drug development program. Cleve Clin Q 51: 293–305.

Pulman J, Hutton JL, Marson AG (2014a). Tiagabine add-on for drug-resistant partial epilepsy. Cochrane Database Syst Rev: Cd001908.

Pulman J, Jette N, Dykeman J et al. (2014b). Topiramate add-on for drug-resistant partial epilepsy. Cochrane Database Syst Rev 2: CD001417.

Raabe A, Schmitz AK, Pernhorst K et al. (2012). Cliniconeuropathologic correlations show astroglial albumin storage as a common factor in epileptogenic vascular lesions. Epilepsia 53: 539–548.

Rathore C, Panda S, Sarma PS et al. (2011). How safe is it to withdraw antiepileptic drugs following successful surgery for mesial temporal lobe epilepsy? Epilepsia 52: 627–635.

Rigamonti D, Hadley MN, Drayer BP et al. (1988). Cerebral cavernous malformations. Incidence and familial occurrence. N Engl J Med 319: 343–347.

Robinson JR, Awad IA, Little JR (1991). Natural history of the cavernous angioma. J Neurosurg 75: 709–714.

Schiller Y, Cascino GD, So EL et al. (2000). Discontinuation of antiepileptic drugs after successful epilepsy surgery. Neurology 54: 346–349.

Schmidt D, Baumgartner C, Loscher W (2004). Seizure recurrence after planned discontinuation of antiepileptic drugs in seizure-free patients after epilepsy surgery: a review of current clinical experience. Epilepsia 45: 179–186.

Shah D, Husain AM (2009). Utility of levetiracetam in patients with subarachnoid hemorrhage. Seizure 18: 676–679.

Shank RP, Gardocki JF, Streeter AJ et al. (2000). An overview of the preclinical aspects of topiramate: pharmacology, pharmacokinetics, and mechanism of action. Epilepsia 41 (Suppl 1): S3–S9.

Shorvon S (2000). Oxcarbazepine: a review. Seizure 9: 75–79.

Sirven JI, Noe K, Hoerth M et al. (2012). Antiepileptic drugs 2012: recent advances and trends. Mayo Clin Proc 87: 879–889.

Stavrou I, Baumgartner C, Frischer JM et al. (2008). Long-term seizure control after resection of supratentorial cavernomas: a retrospective single-center study in 53 patients. Neurosurgery 63: 888–896; discussion 897.

Stockton RA, Shenkar R, Awad IA et al. (2010). Cerebral cavernous malformations proteins inhibit Rho kinase to stabilize vascular integrity. J Exp Med 207: 881–896.

Tomson T, Battino D, Bonizzoni E et al. (2011). Dose-dependent risk of malformations with antiepileptic drugs: an analysis of data from the EURAP epilepsy and pregnancy registry. Lancet Neurol 10: 609–617.

Vajda FJ, O'Brien TJ, Hitchcock A et al. (2004). Critical relationship between sodium valproate dose and human teratogenicity: results of the Australian register of anti-epileptic drugs in pregnancy. J Clin Neurosci 11: 854–858.

White HS, Smith MD, Wilcox KS (2007). Mechanisms of action of antiepileptic drugs. Int Rev Neurobiol 81: 85–110.

Whitehead KJ, Chan AC, Navankasattusas S et al. (2009). The cerebral cavernous malformation signaling pathway promotes vascular integrity via Rho GTPases. Nat Med 15: 177–184.

Witiw CD, Abou-Hamden A, Kulkarni AV et al. (2012). Cerebral cavernous malformations and pregnancy: hemorrhage risk and influence on obstetrical management. Neurosurgery 71: 626–630; discussion 631.

Wyszynski DF, Nambisan M, Surve T et al. (2005). Increased rate of major malformations in offspring exposed to valproate during pregnancy. Neurology 64 961–965.

Yardi R, Irwin A, Kayyali H et al. (2014). Reducing versus stopping antiepileptic medications after temporal lobe surgery. Ann Clin Transl Neurol 1: 115–123.

Index

NB: Page numbers in *italics* refer to figures and tables.

K

L

M

N

O

P

R

S

T

U

V

W

Z

Kolibri 360 GmbH
Ferdinand-Jühlke-Straße 7
99095 Erfurt, Deutschland
produktsicherheit@kolibri360.de